Non-Linear Raman Spectroscopy and Its Chemical Applications

NATO ADVANCED STUDY INSTITUTES SERIES

Proceedings of the Advanced Study Institute Programme, which aims at the dissemination of advanced knowledge and the formation of contacts among scientists from different countries

The series is published by an international board of publishers in conjunction with NATO Scientific Affairs Division

A	Life Sciences	Plenum Publishing Corporation
B	Physics	London and New York
C	Mathematical and Physical Sciences	D. Reidel Publishing Company Dordrecht, Boston and London
D	Behavioural and Social Sciences	Martinus Nijhoff Publishers
E	Engineering and Materials Sciences	The Hague, London and Boston
F	Computer and Systems Sciences	Springer Verlag Heidelberg
G	Ecological Sciences	

Series C – Mathematical and Physical Sciences

Volume 93 – Non-Linear Raman Spectroscopy and Its Chemical Applications

Non-Linear Raman Spectroscopy and Its Chemical Applications

Proceedings of the NATO Advanced Study Institute held at Bad Windsheim, Germany, August 23 - September 3, 1982

edited by

W. KIEFER
Physikalisches Institut, Universität Bayreuth, B.R.D.

and

D. A. LONG
School of Chemistry, University of Bradford, U.K.

D. Reidel Publishing Company

Dordrecht : Holland / Boston : U.S.A. / London : England

Published in cooperation with NATO Scientific Affairs Division

Library of Congress Cataloging in Publication Data
Main entry under title:

Non-linear Raman spectroscopy and its chemical applications.

(NATO advanced study institutes series. Series C, Mathematical and physical sciences ; 93)
"Published in cooperation with NATO Scientific Affairs Division."
Bibliography: p.
Includes index.
1. Raman spectroscopy—Congresses. I. Kiefer, W. (Wolfgang), 1941- II. Long, D. A. (Derek Albert) III. North Atlantic Treaty Organization. Division of Scientific Affairs. IV. Series.
QD96.R34N66 1982 543'.08584 82-11215
ISBN-13: 978-94-009-7914-7 e-ISBN-13: 978-94-009-7912-3
DOI: 10.1007/978-94-009-7912-3

Published by D. Reidel Publishing Company
P.O. Box 17, 3300 AA Dordrecht, Holland

Sold and distributed in the U.S.A. and Canada
by Kluwer Boston Inc.,
190 Old Derby Street, Hingham, MA 02043, U.S.A.

In all other countries, sold and distributed
by Kluwer Academic Publishers Group,
P.O. Box 322, 3300 AH Dordrecht, Holland

D. Reidel Publishing Company is a member of the Kluwer Group

CONTENTS

STIMULATED RAMAN EFFECT

CARS AND APPLICATIONS

RESONANCE CARS

PREFACE

In recent years a number of non-linear Raman spectroscopic techniques have been substantially developed and are now proving to be powerful methods for the solution of many problems not only in spectroscopy but also in chemistry, physics and biology. These techniques include hyper Rayleigh and hyper Raman spectroscopy, coherent anti-Stokes Raman Spectroscopy (CARS), Raman Gain and Inverse Raman Spectroscopy, Photoacoustic Raman Spectroscopy (PARS) and the Raman Induced Kerr Effect (RIKE). Hyper Raman spectroscopy although experimentally difficult is valuable for investigating transitions which are not active in the infrared or in the linear Raman effect; and the other non-linear Raman effects can provide signal strength and resolution which are orders of magnitude higher than those obtainable with linear Raman spectroscopy.

The thirty chapters in this book will form the basis of lectures presented at the NATO Advanced Study Institute in Bad Windsheim, F. R. Germany from August 23 - September 3, 1982. The chapters have been arranged into eight groups. The first group gives state of the art reports on linear Raman spectroscopy and infrared spectroscopy. The second group forms an introduction to the basic theory of non-linear light scattering with particular reference to the polarizability and hyperpolarizability tensors and third order, non-linear susceptibilities. The remaining five sections deal, in detail, with the theory and applications of hyper Rayleigh and hyper Raman spectroscopy, the stimulated Raman effect, coherent anti-Stokes Raman spectroscopy including resonance effects, inverse Raman and Raman gain spectroscopy, photoacoustic Raman spectroscopy and the Raman induced Kerr effect.

We would like to thank the authors for preparing their manuscripts adequately in advance, so that this book could be published in time for use at the Institute. We are much indebted to our secretaries, Miss M. Babendererde at the University of Bayreuth, and Mrs. D. Coultan & Mrs. S. Galloway at the University of Bradford, who have helped in many ways with the planning of

W. Kiefer and D. A. Long (eds.), Non-Linear Raman Spectroscopy and Its Chemical Applications, ix–x.

this Institute and in the preparation of this volume.

We also gratefully acknowledge the financial support of the NATO Scientific Affairs Division in Brussels, and the Universities of Bayreuth, F. R. Germany and Bradford, U.K.

Finally we thank Dr. M. di Lullo and Dr. C. Sinclair, Executive Officers of the NATO Scientific Affairs Division, in Brussels for their interest and guidance.

W. Kiefer
Physikalisches Institut
Universität Bayreuth
D-8580 Bayreuth, FRG

D. A. Long
School of Chemistry
University of Bradford
Bradford, U.K. BD7. 1DP

RAMAN AND INFRARED SPECTROSCOPY

LINEAR RAMAN SPECTROSCOPY: A STATE OF THE ART REPORT

H.W. Schrötter

Sektion Physik
der LMU München
München, West Germany

1. INTRODUCTION

Inelastic light scattering by molecules was discovered in 1928 by C.V. Raman in liquids (1, 2) and soon named after him. The effect had been predicted in 1923 by Smekal (3), when he tried to give a quantum theoretical explanation for dispersion. Energy conservation requires for a quantum mechanical light scattering process

$$M\,v^2/2 + E_m + h\tilde{\nu}_o = M\,v'^2/2 + E_n + h\tilde{\nu}_s\,, \tag{1}$$

where M is the molecular mass, h Planck's constant, v and v' the velocities, E_m and E_n the excitation energies of the molecules, and $\tilde{\nu}_o$ and $\tilde{\nu}_s$ the frequencies of the photons before and after the scattering process, respectively. The change in kinetic energy leads to Doppler broadening and is neglected here, and we can write

$$h\,\tilde{\nu}_o = h\,\tilde{\nu}_s + E_n - E_m \tag{2}$$

The energy difference E_n-E_m corresponds to an eigen frequency $\tilde{\nu}_j$ of the molecule. If $E_n>E_m$, the molecule takes the energy for its excitation from the scattered photon and we speak of a Stokes scattering process; if $E_n<E_m$, the molecule transfers its energy to the photon in an anti-Stokes scattering process. $E_n=E_m$ corresponds to Rayleigh scattering. If we introduce instead of the frequency $\tilde{\nu}_j$ the wavenumber of the Raman shift

W. Kiefer and D. A. Long (eds.), Non-Linear Raman Spectroscopy and Its Chemical Applications, 3–32.

$$\nu_j = \left| \frac{E_n - E_m}{hc} \right| , \tag{3}$$

where c is the velocity of light, we obtain for the wavenumber of the scattered light

$$\nu_s = \nu_o \mp \nu_j \ . \tag{4}$$

In the Stokes Raman spectrum (upper sign) the lines are shifted towards longer wavelengths from the Rayleigh line $\nu_s=\nu_o$, in the anti-Stokes Raman spectrum (lower sign) towards shorter wavelengths. The intensity of the anti-Stokes Raman lines depends on the occupation of the energy level E_m before the scattering process and is therefore lower than the intensity of the Stokes lines.

The molecular energy levels correspond to vibrational and/or rotational excitation, sometimes also electronic excitation (electronic Raman effect). Fig. 1 shows as an example the Stokes Raman spectrum of hydrogen chloride H Cl. At right, next to the exciting line, we see the nearly equidistant lines of the rotational Raman spectrum, originating from $\Delta v=0$, $\Delta J=2$ transitions, at left the strong Q-branch of the vibrational band ($\Delta v=1$, $\Delta J=0$), and next to it the lines of the S-branch ($\Delta v=1$, $\Delta J=+2$) and the O-branch ($\Delta v=1$, $\Delta J=-2$). We see that the vibration-rotation Raman band differs in its appearance from the infrared band, for which the selection rules are $\Delta v=1$, $\Delta J=\pm 1$ (R- and P-branch).

Soon after the first Raman spectra of liquids were published (2), Landsberg and Mandelstam (4) presented the first light scattering spectrum of a crystal, namely quartz, and Ramdas (5) that of ethyl ether in the vapor state. So it was established within a few months after the first discovery, that the Raman effect is a universal phenomenon that occurs in the gaseous, the liquid, and the solid state.

Over 50 years have passed since then and Raman spectroscopy has matured and found wide application, so that it is the subject of bi-annual international conferences (6-9).

The purpose of this article is to give a state-of-the-art report on the achievements and limitations of linear Raman spectroscopy as a basis for a better understanding of the non-linear Raman techniques that are the subject of the school.

In the first part the state of the instrumental development will be described regarding photographic and photoelectric

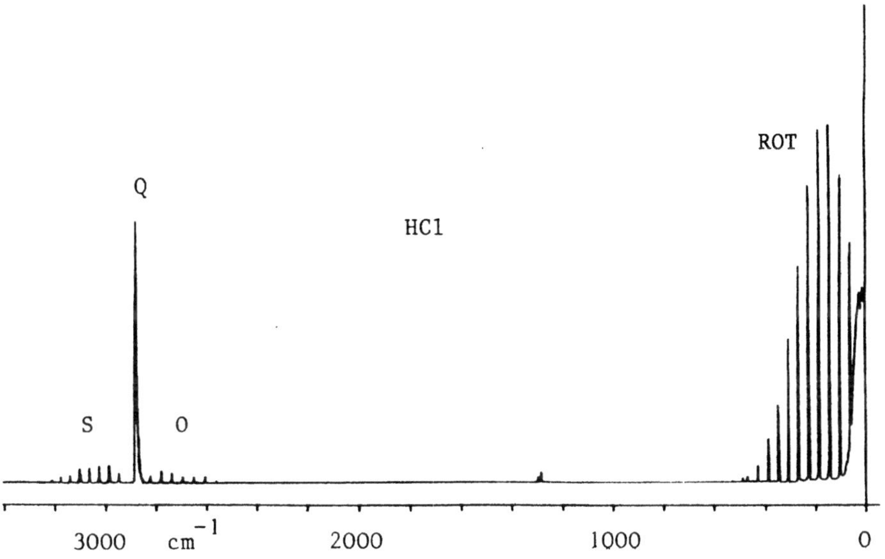

Fig. 1 Raman spectrum of hydrogen chloride HCl. Pressure 1 bar, slit 2 cm^{-1}, time constant 1 s, scanning speed 50 cm^{-1}/min, laser power 800 mW at 514.5 nm. Recorded by J.G. Hochenbleicher.

recording, single-channel and multi-channel detection, short time excitation, micro-sample, rotating sample, and capillary cell techniques. In the second part the experimental methods for accurate measurements of wavenumbers, linewidths, depolarization ratios and Raman scattering cross sections will be reviewed. Some special effects will be treated in the third part: the resonance Raman effect (RRE), surface enhanced Raman scattering (SERS), the measurement of circular intensity differentals (CID), the electronic Raman effect (ERE), and the study of elementary excitations in crystals. Finally we shall turn to examples for applications for the analysis of gas mixtures, remote Raman spectroscopy, the measurement of temperatures, and applications in geology, biology, and medicine.

In view of the contrast between the vast amount of results and the limited time and space that are available, the selection of examples will be rather accidental and is based more on personal preference of the author than on scientific merit on some impartial scale, and apologies are extended to all who are not mentioned.

2. STATE OF INSTRUMENTAL DEVELOPMENT

The introduction of the laser as light source (10-13) has led to an upsurge of interest in Raman spectroscopy and to a rapid development of instrumentation. The present state-of-the-art will be described in the following.

2.1. Photographic Recording

Although most Raman spectra are now recorded by photoelectric techniques, the photographic plate has still its merits for Raman spectra of gases. It allows the accumulation of the available Raman scattered light over long exposure times and a very precise determination of wavenumbers. Six instruments with focal lengths between 1 m and 8 m are used for such investigations in Europe and North America and were described by Brodersen (14) and Weber (15) in their review articles. The best resolution achieved was 0.05 cm^{-1} in the Q-branch of the vibration-rotation band of oxygen (16) by using single mode argon ion laser excitation and compensation of changes of atmospheric pressure, drifts of the laser frequency, and mechanical deformations of the spectrograph (17, 18). The sensitivity of the photographic plates is usually enhanced by pre-baking in a nitrogen or hydrogen atmosphere (19).

From the rotation-vibration Raman bands of several isotopic molecules of nitrogen (20), oxygen (21), and hydrogen (22-24) the equilibrium distances were determined with high precision and are presented in Table 1. Another recent example for photographic recording is the determination of the bond length in the ground state of P_4 as r_o = 222.28 ± 0.05 pm (25). With great care it is also possible to obtain good intensity data by the photographic technique (14), and with computer evaluation of the data it is even possible to separate the purely isotropic part of the Raman intensity, as has been demonstrated for the $2\nu_4$ band of methane (26). But in general precise intensity measurements are easier to perform with the photoelectric technique.

Table 1. Equilibrium Distances of Nuclei in Isotopic Species of Nitrogen, Oxygen, and Hydrogen

Species	r_e/pm	Ref.
$^{14}N_2$	109.7700 ± 0.0007	(20)
$^{14}N^{15}N$	109.7702 ± 0.0004	(20)
$^{15}N_2$	109.7700 ± 0.0008	(20)
$^{16}O_2$	120.754 ± 0.001	(16)
	120.754 ± 0.002	(21)
$^{17}O_2$	120.754 ± 0.002	(21)
$^{17}O^{18}O$	120.754 ± 0.002	(21)
$^{18}O_2$	120.756 ± 0.003	(21)
$^{1}H_2$	74.151 ± 0.002	(24)
$^{1}H^{2}H$	74.154 ± 0.002	(24)
$^{1}H^{3}H$	74.149 ± 0.001	(23)
$^{2}H_2$	74.152 ± 0.001	(24)
$^{2}H^{3}H$	74.151 ± 0.002	(23)
$^{3}H_2$	74.150 ± 0.001	(23)

2.2. Single-Channel Photoelectric Recording

All commercial Raman spectrometers are equipped for photoelectric recording. Most of these instruments consist of a powerful laser (argon or krypton), a sample compartment with provisions for various accessories for different kinds of samples, a double or triple monochromator, often equipped with holographic gratings, a photomultiplier as single-channel detector, and more and more frequently a dedicated computer for spectrometer scanning and data acquisition.

The performance that can be achieved with such an instrument is demonstrated here with examples from the author's laboratory. Fig. 2 shows the pure rotational Raman spectrum of

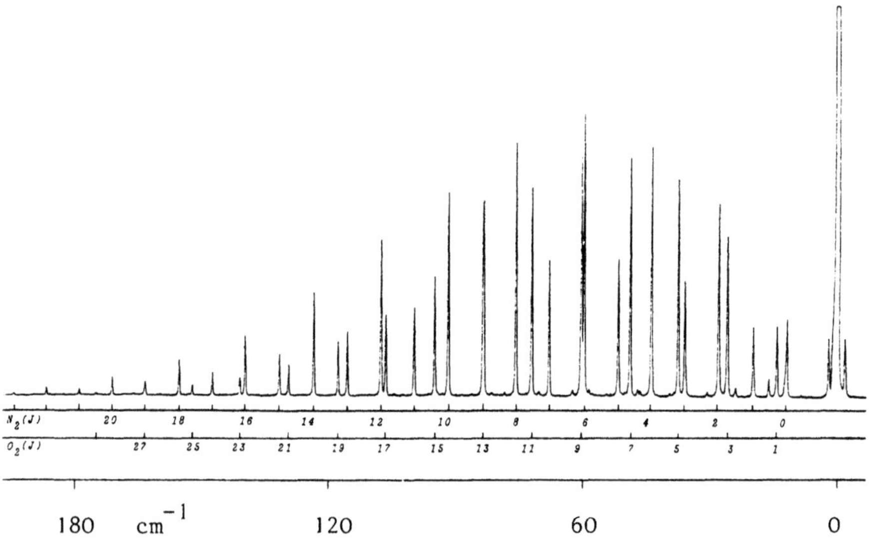

Fig. 2 Pure rotational Raman spectrum of laboratory air. Slit 0.25 cm^{-1}, time constant 0.5 s, scanning speed 4 cm^{-1}/min, laser power 7 W at 514.5 nm (27).

air in the laboratory (27) obtained in a single scan. The two lines of nitrogen and oxygen at 60 cm^{-1} are 0.67 cm^{-1} apart and are clearly resolved using a spectral slit width of 0.25 cm^{-1}, which represents approximately the limit of resolution of commercial Raman spectrometers with a focal length of the order of 1 m. The spin satellites of the Rayleigh line at 2 cm^{-1} due to oxygen (28-30) are well separated from the exciting line, demonstrating the low straylight level of holographic gratings. In CO_2 the weak R-branch lines at 1.5 cm^{-1} could also be resolved (31).

The sensitivity is mainly limited by the reflectivity of the gratings and mirrors in the monochromator and the quantum efficiency and dark current count rate of the photomultiplier. Phototubes with a quantum efficiency of 20 to 40 % and a dark count rate of 1 to 5 cps at -30° C are available. The signal level can be increased by intracavity excitation (12, 32) or multipassing of the laser beam (33) and the signal-to-noise ratio by signal averaging through multiple scanning (34) with a step motor drive or computer control of the spectrometer.

Insufficient reproducibility of the grating drive of a commercial double monochromator for the latter mode of operation can be overcome by starting the multiple scans at the point of steepest slope of a calibration line from a hollow cathode lamp (35) and scanning for a limited range. Fig. 3 shows as example for this kind of operation the Raman spectrum of the combination band $\nu_1+\nu_4$ of acetylene (35), which also contains contributions from the $\nu_3+\nu_5$ and $2\nu_2$ bands. The spectrum was recorded in two parts with multiscanning 30 times each. With an average count rate of 16 cps a resolution of 0.8 cm^{-1} was achieved, using a spectral slit width of 0.5 cm^{-1}.

The photoelectrically recording Raman spectrometer with the best resolution (~0,07 cm^{-1}) was constructed by Fletcher et al. (36, 37) around a monochromator with a focal length of 2 m and a 128 mm x 254 mm ruled grating. The resolution in the Q-branch of oxygen at a pressure of 25o mbar (37) is rather limited by pressure broadening than by the instrument function.

2.3. Multi-Channel Photoelectric Detection

The multiplex advantage of the photographic plate and the much higher quantum efficiency of photoelectric detectors can be combined by the use of image intensifiers and television cameras for the recording of Raman spectra, a technique which has been mainly developed by Delhaye and Bridoux (38-40).

Experiments with continuous and gated image intensifier tubes (41) and with a multichannel image intensifier plate (42, 43) were also performed in the author's laboratory. Fig. 4 shows examples of recordings of the pure rotational Raman spectrum of oxygen at a pressure of 1 bar with different modes of excitation and detection (41). Spectrum A at left was excited with 50 mW of a He-Cd laser at 441.6 nm and recorded for 3s with a 3-stage image intensifier RCA 4549 and a television camera with a SEC

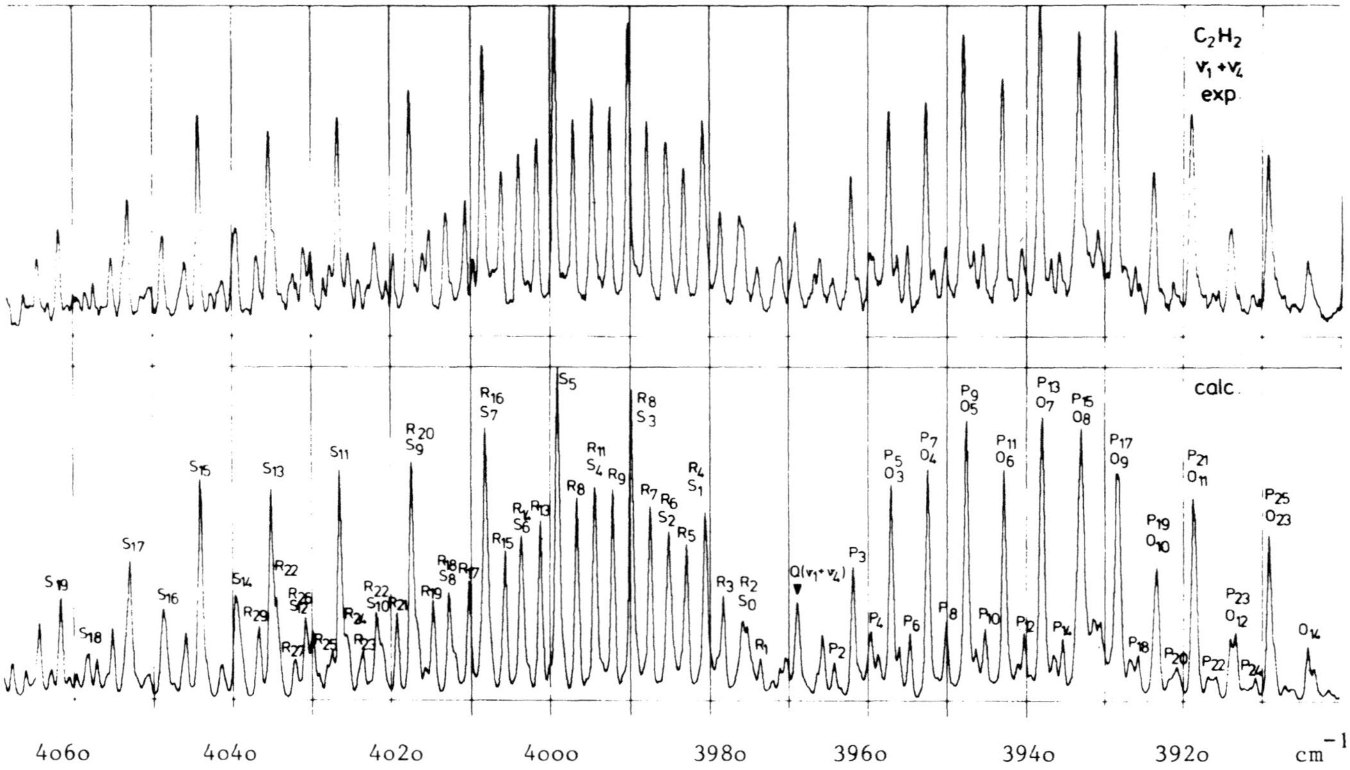

Fig. 3 Measured (exp.) and simulated (calc.) Raman spectrum of the '$\nu_1+\nu_4$' band of acetylene. Pressure 1 bar, slit 0.5 cm^{-1}, 30 scans averaged and computer smoothed, laser power 9 W at 514.5 nm with optical multipass (35).

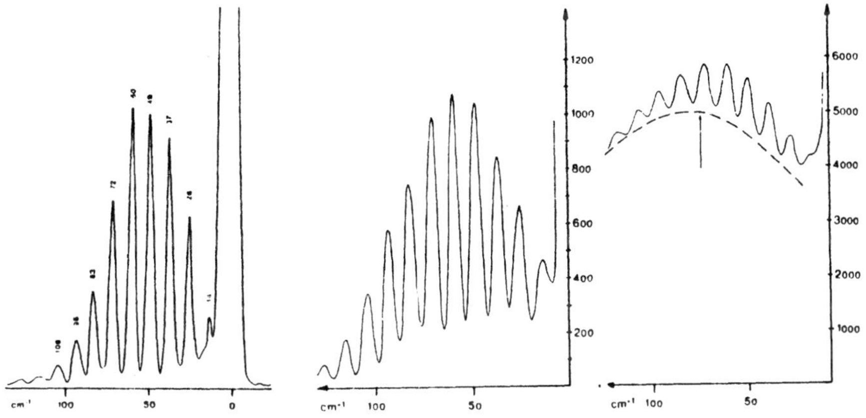

Fig. 4 Pure rotational Raman spectrum of O_2, Pressure 1 bar. A: excited with 150 mJ at 442 nm, B and C 3600 mJ at 532 nm (41). See text for details.

tube. The total laser energy required to produce the spectrum was only 150 mJ. Spectra B in the middle and C at right were excited with 1800 laser pulses of 2 mJ of a frequency doubled Nd-YAG laser at 532 nm with the intentionally low repetition rate of 1 Hz and recorded with a 3-stage image intensifier RCA C 33047 BP1 and the same SEC camera. For spectrum B the image intensifier was gated open for 800 ns for every laser pulse, so that the total sampling time amounts to only 1.4 ms and the dark current was suppressed, whereas for spectrum C the gate remained open for full 30 min. and a strong background accumulated. Through replacement of the SEC camera with a SIT television camera the sensitivity was increased to be equivalent to photon counting with a quantum efficiency of 10 % (41). Further experiments with a multichannel image intensifier plate (L-4263 from Litton), which can be gated easier than the 3-stage intensifier tube, also showed the possibility to reach single photon counting sensitivity (42-44) with these multi-channel detection devices, and a gain in recording time can be realized which is equal to the number of spectral elements that can be resolved by the image intensifier or the television camera, respectively.

Freeman et al. (45) have used an ISIT television camera and demonstrated a gain in recording speed by a factor of 1000 over single-channel recording. Wilbrandt et al. (46, 47) made use of a similar detection system, which is integrated into an electro-optical multichannel spectrometer with computerized data handling (48), to record the resonance Raman spectrum of p-terphenyl in its transient triplet state, which was produced by electron irradiation, with excitation by a single laser pulse.

Solid-state photodiode arrays have now attained such a standard of performance that they can replace television cameras for recording of spectra. Surbeck, Hug et al. have used a 1024 element linear array from Reticon in connection with a concave grating spectrograph to record the Raman spectra of nicotine and toluene (49) in recording times of less than 1 s and pure rotational Raman spectra of some gases (50) in recording times of 14 s with 1 W of argon laser excitation. Comparison of the oxygen spectrum (50) with that in Fig. 4A of this article shows, however, that the photodiode array alone has insufficient sensitivity to provide single photon counting efficiency.

The addition of an image intensifier optically coupled to the diode array enabled Bridoux et al. (51) to record a complete Raman spectrum of P_4S_{10} as polycristalline powder with excitation by a single 25 ps laser pulse of 1 mJ at 532 nm. A double grating foremonochromator with subtractive dispersion as bandpass filter provided the straylight suppression required for solid samples, whereas recording of single picosecond-pulse excited Raman spectra of liquids had already been achieved earlier (52).

2.4. Micro-Sampling Techniques

The smallest sample volume from which a Raman spectrum can be obtained is limited by the spot size to which a laser can be focused, the power density which the sample can withstand, and the sensitivity of the detection system. Delhaye and Dhamelincourt (53, 40, 54) and Rosasco et al. (55, 56) have developed Raman microprobe instruments with which samples of a few μm diameter can be investigated with little difficulty. The commercially available Raman microprobe MOLE allows to investigate a given sample in five successive steps. First the sample is viewed through a microscope in the light of the exciting laser frequency and the proper field of view is selected on a television screen. In a second step the total Raman spectrum of the viewed sample area can be recorded and the components identified through their characteristic Raman lines. Then the laser beam can be focused on small spots, e.g. crystallites or inclusions, and the Raman spectra of the single components are obtained successively. To determine the spatial distribution of a component the spectrometer is tuned to a characteristic Raman line of this compound and the sample area is seen in the light of this line only. Finally the Raman spectrum of a selected area can be recorded in the multi-channel mode over a spectral region of 100 cm^{-1} and temporal changes of the sample composition can be detected and chemical reactions followed. Examples of applications will be given below.

2.5. Special Techniques

For the investigation of resonance Raman spectra of absorbing solutions Kiefer and Bernstein (57) have invented the rotating sample technique. Based on this idea techniques for solid samples, for difference Raman spectroscopy, and for automatic scanning of the depolarization ratio and the reversal coefficient were developed and described by Kiefer (58) in detail. For a combined application of these techniques recently a four-channel photon-counter was developed (59) together with the appropriate four-compartment rotating cell (60). Examples of such applications will be presented below.

A rotating cell for obtaining Raman spectra of micro samples (150 μl) in the backscattering geometry is used by Rogers and Strommen (61) for resonance Raman investigations of biological compounds with UV excitation.

Very intense Raman spectra of liquids with high refractive index can be obtained by the liquid core optical fiber technique of Walrafen and Stone (62). Experimental investigations have so far been made on benzene, where many new overtone and combination bands were detected (62, 63, 58, 41), and tetrachloroethylene (62, 41, 64), where about 30 new second order Raman lines were found.

3. EXPERIMENTAL METHODS

3.1. Wavenumber Measurement

The calibration of the wavenumber scale of Raman spectrometers is usually performed with Fe (65) or Th (66, 67) hollow cathode lamps. Wavelengths measured in air have to be converted to vacuum wavenumbers (68).

For the evaluation of photographically recorded Raman spectra Brodersen and Bentsen (69) use an interpolation for the line positions with a polynomial of 4th order, the three highest coefficients of which were determined as function of the grating angle from 60 spectrograms. The precision that can be achieved is 0.002 to 0.005 cm^{-1} for relative line positons and 0.02 cm^{-1} on an absolute scale (14) and is reflected in the data of N_2 in

Table 1.

Small line shifts due to solvent effects can be accurately determined by difference Raman spectroscopy (70, 58, 71, 72). Any errors arising from the wavenumber scanning are avoided by rapid rotation of a two-compartment sample cell containing solution and solvent, respectively, so that the Raman spectra of both samples are excited nearly simultaneously and the corresponding signals are counted in two channels of a photon counting system and subtracted from each other. Small solvent shifts of Raman lines result in a derivative-like shape in the difference spectrum from which wavenumber shifts can be determined with a precision of up to 0.01 cm^{-1} in favorable cases (71). The technique has been applied to mixtures of CS_2 in benzene, pyridine in benzene, and chloroform in acetone (72). The isotopic shift of the ν_6=261 cm^{-1} Raman band of liquid chloroform upon deuterium substitution was measured to be (1.05 ± 0.08) cm^{-1} (73) and also line shifts upon isotopic dilution were studied in chloroform (74).

3.2. Linewidth Measurement

The line shapes and line widths in Raman spectra are determined by the lifetimes of the molecular states and can therefore be used to study intermolecular forces and interactions. Srivastava and Zaidi (75) have reviewed theoretical and experimental studies on line broadening in gases by collisions. Because the instrumental width has to be significantly narrower than the profile to be measured, linewidths below 2 cm^{-1} can only be determined with interferometric techniques (76, 77). The narrowest linewidth found in a liquid was 0.067 cm^{-1} for liquid nitrogen (76). In low pressure gases Doppler broadening becomes the dominating process for the linewidth and forward scattering has the advantage of the smallest momentum transfer (77, 78). In general, the profile of a Raman band is a Voigt function, a convolution of a Lorentzian and a Gaussian profile (79).

3.3. Depolarization Ratio Measurement

The depolarization ratio of a Raman line ν_j is

$$\rho_j = \frac{3\gamma_j^2}{45a_j^2 + 4\gamma_j^2} , \qquad (5)$$

where a_j, the linear average, and γ_j^2, the anisotropy, are the invariants of the derived polarizability tensor. ρ_j is measured by forming the ratio of the component of the scattered intensity that is polarized perpendicular to the polarization of the incident laser, $I_{j\perp}$, to that polarized parallel to it, $I_{j\parallel}$. Allemand (80) has listed and tested eight different variants of this measurement. When small depolarization ratios are to be measured accurately, divergence errors have to be taken into account (81).

Kiefer and Topp (82) have shown, how $I_\perp$, $I_\parallel$, and ρ can be recorded simultaneously by changing the polarization plane of the incident laser beam with a rotating split polarization rotator, counting the signals alternately into the two channels of a photon counter, and dividing them electronically.

In some cases the polarizability tensor may not be symmetric, e.g. in resonance Raman spectra (83) and the electronic Raman effect (84). Then the depolarization ratio is

$$\rho_j = \frac{3\gamma_j^2 + 5\delta_j^2}{45a_j^2 + 4\gamma_j^2}, \qquad (6)$$

where δ_j^2 is the antisymmetric anisotropy of the derived polarizability tensor. The relative magnitude of the three invariants can now only be determined by an additional measurement of the reversal coefficient for circularly polarized light

$$P_j = \frac{6\gamma_j^2}{45a_j^2 + \gamma_j^2 + 5\delta_j^2}, \qquad (7)$$

which is obtained in a 180° backscattering geometry as the ratio of the co-rotating and the contra-rotating intensities I_{co} and I_{contra}. Kiefer and Nitsch (85) have described, how the reversal coefficient can also be recorded directly.

Often it is useful to obtain separately the isotropic and the anisotropic part of the intensity. This can be done by subtracting $I_{j\perp} \cdot 4/3 \sim 4\gamma_j^2$ from $I_{j\parallel} \sim 45a_j^2 + 4\gamma_j^2$ either in a computer (86) or by direct scanning (87). The latter technique has been applied to the detection of weak Q-branches of isotopic species and hot bands in CO_2 (87-89).

Recently a four-channel photon counting system was developed (59) in order to be able to combine the techniques of direct scanning of the depolarization ratio or the isotropic part of the intensity with the method of difference Raman spectroscopy (90). Fig. 5 shows two examples of its application

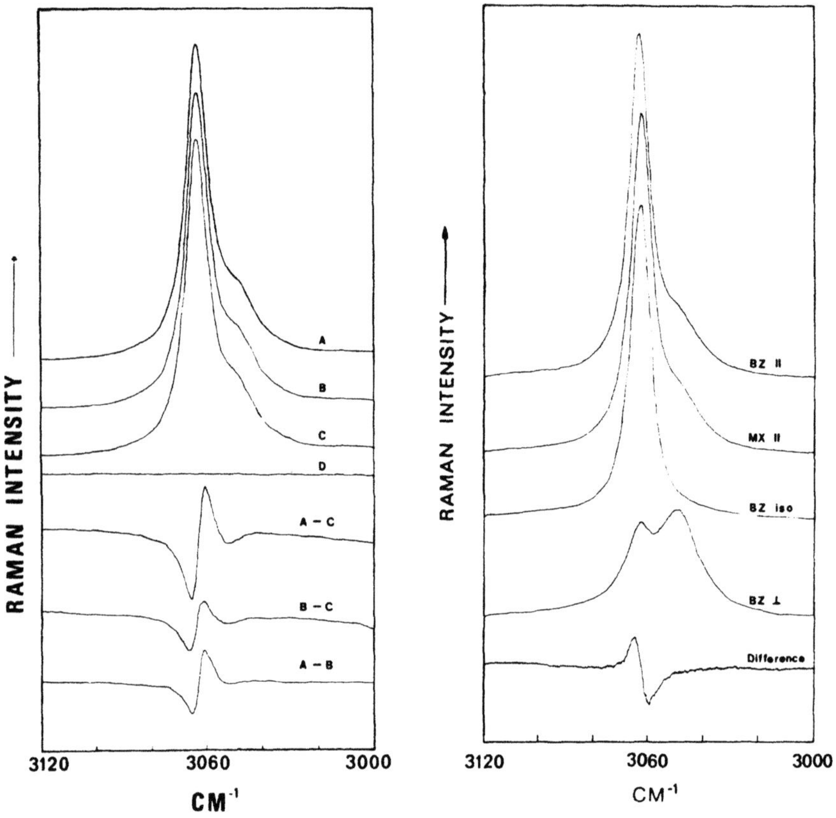

Fig. 5 Raman and difference Raman spectra of the ν_1 and ν_{15} in benzene mixtures. a) benzene and CS_2, b) benzene-d_o and benzene $-d_6$ (90). See text for details.

to a study of the ν_1 and ν_{15} bands of benzene in mixtures (90). Samples A and B in Fig. 5a are mixtures of 0.6 and 0.33 mole fractions of CS_2 in benzene, respectively, while C is pure benzene and D is pure CS_2. From the Raman difference spectra A-C, B-C, and A-B wavenumber shifts of -1.26, -0.54, and -0.68 cm^{-1} are evaluated for ν_1. In Fig. 5b trace BZ∥ is the parallel polarized component of pure benzene $-d_o$, MX∥ the parallel component of benzene $-d_o$ in a 1:1 mixture with benzene $-d_6$, BZ iso the isotropic part of the ν_1 band in pure benzene $-d_o$, BZ⊥ the anisotropic part of the intensity (x2) for pure benzene $-d_o$ clearly showing the depolarized ν_{15} band at 3045 cm^{-1}, and the last trace is the difference spectrum between the isotropic components of the mixture and pure benzene $-d_o$ giving a wavenumber shift of +0.34 cm^{-1} in the mixture.

3.4. Measurement of Raman Scattering Cross Sections

A review of Raman scattering cross sections in gases and liquids has been published recently (91) and only a short summary is given here. The total differential Raman scattering cross section of a vibrational-rotational band ν_j excited by a laser line with the wavenumber ν_o is

$$\frac{d\sigma_j}{d\Omega} = \frac{\pi^2}{45\varepsilon_o^2} \cdot \frac{(\nu_o-\nu_j)^4\, b_j^2}{1-\exp(-hc\nu_j/kT)} \cdot g_j \cdot (45a_j^2+7\gamma_j^2), \tag{8}$$

where $b_j^2=h/(8\pi^2 c\nu_j)$ is the square of the zero-point amplitude, g_j the degree of degeneracy of the normal vibration ν_j, ε_o the permittivity of free space, k Boltzmann's constant, and T the absolute temperature.

The absolute differential Raman scattering cross section of the Q-branch of nitrogen has the well established value (91)

$$\left(\frac{d\sigma}{d\Omega}\right)_{Q,\,N_2} = (5.05 \pm 0.1)\cdot 10^{-48}\,(\nu_o-\nu_{N_2})^4\,\frac{cm^6}{sr} \quad ,$$

where ν_{N_2} is 2331 cm^{-1}. The validity of the ν^4-law was confirmed from 350 to 630 nm within experimental accuracy.

Most scattering cross sections of other Raman bands were measured against the nitrogen Q-branch as standard. Therefore the relative normalized differenctial Raman scattering cross section

$$\Sigma_j = \frac{d\sigma_j/d\Omega}{(d\sigma/d\Omega)_{Q,N_2}} \cdot \frac{(\nu_o-\nu_j)^{-4}}{(\nu_o-\nu_{N_2})^{-4}} \cdot \left(1-\exp\left(-\frac{hc\nu_j}{kT}\right)\right) = \tag{9}$$

$$= \frac{\nu_{N_2}}{\nu_j} \cdot \frac{g_j(45a_j^2 + 7\chi_j\gamma_j^2)}{45a_{N_2}^2 + 7\chi_{N_2}\gamma_{N_2}^2}$$

was defined (91), where χ_j denotes the fraction of the anisotropic scattering which is concentrated in the Q-branch of the Raman band.

Averages of experimentally determined relative normalized differential Raman scattering cross sections for selected gases are given in Table 2. The value for cyanogen (92, 91) had been omitted from the last published compilation (44) and a new value for acetylene (93) has been included in the average. The data from which the averages in Table 2 were calculated agree with one another to within ±2o % with few exceptions, among these the results for benzene, where the results of different measurements vary by as much as ±5o %.

4. SPECIAL EFFECTS IN LINEAR RAMAN SPECTROSCOPY

4.1. Resonance Raman Effect

When the frequency of the exciting radiation is close to an absorption frequency of the scattering system, the transition probability for the two-photon scattering process is enhanced and we speak of a resonance Raman effect (RRE). The first experimental observation was made by Shorygin (94) in 1947. Theories and experimental investigations of the RRE have been the subject of many review articles (e.g. 95-97). When the resonance occurs with a discrete energy level, one speaks of a discrete RRE, when the resonating state is in an absorption continuum, of a continuum RRE (97).

The characteristic features of the RRE can be studied particularly well in the diatomic molecules iodine and bromine. Baierl and Kiefer (98) have recently simplified the theory of the RRE for diatomic molecules by using spherical tensors and Wigner 3-j symbols, and calculated the band profiles of the continuum resonance Raman bands in bromine and iodine in excellent agreement with the experimental data (99). A further simplification and elucidation of the theory of RRE is possible through the application of the δ-approximation or "reflection" method (100) to two-photon processes (101).

In the discrete RRE, which has previously been termed resonance fluorescence, single rotational transitions are enhanced particularly upon exact resonance and the spectra consist of very sharp lines. Very recently Kirillov et al. (102) have resolved two lines in the fundamental band of iodine vapor at 213 cm^{-1} that are 190 MHz ≙ 0.006 cm^{-1} apart, using a pressure of 0.3 mbar.

Table 2. Average Relative Normalized Differential Raman Scattering Cross Sections Σ_j for Q-branches of Vibrational Bands.

Molecule	ν_j (cm^{-1})	Σ_j
N_2	2331	1
O_2	1555	1.04
H_2	4156	3.86
D_2	2986	2.53
F_2	893	0.32
Cl_2	554	2.25
HF	3962	1.34
HCl	2886	3.10
HBr	2558	4.42
HI	2230	5.80
CO	2143	0.93
NO	1877	0.38
CO_2	1388	1.13
	1285	0.75
N_2O	2224	0.50
	1285	1.98
H_2O	3652	3.51
	1595	0.008
H_2S	2611	6.80

Molecule	ν_j (cm^{-1})	Σ_j
O_3	1103	3*
SO_2	1151	3.99*
	519	0.08
C_2H_2	1973	6.20
C_2N_2	2330	11.9†
NH_3	3334	5.83
ND_3	2420	3.06
CH_4	2917	8.55
CCl_4	459	6.45*
SF_6	774	3.60
UF_6	665	12.6
C_2H_4	3020	6.4
	1623	2.12
	1342	2.8
C_2H_3Cl	1632	1.65*
CH_3OH	2846	4.7
C_2H_6	993	1.72
C_6H_6	3070	14.4
	992	11.7
C_6H_{12}	802	4.03

*Average taken only for visible exciting lines

†Cross section for total band

In solutions the RRE leads not so much to very high intensities in the observed spectra because of the competition of absorption, but to the possibility to detect the spectra of absorbing species at very high dilution without serious interference from spectra of other dissolved species and the solvent. A recent example is the investigation of the production of transient radical cations of 1,4-dimethoxybenzene and diazabicyclo-octane in solution at concentrations of 10^{-4} to 10^{-5} mol/l in a stopped-flow experiment with a multi-channel Raman spectrometer using recording times of 20 ms or by rapid mixing and rapid scanning with a single-channel spectrometer using scanning times of 2 s for a range of 400 cm^{-1} (103).

4.2. Surface Enhanced Raman Spectroscopy

When certain polar molecules, such as pyridine, are adsorbed on the rough surface of metals like silver, gold, or copper, the Raman scattering cross section of some normal modes is enhanced by factors of 10^4 to 10^6 compared to scattering from an isolated molecule (104). All attempts to explain this effect, see for instance a number of papers in Ref. (9), have so far failed to reach quantitative agreement with this enormous factor (105, 106). Surface enhanced Raman scattering (SERS) has also been observed on nickel (107) and cadmium (108) surfaces.

4.3. Circular Intensity Differentials

Raman optical activity (109, 110) has been detected in the Raman spectra of chiral molecules. Many of the early investigations of the circular intensity differentials (CID)

$$\Delta = \frac{I_R - I_L}{I_R + I_L} , \qquad (10)$$

where I_R and I_L are the scattered intensities in right and left circularly polarized light, were plagued by optical artefacts caused by imperfections in the polarizing optics, residual birefringence in optical elements, and refractions at angled surfaces. Hug (111) has eliminated most of these artefacts in his multi-channel Raman optical activity spectrometer (112) by collecting the scattered light simultaneously by two collecting lenses in directions of 90° to the incident laser beam and to each other. With this spectrometer vibrational Raman optical activity spectra could be recorded also in perpendicular polarization (112) and investigations on stereochemical problems were carried out (112, 113). Instead of the CID defined in eq. (10)

Hug and coworkers (111, 112) have defined the chirality numbers

$$q_{\parallel,\perp} = \frac{{}^{L}d\sigma_{\parallel,\perp} - {}^{R}d\sigma_{\parallel,\perp}}{({}^{L}d\sigma_{\parallel,\perp} + {}^{R}d\sigma_{\parallel,\perp})/2} , \qquad (11)$$

where $d\sigma$ denotes the differential Raman scattering cross section, in order to achieve better consistence with the notation used in natural and induced optical activity.

4.4. Electronic Raman Effect

The electronic Raman effect (ERE) was discovered as early as 1930 by Rasetti (114) in NO which has a doubly degenerate electronic ground state. The electronic Raman band at 124 cm^{-1} was later investigated with high resolution (115, 116).

Electronic Raman transitions between the low lying energy levels of f-electrons in rare earth ions were studied mainly by Koningstein et al. (117, 118). The ERE is also found for d-electrons in transition metals (118) and unusually high Raman shifts up to 12 600 cm^{-1} were found in Ni^{++} ions (119).

Also the scattering from carriers in semiconductors has been thoroughly investigated (120).

4.5. Elementary Excitations in Crystals

The excitation of the Raman spectra of crystals with the highly collimated laser radiation led to an enormous number of investigations on optical and acoustical phonons, polaritons, plasmons, plasmaritons, magnons, excitons etc. and on phase transitions which have been the subject of a series of conferences (121-123) and of comprehensive books and reviews(124-128). The theory of directional dispersion of phonons and polaritons was recently generalized by Claus et al. (128-132) to include monoclinic and triclinic crystals.

5. APPLICATIONS

Because of limitations in space and time only a few further applications of linear Raman spectroscopy can be mentioned.

5.1. Analysis of Gas Mixtures

Concentrations and the temperature in gas mixtures

including flames can me measured with spatial and temporal resolution (133-135). The concentration measurements are based on known Raman scattering cross sections (91) or on calibration.

Black and Chang (136) used a multi-channel Raman spectrometer for the determination of transient gas concentration profiles in a jet of CF_2Cl_2 emerging from a nozzle, and of temperatures in H_2-air and H_2-O_2 flames. The temperature determination from the CO_2 Raman spectrum poses a special problem because of the influence of Fermi resonance on the Raman intensities (137). Finsterhölzl et al. (138) have calculated the intensities of the hot bands in CO_2 for temperatures up to 3000 K based on parameters obtained from room temperature measurements (27, 88, 89).

The equilibrium between monomeric and dimeric acetic acid in the gas phase has been studied by Gaufrès et al. (139) as a function of temperature and pressure.

Raman spectroscopy with pulsed lasers has also been used for the remote analysis of air pollutants, humidity, and temperature (140, 44). With excitation by a quadrupled Nd-YAG laser at 266 nm water vapor density profiles have been measured in full daylight (141, 142) up to a height of 1000 m.

The pressure and temperature in an expanding molecular-jet-beam have been determined down to 18 mbar and 30 K from the intensity distribution in the pure rotational Raman spectrum of CO_2 excited in the cavity of an argon ion laser (143). Isotope ratio measurements of $^{15}N/^{14}N$ and $^{18}O/^{16}O$ ratios in air samples resulted in good agreement with mass spectrometric determinations (144).

5.2. Applications of Raman Micro Analysis

In bubbles included in minerals Raman lines from $^{12}C^{16}O_2$, $^{13}C^{16}O_2$, and $^{12}C^{16}O^{18}O$ were detected (145). On the basis of relative cross sections of these lines (138) a determination of the isotope ratios $^{13}C/^{12}C$ and $^{18}O/^{16}O$ in samples of geological interest will be possible without destruction of the sample.

In the applications laboratory of Instruments S.A. examples of results obtainable with the Raman microprobe MOLE have been worked out. For instance, dust samples form a super-highway and inclusions in gems and plastic films were analyzed. The identification of wollastonite and the distrinction of calcite and aragonite are also possible.

5.3. Applications in Biology and Medicine

Biological systems can be studied by Raman spectroscopic techniques with relative ease, because water shows only a weak and structureless spectrum in contrast to its strong infrared absorption. Moreover the RRE is helpful for the detection of substances with chromophoric groups as hemoglobin, carotene, chlorophyll, rhodopsin and so on. A great number of systems has been investigated, see papers in Refs. (7-9). The Raman spectroscopic studies of visual pigments and their role in the mechanism of vision has been the subject of a panel session at the last Raman conference (146). A nice example for a biological application is the excitation of the RRE of lobster shell pigments in situ (147).

Indications for an enhancement of the population of vibrational levels of carotenoid molecules in living algae "Chorella pyrenoidosa" above the equilibrium room temperature population have been found by Drissler (148).

The application of Raman spectroscopy in medicine is only just beginning. The influence of oxygen saturation on the resonance Raman spectrum of whole human blood has been investigated by Albrecht and Breitinger (149). The concentration of oxygen, nitrogen, and carbon dioxide in respiratory gases can be measured in a multipass-cell with relatively low laser power employing interference filters for spectral resolution (150). A panel session on biomedical applications is planned for the next Raman conference.

6. CONCLUSION

In this review it has been tried to outline the basic principles of linear Raman spectroscopy and to point out the possibilities and limitations for its application in various fields. The following chapters will treat in much greater detail the new non-linear techniques that have opened new frontiers for Raman spectroscopy.

REFERENCES

1. Raman, C.V., and Krishnan, K.S., Nature 121, 501 (1928).

2. Raman, C.V., Nature 121, 619 (1928); Indian J. Phys. 2, 387 (1928).

3. Smekal, A., Naturwiss. 11, 873 (1923).

4. Landsberg, G., Mandelstam, L., Naturwiss. 16, 557 (1928).

5. Ramdas, L.A., Indian J. Phys. Proc. Indian Assoc. Cultivat. Sci. 3, 131 (1928).

6. Mathieu, J.P., ed., Advances in Raman Spectroscopy, Vol. 1, Heyden, London 1973.

7. Schmid, E.D., Brandmüller, J., Kiefer, W., Schrader, B., and Schrötter, H.W., eds., Proceedings of the 5th Intern. Conf. on Raman Spectroscopy, H.F. Schulz, Freiburg 1976.

8. Schmid, E.D., Krishnan, R.S., Kiefer, W., and Schrötter, H.W., eds., Proceedings of the 6th Intern. Conf. on Raman Spectroscopy, Vol. 1+2, Heyden, London 1978.

9. Murphy, W.F., ed., Proceedings of the 7th Intern. Conf. on Raman Spectroscopy, North-Holland, Amsterdam 1980.

10. Porto, S.P.S., and Wood, D.L., J. Opt. Soc. Amer. 52, 251 (1962).

11. Kogelnik, H., and Porto, S.P.S., J. Opt. Soc. Amer. 53, 1446 (1963).

12. Leite, R.C.C., and Porto, S.P.S., J. Opt. Soc. Amer. 54, 981 (1964).

13. Koningstein, J.A., and Smith, R.G., J. Opt. Soc. Amer. 54, 1061 (1964).

14. Brodersen, S., in Raman Spectroscopy of Gases and Liquids (Weber, A., ed.), Topics in Current Physics, Vol. 11, Springer, Berlin 1979, p. 7.

15. Weber, A., ibid., p. 71.

16. Loëte, M., and Berger, H., J. Mol. Spectrosc. 68, 317 (1977).

17. Berger, H., Faivre, M., Champion, J.P., and Moret-Bailly, J., J. Mol. Spectrosc. 45, 298 (1973).

18. Champion, J.P., and Berger, H., J. Mol. Spectrosc. 55, 15 (1975).

19. Schlupf, J., and Weber, A., Appl. Opt. 10, 2373 (1971).

20. Bentsen, J., J. Raman Spectrosc. 2, 133 (1974).

21. Edwards, H.G.M., Long, D.A., Najm, K.A.B., and Thomsen, M., J. Raman Spectrosc. 10, 60 (1981).

22. Edwards, H.G.M., Long, D.A., and Mansour, H.R., J. Chem. Soc. Faraday II, 74, 1203 (1978).

23. Edwards, H.G.M., Long, D.A., Mansour, H.R., and Najm, K.A.B., J. Raman Spectrosc. 8, 251 (1979).

24. Edwards, H.G.M., Farwell, D.W., Gorvin, A.C., and Long, D.A., private communication 1982.

25. Brassington, N.J., Edwards, H.G.M., and Long, D.A., J. Raman Spectrosc. 11, 346 (1981).

26. Lolck, J.-E., Brodersen, S., and Robiette, A.G., J. Raman Spectrosc. 12, 49 (1982).

27. Finsterhölzl, H., Klöckner, H.W., Srinivasan, K., Schrötter, H.W., and Brandmüller, J., Indian J. Pure Appl. Phys. 16, 370 (1978).

28. Jammu, K.S., St. John, G.E., and Welsh, H.L., Can. J. Phys. 44, 797 (1966).

29. Lepard, D., Can. J. Phys. 48, 1664 (1970).

30. Altmann, K., Strey, G., Hochenbleicher, J.G., and Brandmüller, J., Z. Naturforsch. 27a, 56 (1972).

31. Altmann, K., Klöckner, H.W., and Strey, G., Z. Naturforsch. 31a, 1311 (1976).

32. Barrett, J.J., and Adams III, N.I., J. Opt. Soc. Amer. 58, 311 (1968).

33. Kiefer, W., Bernstein, H.J., Wieser, H., and Danyluk, M., J. Mol. Spectrosc. 43, 393 (1972).

34. Ziegler, E., and Hoffmann, E.G., Österr. Chem. Z. 68, 319 (1967).

35. Finsterhölzl, H., Schrötter, H.W., and Strey, G., J. Raman Spectrosc. 11, 375 (1981).

36. Fletcher, W.H., and Rayside, J.S., J. Raman Spectrosc. 2, 3 (1974).

37. Fletcher, W.H., Rayside, J.S., and Mc. Lendon, W.B., J. Raman Spectrosc. 7, 205 (1978).

38. Delhaye, M., Appl. Opt. 7, 2195 (1968).

39. Bridoux, M., Delhaye, M., Nouv. Rev. d'Optique appl. 1, 23 (1970).

40. Bridoux, M., Delhaye,M., in Advances in Infrared and Raman Spectroscopy (Clark, R.J.H., and Hester, R.E., eds.), Vol. 2, Heyden, London 1976, p. 140.

41. Schmid, W.J., Dissertation, LMU München 1980.

42. Klöckner, H.W., Dissertation, LMU München 1977.

43. Klöckner, W., Schmid, W.J., and Schrötter, H.W., in Ref. (7), p. 766.

44. Schrötter, H.W., in Advances in Infrared and Raman Spectroscopy (Clark, R.J.H., and Hester, R.E., eds.), Vol. 8, Heyden, London 1981, p. 1.

45. Freeman, J.J., Heaviside, J., Hendra, P.J., Prior, J., and Reid, E.S., Appl. Spectrosc. 35, 196 (1981).

46. Wilbrandt, R., Jensen, N.H., Pagsberg, P., Sillesen, A.H., and Hansen, K.B., in Ref.(8), p. 510.

47. Wilbrandt, R., Jensen, N.H., Pagsberg, P., Sillesen, A.H., Hansen, K.B., and Hester, R.E., Chem. Phys. Lett. 60, 315 (1979).

48. Hansen, K.B., Wilbrandt, R., and Pagsberg, P., Rev. Sci. Instrum. 50, 1532 (1979).

49. Surbeck, H., Hug, W., Gremaud, M., Bridoux, M., Deffontaine, A., and Da Silva, E., Opt. Commun. 38, 57 (1981).

50. Hug, W., and Surbeck, H., J. Raman Spectrosc. (to be published).

51. Bridoux, M., Deffontaine, A., Delhaye, M., Rose, B., and Da Silva, E., J. Raman Spectrosc. 11, 515 (1981).

52. Bridoux, M., in Ref. (8), Vol. 1, p. 381.

53. Delhaye, M., and Dhamelincourt, P., J. Raman Spectrosc. 3, 33 (1975).

54. Dhamelincourt, P., in Ref. (8), Vol. 1, p. 399

55. Rosasco, G.J., Etz, E.S., and Cassatt, W.A., Appl. Spectrosc. 29, 396 (1975).

56. Rosasco, G.J., in Ref. (8), Vol. 1, p. 389.

57. Kiefer, W., and Bernstein, H.J., Appl. Spectrosc. 25, 501 (1971).

58. Kiefer, W., in Advances in Infrared and Raman Spectroscopy (Clark, R.J.H., and Hester, R.E., eds.), Vol. 3, Heyden, London 1977, p. 1.

59. Eichele, H., Kolb, G., Laane, J., Beckmann, A., and Kiefer, W., J. Phys. E, Sci. Instrum. 14, 1397 (1981).

60. Laane, J., and Kiefer, W., Appl. Spectrosc. 35, 428 (1981).

61. Rodgers, E.G., and Strommen, D.P., Appl. Spectrosc. 35, 215 (1981).

62. Walrafen, G.E., and Stone, J., Appl. Spectrosc. 26, 585 (1972).

63. Walrafen, G.E., Phys. Bl. 30, 540 (1974).

64. Schmid, W.J., and Schrötter, H.W., J. Raman Spectrosc. 10, 212 (1981).

65. Crosswhite, H.M., J. Res. Natl. Bur. Stands. 79A, 17 (1975).

66. Giachetti, A., Stanley, R.W., Zalubas, R., J. Opt. Soc. Amer. 60, 474 (1970).

67. Valero, F.P.J., J. Opt. Soc. Amer. 60, 1675 (1970).

68. Strey, G., Spectrochimica Acta 25A, 163 (1969).

69. Brodersen, S., and Bentsen, J., J. Raman Spectrosc. 3, 207 (1975).

70. Kiefer, W., Appl. Spectrosc. 27, 253 (1973).

71. Laane, J., and Kiefer, W., J. Chem. Phys. 72, 5305 (1980).

72. Laane, J., and Kiefer, W., Appl. Spectrosc. 35, 267 (1981).

73. Laane, J., Eichele, H., Hohenberger, H.P., and Kiefer, W., J. Mol. Spectrosc. 86, 262 (1981).

74. Laane, J., and Kiefer, W., J. Chem. Phys. 73, 4971 (1980).

75. Srivastava, R.P., and Zaidi, H.R., in Raman Spectroscopy of Gases and Liquids (Weber, A., ed.), Topics in Current Phys Physics, Vol. 11, Springer, Berlin 1979, p. 167.

76. Clements, W.R.L., and Stoicheff, B.P., Appl. Phys. Lett. 12, 246 (1968).

77. Clements, W.R.L., and Stoicheff, B.P., J. Mol. Spectrosc. 33, 138 (1970).

78. Weber, A., in The Raman Effect (Anderson, A., ed.), Vol. 2, Marcel Dekker, New York, 1973, p. 543.

79. Asthana, B.P., and Kiefer, W., Verhandl. DPG (VI) 17, 390 (1982); Appl. Spectrosc. (to be published).

80. Allemand, C.D., Appl. Spectrosc. 24, 348 (1970).

81. Hess, W.R., Hacker, H., Schrötter, H.W., and Brandmüller, J., Z. angew. Phys. 27, 233 (1969).

82. Kiefer, W., and Topp, J.A., Appl. Spectrosc. 28, 26 (1974).

83. Spiro, T.G., and Strekas, T.C., Proc. Natl. Acad. Sci. USA 69, 2622 (1972).

84. Mortensen, O.S., and Koningstein, J.A., J. Chem. Phys. 48, 3971 (1968).

85. Kiefer, W., and Nitsch, W., Appl. Spectrosc. 31, 50 (1977).

86. Scherer, J.R., Kint, S., and Bailey, G.F., J. Mol. Spectrosc. 39, 146 (1971).

87. Klöckner, W., Kiefer, W., and Schrötter, H.W., Appl. Spectrosc. 31, 52 (1977).

88. Srinivasan, K., Finsterhölzl, H., Klöckner, H.W., Illig, D., and Schrötter, H.W., Z. Naturforsch. 32a, 1070 (1977).

89. Klöckner, H.W., Finsterhölzl, H., Srinivasan, K., and Schrötter, H.W., Appl. Spectrosc. 32, 401 (1978).

90. Laane, J., and Kiefer, W., Appl. Spectrosc. 35, 428 (1981).

91. Schrötter, H.W., and Klöckner, H.W., in Raman Spectroscopy of Gases and Liquids (Weber, A., ed.), Topics in Current Physics, Vol. 11, Springer, Berlin 1979, p. 123.

92. Warsop, P.A., and Montero, S., J. Raman Spectrosc. 7, 115 (1978).

93. Orduna, F., Domingo, C., Montero, S., and Murphy, W.F., Mol. Phys. 45, 65 (1982).

94. Shorygin, P.P., Zh. Fiz. Khim. SSSR 21, 1125 (1947).

95. Behringer, J., in Molecular Spectroscopy (Barrow, R.F., Long, D.A., and Millen, D.J., eds.), Vol. 2, The Chemical Society, London 1974. p. 100.

96. Behringer, J., ibid., Vol. 3, 1975, p. 163.

97. Rousseau, D.L., Friedman, J.M., and Willians, P.F., in Raman Spectroscopy of Gases and Liquids (Weber, A., ed.), Topics in Current Physics, Vol. 11, Springer, Berlin 1979, p. 203.

98. Baierl, P., and Kiefer, W., J. Raman Spectrosc. 10, 197 (1981).

99. Baierl, P., and Kiefer, W., J. Raman Spectrosc. 11, 393 (1981).

100. Condon, E.V., Phys. Rev. 32, 858 (1928).

101. Baierl, P., and Kiefer, W., J. Chem. Phys. (in print).

102. Kirillov, D., Lam, L.K., and Hellwarth, R.W., J. Mol. Spectrosc. 91, 269 (1982).

103. Merlin, J.C., Lorriaux, J.-L., and Hester, R.E., J. Raman Spectrosc. 11, 384 (1981).

104. Van Duyne, R.F., in Chemical and Biochemical Applications of Lasers (Moore, C.B., ed.), Vol. 4, Academic Press, New York 1979.

105. Furtak, T.E., and Reyes, J., Surface Sci. 93, 351 (1980).

106. Lee, T.-K., and Birman, J.L., J. Raman Spectrosc. 10, 141 (1981).

107. Krasser,W., and Renouprez, A.J., J. Raman Spectrosc. 11, 425 (1981).

108. Loo, B.H., J. Chem. Phys. 75, 5955 (1981).

109. Barron, L.D., in Ref.(7), p. 677.

110. Barron, L.D., in Advances in Infrared and Raman Spectroscopy (Clark, R.J.H., and Hester, R.E., eds.), Vol. 4, Heyden, London 1978, p. 314.

111. Hug, W., Appl. Spectrosc. 35, 115 (1981).

112. Hug, W., and Surbeck, H., Chem. Phys. Lett. 60, 186 (1979).

113. Hug, W., Kamatari, A., Srinivasan, K., Hansen, H.-J., and Sliwka, H.-R., Chem. Phys. Lett. 76, 469 (1980).

114. Rasetti, F., Z. Physik 66, 646 (1930).

115. Fast, H., Welsh, H.L., and Lepard, D.W., Can. J. Phys. 47, 2879 (1969).

116. Rich, N.H., and Welsh, H.L., J. Opt. Soc. Amer. 61, 977 (1971).

117. Koningstein, J.A., and Mortensen, O.S., in The Raman Effect (Anderson, A., ed.), Vol. 2, Marcel Dekker, New York 1972, p. 519.

118. Koningstein, J.A., in Molecular Spectroscopy (Barrow, R.F., Long, D.A., and Sheridan, J., eds.), Vol. 4, The Chemical Society, London 1976, p. 196.

119. Abdalian, A.T., and Moch, P., in Ref. (8), Vol. 2, p. 410.

120. Klein, M.V., in Light Scattering in Solids (Cardona, M., ed.), Topics in Applied Physics, Vol. 8, Springer, Berlin 1975, p. 147.

121. Wright, G.B., ed., Light Scattering Spectra of Solids, Springer, New York 1969.

122. Balkanski, M., ed., Proc. 2nd Int. Conf. on Light Scattering in Solids, Flammarion, Paris 1971.

123. Balkanski, M., Leite, R.C.C., and Porto, S.P.S., eds. Proc. 3rd Int. Conf. on Light Scattering in Solids, Flammarion, Paris 1976.

124. Poulet, H., Mathieu, J.P., Spectres de Vibration et Symétrie des Cristaux, Gordon and Breach, Paris 1970.

125. Brandmüller, J., and Schrötter, H.W., Topics in Current Chemistry, Vol. 36, Springer, Berlin 1973, p. 85.

126. Cardona, M., ed., Light Scattering in Solids, Topics in Applied Physics, Vol. 8, Springer, Berlin 1975.

127. Claus, R., Merten, L., and Brandmüller, J., Light Scattering by Phonon-Polaritons, Springer Tracts in Modern Physics, Vol. 75, Springer, Berlin 1975.

128. Claus, R., phys. stat. sol. (b) 100, 9 (1980).

129. Claus, R., phys. stat. sol. (b) 88, 683 (1978).

130. Claus, R., and Borstel, G., phys. stat. sol. (b) 88, K 123 (1978).

131. Merten, L., and Claus, R., phys. stat. sol. (b) 89, 159 (1978).

132. Claus, R., J. Raman Spectrosc. 10, 183 (1981).

133. Lapp, M., and Penney, C.M., eds., Laser Raman Gas Diagnostics, Plenum Press, New York 1974.

134. Lapp, M., and Penney, C.M., in Advances in Infrared and Raman Spectroscopy (Clark, R.J.H., and Hester, R.E., eds.), Vol. 3, Heyden, London 1977, p. 204.

135. Lapp, M., in Ref. (8), Vol. I, p. 219.

136. Black, P.C., and Chang, R.K., AIAA J. 16, 295 (1978).

137. Howard-Lock, H.E., and Stoicheff, B.P., J. Mol. Spectrosc. 37, 321 (1971).

138. Finsterhölzl, H., Illig, D., and Schrötter, H.W., Verhandl. DPG (VI) 17, 391 (1982).

139. Gaufrès, R., Maillols, J., and Tabacik, V., J. Raman Spectrosc. 11, 442 (1981).

140. Inaba, H., in Laser Monitoring of the Atmosphere (Hinkley, E.D., ed.), Topics in Appl. Phys., Vol. 14, Springer, Berlin 1976, p. 153.

141. Renaut, D., Pourny, J.C., and Capitini, R., Optics Letters 5, 233 (1980).

142. Capitini, R., Renaut, D., and Pourny, J.C., in Ref. (9), p. 238.

143. Silvera, I.F., and Tommasini, F., Phys. Rev. Lett. 37, 136 (1976).

144. Bloom, S.D., Harney, R.C., and Milanovich, F.P., Appl. Spectrosc. 30, 64 (1976).

145. Dhamelincourt, P., Beny, J.-M., Dubessy, J., Poty, B., Bull. Minéral. 102, 600 (1979) and private communication.

146. See Ref. (9), pp. 532-562.

147. Nelson, W.H., and Carey, P.R., J. Raman Spectrosc. 11, 326 (1981).

148. Drissler, F., Collective Phenomena 3, 147 (1981).

149. Albrecht, H., and Breitinger, D., in Ref. (7), p. 212.

150. Albrecht, H., Müller, G., and Schaldach, M., in Ref. (8), Vol. 2, p. 526.

INFRARED SPECTROSCOPY: A STATE OF THE ART REPORT

J. R. Durig and J. F. Sullivan

Department of Chemistry
University of South Carolina
Columbia, South Carolina, U.S.A.

1. INTRODUCTION

Due to the complementary nature of infrared and Raman spectroscopy for studying the vibrational and rotational energies of molecules, it is interesting to compare the two techniques in terms of sample handling, instrumentation, and applications. This information, which is listed in Table 1, is taken from an article written by H. J. Sloan [1] and represents state-of-the-art commercial Raman instrumentation and its dispersive infrared counterparts as of ten years ago. Of course, much of this information must now be modified in light of the recent innovations in both Raman and infrared spectroscopy, and, in particular, Fourier transform infrared spectroscopy. The recent marketing of a table-top Fourier transform infrared spectrometer by Nicolet Instruments, Inc., with a pricetag competitive with the computerized grating instruments available, establishes Fourier transform infrared spectroscopy as the state-of-the-art technique for the applied chemist who does a variety of sophisticated research. In this chapter, we will attempt to summarize the state-of-the-art of infrared spectroscopy at the present time in terms of instrumentation, applications and techniques.

2. HISTORY

In order to present the state-of-the-art of infrared spectroscopy in its proper perspective, a brief review of the history of infrared spectroscopy should be given; however, a more detailed and candid historical review has recently been pre-

W. Kiefer and D. A. Long (eds.), Non-Linear Raman Spectroscopy and Its Chemical Applications, 33–50.

Table 1 Comparison of Infrared and Raman Spectroscopy[a] in 1971

INSTRUMENTATION	RAMAN	INFRARED
Relative complexity	Moderate	Slightly greater
Source	Laser	Blackbody
Detector	Photomultiplier tube	Thermal
Resolution	ca. 0.25 cm^{-1}	ca. 0.20 cm^{-1}
Principal limitation	Energy	Energy
Wavenumber range	? 0-4000 Δcm^{-1}	180-4000 cm^{-1} (one instrument) 10-400 cm^{-1} (second instrument)
Purge requirement	No	Yes
Photometry	SB emission	DB absorption
SAMPLE HANDLING	**RAMAN**	**INFRARED**
General applicability	95%	99%
Sample limitations	Color; fluorescence	Single crystals; metals; aqueous solutions
Ease of sample preparation	Very simple	Variable
(1) Liquids	Very simple	Very simple
(2) Powders	Very simple	More difficult
(3) Single crystals	Very simple	Very difficult
(4) Polymers	Very simple	More difficult
(5) Single fibers	Possible	Very difficult
(6) Gases and vapors	Now possible	Simple
Cells	Very simple	More complex
Micro work	Good (<10 μg)	Good (<10 μg)
Trace work	Sometimes	Sometimes
High and low temperature	Moderately simple	Moderately simple
APPLICATIONS	**RAMAN**	**INFRARED**
Fingerprinting	Excellent	Excellent
Best vibrations	Symmetric	Asymmetric
Assignment work	Excellent	Very good
Group frequencies	Excellent	Excellent
Aqueous solutions	Very good	Very difficult
Quantitative analysis	More difficult	Good
Low frequency modes	Excellent	Difficult

[a]Taken from Ref. [1].

sented by R. N. Jones [2]. The existence of infrared radiation was first recognized in 1800 by Sir William Herschel [3] who, with a glass prism and blackened thermometers, observed an increase in temperature in the region beyond the red end of the projected solar spectrum. The time period from 1830 to 1870 included the design of thermopiles by Nobili, the construction of the first mid-infrared spectrometer upon the discovery of the transparency of sodium chloride to infrared radiation by Melloni, and progress in emission spectroscopy in the visible. Additionally, the invention of the bolometer by Langley in 1878 enabled infrared spectrometers with much higher resolution to be designed.

In the late 1800's, Michelson's expertise in optics culminated in the design of an instrument, termed an interferometer, in which the relative path differences of two interfering beams could be easily and precisely varied to produce an interference pattern [4,5]. However, it was not until 1911 that the first true infrared interferogram was published by Rubens and Wood [6]. Although Ruben's interferometer differed slightly from the Michelson design, he suffered the same major problem that Michelson faced, i.e., the lack of computational resources with which to transform his interferogram into a spectrum.

In the meantime, dispersive spectroscopy had greatly advanced as evidenced by the publication in 1905 [7] of the infrared spectra of hundreds of compounds, both organic and inorganic, by W. W. Coblentz. The establishment of quantum physics and the development of molecular spectroscopy came about in the 1920's and 1930's, and commercial dispersive infrared spectrometers were produced in the early 1940's due to wartime needs. An additional outcome of this period was the advancement in the science of electronics.

The first numerical Fourier transform calculation by Fellgett [8] in 1949 marked a new era in the potential uses of interferometry. In the early 1950's, J. D. Strong assembled a group of investigators, including H. A. Gebbie, E. V. Loewenstein, and G. Vanasse, who showed that the advantages of multiplexing and throughput claimed for interferometry by Fellgett [9] and Jacquinot and Dufour [10] could be realized. By this time, it was possible to calculate the necessary Fourier transforms; however, it remained a laborious and time consuming undertaking. At this point, the investigations in multiplex spectroscopy diverged into the development of different types of interferometers. Gebbie, in conjunction with G. W. Chantry, J. E. Chamberlain, and J. E. Gibbs, turned to Michelson interferometers, while Strong and Vanasse developed the lamellar grating interferometer.

At the onset of the 1960's, interest in interferometric spectroscopy was growing, and that decade was to see many advances in the theory of interferometric measurements and its application to physical systems, as well as great strides in equipment and technology. Some of the researchers associated with this period who deserve recognition include P. and J. Connes with their astronomical spectra, E. E. Bell and R. Sanderson's work in asymmetric Fourier transform spectroscopy, J. M. Dowling and C. M. Randall, who studied atmospheric gases, and L. Mertz, who contributed greatly to the design of current instrumentation.

The key developments of this period did not come from spectroscopists, however. The availability of inexpensive but powerful minicomputers coupled with the publication of the Cooley-Tukey algorithm for fast Fourier transformation, which was a real breakthrough for mid-infrared Fourier transform spectroscopy, removed the last major drawback to interferometric spectroscopy. Gebbie [11] has commented that computers, like photographic plates, are wonderfully sophisticated and essential spectroscopic tools that, fortunately, the spectroscopist did not have to invent himself.

In the early 1970's it became possible to purchase commercial interferometric instruments, and the field has blossomed as the techniques have become available to those without extensive machining and design capabilities.

3. INSTRUMENTATION

Even to a spectroscopist who has a well-versed knowledge of infrared instrumentation, the diversity of infrared spectrometers available can be confusing. A. Lee Smith [12] has presented a decipherable classification of infrared spectrometers (Fig. 1) which he adapted from a detector approach scheme by Winefordner et al. [13]. A sequential device uses a single detector and its characterization lies in the fact that the information is collected sequentially, i.e., each spectral element is scanned at one time. Spatial devices rely on multiple detectors which collect the spectral element information simultaneously and independently. Multiplex devices yield individual spectral element information through the use of a single detector which receives signals simultaneously from different elements of the spectrum.

The Michelson interferometer is the basic instrument for Fourier transform infrared spectroscopy today. As can be seen from Fig. 2, the radiation from the source passes onto a beamsplitter (an ideal but non-existent beamsplitter has zero

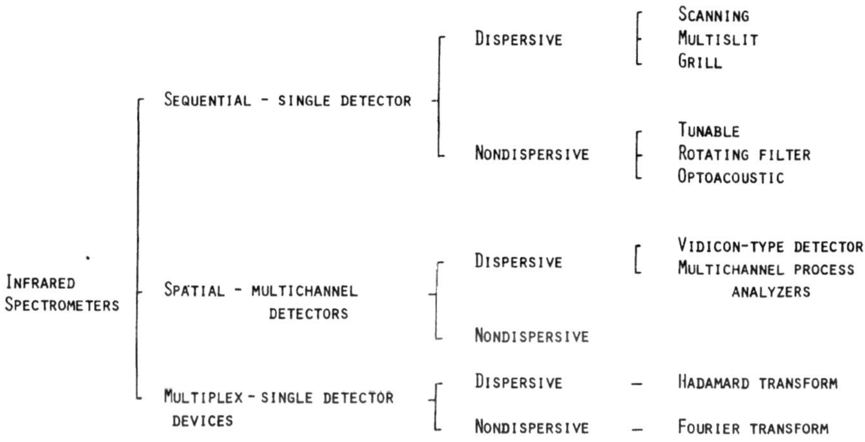

Fig. 1 Classification of infrared spectrometers (used by permission, Ref. [12]).

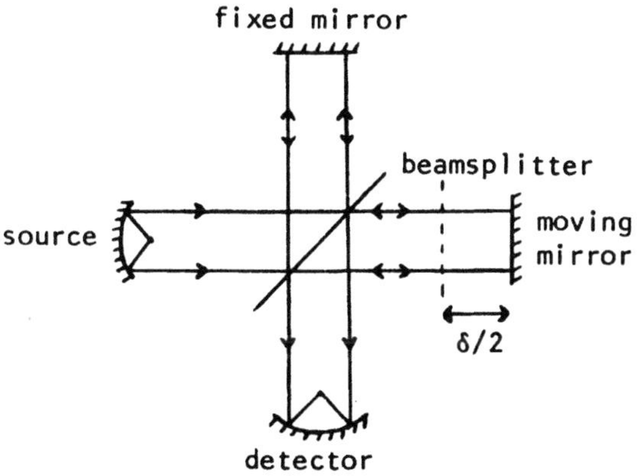

Fig. 2 Basic design of a Michelson interferometer.

absorption and 50% transmittance and reflectance) which divides the radiation into two beams, one of which travels to a fixed mirror and the other to an oscillating mirror. After being reflected, the two beams recombine and interfere at the beamsplitter and 50% of the beam which was reflected from the oscillating mirror is transmitted back in the direction of the source, whereas the other 50% is reflected by the beamsplitter towards the detector. Fifty percent of the beam which was reflected from the fixed mirror is reflected towards the source and the other 50% is transmitted through the beamsplitter towards the detector. When the mirror oscillates at a constant speed, the intensity of each spectral element is modulated sinusoidally at a frequency which depends on the frequency of the spectral information and the mirror speed, and therefore the detector output is a sum of sinusoidal functions. The spectrum can thus be obtained by computing the cosine Fourier transform of the interferogram which is the integral of the contributions from all wavenumbers in the spectrum. For a more detailed discussion of Fourier transform infrared spectroscopy, the reader is referred to recent monographs by P. R. Griffiths [14] and A. E. Martin [15].

A. Sources

The most commonly used source for the far infrared spectral region is a mercury lamp with silica envelope whose chief advantage is that it radiates a greater proportion of its total energy at longer wavelengths and less at shorter wavelengths. Additionally, at frequencies below 100 cm^{-1}, the silica envelope transmits a useful amount of energy from the plasma itself. By surrounding the mercury lamp with a water cooled jacket, the proportion of energy radiated at shorter wavelengths is reduced, the stability of the arc is improved, and a large temperature gradient in the interferometer is avoided. Since there is no shortage of energy at higher frequencies (500 to 10,000 cm^{-1}), a Globar or Nichrome refractory glower may be used as sources for these regions.

B. Beamsplitters

As stated previously, an ideal beamsplitter is one which transmits 50% of the radiation and reflects 50% of the radiation. In the far infrared region the most commonly used beamsplitter material is polyethylene terephthalate (Mylar in the U.S., Melinex in Europe) whose refractive index of ~1.7 is high and therefore a sufficiently high reflectance is attained without the application of a reflective coating. As a result of multiple reflections occurring within the beamsplitter, the efficiency of the beamsplitter varies with frequency and therefore to obtain a good signal over a wide spectral range, a number of beamsplitters with varying thicknesses are required to cover the 500 to 10 cm^{-1} frequency region. Another type of beamsplitter which can be used for the far infrared spectral region is a wire grid, the efficiency of which at a particular wavelength depends on the grating constant of the mesh, and whose reflectance is strongly frequency dependent. This type of beamsplitter may be superior to Mylar but it has not been extensively used, perhaps due to its manufacturing cost.

For higher frequencies, flat plates of transparent optical materials are used. For the near infrared region, calcium fluoride or quartz, coated with a thin film of iron oxide, are most commonly used whereas for the mid-infrared spectral region germanium and silicon films are generally used. These high refractive index materials are brittle and therefore must be supported on a transmitting substrate (e.g., CsI, CsBr, NaCl, or KBr), the choice of which depends on the desired spectral range. These substrates should have the characteristics of a low refractive index, a high transmittance, a high degree of flatness, and insensitivity to atmospheric components, e.g., water vapor. With the use of these thick plates, a compensator plate of equal thickness of the same material is placed on the other side of the film or coating to prevent dispersion effects.

C. Detectors

The most commonly used detectors for Fourier transform infrared spectroscopy are triglycine sulfate (TGS) or deuterated triglycine sulfate (DTGS) pyroelectric bolometers because they have a good response to a wide range of infrared frequencies (5000 to 33 cm^{-1}), and they can be used at high chopping frequencies, although there is some loss of efficiency at chopping frequencies of >1000 Hz.

If the desired spectral range is the mid-infrared region (5000-400 cm^{-1}), the mercury cadmium telluride (MCT) detector, which can be conveniently operated at liquid nitrogen temperature, offers fast response and increased sensitivity. Helium-cooled bolometers also offer high sensitivity in both the far and mid-infrared regions, but have lacked general usage. However, as energy-inefficient sampling techniques become more commonplace, the popularity of these detectors will probably increase.

3. APPLICATIONS AND TECHNIQUES

In the preceding discussion, we have only briefly treated the instrumentation used in Fourier transform spectroscopy. In the space allotted, it is impossible to comment on the theoretical and optical considerations of interferometry, the details of interferometer hardware, and data systems. In the present section we will attempt to summarize several aspects of state-of-the-art infrared spectroscopy in terms of applications and techniques. By no means will this be an all-encompassing review, but rather an indication that many of the problems previously encountered in infrared spectroscopy in terms of sample handling and applications are now being overcome.

A. Gas Chromatography

Gas chromatography Fourier transform infrared spectroscopy (GC/FT-IR) has become a routine and powerful analytical technique over the past ten years. Since the first description of the experiment using a rapid-scanning interferometer in 1967 [16], researchers in this area have striven to optimally design a system which offers adequate chromatographic and spectral resolution, spectral sensitivity, and sophisticated data handling. Although GC/IR analyses can be made by using a rapid-scan dispersive spectrometer, the interferometer gives superior results mainly because of its energy advantage. Additionally, the detection limits of this technique have decreased from microgram to nanogram quantities in the past few years.

The chromatographic resolution and spectral sensitivity of GC/FT-IR experiments depend on such factors as the dimensions of the light pipe [17], i.e., a long, narrow gas cell with highly reflective walls through which the effluent from the GC column is continuously passed; the flow rate; the type of column employed; and the spectral acquisition rate. Increased separation capabilities through the use of capillary GC/FT-IR are currently being explored [18] where the most important factor in its development is increasing the sensitivity of the technique.

The vast amount of data generated in a GC/FT-IR experiment necessitates continual development and improvement in data handling. Until very recently, the time required to transform the interferogram exceeded the time needed to collect the interferogram. However, array processors now provide the necessary speed for computations and this aspect will undoubtedly lead to many alternatives in data handling [19]. As pointed out by P. R. Griffiths [20], the detection limits of GC/FT-IR will never equal those using the GC/MS technique, but a combined GC/FT-IR/MS system may provide the greatest amount of information for rapid and precise identification of components in a complex mixture.

B. Diffuse Reflectance Infrared Spectroscopy

The infrared spectra of difficult solid materials may be obtained in several ways. Attenuated total reflection (ATR) is probably the method of choice for soft materials such as gums or self-adhering materials, and fibers. However, for the ATR technique to be effective and reliable, the sample must afford good contact with the ATR plate.

Diffuse reflectance has recently emerged as a viable technique for identifying solid materials in pure powdered form or as trace quantities in nonabsorbing matrices. This technique has generally been restricted to Fourier transform interferometers because the spectra can be measured with a very high signal-to-noise ratio, particularly when an MCT detector is used [21]. The drawbacks of this technique include the inversion of bands of high absorptivity and irreducible intensities, due to the presence of specularly reflected radiation. However, band ratios may be used for quantitative analyses at low concentrations, and at high concentrations if suitable corrections are made [22]. Nevertheless, diffuse reflectance FT-IR spectroscopy (DRIFT) is emerging as a superior and convenient method for microsampling [23] and for use in surface chemistry analyses.

C. Photoacoustic Infrared Spectroscopy

Photoacoustic Fourier transform infrared spectroscopy (PAS) is another very recently developed technique for the study of solid samples [24], as well as liquids and gases. The attractiveness of this technique, particularly to the analytical chemist, lies in the fact that the morphology of the sample presents no formidable preparation. Although photoacoustic spectra have a much lower signal-to-noise ratio as compared to diffuse reflectance spectra, several of the disadvantages of DRIFT are overcome by the use of PAS [25]. For example, for samples at high concentration or for bands with high absorptivities, no spectral anomalies have been observed in PAS spectra; however, quantitative analyses of PAS spectra require linearization procedures. Yeboah et al. [25] have compared several aspects of both the PAS and DRIFT techniques and concluded that the techniques are indeed complementary.

D. Time Resolved Interferometry

Time resolved spectroscopy appears to have potential for the study of reaction intermediates and products of gas-phase reactions in that it may provide sufficient information for the examination of the band shapes of materials, and energy distributions within the molecules. The main difference between time resolved spectroscopy and conventional FT-IR spectroscopy is that the source intensity, which may be in either the emission mode or absorption mode, is time dependent rather than time independent.

The utilization of FT-IR methods in time resolved spectroscopy was originally developed for step-scan interferometry [26,27] and recently A. Mantz has adapted this technique for rapid-scanning interferometers [28,29]. However, several problems associated with the control of experimental conditions have led to misinterpreted results, but they are now being resolved [30,31]. Most recently, time resolved Fourier transform spectroscopy has been applied to the study of the stretching-relaxation of polymers [32], and, as pointed out by Mamantov and coworkers [30], FT-IR time resolved spectroscopy should have its greatest impact in the field of relaxation and energy transfer as equipment which has better time resolution is developed.

E. Vibrational Circular Dichroism

The routine measurements of vibrational optical activity (VOA) spectra, which include infrared vibrational circular dichroism (VCD) and Raman optical activity (ROA), has taken place only over the past several years. A comprehensive review of both of these techniques has recently been given by L. A. Nafie [33], who has pointed out that optical activity is a unique tool in the study of molecular structure since it is able to probe the three-dimensional stereochemistry of molecules in disordered phases; the only other probe capable of distinguishing optical isomers is single crystal X-ray diffraction.

While the VCD and ROA techniques have many common features such as yielding initial spectra which register the difference between left and right circularly polarized intensities, the measurement of the intensities in vibrational transitions, the use of the same vibrational mode analysis in the calculations, and about the same level of detection, they are also complementary techniques, and perhaps will become more so than classical infrared and Raman spectroscopy.

To fully comprehend the experimental difficulties in obtaining VCD spectra, one only has to realize that the experimentalist is measuring a change in the band intensity which is generally 10^{-4} to 10^{-6} of the magnitude of the band itself. Therefore, this fact, coupled with the fact that the VCD spectral features are four to five orders of magnitude weaker than ordinary absorbance spectra, has necessitated improvements in signal quality, ease of measurement, and extension of the spectral range below ~1600 cm^{-1}. It is not surprising then that Nafie and coworkers have extended their studies from dispersive VCD spectroscopy to the use of Fourier transform spectroscopy in conjunction wih a double modulation approach [34]. Their continual instrumental improvements have thus far resulted in reduced noise levels with reduced scanning time and the measurement of high quality VCD spectra to as low as 1280 cm^{-1} [35]. Further developments may include the use of tunable infrared lasers as sources, as well as the development of the theoretical methods [33].

F. Conventional Fourier Transform Infrared Spectroscopy

Throughout the preceding sections, we have discussed, albeit briefly, the analytical aspects of state-of-the-art infrared spectroscopy. On the more traditional level, however, research is continually being conducted in the areas of vibrational assignments, skeletal bending modes including ring vibrations, torsional vibrations, and conformational analyses, all of which comprise the authors' own research interests. Most recently, the majority of our research efforts have been directed towards conformational analyses in which one of the major goals is the calculation of the energy difference between the two conformers, ΔE, as well as the energy necessary for interconversion. Once these data are available, calculation of the thermodynamic functions, G, H, S and C_p, is possible. The calculation of these energy quantities is facilitated by using a potential function which describes the vibrational motion, or internal rotation (torsion), as a function of the dihedral angle, α. The potential function is called asymmetric because both the frame and the top portions of the molecule have no symmetry element higher than a plane. Fourier transform far infrared spectroscopy can provide a wealth of information on the torsional transitions of a large number of asymmetric rotors. Usually the spectra are very "rich" with many "hot bands" and frequently overtones with accompanying "hot bands" can also be observed. The transitions may yield very sharp Q-branches when the asymmetric torsions give rise to C-type band contours. When two or more conformers are present in appreciable quantities it may be difficult to distinguish which Q-branch series should be assigned to which conformer. In many cases the Raman spectrum of the vapor can provide the polarization data necessary for making the correct assignment. For example, if one of the conformers has a symmetry plane and the asymmetric torsional mode is an A" vibration giving rise to a depolarized Raman line, and if the other conformer has no symmetry so that the asymmetric torsion gives rise to a polarized Raman line, then one can readily distinguish between the two torsions in the Raman effect. Also, for some molecules, the asymmetric torsional overtones are observed in the Raman effect; these data can provide a check of the assignments made on the basis of the infrared data. Nevertheless we have found that the far infrared spectra usually contain many more transitions than the Raman spectra and that they are much more valuable for studying asymmetric torsions. From these torsional transitions it is not only possible to obtain the potential function governing the internal rotation but it is also possible to obtain the energy differences between conformers and the various barriers to interconversion. Additionally, under favorable circumstances, it may be possible to obtain the dihedral angle and identify the symmetry of the most stable conformer. Only recently has far infrared FT spectroscopy been used to study asymmetric torsions.

As an example, all of the asymmetric torsional data for gaseous propanal [36] arose only from the far infrared spectrum

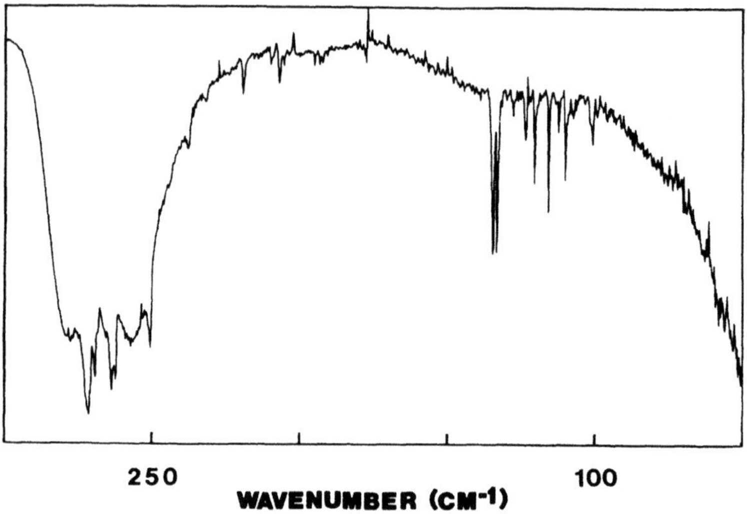

Fig. 3 Far infrared spectrum of gaseous propanal (used by permission, Ref. [36]).

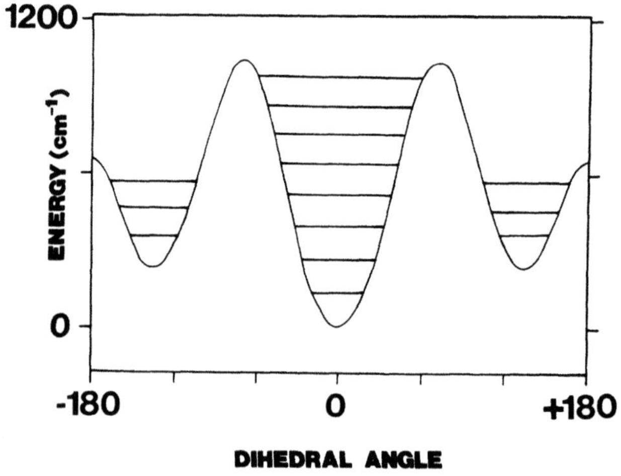

Fig. 4 Torsional potential function, including observed energy levels, for the asymmetric torsion of propanal. The dihedral angle of zero corresponds to the <u>s-cis</u> conformer. The <u>gauche</u> levels shown are doubly degenerate, with the splittings between these levels too small to be apparent on the scale used (used by permission, Ref. [36]).

Table 2 Observed Far Infrared Frequencies and Proposed Assignment for the Asymmetric Torsion of Gaseous Propanal[a]

$\nu(cm^{-1})$	Rel. Int.	Assignment		Obs-Calc[b]
271.5	vs	g	skeletal bend, ν_{15}	
269.3	s	c	$2 \leftarrow 0$	2.0[c]
264.0	s	c	skeletal bend, ν_{15}	
262.5	s	c	$3 \leftarrow 1$	-2.0[c]
250.6	m	c	$4 \leftarrow 2$	-7.1[c]
244.4	vw	c	$5 \leftarrow 3$	-3.5[c]
238.0	w	c	$6 \leftarrow 4$	3.6[c]
232.0	vw			
135.1	vs	c	$1 \leftarrow 0$	1.2
133.8	vs	c	$2 \leftarrow 1$	0.3
128.3	vw	c	$3 \leftarrow 2$	-2.6
124.3	w	c	$4 \leftarrow 3$	-2.6
121.0	m	c	$5 \leftarrow 4$	-0.1
116.5	s	c	$6 \leftarrow 5$	3.2
113.1	w	g	$1\pm \leftarrow 0\mp$	-2.6
110.7	m	c	$7 \leftarrow 6$	0.1
108.1	w	g	$2\pm \leftarrow 1\mp$	2.1
102.2	vw	g	$3- \leftarrow 2+$	0.5[c]
101.5	vw	c	$8 \leftarrow 7$	6.8[c]

[a]Data taken from Ref. [36]. Abbreviations used: v, very; s, strong; m, moderate; w, weak; g, gauche; c, s-cis.

[b]Calculated using the following potential coefficients: V_2 = 559, V_3 = 651, V_4 = -105, and V_6 = -46 cm^{-1}.

[c]Not used in the final fit.

which is shown in Fig. 3. The strong band at 271.5 cm^{-1} had previously been assigned [37,38] to the s-cis in-plane bending mode ν_{15} whereas the s-cis asymmetric torsion ν_{24} was assigned [37] to an infrared absorption at 133 cm^{-1}. These two bands were found to remain in the spectrum of the annealed solid phase [36]. From Fig. 3, one can see that the asymmetric torsional mode at 133 cm^{-1} is not one Q-branch, but a series of Q-branches which do not follow the normal trend for relative intensities expected from Boltzmann statistics, and very careful analysis was necessary in order to assign these bands to the correct torsional transitions. The observation of a number of weak, sharp bands, similar in frequency to the skeletal bending modes, enabled a definitive assignment to be made, since these bands turned out to be the two-quantum transitions (double jumps) of the asymmetric torsion. The complete assignment of the observed far infrared bands, excluding those bands which were assigned to the methyl torsional modes, is presented in Table 2, and the potential function is shown in Fig. 4. The poor fit of the s-cis 8 ← 7 transition results from the proximity of the eighth level to the barrier, being the last transition observed below the barrier. Because of this, the level is very sensitive to small changes in the potential function. For a more detailed treatment of conformational analyses and asymmetric potential functions, the interested reader is referred to a recent review article on this subject [39].

G. Tunable Infrared Laser Sources

Throughout this chapter, we have discussed the state-of-the-art of Fourier transform infrared spectroscopy and, as we have seen, it is one of the most versatile techniques available to the spectroscopist and analytical chemist. Equally exciting though is the development, over the past decade, of practical tunable infrared laser sources with which high resolution absorption spectra can be obtained. An excellent review on this subject has recently appeared [40]. The major difference between lasers and the conventional thermal sources is not in the total power, but in the high spectral radiance which is a result of their narrow linewidths and spatial coherence. In general, and under similar conditions, the spectral radiance of a blackbody is approximately thirteen orders of magnitude weaker than many tunable lasers. Several types of lasers which have been used as tunable infrared sources include semiconductor diode lasers which have been the most commonly employed, atomic and molecular gas lasers whose ranges may be too restricted for general spectroscopic purposes unless they are operated at high pressures or Zeeman-tuned, spin-flip Raman lasers, polariton lasers whose major drawback in high resolution spectroscopy is the breadth of the output linewidth, Raman frequency shifting devices whose applications may lie in their high output powers, optical parametric oscillators, several types of nonlinear optical mixing techniques, color-center lasers which have found use in the generation of tunable radiation in the near infrared, and other sources which are in their preliminary stages of development, i.e., free-electron lasers, Landau emission and vibronic transition lasers.

We have only presented an extremely brief overview of the status of tunable infrared laser sources; the interested reader is urged to consult the review article by R. S. McDowell [40] and the references therein. The full extent of the applications of tunable laser spectroscopy has not yet been realized; however, there are a few applications such as isotopic analyses, chemical reactions and kinetics, vibrational energy transfer rates, pollution monitoring, and trace-gas analyses, which are very promising. Although there may be great potential for tunable infrared laser spectroscopy, it is still in its infancy and is, at the present time, a laboratory curiosity.

5. SUMMARY AND CONCLUSIONS

The purpose of this chapter is to expose the reader to various state-of-the-art applications and techniques in infrared spectroscopy. However, the experimental accessories, techniques and applications in many of these areas are developing so rapidly that one must stay abreast of the literature. There are several areas of Fourier transform infrared spectroscopy which have not been covered, such as the biological applications reported by R. J. Jakobsen and his coworkers at Battelle-Columbus Laboratories, the measurement of gaseous pollutants and atmospheric components, hardware and software developments, and the tremendous strides being made in data processing routines. There is little doubt that when given the choice between Fourier transform infrared spectroscopy and dispersive infrared spectroscopy, the majority of researchers choose the former mainly because of its energy advantage.

If we now look back at Table 1 which is a comparison of dispersive infrared and Raman spectroscopy in 1971, we can see that in the area of infrared applications, aqueous solutions are no longer "very difficult" since the intense water bands can be removed by spectral subtraction if a suitable sample pathlength is used. However, it should be added that sufficiently thin spacers for liquid cells can present problems. Also the cost of the window materials can be quite high compared to the simple Pyrex capillary cells used in Raman spectroscopy and, additionally, it is necessary to use more than one window material to cover the infrared spectral region from 100 to 4000 cm^{-1}. Thus, it may no longer be considered as "very difficult" to obtain infrared spectra of aqueous solutions, but it is still not as convenient as obtaining the corresponding Raman spectra over this spectral range. From the authors' own research accounts, it is evident that the observation and resolution of low frequency modes in the infrared spectra is no longer difficult, especially if very sensitive detectors such as helium-cooled bolometers are used. However, let us hasten to add that obtaining good quality far infrared spectra still requires considerable effort to remove the water vapor from the interferometer housing by strong flushing with dry nitrogen or a vacuum housing which means that the observation of low frequency modes cannot be classified as routine. Perhaps the major revisions in the Table occur under the heading "Sample Handling" because we have seen that many of the state-of-the-art infrared techniques such as PAS and DRIFT can be applied to samples which require little or no preparation. Nevertheless, it is still no small chore to obtain polarized infrared spectra of single crystals particularly in the lattice spectral region. Clearly Raman spectroscopy has a decided edge in the obtaining of low frequency spectra of solids. Modifications must also be made to the Instrumentation Section where MCT detectors used in FT-IR are comparable in sensitivity to the thermocouples in dispersive infrared. Additionally, only one Fourier transform infrared interferometer is required to cover the spectral range of 4000 to 10 cm^{-1}, if accessories are used. In fact, some of the newly designed interferometers have automatic beamsplitter changers which do not require the opening of the interferometer housing. What remains as the principal limitation in infrared spectroscopy is, of course, energy. Infrared spectroscopy has come a long way in the past decade and it is expected that it will even become more popular in other scientific disciplines as the low cost interferometers become available to the industrial scientists.

REFERENCES

1. H. J. Sloan, Appl. Spectrosc., 25, 430 (1971).

2. R. N. Jones, Can. J. Spectrosc., 26, 1 (1981).

3. W. Herschel, Phil. Trans. Roy. Soc. (London), 90, 284 (1800).

4. A. A. Michelson, Phil. Mag. Ser. 5, 31, 256 (1891).

5. A. A. Michelson, Phil. Mag. Ser. 5, 34, 280 (1892).

6. H. Rubens and R. W. Wood, Phil. Mag. Ser. 6, 21, 249 (1911).

7. W. W. Coblentz, "Investigations of Infrared Spectra, Parts I and II," Carnegie Institute of Washington, Publication No. 35 (1905).

8. P. Fellgett, Aspen Int. Conf. on Fourier Spect., (G. A. Vanasse, A. T. Stair, and D. J. Baker, Eds.), AFCRL-71-0019, p. 139 (1970).

9. P. Fellgett, J. Phys. Radium, 19, 187 (1958).

10. P. Jacquinot and C. J. Dufour, J. Rech. C.N.R.S., 6, 91 (1948).

11. H. A. Gebbie, Appl. Opt., 8, 501 (1969).

12. A. L. Smith, Applied Infrared Spectroscopy, (P. J. Elving and J. D. Winefordner, eds.), Wiley, New York, 1979.

13. J. D. Winefordner, J. J. Fitzgerald, and N. Omenetto, Appl. Spectrosc., 29, 369 (1975).

14. P. R. Griffiths, Chemical Infrared Fourier Transform Spectroscopy, Wiley, New York (1975).

15. A. E. Martin, Infrared Interferometric Spectrometers, (J. R. Durig, ed.), Elsevier, Amsterdam, 1980.

16. M. J. D. Low and S. K. Freeman, Anal. Chem., 39, 194 (1967).

17. P. R. Griffiths, Appl. Spectrosc., 31, 284 (1977).

18. D. Kuehl, 1981 Int. Conf. Fourier Transform Infrared Spectrosc., Proc. SPIE, 289, 140 (1981).

19. S. C. Simonoff, M. L. Olson, D. Kuehl, and R. B. Turner, 1981 Int. Conf. Fourier Transform Infrared Spectrosc., Proc. SPIE, 289, 135 (1981).

20. P. R. Griffiths, 1981 Int. Conf. Fourier Transform Infrared Spectrosc., Proc. SPIE, 289, 446 (1981).

21. M. P. Fuller and P. R. Griffiths, Anal. Chem., 50, 1906 (1978).

22. P. R. Griffiths and M. P. Fuller, in Advances in Infrared and Raman Spectroscopy, (R.J.H. Clark and R. E. Hester, eds.), Vol. 9, Heyden, London, 1981.

23. M. P. Fuller and P. R. Griffiths, Appl. Spectrosc., 34, 533 (1980).

24. D. W. Vidrine, Appl. Spectrosc., 34, 314 (1980).

25. S. A. Yeboah, P. R. Griffiths, K. Krishnan, and D. Kuehl, 1981 Int. Conf. Fourier Transform Infrared Spectrosc., Proc. SPIE, 289, 105 (1981).

26. R. E. Murphy, F. Cook, and H. Sakai, J. Opt. Soc. Am., 65, 600 (1975).

27. H. Sakai and R. E. Murphy, Appl. Opt., 17, 1342 (1978).

28. A. W. Mantz, Appl. Spectrosc., 30, 459 (1976).

29. A. W. Mantz, Appl. Opt., 17, 1347 (1978).

30. A. A. Garrison, R. A. Crocombe, G. Mamantov, and J. A. DeHaseth, Appl. Spectrosc., 34, 399 (1980).

31. J. A. DeHaseth, 1981 Int. Conf. Fourier Transform Infrared Spectrosc., Proc. SPIE, 289, 34 (1981).

32. D. E. Honigs, R. M. Hammaker, W. G. Fateley, and J. L. Koenig, in Vibrational Spectra and Structure, (J. R. Durig, ed.), Vol. 11, Elsevier, Amsterdam, in press.

33. L. A. Nafie, in Vibrational Spectra and Structure, (J. R. Durig, ed.), Vol. 10, Elsevier, Amsterdam, 1981.

34. L. A. Nafié and M. Diem, Appl. Spectrosc., 33, 130 (1979).

35. L. A. Nafie, E. D. Lipp, and C. G. Zimba, 1981 Int. Conf. Fourier Transform Infrared Spectrosc., Proc. SPIE, 289, 457 (1981).

36. J. R. Durig, D. A. C. Compton, and A. Q. McArver, J. Chem. Phys., 73, 719 (1980).

37. S. G. Frankiss and W. Kynaston, Spectrochim. Acta, 28A, 2149 (1972).

38. A. Chapput, B. Roussel, and G. Fleury, J. Raman Spectrosc., 2, 117 (1974).

39. D. A. C. Compton, in Vibrational Spectra and Structure, (J. R. Durig, ed.), Vol. 9, Elsevier, Amsterdam, 1980.

40. R. S. McDowell, in Vibrational Spectra and Structure, (J. R. Durig, ed.), Vol. 10, Elsevier, Amsterdam, 1981.

SOME CHEMICAL APPLICATIONS OF GAS PHASE LINEAR RAMAN SPECTROSCOPY

J. R. Durig and J. F. Sullivan

Department of Chemistry
University of South Carolina
Columbia, South Carolina, U.S.A.

1. INTRODUCTION

Almost as soon as the Raman effect was discovered for liquids [1] and solids [2], the Raman spectra of several gases were reported [3-5]. Since the number of scattering molecules per unit volume is considerably less in the gas phase than in the other phases, weak signal intensity initially prohibited observation of all but the stronger Raman lines of gases. Raman band contours of nontotally symmetric vibrations are ordinarily very weak and broad, and frequently little information can be obtained from them even in terms of the band center. However, it should be pointed out that the band width is as good an indicator of a nontotally symmetric vibration as the depolarization ratios [6]. Nevertheless, with the advent of high-powered lasers and vastly improved detection systems, observation of weak Raman peaks has become routine, and the techniques of gas phase Raman spectroscopy have become more and more commonplace. The current uses include, but are not limited to, the completion of vibrational assignments, structural determinations, conformational analyses, and the study of anharmonic vibrations such as internal torsions, ring-puckering modes, and the low frequency bending motions of quasi-linear molecules. We shall briefly touch on each of these subjects using mainly illustrations from our own studies but this limitation should not be taken as an indication that our laboratory is the only laboratory utilizing gas-phase Raman data. We have simply taken data from our own studies for the most part because it is readily available and we have had a limited time to prepare the current manuscript. The reader should keep these limitations in mind as we review the various applications of gas-phase Raman spectroscopy.

W. Kiefer and D. A. Long (eds.), Non-Linear Raman Spectroscopy and Its Chemical Applications, 51–80.

2. VIBRATIONAL ASSIGNMENTS

For several years we have been using gas-phase Raman spectroscopy for completing vibrational assignments of small molecules. As examples of such studies, we shall review our studies on hydrazine [7], disilylsulfide [8], and phenol [9]. Prior to our studies on hydrazine [7], all of the earlier studies were made on either the solid or liquid states with the exception of the work of Kotov and Tatevskii [10] who recorded the Raman spectrum of hydrazine vapor with Hg arc excitation. Since hydrazine exhibits a considerable amount of hydrogen bonding in the condensed states, it is expected that many of the frequencies are appreciably shifted with condensation and provide only poor estimates of the normal modes for the isolated molecule. Therefore, we recorded [7] the Raman spectra of gaseous N_2H_4 and N_2D_4 using laser excitation with the hope of obtaining frequencies for several of the normal symmetric vibrations. The Raman spectrum of gaseous N_2H_4 (see Fig. 1A) has five pronounced, polarized lines at 3398, 3329, 3302, 3266 and 1076 cm^{-1}. For a molecule of C_2 symmetry there should be two

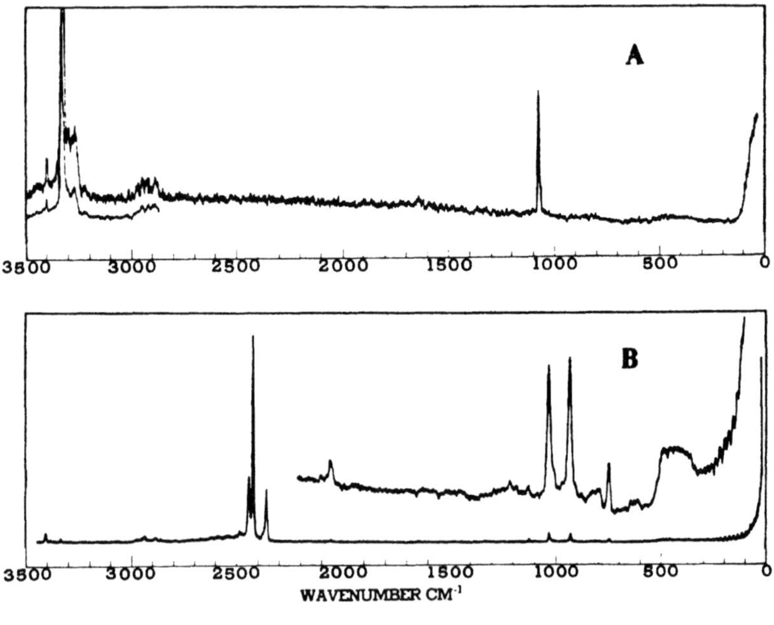

Fig. 1 Raman spectra of gaseous (A) N_2H_4 and (B) N_2D_4 recorded with SBW = 6 cm^{-1} (used by permission, Ref. [7]). The very broad apparent band in the 500-350 cm^{-1} region in spectrum (B) is due to the sample cell and not to the compound.

N-H stretching vibrations which give rise to polarized Raman bands. Therefore, the NH_2 antisymmetric stretch of symmetry "a" was assigned to the Raman line at 3398 cm^{-1} and the symmetric motion of the corresponding symmetry was assigned to the Raman line at 3329 cm^{-1}, the strongest line in the Raman spectrum. The N-H antisymmetric stretch of symmetry species "a" had been assigned in previous studies to bands of considerably lower frequency but the Raman spectrum of the gas led to a definite assignment of the 3398 cm^{-1} line for this normal mode.

The N-N stretch for the N_2H_4 molecule can be clearly seen at 1076 cm^{-1} (see Fig. 1A) in the Raman spectrum of the gas. In an earlier infrared study [10] of this molecule, this band had been assigned to the NH_2 wagging motion, but the intensity of this band in the Raman effect precludes that assignment. The two Raman polarized lines at 3302 and 3266 cm^{-1} were assigned as the overtones of the two NH_2 deformations. There is a very weak line at 1642 cm^{-1} of undetermined polarization which was tentatively assigned to the NH_2 antisymmetric deformation.

The Raman spectrum of gaseous N_2D_4 is shown in Fig. 1B. There are seven easily discernable Raman lines in the "survey" spectrum for this molecule. The assignment of the Raman lines in the N-D stretching region followed directly from that given for the corresponding region in the spectrum of the "light" compound. The Raman lines at 2442, 2422 and 2358 cm^{-1} were assigned to ν_1, ν_2 and $2\nu_3$, respectively, with ν_2 having a somewhat higher frequency than expected because of its Fermi resonance with $2\nu_3$.

The appearance of three easily discernable Raman bands in the general region of 1000 cm^{-1} is the result of more extensive mixing of the N-D bending modes with the N-N stretch. The Raman lines at 1031 and 930 cm^{-1} were assigned to the N-N stretch and ND_2 wag, respectively. The ND_2 rock is clearly observed at 749 cm^{-1} and one of the ND_2 deformations appears at 1126 cm^{-1} with the other one probably falling at 1218 cm^{-1}. The shift factor of the 1126 cm^{-1} line seems large but this deformation can mix with the N-N stretch which could result in a lower frequency for this normal mode. The fact that the 1126 cm^{-1} line is depolarized does not preclude its assignment to a normal mode of "a" symmetry since the polarization might not be experimentally detectable.

These Raman data on the vapor clearly show that the NH_2 antisymmetric stretch in the gas phase is much higher in frequency than it had been previously thought to be. Also, the assignment of the N-N stretch was unambiguous for the "light" molecule. In the earlier infrared study [11] of N_2D_4 vapor, a series of Q-branches was observed beginning at 637 and running

Table 1 Observed[a] Raman Frequencies of Gaseous N_2H_4 and N_2D_4

N_2H_4			N_2D_4			
$\Delta\nu(cm^{-1})$	rel. int.	pol.	$\Delta\nu(cm^{-1})$	rel. int.	pol.	Assignment and Description
3398	m	p	2442	s	p	ν_1(a)$NH_2(ND_2)$ antisym. str.
3329	vvs	p	2422	vvs	p	ν_2(a)$NH_2(ND_2)$ sym. str.
3302	m	p	2358[b]	s	p	$2\nu_{10}$(2x~661) overtone of $NH_2(ND_2)$ def.
3266	m	p				$2\nu_3$(2x1642) overtone of $NH_2(ND_2)$ def.
2958[c]			3072[c]			
2925	mw	?	2939	w	p	impurity?
2890			2885			
			2054	m	p	$2\nu_5$(2x1031=2062)
			1218	vw	?	ν_{10}(a)ND_2 def.
1642	vvw	?	1126	m,br	dp	ν_3(b)$NH_2(ND_2)$ def.
1076	s	p	1031	s	p	ν_5(a)N-N str. (mixed with ND_2 wag)
			1009	w,sh	dp	ν_{11}(b)ND_2 wag (?)
			930	s	p	ν_4(a)ND_2 wag (mixed with N-N str.)
			~805	w	dp	ν_{12}(b)ND_2 rock
			749	m	p	ν_6(a)ND_2 rock

[a]Data taken from Ref. [7]. Abbreviations used: v, very; s, strong; m, medium; w, weak; sh, shoulder; br, broad; p, polarized; dp, depolarized.

[b]$2\nu_3$ in the N_2D_4 would be predicted around 2416 cm^{-1} but it is probably considerably lower because of Fermi resonance with ν_2.

[c]These lines varied in intensity with purification and therefore are concluded to be due to impurities.

to 800 cm^{-1} with various spacings and intensities. A rather strong Q-branch at 722 cm^{-1} was chosen as the center of the "b" species fundamental, ν_{12}, whereas the corresponding rocking mode of the "a" species was assigned to the "washed-out" portion of the fine structure at 688 cm^{-1}. This latter assignment is certainly not consistent with the Raman spectrum of N_2D_4 vapor where a polarized line was observed at 749 cm^{-1}. Thus, it appears that the center of the infrared band with the fine structure is around 750 cm^{-1} which leaves unexplained why the Q-branch series extends to much lower frequencies from the band center than to higher frequencies. This assignment also leaves unexplained the pronounced Q-branch in the infrared spectrum at 722 cm^{-1}. However, it should be mentioned that a similar problem also exists with the fine structure on the infrared absorption between 950 and 1000 cm^{-1}. In the earlier study [11] the band center for the ND_2 wag was chosen at 1009 cm^{-1} which corresponds well with the depolarized Raman line of the same frequency. A Q-branch at 985 cm^{-1} was assigned as the other ND_2 wagging mode whereas the relatively strong polarized Raman line at 930 cm^{-1} is clearly the ND_2 wag of the "a" species. Thus, the Q-branch at 985 cm^{-1} is unexplained with this assignment. These results are summarized in Table 1.

As the second example of the utility of gas-phase Raman data for vibrational assignment purposes we have chosen the disilylsulfide molecule [8]. The observed spectra of both $(SiH_3)_2S$ and $(SiD_3)_2S$ are shown in Fig. 2 and the observed frequencies are listed in Table 2. Prior to our study there had been two previous vibrational studies [12,13] of this molecule but neither study included the Raman spectra of the gas nor solid and the Raman spectrum of the liquid did not include quantitative depolarization values. Because of the lack of these data, a detailed vibrational analysis was not presented.

The symmetric Si-S-Si stretch is found as a medium-strong, polarized line in the Raman spectrum of the gas at 487 cm^{-1} and deuteration shifts this mode to 448 cm^{-1} which indicates some degree of coupling with the SiD_3 motions. The Raman line at 514 cm^{-1} in the spectrum of gaseous $(SiH_3)_2S$ had previously been attributed [12,13] to the antisymmetric skeletal stretch (B_1); however, this line is clearly polarized in the spectra of both the gas and liquid (507 cm^{-1}). We [8] were unable to adequately explain the presence of this line in terms of a non-fundamental process or a reasonable impurity so its origin is unclear at this point. The Si-S-Si skeletal deformation is found at 141 and 134 cm^{-1} in the Raman spectra of gaseous $(SiH_3)_2S$ and $(SiD_3)_2S$, respectively.

The A_1 SiD_3 symmetric deformation appears as a polarized Raman line at 712 cm^{-1} in the spectrum of the gas; however, the

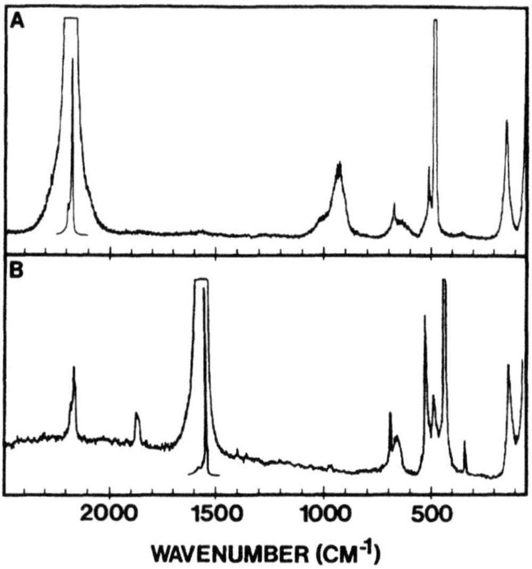

Fig. 2 Raman spectra of gaseous (A) disilylsulfide-d_0 and (B) disilylsulfide-d_6 recorded with SBW = 3 cm^{-1} (used by permission, Ref. [8]).

Raman spectrum of gaseous $(SiH_3)_2S$ does not show a similar line in the silyl deformation region. This result had previously been observed for related compounds containing SiH_3 groups [14,15]. The two SiH_3 antisymmetric deformations were observed at 940 (ν_9) and 928 cm^{-1} (ν_{19}) as weak broad Raman lines which shifted to 689 and 677 cm^{-1}, respectively, with deuteration. The SiH_3 rock of A_1 symmetry appears as a pronounced Raman line at 676 cm^{-1} and shifts to 540 cm^{-1} with deuteration; there is a considerable intensity enhancement because of the increased mixing (approximately 50% in the deuterium compound but only 15% in the light molecule) of the SiD_3 rock with the Si-S-Si stretch in the deuterium compound. The SiH_3 antisymmetric stretch can be confidently assigned at 2188 cm^{-1} because of the polarized nature of this line and it is clearly resolved from the SiH_3 symmetric stretch at 2177 cm^{-1}. The assignments of the remaining observed Raman lines for both the "light" and "heavy" molecules are summarized in Table 2. Again it should be emphasized that the confident assignment of several of the normal modes was made possible by the availability of the Raman spectra of the gaseous $(SiH_3)_2S$ and $(SiD_3)_2S$ molecules.

The final example in this section will be the Raman spectrum of phenol [9]. The vibrational spectrum of phenol had been previously examined using a number of spectroscopic techniques

Table 2 Observed[a] Raman Frequencies of Gaseous $(SiH_3)_2S$ and $(SiD_3)_2S$

$(SiH_3)_2S$			$(SiD_3)_2S$			
$\Delta\nu(cm^{-1})$	rel. int.	pol.	$\Delta\nu(cm^{-1})$	rel. int.	pol.	Assignment and Description
2188	s	p	1584	s	p	ν_1 $SiH_3(SiD_3)$ antisym. str. (A_1)
2177	vvs	p	1567	vvs	p	ν_2 $SiH_3(SiD_3)$ sym. str. (A_1)
1000	w,br	dp				$\nu_6 + \nu_{17} = 1004$ (B_1)
			984	vw	p	$\nu_5 + \nu_6 = 988$ (A_1)
			712	mw	p	ν_3 SiD_3 sym. def. (A_1)
940	w,br	dp	689	w	dp	ν_9 $SiH_3(SiD_3)$ antisym. def. (A_2)
928	w,br	dp	677	w	dp	ν_{19} $SiH_3(SiD_3)$ antisym. def. (B_2)
			665	w,sh	dp	ν_{15} SiD_3 sym. def. (B_1)
676	w	p	540	m	p	ν_5 SiH_3 (SiD_3) rock (A_1)
625	vw,br	dp				ν_{20} SiH_3 rock (B_2)
			521	w	dp	ν_{17} SiSSi str. (B_1)
514	w	p				Impurity (see Text)
			465	w,sh	dp	ν_{10} SiD_3 rock (A_2)
487	ms	p	448	ms	p	ν_6 SiSSi str (A_1)
141	w	p	134	mw	p	ν_7 SiSSi bend (A_1)

[a]Data taken from Ref. [8]. Abbreviations used: see Table 1.

which included uv-vis [16], the infrared spectrum of the vapor [17], and the infrared and Raman spectra of both the liquid and solid (see Ref. [9] for a summary of the references for the earlier work). However, the Raman spectrum of gaseous phenol had not been reported and such data should be quite useful in assigning the in-plane fundamentals of this molecule. Therefore, we [9] examined the medium resolution Raman spectrum of phenol vapor and found that a few of the features of the spectrum warranted comment, in light of the generally accepted vibrational assignment for this molecule [16]. Typical survey spectra covering much of the range from 1800 cm^{-1} to about 100

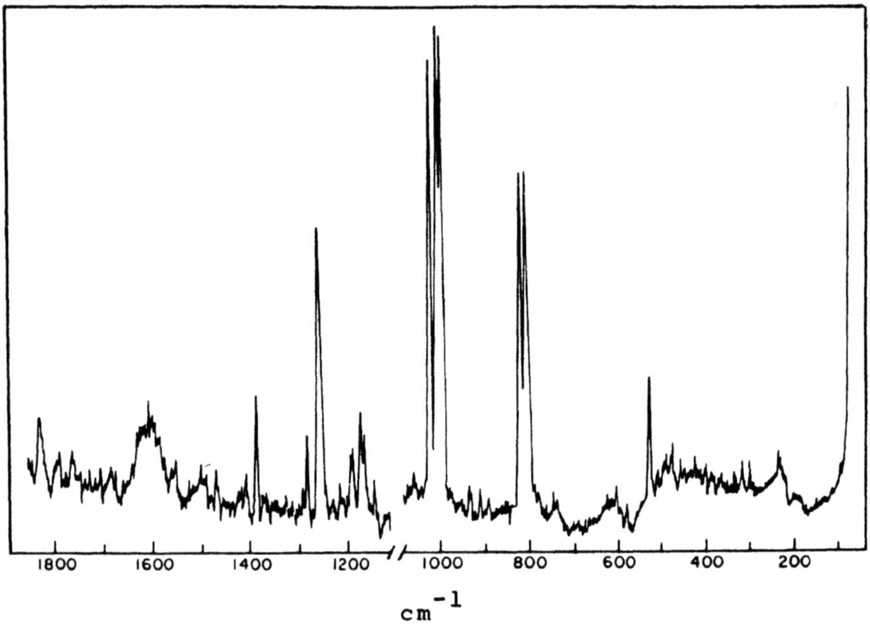

Fig. 3 Raman spectrum of gaseous phenol (used by permission, Ref. [9]).

cm^{-1} are shown in Fig. 3. The spectra were recorded at a spectral slitwidth of about 2.5 cm^{-1} on a Cary Model 82 Raman spectrophotometer equipped with a CRL 53Å argon ion laser which delivered about 3.3 W of 5145 Å radiation at the sample. Sample pressures of about 12 torr were obtained at 75°C in a heated Cary multipass accessory. In Table 3 the observed Raman lines for the vapor are tabulated along with their commonly accepted assignments. Many of the features can be routinely interpreted on the basis of previous vibrational assignments [16]; however, several lines require some comments.

Phenol can assume two symmetrically equivalent configurations by undergoing internal rotation around the CO bond axis. The degenerate pairs for the phenyl moiety are split slightly in the non-rigid phenol molecule; however, the C_s point group is not completely able to explain or distinguish all of the vibrational sublevels that result. Brand and co-workers [16] have defined a molecular group G_4 of order 4 and isomorphous with the C_{2v} point group in their excellent treatment of the UV spectrum of phenol, but we shall retain the C_s point group designations since most of the infrared and Raman gas (g) features can be interpreted using the simpler point group.

Table 3 Observed Raman Frequencies and Assignments for Gaseous Phenol[a]

Frequency[b] (cm^{-1})	Rel. Int.[c]	Assignment[d]	Approximate Description
3656(3656)	>20		OH stretch
3073(3063)	>20	$\nu_2(a_1)$	CH stretch
1833	3	$\nu_{18a}(a_1)+\nu_{12}(a_1)=1849(A_1)$	
1612(1610)	4	$\nu_{8b}(b_2)$	CC stretch
1602(1603)	4	$\nu_{8a}(a_1)$	CC stretch
1504(1501)	1	$\nu_{19a}(a_1)$	CC stretch
1470(1472)	1	$\nu_{19b}(b_2)$	CC stretch
1388(1343)	6	$\nu_{14}(b_2)$	CC stretch
1285(1277)[e]	4	$\nu_3(b_2)$	CC stretch
1264(1261)	12	$\nu_{7a}(a_1)$	CO stretch
1196	3		?
1176(1176)	4		OH bend
1168(1168)	3	$\nu_{9a}(a_1)$	CH bend
1149(1150)	1	$\nu_{9b}(b_2)$	CH bend
1025(1025)	19	$\nu_{18a}(a_1)$	CH bend
1007	22	$2\nu_{16b}(b_1)=1006(A_1)$	
999(999)	20	$\nu_1(a_1)$	ring deformation
824(823)	15	$\nu_{12}(a_1)$	X-sensitive ring
810	15	$2\nu_{16a}(a_2)=817(A_1)$	
528(526)	7	$\nu_{6a}(a_1)$	X-sensitive ring
235(244)	3	$\nu_{11}(b_1)$	X-sensitive ring

[a] Data taken from Ref. [9].

[b] The numbers in parentheses denote previously assigned infrared frequencies from the vapor [16].

[c] Based on the 1007 cm^{-1} line having an intensity of 22.

[d] Wilson's notation (Phys. Rev. 45, 706 (1934)) and symmetry species for the G_4 group.

[e] Estimated from the UV spectrum [16].

A weak to medium intensity band at 1388 cm^{-1} is about 45 cm^{-1} higher in frequency than the peak assigned to the C=C stretching mode, ν_{14} (1343 cm^{-1}), in the infrared (g) spectrum [16]. Both the Raman (g) and infrared (g) bands can be accounted for by an appropriate choice of combination and/or overtone bands. It is, therefore, unclear as to which of them is actually the ν_{14} C=C stretching fundamental. The infrared (g) band has a distinct type A band contour. Since overtone and combination bands are usually quite weak in the Raman effect unless they are in Fermi resonance with a fundamental, it is much more reasonable to assign the 1388 cm^{-1} Raman line as the ν_{14} fundamental. This necessitates the reassignment of the 1343 cm^{-1} infrared band as an overtone or a combination band. Such bands are frequently found with considerable intensity in the infrared spectrum.

The highest in-plane CH bending mode has been assigned at 1277.4 cm^{-1} primarily from the UV spectrum of the vapor [16]. There is no evidence of this band in the infrared (g) spectrum and the nearest feature in the infrared spectrum of the condensed phase is at about 1292 cm^{-1}. The Raman band at 1285 cm^{-1} lies well within the range found for ν_3 in other benzene derivatives [18] and we therefore proposed assigning the Raman (g) band to this CH bending fundamental. The CO stretch, ν_{7a}, has been assigned at 1261 cm^{-1} in the infrared spectrum [16] and the corresponding Raman line was found at 1264 cm^{-1}.

One of the two unusual features in the Raman (g) spectrum is the appearance of the strongest a' band below 2000 cm^{-1} at 1007 cm^{-1}, close to the CH in-plane bending mode ν_{18a} at 1025 cm^{-1} and the in-plane ring deformational vibration ν_1 at 999 cm^{-1}. Two possibilities exist for the assignment of the 1007 cm^{-1} line. The most likely is that the band is an overtone of the out-of-plane X-sensitive ring deformational vibration ν_{16b} which has been assigned at 503 cm^{-1} in the infrared (g) spectrum [16]. The presence of two in-plane fundamentals near the a' overtone likely results in strong Fermi resonance which accounts for its strength. Otherwise, it is necessary to reassign one of the in-plane CH bending modes at 1007 cm^{-1} and there are already sufficient bands above 1000 cm^{-1} to accommodate these motions fully. In the infrared (g) spectrum recorded with relatively high pressure (~65 torr), there is only very weak evidence of a band at 1007 cm^{-1}, but the fundamental at 999 cm^{-1} is observed in the infrared spectrum only at very long pathlengths. Therefore this overtone would not be expected to have any appreciable intensity in the infrared spectrum.

There are two strong a' bands at 824 and 810 cm^{-1}. In previous assignments the bands have been invariably interpreted as having different symmetries. Opinions as to which is a' and which is a" have, however, varied. Quite clearly the Raman (g) spectrum shows both lines belong to the totally symmetric symmetry species a'. The reassignment of one or the other therefore effects the interpretation of both the in-plane and out-of-plane vibrational groups in phenol. One of the lines is clearly the in-plane deformation ν_{12}, which has been previously assigned at 823 cm^{-1}, and we concur with this assignment [16]. The assignment of the other band was more difficult. The most reasonable assignment of the 810 cm^{-1} line is to the overtone $2\nu_{16a}$, which gives a frequency of 817 cm^{-1}. This overtone would be in resonance with ν_{12} so that ν_{12} moves up to 824 cm^{-1} and the overtone moves to 810 cm^{-1}; this resonance could then be used to explain the nearly equal intensity of these two Raman lines. This resonance has been proposed [16] from UV studies. It seems clear that the 810 cm^{-1} line is not ν_{10a} (an a_2 mode under the G_4 symmetry group) as proposed from infrared studies, and the infrared contour is consistent with an in-plane motion but not an out-of-plane motion. It is possible that ν_{10a} may fall in this same region but it cannot be assigned to the infrared or Raman bands at 810 cm^{-1}.

In conclusion we [9] were able to reassign the ν_2, ν_{14} and ν_3 fundamentals of phenol to Raman lines at 3073, 1388 and 1285 cm^{-1}. These frequencies differ appreciably from the proposed assignments for these normal modes based on the infrared spectrum [16] of gaseous phenol. We also observed two very strong lines at 1007 and 810 cm^{-1} which appear to be overtones in Fermi resonance with nearby fundamentals. It is possible that the 810 cm^{-1} line is the ν_{6a} fundamental and further studies of phenol-d_5 will be needed to determine if the 810 cm^{-1} line is an overtone or a fundamental. The infrared band at 810 cm^{-1} which has a contour similar to the $\nu_{12}(a_1)$ fundamental must correspond to the Raman line at this same frequency and it cannot, therefore, correspond to the ν_{10a} fundamental as previously assigned since ν_{10a} is an out-of-plane CH bending motion. Further infrared data under higher resolution are needed before the CH stretching motions can be confidently assigned since it was necessary to reassign the frequency for the ν_2 fundamental [9].

It is clear from the above three examples that the Raman spectra of gases can be invaluable for making complete assignments for a large variety of molecules but, for the most part, it has been neglected mainly because of the need for reasonably high powered laser sources for collection of the data. Frequently such source intensity will cause the sample to decompose or the focused beam will "burn" holes in the mirror or sample windows which of course makes the recording of such data somewhat difficult. Nevertheless one usually finds that the Raman data for the gaseous phase is of sufficient value to warrant the effort to collect it.

3. STRUCTURAL DETERMINATIONS

The next area which we would like to review is the utility of gas-phase Raman data for structural determinations which can be conveniently divided into the use of low resolution data for symmetry determinations and high resolution rotational spectra for the determination of bond distances and angles. In the first case it is usually necessary to use the Raman data in conjunction with the corresponding infrared data, as well as the spectral data for the solid state. In many of the early studies on symmetry determinations, the Raman data for the liquid was compared to the infrared spectrum of the vapor and conclusions were made which later proved incorrect. Therefore, it is necessary to compare the infrared and Raman data for the same physical states if one wants to determine the presence of a symmetry center or the selection rules being followed by the molecule. Since overtone and combination bands are usually very weak in the Raman effect, it is frequently possible to determine, from only the number of observed Raman lines, some idea of the molecular symmetry of the molecule under investigation. Additionally the polarization of a given normal mode may also provide a clue to the molecular symmetry. Nevertheless the most confident determination can usually be made by a comparison of both the infrared and Raman data for both the fluid and solid states.

As an example of symmetry determination let us consider some molecules of general formula X_2Y_4 which exhibit a wide variety of molecular symmetries. Five possible structures for these X_2Y_4 molecules will be considered which include the planar (D_{2h}), staggered with a twist angle of 90° (D_{2d}), and three resulting from a nonplanar arrangement of the XXY_2 moiety: trans (C_{2h}), gauche (C_2), and cis (C_{2v}). These five symmetries are illustrated in Fig. 4 and the representations for the normal modes along with their infrared and Raman activity for each molecular symmetry group are given in Table 4.

The first molecule to be considered is B_2F_4 for which the Raman spectra of the gas and solid were reported by Durig et al. [19]. Major Raman lines for the gas were reported at 1404 (w,p), 1368 (w,br,dp), 673 (vs,p), 380 (w,dp?), 326 (w,sh,dp), and 318 cm^{-1} (s,p) with corresponding lines in the spectrum of the solid at nearly the same frequencies except for the 1368 cm^{-1} line which shifted to 1308 cm^{-1}. These six lines are exactly the number expected for the planar molecule of D_{2h} symmetry where the three polarized lines in descending frequencies are the B-B stretch, the B-F stretch, and the BF_2 deformation, respectively. The three depolarized lines, again in descending frequencies, are the BF_2 stretch (B_{1g}), the BF_2 wag (B_{2g}), and the BF_2 rock (B_{1g}), respectively. A comparison

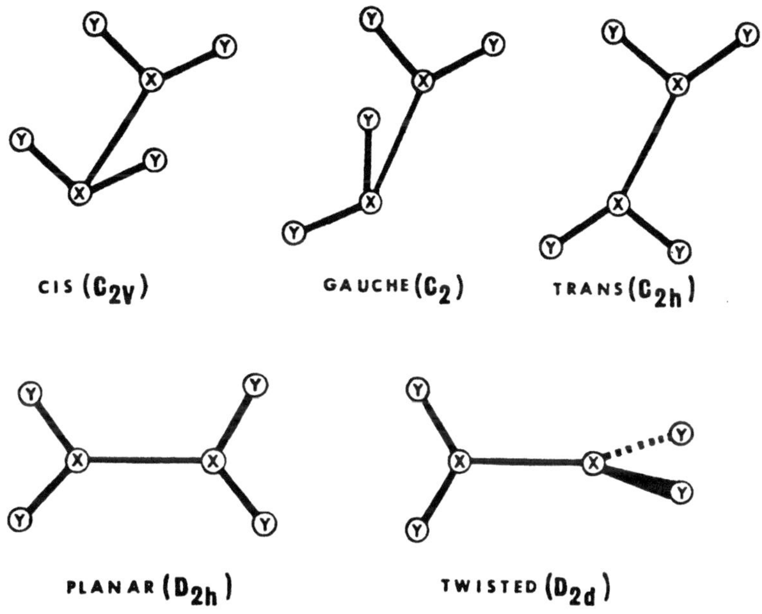

Fig. 4 Five possible structures for X_2Y_4 molecules.

Table 4 Representations of Normal Modes and their Activity for Molecules with Molecular Formula X_2Y_4

D_{2h}	$3A_g(R,p) + 2B_{1g}(R,dp) + 1B_{2g}(R,dp) + 1A_u(-) + 1B_{1u}(IR) + 2B_{2u}(IR) + 2B_{3u}(IR)$
D_{2d}	$3A_1(R,p) + 1B_1(R,dp) + 2B_2(R,dp,IR) + 3E(R,dp,IR)$
C_{2h}	$4A_g(R,p) + 2B_g(R,dp) + 3A_u(IR) + 3B_u(IR)$
C_2	$7A(R,p,IR) + 5B(R,dp,IR)$
C_{2v}	$4A_1(R,p,IR) + 3A_2(R,dp) + 2B_1(R,dp,IR) + 3B_2(R,dp,IR)$

of these data with the infrared frequencies clearly showed that the rule of mutual exclusion is retained in the spectra for both the fluid and solid states and the molecule has D_{2h} symmetry in all physical states. No evidence could be found for the non-planar D_{2d} form in the gas as had been previously reported from an infrared study [20].

The second molecule which we shall consider in this section is tetramethylhydrazine. The group frequencies for the CH_3 groups are well known and they do not interfere with the analysis of the spectral data for the skeletal modes.

We recorded [21] the Raman spectra of the gas, liquid, and solid and measured the depolarization values. Only the skeletal stretching and bending frequencies were expected to show significant variations with changes in molecular structure. The Raman spectrum of solid tetramethylhydrazine shows eleven distinct lines between 300 and 1050 cm^{-1}. This is exactly the number expected for the skeletal modes, excluding the torsion, if the molecule has C_2 symmetry. If the trans form were present only six lines would be expected and mutual exclusion would be operative. However, excellent correspondence between the infrared and Raman bands in the solid was found. Therefore, the molecule has C_2 symmetry in the solid. There were no bands observed in the liquid which disappear with solidification. In the Raman spectrum of the gas, the Q-branches for the symmetric modes were well pronounced and the broad lines of the antisymmetric modes could still be identified. Thus, it was concluded that the tetramethylhydrazine molecule exists in only one isomeric form in all states and this form is the gauche (C_2) isomer. These two examples clearly show the utility of Raman spectral data for symmetry determinations.

As indicated earlier, high resolution rotational Raman spectra can be used to obtain bond distances and angles. The pure rotational Raman spectra of cyclohexane-d_0 and cyclohexane-d_{12} were recorded [22] under high resolution with an argon laser as the exciting source. The spectra consisted of reasonably sharp lines which formed well developed R- and S-branches and, from the analysis of these spectra, the following molecular constants were obtained: for C_6H_{12}, B_0 = 0.143429 ± $2x10^{-6}$ cm^{-1}, D_J = (1.56 ± 0.03)$x10^{-8}$ cm^{-1} and for C_6D_{12}, B_0 = 0.109469 ± $1x10^{-6}$ cm^{-1} and D_J = (0.61 ± 0.01)$x10^{-8}$ cm^{-1}. From these two experimental moments of inertia it is not possible to obtain the four unknown structural parameters, i.e., two bond lengths and two interbond angles, but by assuming a value for ∢CCC and calculating a value for r(C-H) from the correlation between this bond length and the average frequency of its stretching vibration [23], it is possible to obtain the other structural parameters. The results of this study [22] are given in Table 5 along with values for the structural parameters obtained from three different electron diffraction studies [24-26]. The results compare very favorably and show that excellent structural parameters can be obtained in favorable cases.

Table 5 Structural Parameters[a] for Cyclohexane

	∢C-C-C,deg	r(C-C),Å	∢H-C-H,deg	r(C-H),Å
Electron Diffraction				
Ref. [24]	111.55±0.15°	1.528±0.005	tetrahedral	1.104±0.005
Ref. [25]	111.05±0.12°	1.528±0.003	110.0±2.5°	1.119±0.004
Ref. [26]	111.4 ±0.2°	1.535±0.002	107.2±1°	1.121±0.004
Raman Effect				
Ref. [22]	111.4±0.3°	1.535_1±0.001_3	110.0±0.3°	1.102±0.001

[a]Value of ∢C-C-C is assumed. Value of r(C-H) determined from bond length-bond frequency correlation [23].

4. CONFORMATIONAL ANALYSIS

The utility of Raman spectral data from the gas phase for conformational analyses has been pretty much limited to very recent studies but a large number of papers on the temperature studies of the Raman spectra of liquids have been published over the past thirty years. Therefore we shall first briefly review this latter use of Raman data for conformational analyses and then review the more recent gas-phase Raman studies on molecules with asymmetric torsional modes and more than one conformer present at ambient temperature. In principle, temperature studies on the Raman spectra of gases could provide ΔH values but such studies are relatively difficult to carry out. In general the spectral lines due to the different conformers are identified in the gas phase, then temperature studies are carried out on the liquid phase, and finally the spectrum of the solid is recorded to verify that the high energy conformer lines do disappear in the solid. However, it should be pointed out that the high energy conformer may be more stable in the solid state because of packing factors but, in general, the more thermodynamically favored conformer is the one remaining in the crystalline solid.

As an example of the identification of a second conformer in the gaseous state by Raman spectroscopy, the Raman spectrum [27] of $(CH_3)_2PP(CH_3)_2$ is shown in Fig. 5. For this molecule only the skeletal stretching and bending motions are expected to show significant frequency variations with changes in the molecular structure. Since all of the carbon-hydrogen motions except the methyl torsions fall above 800 cm^{-1} and the frequencies of the skeletal modes are all below 800 cm^{-1}, it was

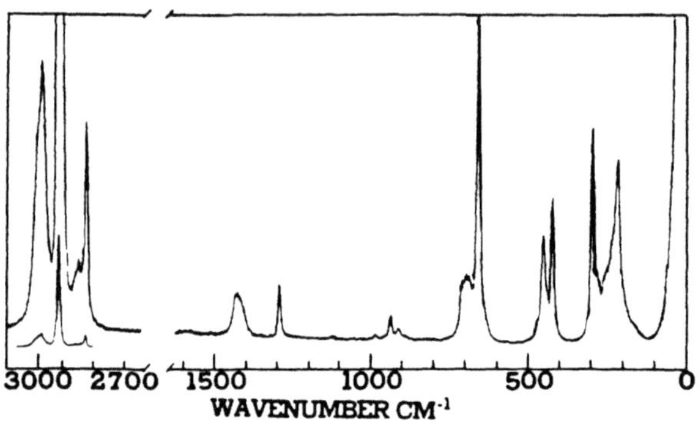

Fig. 5 Raman spectrum of gaseous $(CH_3)_2PP(CH_3)_2$ at 80°C (used by permission, Ref. [27]).

quite easy to assign the fundamental frequencies for the normal modes of the heavy atom skeleton. In the spectral region between 400 and 500 cm^{-1} there are two lines at 426 and 457 cm^{-1} which have corresponding frequencies of 429 and 455 cm^{-1} in the spectrum of the liquid [28] and these lines must be assigned as the P-P stretching modes of the gauche and trans conformers, respectively, since these are the only normal modes expected in this region. The 426 cm^{-1} line disappeared with solidification and only three Raman lines at 314, 260, and 231 cm^{-1} remained in the PC_2 bending region which is exactly the number expected for the trans conformer. Further evidence for the presence of only the trans conformer in the solid state was obtained from a comparison of the infrared and Raman frequencies for the skeletal modes from which their alternate forbiddenness provided conclusive evidence that solid tetramethylbiphosphine exists as the trans isomer. Additional evidence for the presence of the gauche isomer in the gas phase can be found in the 180-300 cm^{-1} region where lines were observed at 299 (301), 284 (287), 261, 254 (258), 241 (243), 233, 222 (226), and 186 (190) cm^{-1} with the corresponding frequencies for the liquid listed in parentheses. Since only three lines were observed in this spectral region for the trans isomer in the solid state, these additional lines had to be attributed to PC_2 bending modes of the gauche isomer [28]. The relative intensities of the 426 and 457 cm^{-1} lines were studied as the sample temperature was varied from ambient to 80°C but the relative intensities remained essentially the same which indicates a very small energy difference between the two isomers. Therefore, on a statistical basis, the gauche should be in approximately a 2 to 1 ratio which is consistent with the approximately 60% concentration estimated from the Raman spectrum of the liquid [27]. These results are in complete disagreement with the conclusions from the electron diffraction study [29] of this molecule where it was concluded that the molecule exists in only the trans conformer in the gas phase.

For the second example in this section we shall use the Raman spectrum of gaseous n-butane. In this case we shall show how the determination of the asymmetric potential function can be used to obtain the energy difference between the two conformers.

In order to be able to characterize the asymmetric potential function, four types of information are required: (1) the approximate dihedral (torsional) angle of each conformer, because the number of torsional energy levels is directly related to the number of potential minima; (2) the approximate relative enthalpy difference between the high and low energy conformers, since this is one of the constraints defining the potential function; (3) the change in molecular kinetic energy as a function of torsional angle; and (4) the accurate observation and assignment of the torsional transition frequencies. Previous vibrational studies [30,31] of n-butane had shown that the molecule exists in the gas phase as a mixture of the s-trans and gauche conformers. The s-trans asymmetric torsion is expected to show sharp Q-branches in the infrared spectrum but only one broad ill-defined band, centered at 121 cm^{-1}, was observed [30]. Compton et al. [32] recorded the low frequency Raman spectrum of this molecule at relatively high sensitivity (see Fig. 6) and they were able to resolve four bands due to the asymmetric torsion of the gauche conformer. Additionally, they observed several bands between 260 and 210 cm^{-1} of which four were assigned to the overtones of the asymmetric torsion of the s-trans conformer (see Table 6). From these data, potential coefficients of V_1 = 395±2, V_3 = 1166±5, V_5 = 26±1 and V_6 = -34±2 cm^{-1} were obtained. The resulting potential function (Fig. 7) has an enthalpy difference of 311±10 cm^{-1} and a gauche dihedral angle of 118±1°. This value of 311 cm^{-1} for ΔH of the gas is considerably larger than the values of 189 cm^{-1} and 195 cm^{-1} obtained from the temperature and pressure dependence of the Raman spectrum of the liquid [33,34]. Such a large difference between the ΔH values for the two phases is not surprising for molecules which have considerably different shapes for their two conformers and the more symmetrical one can be accommodated more readily in the liquid phase.

5. LOW FREQUENCY, ANHARMONIC VIBRATIONS

For several years, there has been considerable interest in large amplitude, anharmonic vibrations which usually occur in the low frequency spectral region. The anharmonic vibrations

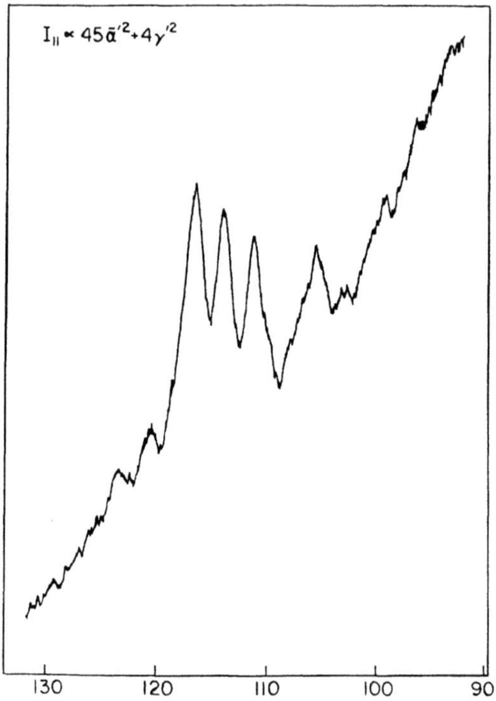

Fig. 6 Low frequency Raman spectrum of gaseous n-butane recorded with SBW = 0.7 cm^{-1} (used by permission, Ref. [32]).

which will be considered are the ring bending modes of "saturated" five-membered rings, bending modes of quasi-linear molecules, and internal torsional oscillations. Since these vibrations are low in frequency, several excited vibrational states are ordinarily populated at ambient temperature, and upper state transitions are usually observed. Correct assignments of these "hot bands" can provide a knowledge of the vibrational potential function and barriers to interconversion between molecular forms. For example, the most stable conformation, as well as the path for interconversion, may often be determined for five-membered rings; for quasi-linear molecules, the barrier to linearity may frequently be obtained. For molecules with two C_{3v} rotors, not only is the potential barrier obtained, but both the sine-sine and cosine-cosine coupling terms may be evaluated. In order to best study anharmonic vibrations, the molecules must be in the gas phase, where they are free from the interactions present in the liquid or solid phases. Initially, these vibrations were studied by the far infrared technique and, with the development of FT-IR spec-

Table 6 Observed[a] Raman Frequencies of Gaseous n-Butane below 300 cm^{-1}

freq, cm^{-1}	rel.int.	assignment[b]	obs-calc[c]
254.9	m	t	
248.2	w,sh	t	
244.2	s	t 2 ← 0	-0.4
239.1	m	t 3 ← 1	0.6
232.0	m	t 4 ← 2	-0.1
230.5	vw,sh	t	
224.9	w	t 5 ← 3	-0.1
123	vw,br		
116.6	s	g 1± ← 0±	0.1
114.1	m	g 2± ← 1±	0.0
111.3	m	g 3± ← 2±	-0.1
105.6	w	g	

[a]Data taken from Ref. [32]. Abbreviations used: s, strong; m, medium; w, weak; sh, shoulder; br, broad; v, very.

[b]t, s-trans; g, gauche.

[c]Calculated using the following potential coefficients: $V_1 = 395.1$, $V_3 = 1166.0$, $V_5 = 26.3$, and $V_6 = -34.0$ cm^{-1}.

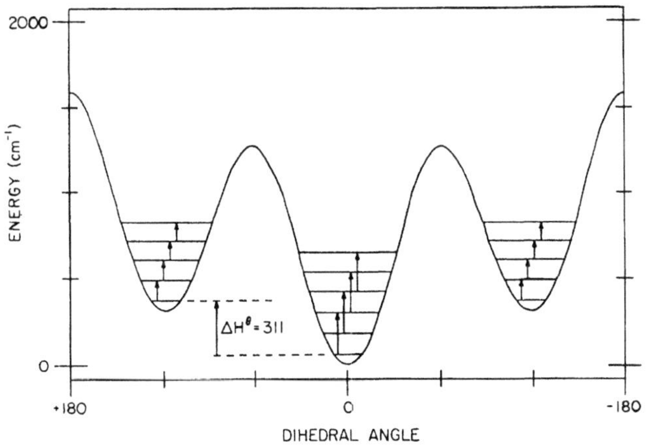

Fig. 7 The asymmetric potential function of n-butane showing the observed torsional transitions. The dihedral angle of zero corresponds to the s-trans conformer (used by permission, Ref. [32]).

troscopy, considerable research effort has been expended in this area. However, until recently, little attention has been given to Raman spectroscopy which is the other technique for directly observing anharmonic vibrations. The fundamentals of these large amplitude vibrations, if observed at all, often yield broad, weak band contours in the gas-phase Raman spectrum but the overtones are of the proper symmetry to yield the desired Q-branches. Because of both the electrical and the mechanical anharmonicity, these vibrational overtones are allowed and they usually have sufficient intensity to be observed, even when the fundamentals may not be observable. Thus, gas-phase Raman spectroscopy offers another method of directly observing these anharmonic vibrations. It has its own set of advantages and disadvantages as an adjunct to far infrared spectroscopy, or as the primary technique for studying these large amplitude motions.

A. Five-membered Rings

In five-membered ring molecules, there are two low frequency out-of-plane ring motions. If the five-membered ring contains no endocyclic double bonds, the two out-of-plane ring motions can usually couple and the Hamiltonian for this motion is generally written as a two-dimensional problem in Cartesian coordinates. This two-dimensional Cartesian Hamiltonian is then transformed to polar coordinates, q and ϕ. There have been several recent investigations carried out on the radial modes of saturated five-membered rings by Raman spectroscopy in order to observe the effects of a finite barrier to planarity and to calculate the two-dimensional potential surface for the two low frequency out-of-plane modes.

We recorded [35] the Raman spectrum of gaseous silacyclopentane in the region of the ring-twisting mode and observed a series of Q-branches at 264.5, 259.9 and 255.3 cm^{-1} (Fig. 8). These Q-branches were assigned as the fundamental and hot bands of the radial vibration of a hindered pseudorotator. The dimensionless potential constants were varied in a least-squares manner in order to simultaneously fit the transitions for the ring-twisting and the previously reported [36,37] ring-puckering modes of silacyclopentane. The following potential constants were determined: A = 2.2056; B = -90.1340; C = 83.4345; D = -2.3774; and E = 0.2494 cm^{-1}. Because of the limitations of the size of the basis set, only two puckering mode transitions in the first excited twisted state could be calculated with sufficient accuracy (Table 7). The potential equation utilizing the above dimensionless constants reduces to $V(cm^{-1}) = 0.0776\, q_1^4 - 6.6990\, q_1^2$. The result is a double minimum potential along the ring-puckering coordinate with the bent configuration being 144.6 cm^{-1} lower in energy than the planar configuration.

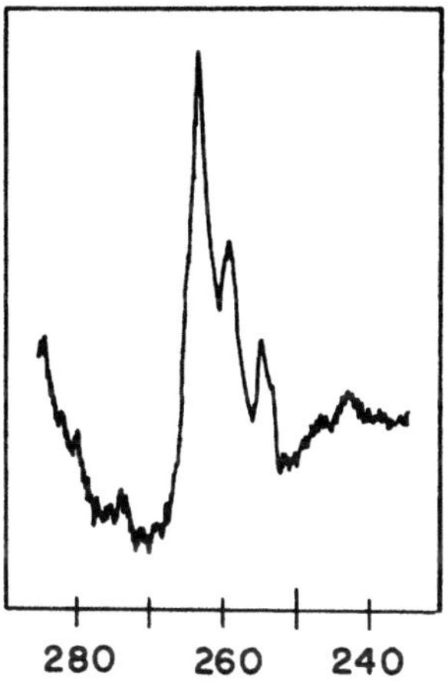

Fig. 8 Raman spectrum of the twisting mode of silacyclopentane in the gas phase at its ambient vapor pressure (used by permission, Ref. [35]).

A similar slice along the ring-twisting coordinate at $\phi = 90°$ ($q_1 = 0$) leads to a potential of $V(cm^{-1}) = 4.8324\ q_2^4 - 173.6\ q_2^2$ which represents a double minimum potential function. However the twisted conformation is 1558.6 cm^{-1} lower in energy than the planar conformation. According to this two-dimensional potential function [35], the interconversion between the two stable twisted forms takes place through a slightly puckered conformation which is 144.6 cm^{-1} more stable than the planar conformation. The path for the interconversion between the stable twisted conformers is hindered by a barrier of 1414 cm^{-1}.

In summary it is clear that by a combination of far infrared data ($\Delta v = 1$) for the ring-puckering (pseudorotational) modes and the $\Delta v = 1$ transitions for the ring-twisting (radial) modes obtained from the Raman effect for saturated five-membered rings, it is possible to observe the effects of a finite barrier to planarity and to calculate the two-dimensional potential surface for the low frequency out-of-plane modes for these molecules. On the basis of a very limited number of cases it would seem that the one-dimensional model breaks down as the path to interconversion approaches one which involves a planar intermediate (Table 8). Unfortunately, it is usually not possible to determine a priori if a twisted molecule interconverts through a planar or puckered intermediate. In cases where the barrier to planarity is very large compared to the barrier to interconversion or pseudorotation, the second-order

Table 7 Observed[a] and Calculated Frequencies of the Low Frequency Skeletal Ring Modes of Silacyclopentane

Transition[b] $n_2,n_1 - n_2',n_1'$	Observed (cm^{-1})	Calculated[c] (cm^{-1})	Δ
0,0 - 0,1	101.7	101.9	-0.2
0,1 - 0,2	99.6	99.5	0.1
0,2 - 0,3	97.3	97.4	-0.1
0,3 - 0,4	95.0	94.4	0.6
0,4 - 0,5	92.6	93.7	-1.1
1,0 - 1,1	100.7	101.7	-1.0
1,1 - 1,2	98.7	97.7	1.0
0,0 - 1,0	264.5	264.5	0.0
1,0 - 2,0	259.9	259.6	0.3
2,0 - 3,0	255.3	d	---

[a]The bending frequencies are taken from Refs. [35-37].

[b]n_1 and n_2 are quantum numbers of the bend and twist, respectively.

[c]Calculated using a reduced potential function with an isotropic harmonic oscillator basis set of size 121 x 121 for the ℓ even levels and 110 x 110 for the ℓ odd levels.

[d]The basis set is not large enough for convergence at this level so the calculated frequency is omitted.

Table 8 Comparison of Barriers (cm^{-1}) between the One-dimensional and Two-dimensional Treatment of the Two Low Frequency Bending Modes of Some Five-membered Rings

Molecule	One-dimensional barrier to interconversion	Two-dimensional barrier to planarity	Barrier to interconversion	Ref.
cyclopentane	----	1824	----	38
silacyclopentane	1390	1558.6	1414	35-37
germacyclopentane	2043	1474.4	1454	38,39
selenacyclopentane[a]	1882	1693	1693	40,41
cyclopentanone[a]	1303;1687	750	----	41-43

[a]The interconversion for these molecules is through the planar conformation whereas the pathway for interconversion for the other molecules takes place through a slightly puckered configuration.

perturbation calculation of Ikeda and Lord [42] can be useful in calculating the barrier values. Neither silacyclopentane nor germacyclopentane fit these criteria, and it seems clear that the detailed multidimensional treatments which require a great deal of data provide the most reliable means of understanding the dynamics of saturated five-membered ring molecules.

B. Quasi-linear Molecules

Molecules whose heavy atom skeletons have a moderate or low barrier at a linear configuration are termed "quasi-linear". This type of low frequency bending mode is generally a large amplitude, anharmonic vibration. Until very recently, electron diffraction studies had yielded most of the structural information concerning these molecules. However, electron diffraction results yield "effective values" for the skeletal angles and bond distances rather than equilibrium values and hence might predict a molecule to have a lower symmetry than would be expected for the equilibrium configuration. This "shrinkage" effect is particularly problematic in molecules which have perpendicular vibrations as in three- (X=Y=Z) or four-membered (M-X=Y=Z) chains or when a low frequency, large amplitude vibration is involved. Corrections can be made for this "shrinkage" effect, but the nature of the potential function which governs the low frequency bending mode must be known. Additionally, if the potential function can be determined, it is possible to obtain the equilibrium value for the skeletal angle if the reduced mass for the normal vibration is known.

Gas-phase Raman spectroscopy is the primary technique for investigation of these types of molecules. The assignment of the observed transitions in the Raman spectrum is less difficult because the transitions are of the type $\Delta v = 2$ and hence the Q-branches are much more intense than the rotational branches. Additionally, the selection rules for this bending vibration differ in the infrared and Raman effect. The infrared spectrum is far more complicated because the allowed transitions are $\Delta v = 1$, $\Delta \ell = \pm 1$, whereas the Raman selection rules ($\Delta v = 2$, $\Delta \ell = 0$) lead to $(v + 2)/2$ transitions originating from even v levels and $(v + 1)/2$ for levels of odd v. Thus the Raman spectral data are essential for the initial assignment of the observed transitions, and the observed infrared transitions must be in agreement with the calculated energy levels obtained from the Raman data.

We have successfully obtained the potential functions governing the low frequency bending modes in carbon suboxide (C_3O_2), disiloxane ($(SiH_3)_2O$) and silylisocyanate (SiH_3NCO). From the assignment of the Q-branches observed in the far infrared and low frequency Raman spectra and the subsequent determi-

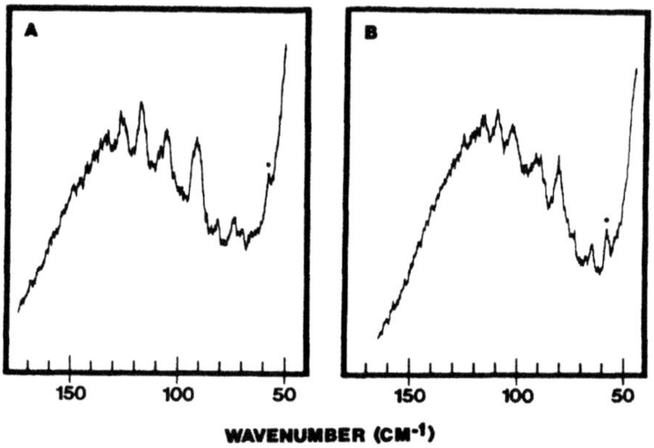

Fig. 9 The low frequency Raman spectra of gaseous (A) disiloxane-d_0 and (B) disiloxane-d_6 in the $2\nu_{11}$ region recorded with SBW = 1.0 cm^{-1}. The asterisks mark the Δ59 cm^{-1} line of air, present as a minor impurity (used by permission, Ref. [14]).

nation of the potential function [44], it was concluded that carbon suboxide is linear since the lowest vibrational energy level at 19.7 cm^{-1} lies above the barrier of 14±2 cm^{-1}. Contrary to this, it has been established [14] that disiloxane is quasi-linear. The series of Q-branches observed in the $2\nu_{11}$ region in the Raman spectrum of gaseous disiloxane (Fig. 9) was assigned, and it was found from the determination of the potential function that four vibrational levels lie below the barrier of 112 cm^{-1} but that a significant number of molecules occupy states with a linear configuration (Table 9). The Si-O-Si equilibrium angle was calculated to be 149±2° which is in fair agreement with the angle of 144.1±0.8° reported in an electron diffraction study [45]. The Raman spectrum of gaseous SiH_3NCO has also been recorded [46] and a series of Q-branches was observed. These Q-branches were attributed to the double jumps of the anharmonic, low frequency SiNC bending mode. The potential function was determined and a barrier of 45 cm^{-1} at the linear configuration was found. Only two vibrational levels lie below this barrier; the 1,1 ← 0,0 transition was calculated to be 15.3 cm^{-1}. The equilibrium Si-N=C angle is 153±2° which is in agreement with the angle of 151.7° reported by Glidewell et al. [47] who used an anharmonic potential with a barrier of 20 cm^{-1} to fit the observed shrinkages. In Table 10, the potential constants (A and B), barriers to linearity, and the frequencies of the 1 ← 0 and 2 ← 0 transitions are listed for these three molecules.

Table 9 Observed and Calculated Frequencies (cm^{-1}) for $(SiH_3)_2O$ in the $2\nu_{11}$ Region

Transition $v'\|\ell'\| \leftarrow v\|\ell\|$	Obs.	Calc.[a]	Δ	Weight
2,0 ← 0,0	76.0	75.4	0.6	1.0
4,0 ← 2,0	83.6	83.2	0.4	1.0
3,1 ← 1,1	93.2	93.3	-0.1	1.0
4,2 ← 2,2	107.5	108.1	-0.6	0.5
5,1 ← 3,1	107.5	104.0	3.5	0.5
6,0 ← 4,0	110.9	109.5	1.4	1.0
5,3 ← 3,3	119.5	120.6	-1.1	0.4
6,2 ← 4,2	119.5	118.8	0.7	0.4
7,1 ← 5,1	119.5	122.0	-2.5	0.4
8,0 ← 6,0	127.9	129.2	-1.3	1.0
6,4 ← 4,4	130.2	131.4	-1.2	0.2
7,3 ← 5,3	130.2	130.8	-0.6	0.2
9,1 ← 7,1	130.2	132.0	-1.8	0.2
8,2 ← 6,2	135.7	133.2	2.5	0.5
7,5 ← 5,5	139.0	141.1	-2.1	0.25
8,4 ← 6,4	139.0	141.0	-2.0	0.25

[a]Using the reduced potential function $V(cm^{-1}) = 1.07q^4 - 21.9q^2$ with a two-dimensional isotropic harmonic oscillator basis set of size 121 x 121 for the even ℓ levels and 110 x 110 for the odd ℓ levels (max. ℓ = 20) with γ = 3.09 amu^{-1} $Å^{-2}$ rad^{-2}.

Table 10 Potential Constants, Barriers to Linearity, and Transitions Determined in Some Quasi-linear Molecules

Molecule	A(cm^{-1})	B(cm^{-1})	Barrier (cm^{-1})	1 ← 0 (cm^{-1})	2 ← 0 (cm^{-1})	Ref.
carbon suboxide	0.728	-6.40	14±2	22.2	61.0	44
disiloxane	1.07	-21.9	112±5	8.7	75.4	14
silylisocyanate	0.858	-12.5	45±3	15.3	63.2	46

It is clear from the previous discussion that in many cases the gas-phase Raman spectrum provides the necessary information in determining the potential function which governs the low frequency bending modes in quasi-linear molecules. Additionally, if the reduced mass of the normal vibration is known, one can calculate the equilibrium angle in the ground vibrational state.

C. Internal Torsional Modes

In a series of studies [48-50] we have shown that gas-phase Raman spectroscopy can be applied successfully to the study of torsional overtones ($\Delta v = 2$). While it was shown that mechanical anharmonicity contributes in part to the Raman intensity of these torsional overtones, it is primarily the large electrical anharmonicity, $\partial^2\alpha/\partial Q^2$, which for the most part causes these double quantum transitions to be observed at all. Evidence for this argument is the observed intensity ratios of the $2 \leftarrow 0$ and higher transitions. In the ethyl halides [48], the $3 \leftarrow 1$ transition is at least as intense as the $2 \leftarrow 0$, and quite often more intense. Also, the intensities of higher torsional transitions deviate even more radically from simple Boltzmann intensity predictions. Thus, the higher vibrationally excited torsional states cause a large electrical anharmonicity as a result of the large amplitude of vibration, and this accounts for the anomalous intensities observed in the Raman spectra.

As a result of this large electrical anharmonicity making two-quantum torsional transitions observable in the Raman effect, several advantages are obtained over standard techniques of determining torsional barriers. For example, in the haloethanes, gas-phase Raman spectroscopy offers an immediate advantage over microwave spectroscopy for the calculation of torsional barriers in that molecules which contain several quadrupolar atoms may be studied. Also, gas-phase Raman spectroscopy may be used in some cases where the infrared spectrum of the vapor could not be obtained because of lack of sufficient sample vapor pressure, poor band contours, or insignificant dipole moment change.

However, there are two limitations in using gas-phase Raman spectroscopy as a tool for determining torsional barriers. One is that insufficient vapor pressure, even at elevated temperatures, precludes the observation of torsional overtones due to the lack of enough scattering molecules in the laser beam to produce a detectable signal. Additionally, the $\Delta v = 2$ torsional transitions quite often fall in the region of fundamental molecular vibrations. This is particularly true for the haloethanes, where the carbon-halogen stretching and bending modes fall in the torsional overtone region and are ordinarily quite intense in the Raman effect. Another minor limitation to using gas-phase Raman spectroscopy is that the higher torsional energy levels fall in a region where the possibility exists for interaction with other molecular energy levels, i.e., Fermi resonance. This has been observed [49] in several instances but fortunately deuterium substitution can quite often circumvent this problem.

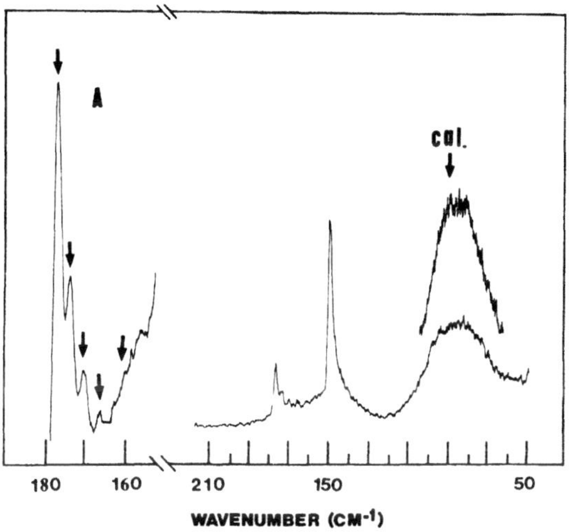

Fig. 10 The low frequency Raman spectrum of gaseous CF_3CH_2Br. Insert (a) shows the torsional overtone region recorded with SBW = 1 cm^{-1}. The 1 ← 0 transition was calculated at 88.3 cm^{-1} (used by permission, Ref. [50]).

To illustrate the observation of the torsional overtones we are showing the Raman spectrum of gaseous CF_3CH_2Br in the region from 50 to 210 cm^{-1} (Fig. 10). Whereas the fundamental leads to a very broad Raman line with an ill-defined center, the overtone region has four well defined lines at 176.5, 173.0, 170.1 and 165.9 cm^{-1} which have been assigned [50] as the 2 ← 0, 3 ← 1, 4 ← 2, and 5 ← 3 transitions, respectively. From these data the barrier to internal rotation was calculated to have a value of 4.39±0.07 kcal/mol and a V_6 term of 103±25 cal/mol. All four of these transitions were fit to better than a wavenumber with this potential function. Many other barriers have been determined [48-50] by Raman spectral studies of gases, and therefore the direct observation of torsional frequencies should be added as another application of gas-phase Raman spectroscopy. The succeeding double jumps are much more intense than one would expect from the Boltzmann factors, but the electrical and mechanical anharmonicity of the large amplitude torsional motions apparently accounts for the unusual intensity of the second, third, and fourth "hot bands."

ACKNOWLEDGEMENT

The authors gratefully acknowledge the financial support for much of the work cited herein by the National Science Foundation through Grant CHE-79-20763.

REFERENCES

1. C. V. Raman and K. S. Krishnan, Nature, 121, 501 (1928).

2. G. Landsberg and L. Mandelstam, Naturwissenschaften, 16, 557 (1928).

3. L. A. Ramdas, Nature, 122, 57 (1928).

4. C. V. Raman, Indian J. Phys., 2, 387 (1928).

5. R. S. Krishnan and R. K. Shankar, J. Raman Spectrosc., 10, 1 (1981).

6. G. Herzberg, Molecular Spectra and Molecular Structure II. Infrared and Raman Spectra of Polyatomic Molecules, D. Van Nostrand Co., Inc., Princeton, New Jersey, 1945.

7. J. R. Durig, M. G. Griffin, and R. W. MacNamee, J. Raman Spectrosc., 3, 133 (1975).

8. J. R. Durig, M. J. Flanagan, and V. F. Kalasinsky, Spectrochim. Acta, 34A, 63 (1978).

9. H. W. Wilson, R. W. MacNamee, and J. R. Durig, J. Raman Spectrosc., 11, 32 (1981).

10. Y. I. Kotov and V. M. Tatevskii, Opt. Spectrosc., USSR (English Transl.), 15, 65 (1963).

11. J. R. Durig, S. F. Bush, and E. E. Mercer, J. Chem. Phys., 44, 4238 (1968).

12. H. R. Linton and E. R. Nixon, J. Chem. Phys., 29, 921 (1958).

13. E. A. V. Ebsworth, R. Taylor, and L. A. Woodward, Trans. Faraday Soc., 55, 211 (1959).

14. J. R. Durig, M. J. Flanagan, and V. F. Kalasinsky, J. Chem. Phys., 66, 2775 (1977).

15. J. R. Durig, K. S. Kalasinsky, and V. F. Kalasinsky, J. Mol. Struct., 35, 201 (1976).

16. H. D. Bist, J. C. D. Brand, and D. R. Williams, J. Mol. Spectrosc., 24, 402 (1967); 24, 413 (1967); 21, 76 (1966).

17. H. W. Wilson, Anal. Chem., 46, 962 (1974).

18. G. Varsanyi, Vibrational Spectra of Benzene Derivatives, Academic Press, New York, 1969.

19. J. R. Durig, J. W. Thompson, J. D. Witt, and J. D. Odom, J. Chem. Phys., 58, 5339 (1973).

20. J. N. Gayles and J. Self, J. Chem. Phys., 40, 3530 (1964).

21. J. R. Durig, R. W. MacNamee, L. B. Knight, and W. C. Harris, Inorg. Chem., 12, 804 (1973).

22. R. A. Peters, W. J. Walker, and A. Weber, J. Raman Spectrosc., 1, 159 (1973).

23. D. C. McKean, J. L. Duncan, and L. Batt, Spectrochim. Acta, 29A, 1037 (1973).

24. M. Davis and O. Hassel, Acta Chem. Scand., 17, 1181 (1963).

25. H. J. Geise, H. R. Buys, and F. C. Mijlhoff, J. Mol. Struct., 9, 447 (1971).

26. O. Bastiansen, L. Fernholt, H. M. Seip, H. Kambara, and K. Kuchitsu, J. Mol. Struct., 18, 163 (1973).

27. J. R. Durig and R. W. MacNamee, J. Mol. Struct., 17, 426 (1973).

28. J. R. Durig and J. S. DiYorio, Inorg. Chem., 8, 2796 (1969).

29. A. McAdam, B. Beagley, and T. G. Hewitt, Trans. Faraday Soc., 60, 2732 (1970).

30. J. R. Durig and D. A. C. Compton, J. Phys. Chem., 83, 265 (1979).

31. I. Harada, H. Takeuchi, M. Sakakibara, H. Matsuura, and T. Shimanouchi, Bull. Chem. Soc. Jpn., 50, 102 (1977).

32. D. A. C. Compton, S. Montero, and W. F. Murphy, J. Phys. Chem., 84, 3587 (1980).

33. J. Devaure and J. Lascombe, Nouv. J. Chim., 3, 579 (1979).

34. S. Kint, J. R. Scherer, and R. G. Snyder, J. Chem. Phys., 73, 2599 (1980).

35. J. R. Durig, W. J. Natter, and V. F. Kalasinsky, J. Chem. Phys., 67, 4756 (1977).

36. J. Laane, J. Chem. Phys., 50, 1946 (1969).

37. J. R. Durig and J. N. Willis, Jr., J. Mol. Spectrosc., 32, 320 (1969).

38. J. R. Durig, Y. S. Li, and L. A. Carreira, J. Chem. Phys., 58, 2393 (1973).

39. J. R. Durig and J. N. Willis, Jr., J. Chem. Phys., 52, 6108 (1970).

40. J. R. Durig and W. J. Natter, J. Chem. Phys., 69, 3714 (1978).

41. W. H. Green, A. B. Harvey, and J. A. Greenhouse, J. Chem. Phys., 54, 850 (1971).

42. T. Ikeda and R. C. Lord, J. Chem. Phys., 50, 2438 (1969).

43. J. R. Durig, G. L. Coulter, and D. W. Wertz, J. Chem. Phys., 49, 675 (1968).

44. L. A. Carreira, R. O. Carter, J. R. Durig, R. C. Lord and C. C. Milionis, J. Chem. Phys., 59, 1028 (1973).

45. A. Almenningen, O. Bastiansen, V. Ewing, K. Hedberg, and M. Traetteberg, Acta Chem. Scand., 17, 2455 (1963).

46. J. R. Durig, K. S. Kalasinsky, and V. F. Kalasinsky, J. Chem. Phys., 69, 918 (1978).

47. C. Glidewell, A. G. Robiette, and G. M. Sheldrick, Chem. Phys. Lett., 16, 526 (1972).

48. J. R. Durig, W. E. Bucy, L. A. Carreira, and C. J. Wurrey, J. Chem. Phys., 60, 1754 (1974).

49. J. R. Durig, W. E. Bucy, and C. J. Wurrey, J. Chem. Phys., 60, 3293 (1974).

50. A. D. Lopata and J. R. Durig, J. Raman Spectrosc., 6, 61 (1977).

INTRODUCTION TO NON-LINEAR RAMAN SPECTROSCOPY

GENERAL INTRODUCTION TO NON-LINEAR RAMAN SPECTROSCOPY

W. Kiefer

Physikalisches Institut der Universität Bayreuth
D-8580 Bayreuth, F.R. Germany

I. HISTORICAL INTRODUCTION

The introduction of lasers into Raman spectroscopy has stimulated this traditional field of molecular spectroscopy in various aspects. In conventional Raman spectroscopy a renaissance began by simply replacing the light source.Many new techniques,specially adapted to the coherent emission of the new Raman excitation source, were developed which permit the molecular spectroscopist to record Raman spectra of a great variety of compounds, from deeply coloured or even black materials, to highly fluorescent molecules. Parallel to these achievements there has been a rapid development of fundamentally new methods. These are based on the contributions of the non-linear part of the induced dipole moment or the induced polarization to the intensity of the frequency-shifted light.

Soon after the discovery of the laser it was demonstrated that the intense electric field of a laser can produce nonlinear optical effects in various media. The first such demonstration was by Franken et al. [1]. The beam of a ruby laser was focussed into a quartz crystal and radiation at the second harmonic of the ruby laser was generated in this experiment. At about the same time Kaiser and Garrett [2] observed another nonlinear optical effect produced by a laser beam two-photon absorption.

The coupling of high intense electromagnetic fields with molecular (mechanical) vibrations naturally was a consequent step in this new field of non-linear optics. However, it should be mentioned, that the discovery of the stimulated Raman effect by Woodbury and Ng [3] occured rather accidentally. They were engaged

W. Kiefer and D. A. Long (eds.), Non-Linear Raman Spectroscopy and Its Chemical Applications, 83–97.

in Q-switching a ruby laser with a Kerr cell as an electrooptical shutter. It was noticed that for sufficiently high laser-pulse intensity less red light was emitted. The authors traced this unexpected result to enhanced emission in the infrared, corresponding to Stokes-shifted Raman scattering in nitrobenzene. The correct interpretation as stimulated Raman scattering in nitrobenzene was given shortly later by Eckhardt et. al. [4], who also reported the observation of this new effect for a variety of other liquids. Although most studies on the stimulated Raman effect were devoted to either generation of new coherent light sources or to applications particularly to time-resolved spectroscopy it must be stated that the stimulated Raman effect certainly is the fundamental process inherent in many other non-linear Raman processes discussed here. As example, Terhune [5] and Maker and Terhune [6] discovered in the years 1963-1965 that by three- or four-wave mixing, stimulated Raman emissions coupled with the original laser light to produce coherent radiation at anti-Stokes frequency. This well-known paper [6] in which several different non-linear optical effects arising from an induced optical polarization third order in the electric field strength have been described led to the early foundations of a new kind of Raman spectroscopy, i.e. CARS or coherent anti-Stokes Raman spectroscopy. Of all the non-linear coherent Raman techniques, CARS has probably received the most attention over the past few years. Consequently this technique will be treated extensively in this Advanced Study Institute.

In 1964 another kind of non-linear Raman spectroscopy was introduced [7]. Up to that time, all investigations of Raman processes, whether linear or non-linear, have been concerned with the emission of the scattered radiation and its spectrum. However, according to the dispersion theories of Kramers and Heisenberg [8], and Dirac [9], the occurence of scattering of radiation should manifest itself both in emission and in absorption. Jones and Stoicheff [7] have been the first to demonstrate this. In their experiments, the scattering medium was irradiated simultaneously with intense monochromatic light of frequency ω_L and with an intense continuum. Under this excitation the molecules are stimulated to emit radiation at ω_L and at the same time to absorb radiation at $\omega_L + \omega_R$ or $\omega_L - \omega_R$ from the continuum, the net effect being to change their energy states by $+\hbar\omega_R$ or $-\hbar\omega_R$, respectively (ω_R being the molecular frequency). This so-called inverse Raman effect has been recently applied very successfully by Owyoung [10] for high-resolution molecular spectroscopy. In his experiments he has replaced the continuum by tunable laser sources.

The method called "Stimulated Raman gain spectroscopy" is in principle based on the same physical phenomenon as the inverse Raman effect. Instead of observing the anti-Stokes absorption induced in a Raman active medium ("Raman loss") by a strong laser beam via the inverse Raman effect, one can also measure the stimulated Raman gain, e.g. at Stokes frequency. The high-resolution spectroscopic capability of this stimulated Raman gain technique was first demonstrated by Lallemand et al. [11] already in 1966 by using a pressure-tuned H_2 Raman oscillator to provide the narrow-band source for probing the backward Raman gain profile induced in a second cell of H_2 at constant pressure. Again, by employing tunable lasers, Owyoung [12] was able to develope this "stimulated Raman gain" technique to a high standard.

Another new Raman spectroscopic technique, which is closely related to stimulated Raman gain spectroscopy is photoacoustic Raman spectroscopy (PARS). Pioneering work by Kerr and Atwood [13] and by Kreuzer [14] on the use of photoacoustic methods to measure the absorption of laser radiation in gases samples has revived interest in a spectroscopic technique that was discovered nearly a century ago by Bell [15] and Tyndall [16]. The combination of this technique with non-linear Raman methods was first reported by Barrett and Berry [17] in 1978.

The introduction of the Raman-induced Kerr Effect (RIKE) by Heiman et al. [18] in 1976 and subsequent variations of this technique by Eesley et al. [19] provided the "polarization spectroscopy" forms of stimulated Raman spectroscopy. In this new effect, a laser pulse can be made to induce a Kerr effect only at Raman-shifted frequencies.

The non-linear Raman methods mentioned above are all coherent techniques which means, that the signal is contained in a coherent laser beam. Furthermore, they are all associated with the third-order non-linear susceptibility $\chi^{(3)}$. The latter is important for these phenomena, since an ensemble of molecules all contributing to the suscpeptibility are involved in the non-linear process. As the roots of this Advanced Study Institute are to be found in the field of molecular spectroscopy and its applications in chemistry, we will also include non-coherent effects that are associated with the second-order non-linear polarizability of a single molecule. Elastic and inelastic light scattering at about double incident laser frequency are associated with this second-order non-linear polarizability and they are called hyper-Rayleigh and hyper-Raman scattering, respectively. The possibility of hyper-Raman scattering was first explicitly recognized by Decius and Rauch [20] in 1959 and the first experimental observation of hyper Rayleigh and hyper Raman scattering was made six years later by Terhune et al. [21]. Since in hyper Raman scattering due to different selection rules silent modes can be detected this technique is complementary to linear Raman and infrared spectroscopy and hence can provide valueable spectroscopic information.

In the sections to follow we will outline only the principal aspects of the methods mentioned above under a common point of view. A more detailed discussion on each topic will be given in the various chapters of this book.

2. HYPER RAYLEIGH AND HYPER RAMAN SCATTERING

Let us first briefly discuss spontaneous non-linear Raman spectroscopy on an isolated single molecule. Generally the induced dipole moment $\vec{p}$ in a molecular system is written in the form

$$\vec{p} = \alpha\vec{E} + \beta\vec{E}\vec{E} + \gamma\vec{E}\vec{E}\vec{E} + \ldots \quad (1)$$

where α is the polarizability, β the hyperpolarizability and γ the second hyperpolarizability. The α,β,γ, etc. are tensors of rank 2,3,4, etc., respectively. $\vec{E}$ is the electric field, $\vec{E}\vec{E}$ is the dyadic product of two electric fields, etc. The non-linear terms in Equation 1 are usually small compared to the linear term which gives rise to normal, linear Raman scattering. However, when the electric field is sufficiently large as is the case when for instance a Q-switched laser is focused on the sample contributions from the second term in Equation 1 is sufficiently intense to be detected. This scattering is at a frequency $2\omega_L \pm \omega_R$, where ω_L is the frequency of the exciting laser beam and $-\omega_R$ and $+\omega_R$ are the Stokes and anti-Stokes hyper Raman displacements, respectively. Also scattering at frequency $2\omega_L$ may occur, depending on the symmetry of the molecule. Scattering at frequencies $2\omega_L$ and $2\omega_L \pm \omega_R$ are called hyper-Rayleigh and hyper Raman scattering, respectively.

The hyper Raman effect is a three-photon process involving two virtual states of the scattering system. The level scheme for hyper Rayleigh and hyper Raman scattering is presented in Figure 1.

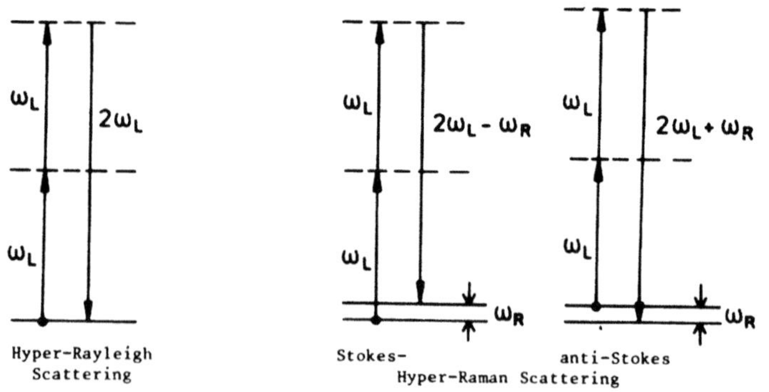

Fig. 1 Schematic level diagram for hyper-Rayleigh- and hyper-Raman scattering.

Two laser photons of frequency ω_L are simultaneously scattered to give a photon at frequency $2\omega_L$ (hyper-Rayleigh), at frequency $2\omega_L - \omega_R$ in the Stokes hyper Raman case, and $2\omega_L + \omega_R$ when an excited molecular vibration (or rotation) is destroyed (anti-Stokes hyper-Raman). Thus, hyper-Raman lines are found in the spectral neighbourhood of the second harmonic ($2\omega_L$), of the incident laser radiation. It should be mentioned, that Rayleigh scattering always occurs, but hyper Rayleigh scattering occurs only if the scattering material does not have a centre of symmetry. Frequency doubled radiation consists of hyper Rayleigh scattering from a pure crystal and consequently a crystal used for frequency doubling must not have a centrosymmetric unit cell.

The importance of the hyper-Raman effect as a spectroscopic tool results from its symmetry selection rules. These are determined by products of three dipole moment matrix elements relating the four levels indicated in Figure 1. It turns out that all infrared active modes of the scattering system are also hyper-Raman active. In addition, as already mentioned above, the hyper Raman effect allows the observation of "silent" modes, which are accessible neither by infrared nor by linear Raman spectroscopy. Hyper Raman spectra have been observed for the gaseous, liquid and solid state. A full description of theory and practise of hyper-Raman spectroscopy is given by Long in this book [22].

3. STIMULATED RAMAN EFFECT

The stimulated Raman process is schematically represented in Figure 2. A light wave at frequency ω_S is incident on the material system simultaneously with a light wave at frequency ω_L. While the incident light beam loses a quantum ($\hbar\omega_L$) and the

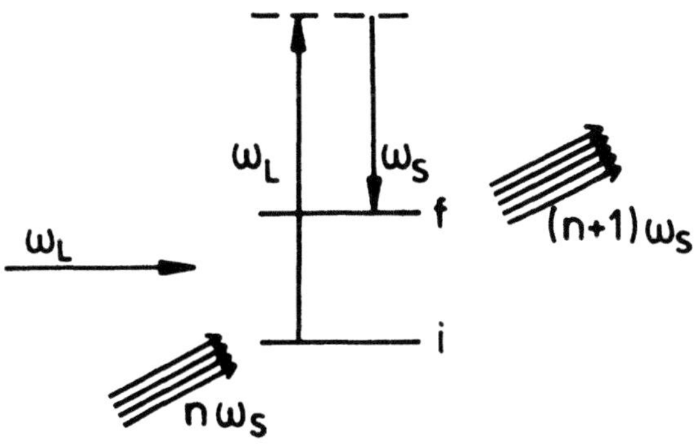

Fig. 2 Schematic diagram for stimulated Raman scattering as a quantum process (adapted from Ref. 23).

material system is excited by a quantum $\hbar(\omega_L - \omega_S)$, a quantum $\hbar\omega_S$ is added to the wave at frequency ω_S, which consequently becomes amplified.

A theoretical description of the stimulated Raman effect in terms of a nonlinear Raman susceptibility was presented by Bloembergen [23] soon after the experimental discovery of this effect. He showed that a polarization at Stokes frequency ω_S is generated via the third-order non-linear susceptibility $\chi^{(3)}$:

$$P(\omega_S) = \chi^{(3)}(-\omega_S, \omega_S, \omega_L, -\omega_L)\ E\ (\omega_S)\,|E(\omega_L)|^2 \tag{2}$$

For the case that pump frequency ω_L and Stokes frequency ω_S differ by a molecular eigenfrequency ω_R, i.e. $\omega_L - \omega_S = \omega_R$, the third-order non-linear susceptibility has a negative imaginary part. This leads to a negative absorption coefficient or an exponential gain at the Sotkes frequency ω_S proportional to the laser intensity at pump frequency ω_L [25]. The gain factor g, which describes this exponential growth has been shown to depend on the differential Raman cross section $d\sigma/d\Omega$ as well as on the linewidth Γ of the molecular transition (ω_R) in the following way [24]

$$g(\omega_S) \sim \left(\frac{d\sigma}{d\Omega}\right) \cdot \frac{\Gamma}{(\omega_L - \omega_S - \omega_R)^2 + \Gamma^2} \tag{3}$$

Optimum gain is found at the center of the Raman line where $\omega_R = \omega_L - \omega_S$. There, the gain constant for stimulated Raman scattering at Stokes frequency is given by

$$g_S \sim \left(\frac{d\sigma}{d\Omega}\right) \cdot \frac{1}{\Gamma} \tag{4}$$

From Equation 4 we immediately recognize that in stimulated Raman scattering processes where only <u>one</u> input laser field with frequency ω_L is employed a coherent Stokes wave is generated for those Raman modes which have the highest ratio between differential Raman cross section and linewidth Γ. The latter corresponds to the dephasing time T_2 of the physical system, $\Gamma = 1/T_2$ and reflects the damping of the system.

In this classical description of stimualted Raman scattering a coherent light beam at Stokes frequency ω_S is produced by a pump field at $\omega_L = \omega_S + \omega_R$. The distinctive feature of stimulated Raman scattering is that an assemblage of coherently driven molecular vibrations provides the means of coupling the two light waves at frequency ω_L and ω_S by modulating the nonlinear susceptibility.

From the fact that the Stokes beam interacting with the pump laser beam automatically establishes correct phasing of the molecular oscillators so as to ensure propagation of the Stokes beam, it follows that Raman gain is possible for any direction of propagation of the Stokes beam with respect to the pump beam.

However, the greatest Stokes intensity should be observed from a Raman cell along the direction of its greatest illuminated length, unless some other considerations are introduced.

In the ordinary Raman effect, few molecules are found in their excited vibrational state. The strong pumping action of a laser beam changes this situation drastically, so that an appreciable fraction of all molecules in the laser beam are soon made available for anti-Stokes emission. Classically, the anti-Stokes radiation is generated by the interaction of the laser beam with molecular vibrations, but the phase of the latter is established by the still more intense Stokes radiation. As a consequence, an index-matching requirement exists for this three-wave process. The direction of emission of the anti-Stokes intensity is given by the momentum matching condition [24]

$$\vec{k}_{AS} = 2\,\vec{k}_L - \vec{k}_S \tag{5}$$

Since in condensed phases there is curvature of the dispersion of the linear index of refraction, $n = n(\omega)$ and because

$$|\vec{k}| = \frac{n(\omega)\cdot\omega}{c} \tag{6}$$

one immediately recognizes that a collinear arrangement between pump, Stokes and anti-Stokes beams is not possible. However, a direction can be found which satisfies the vector equation (5), but it confines the S and AS photon pairs to conical beams coaxial with the laser beam. Figure 3 shows the required construction, the cones being generated by rotating the plane of the figure around $\vec{k}_L$.

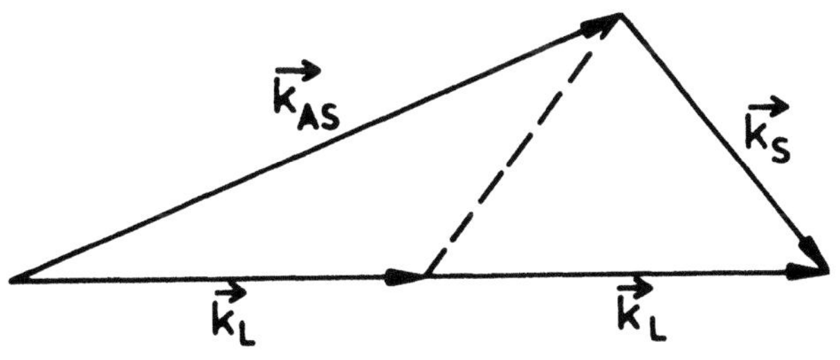

Fig. 3 Momentum conservation for stimulated anti-Stokes Raman scattering (Representation of Equation 5).

Similarly to the generation of coherent fundamental Stokes and anti-Stokes radiation, higher-order stimulated Stokes and anti-Stokes emission can be produced when high pump intensities are employed.

This very short treatment on stimulated Raman effect should serve to illustrate only some principal aspects of this type of non-linear Raman scattering. In the articles by Laubereau [26] and Kaiser and Zinth [27,28] more information on stimulated Raman scattering as well as its application in picosecond spectroscopy is given in this book. We also refer to review articles and books listed in References 29 - 41.

4. NON-LINEAR RAMAN SPECTROSCOPY BASED ON THIRD-ORDER SUSCEPTIBILITIES

From the foregoing short discussion on stimulated Raman scattering it became clear that during this non-linear process coherently driven molecular vibrations are generated and that there is strong coupling between the mechanical waves and the light waves when the frequency difference between pump and Stokes beam matches the molecular vibrational frequency. In what usually is named as stimulated Raman effect only one input laser field (ω_L) is used for this type of excitation. We have seen, in section 3, that only particular Raman modes, i.e. those with highest gain factors, give rise to stimulated Stokes emission. Thus, for molecular spectroscopy in which the interest dominates to determine all Raman active modes, excitation with one strong laser field would not serve the purpose, although it would provide very high signals in form of a coherent beam, but unfortunately, only at one particular vibrational frequency. However, the advantages of stimulated Raman scattering, being high signal strength and coherent radiation, can be fully exploited by a very simple modification of the type of excitation. The trick is simply to provide the molecular system with an intense external Stokes field by using a second laser beam at Stokes frequency ω_S instead of having initially the Stokes field produced in the molecular system by conversion of energy from the pump field. Thus, by keeping one of the two lasers, e.g. the laser with frequency ω_L, fixed in frequency and the other one, e.g. the laser beam at Stokes frequency ω_S tunable, one is now able to excite selectively coherent molecular vibrations at any desired frequency ω_R. A variety of "new" Raman techniques based on this idea has been developed in the last decade. I have put the word new between quotation marks, since only the various techniques are new, but not their common physical basis, which is the stimulated Raman process. Several names for these non-linear Raman spectroscopic techniques based on the third-order non-linear susceptibility have been used in the literature:
"Coherent Active Raman Spectroscopy"[42,43], "Coherent Raman Spectroscopy" [44] or "Active Raman Spectroscopy"[45,46]. These various names stand for methods which combine the wide spectroscopic potentials of spontaneous Raman spectroscopy and the high efficiency of scattering, strong excitation and phasing of molecular

vibrations in a macroscopic volume of substance, that is, as already emphasized the features inherent in stimulated Raman scattering. Certainly, progress in these techniques was connected mainly with the achievements in the development of tunable lasers. The following acronyms of some of these non-linear (coherent) Raman techniques have been widely used [47]: CARS (Coherent anti-Stokes Raman Spectroscopy), CSRS (Coherent Stokes Raman Spectroscopy), PARS (Photoacoustic Raman Spectroscopy), RIKE (Raman Induced Kerr Effect), SRGS (Stimulated Raman Gain Spectroscopy), IRS (Inverse Raman Scattering), ASTERISK (a four colour technique, so named because of the arrangement of the planes of polarization in the experiment), and HORSES (Higher-Order Raman Spectral Excitation Studies). Most of these techniques will be discussed extensively in this NATO ASI. It is the prupose of this introductory chapter to very shortly illuminate the basic principles of these techniques under a common point of view. This will be done next.

A schematic diagram for some of the methods mentioned above is illustrated in Figure 4. The common physical aspect is the excitation of Raman-active molecular vibrations (or rotations) in the field of two laser beams

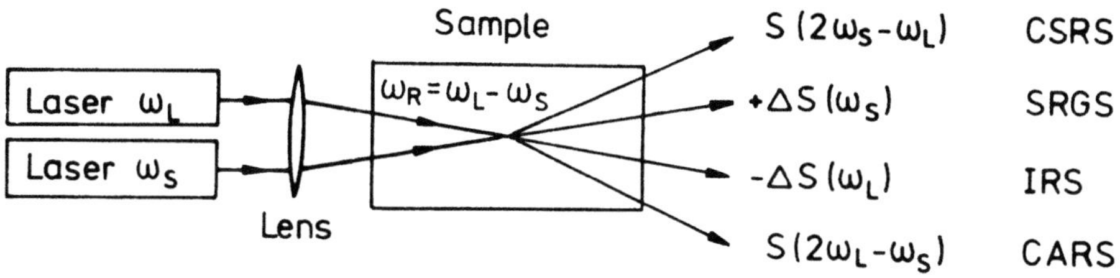

Fig. 4 Schematic diagram for a few techniques in non-linear (coherent) Raman spectroscopy (CSRS: Coherent Stokes Raman Spectroscopy; SRGS: Stimulated Raman gain Spectroscopy; IRS: Inverse Raman Spectroscopy; CARS: Coherent anti-Stokes Raman Spectroscopy)

with frequencies ω_L and ω_S in such a way that their difference corresponds to the frequency of the molecular vibration ω_R ($\omega_L-\omega_S=\omega_R$). The strong coupling between the generated coherent molecular vibrations with the input laser fields via the third-order non-linear susceptibility opens the possibility for the various techniques. As mentioned in section 3 an anti-Stokes signal at frequency $\omega_{AS} = \omega_L + \omega_R = 2\omega_L - \omega_S$ is generated when the momentum matching condition (Equation 5, Figure 3) is fullfilled. However, in contrary to stimulated Raman scattering excited with only one laser (ω_L), the anti-Stokes intensity is not scattered into a cone as described above, instead it is scattered in one direction, which lies in the plane given by the two laser directions ω_L and ω_S and which is determined by the momentum vector diagram shown in Figure 3. Therefore, this so-called CARS-spectroscopy is simply performed by measuring the signal $S(2\omega_L-\omega_S) = S(\omega_L+\omega_R)$, which is a coherent beam emitted into a certain direction. These coherent signals with anti-Stokes frequencies are generated each time the frequency difference of the input laser fields matches a molecular frequency.

The mixing of the two laser fields can also produce radiation on the Stokes side of the ω_S-laser. The direction of this coherent Stokes Raman signal (CSRS) is again determined by a corresponding momentum conservation diagram, which leads to a different direction (see Figure 4, labelled by $S(2\omega_S-\omega_L)$). From the symmetry of this two generated coherent beams, there seems to be no reason to favour the CSRS or CARS technique but, since $2\omega_S-\omega_L = \omega_S- \omega_R$ is to low wavenumber of both ω_L and ω_S there is strong tendency for CSRS to be overlapped by fluorescence. Furthermore, the CSRS signal is in principle weaker than the CARS signal. For these reasons the CARS technique is more frequently used.

The interaction of the electric fields of the two input lasers with frequencies ω_L and ω_S with the electric field associated with the coherent molecular vibrations can yield also to a gain or a loss in the power of the lasers. The method where the gain at the Stokes frequency (labelled in Figure 4 by $+\Delta S(\omega_S)$) is measured is generally referred to as "Stimulated Raman Gain Spectroscopy" whereas the "Inverse Raman Scattering" is the terminology commonly used to designate the induced loss a the pump laser frequency (Figure 4,$-\Delta S(\omega_L)$).

From an experimental point of view, it is quite evident, that for the four non-linear (coherent) Raman techniques discussed until now, one either measures the radiation generated at anti-Stokes frequency (CARS, $\omega_{AS} = 2\omega_L-\omega_S$) or at Stokes frequency (CSRS, $2\omega_S-\omega_L$) or one determines the change ΔS in the laser beam power (ω_L: IRS; ω_S: SRGS). In order to get full Raman information of the medium, it is necessary to tune the frequency difference $\omega_L-\omega_S$; then, successively all Raman-active vibrations (or rotations, or rotation-vibrations) will be excited and a complete "Raman spectrum" is obtained.

In Photoacoustic Raman Spectroscopy (PARS) the sample is illuminated again by the two laser beams ω_L,ω_R the same way as shown in Figure 4, i.e. with spatially and temporally overlapping. Due to the stimulated Raman effect, a population of a particular energy level (ω_R) of the sample is achieved. As the vibrationally excited molecules relax by means of collisions, a pressure wave is generated in the sample and this acoustic signal is detected by a sensitive microphone. In the PARS technique one usually modulates either one or both of the laser beams. When frequency matching is achieved ($\omega_R = \omega_L - \omega_S$) pressure variations are generated in the sample and an acoustic signal at the laser modulation frequency is detected. Again, by tuning one of the two laser frequencies one can obtain complete Raman information from the sample.

As already mentioned above, the introduction of the Raman induced Kerr effect (RIKE) provided the "polarization spectroscopy" forms of stimulated Raman spectroscopy. In RIKE [18], again two laser bemas with frequencies ω_L and ω_S (pump and probe beam) are made spatially overlapped in a sample similar as shown in Figure 4. In addition, however, the probe beam is linearly polarized, whereas the pump beam is circularly polarized. The probe beam is normally blocked behind the sample through a polarizer with perpendicular polarization in respect to the laser polarization. The circularly polarized pump beam at frequency ω_L now induces complex, anisotropic changes in the refractive indices of the sample. These changes exhibit resonances when $\omega_L - \omega_R$ is near the frequency of a Raman-active vibration in the scattering medium. Being anisotropic, these changes caùse the exciting polarized probe beam to be partially transmitted through the "crossed" polarizer behind the sample. By tuning the frequency $\omega_L - \omega_S$ this technique permits the observation of a Raman spectrum.

One of the major advantages of these non-linear (coherent) Raman techniques over conventional Raman spectroscopy is the ability to obtain Raman spectra with very high resolution which is solely determined by the line width of the input sources. The experimental requirements for high resolution measurements utilizing these techniques preclude the necessity of using monochromators, interferometers, or other apparatus usually used in conventional Raman spectroscopy. Since precisely controlled, tuneable, narrow line-width laser sources with appreciable power are available commercially, high resolution molecular spectroscopy, which for quite a long time was only possible by infrared absorption, has become feasible now also in the visible region through these non-linear techniques.

The advantages of these techniques, namely high signal strength and simultaneously high resolution is clearly demonstrated in the example shown in Figures 5 and 6. In Figure 5 a conventional (linear) Raman spectrum of NO at a pressure of 100 Torr obtained with a commercial Raman 0.85m-double monochromator and excitation with an argon-ion laser of 1.5 Watt power is displaced [48].

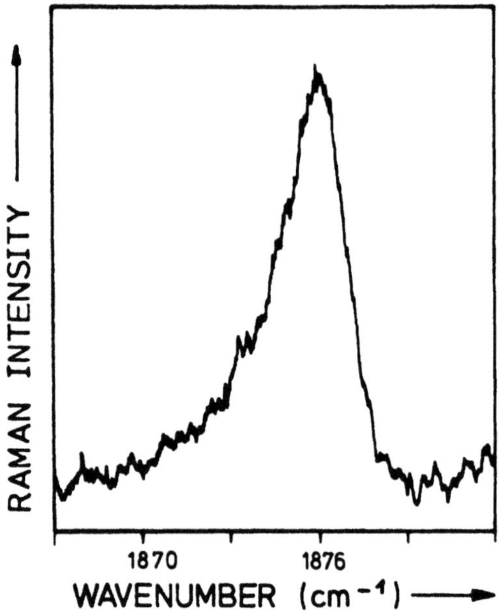

Fig. 5 Conventional (linear) Raman spectrum of the Q-branch of NO at 100 Torr pressure. Excitation: 514.5 nm, 1.5 Watt; slits: 2 cm^{-1} [48].

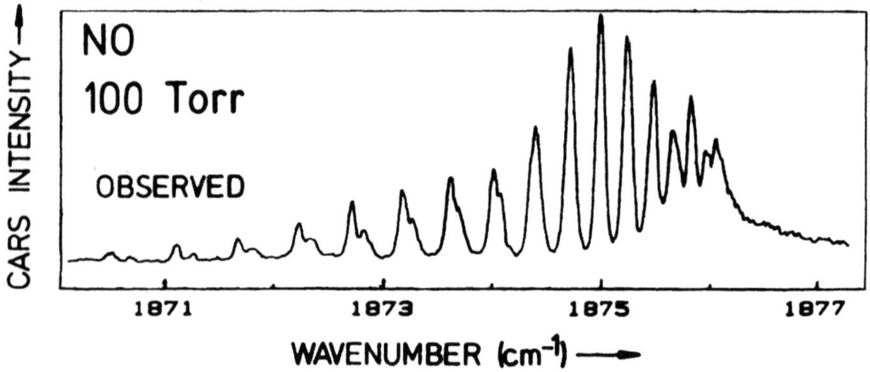

Fig. 6 CARS spectrum of the Q-branch of NO at 100 Torr pressure. Pulsed excitation: power of ω_L-Laser: 10 kW, power of ω_S-Laser: 5kW; pulse widths 5 ns; repetition rate 10 Hz. (Adapted from Reference 49)

The resolution was about 2 cm^{-1}. Figure 6 shows the CARS spectrum excited with a nitrogen-laser pumped dye-laser system [49] but with a resolution of 0.05 cm^{-1}. The peak power of the pulsed dye lasers (ω_L,ω_S) have been 10 kW and 5 kW, respectively, the pulse width was about 5 ns and the repetition rate was 10 Hz. From these values one can calculate a time-averaged total power of 0.75 mW only. Although the total average power has been reduced by a factor of 2000 in CARS excitation, the improvement in resolution is about 40. As will be shown in this Advanced Study Institute, even higher resolution is possible.

REFERENCES

1. Franken, P.A., Hill, A.E., Peters, C.W., Weinreich, G., Phys. Rev. Letters 7, 118 (1961).
2. Kaiser, W., and Garrett, C.G.B., Phys. Rev. Letters 7, 229 (1961).
3. Woodbury, E.J., Ng, W.K., Proc. IRE 50, 2367 (1962).
4. Eckhardt, G., Hellwarth, R.W., McClung, F.J., Schwarz, S.E., and Woodbury, E.J., Phys. Rev. Letters 9, 455 (1962) .
5. Terhune, R.W., Solid State Design 4, 38 (1963).
6. Maker, P.D. and Terhune, R.W., Phys. Rev. 137, A 801 (1965).
7. Jones, W.J., and Stoicheff, B.P., Phys. Rev. Letters 13,657 (1964).
8. Kramers, H.A., and Heisenberg, W., Z. Physik 31, 681 (1925).
9. Dirac, P.A.M., Proc. Roy. Soc. (London) A114, 247, 710 (1927).
10. Owyoung, A. in Laser Spectroscopy IV, Walther H. and Rothe, K.W., eds., Springer Series in Optical Sciences, Vol. 21, 1979, p. 175.
11. Lallemand, P., Simova, P., and Bret, G., Phys. Rev. Letters 17, 1239 (1966).
12. Owyoung, A., IEEE J. Quantum Electron. 14, 192 (1978).
13. Kerr, E.L.,Atwood, J.G., Appl. Opt. 7, 915 (1968).
14. Kreuzer, L.B., J. Appl. Phys. 42, 2934 (1971).

15. Bell, A.G., Proc. Am. Assoc. Adv. Sci. 29, 115 (1880) and Philos. Mag. 11, 510 (1881).

16. Tyndall, J., Proc. Roy. Soc. (London) 31, 307, 478 (1881).

17. Barrett, J.J., and Berry, M.J. in Proceed. Sixth Int. Conf. Raman Spectrosc., Schmid, E.D., Krishnan, R.S., Kiefer, W., and Schrötter, H.W., eds., Heyden, London, 1978, Vol. 2, page 466.

18. Heiman, D., Hellwarth, R.W., Levenson, M.D., and Martin, G., Phys. Rev. Letters 36, 189 (1976).

19. Eesley, G., Levenson, M.D., and Tolles, W.M., IEEE J. Quantum Electron. 14, 45 (1978).

20. Decius, J.C. and Rauch, J.E., Ohio State Symp. Molec. Spectr Columbus, Paper 48, 1959.

21. Terhune, R.W., Maker, P.P., and Savage, C.M., Phys. Rev. Letters 14, 681 (1965).

22. Long, D.A., this book, page 165.

23. Bloembergen, N., Proc. IEEE 51, 124 (1963).

24. Bloembergen, N., Am. J. Physics 35, 989 (1967).

25. Bloembergen, N., in Proceedings of the Sixth Int. Conf. Raman Spectrosc., Vol. 1, Schmid, E.D., Krishnan, R.S., Kiefer, W., and Schrötter, H.W., eds., Heyden, London, 1978, page 335.

26. Laubereau, A., this book, page 183.

27. Kaiser, W., and Zinth, W., this book, page 211.

28. Zinth, W., and Kaiser, W., this book, page 227.

29. Bloembergen, N., Nonlinear Optics, Benjamin, New York, 1965.

30. Shen, Y.R., and Bloembergen, N., Phys. Rev. 137, A 1787 (1965).

31. Giordmaine, J.A., and Kaiser, W., Phys. Rev. 144, 676 (1966).

32. Maier, M., Kaiser, W., and Giordmaine, J.A., Phys. Rev. 177, 580 (1969).

33. Pantell, R.H., and Puthoff, H.E., Fundamentals of Quantum Electronics, Wiley, New York, 1969.

34. Lallemand, P., in The Raman Effect, Anderson, A., ed., Vol. 1, p. 287, Dekker, New York, 1971.

35. Kaiser, W., and Maier, M., in Laser Handbook, Arecchi, F.T., and Schulz-Dubois, E.O., ed.,Vol. 2, p. 1077, North-Holland Publ., Amsterdam, 1972.

36. Shen , Y.R., in Light Scattering in Solids, Cardona, M., ed., Topics in Applied Physics, Vol. 8, p. 275, Springer, Berlin, 1975.

37. Wang, C.S., in Quantum Electronics: A treatise, Rabin, H. and Tang, C.L., eds, Vol. 1, p. 447, Academic Press, New York, 1975.

38. Yariv, A., Quantum Electronics , Wiley, New York, 1975.

39. Maier, M., Appl. Phys. 11, 209 (1976).

40. Ducuing, J. in Nonlinear Optics, Harper, P.G., and Wherrett, B.S., eds., p. 11, Academic Press, New York, 1977.

41. Penzkofer, A., Laubereau, A., and Kaiser, W., Progr. Quant. Electron. 6, 55 (1979).

42. Akhmanov, S.A., Dimitriov, V.G., Kovrigin, A.I., Koroteev, N.I., Tunkin, V.G., Kholodnykh, A.I., JETP Lett., 15, 425 (1972).

43. Akhmanov, S.A. in Nonlinear Spectroscopy, Bloembergen, N., ed., North-Holland Publ. Comp., Amsterdam, 1977, p. 217.

44. Eesley, G.L. Coherent Raman Spectroscopy, Pergamon Press, Oxford, 1981.

45. Kiefer, W., J. Mol. Struct. 59, 305 (1981).

46. Kiefer, W., in Spectroscopy in Chemistry and Physics: Modern Trends, Comes, F.J., Müller, A., and Orville-Thomas, W.J., eds., Elsevier, Amsterdam, 1980, p. 305.

47. Harvey, A.B., Chemical Applications of Nonlinear Raman Spectroscopy, Academic Press, New York, 1981.

48. Fietz, H., and Kiefer, W., unpublished.

49. Beckmann, A., Fietz, H., Kiefer, W., and Laane, J., Phys. Rev. 24A, 2518 (1981).

THE POLARIZABILITY AND HYPERPOLARIZABILITY TENSORS

D. A. Long

School of Chemistry
University of Bradford
Bradford, West Yorkshire
United Kingdom

1. INTRODUCTION

This chapter is concerned with the properties of the polarizability and hyperpolarizability tensors with particular reference to their roles in Rayleigh and Raman scattering and in hyper Rayleigh and hyper Raman scattering, respectively. Before embarking on the detailed mathematical treatments of these tensors it seems appropriate to indicate briefly the nature of these forms of scattering and why an understanding of these tensors is necessary.

The next chapter will deal with the theory of hyper Rayleigh and hyper Raman scattering and with applications of these effects.

2. RAYLEIGH AND RAMAN SCATTERING

When monochromatic radiation of normal irradiance and wavenumber $\tilde{\nu}_0$ is incident on systems like dust-free, transparent gases and liquids, or optically perfect, transparent, solids, most of it is transmitted without change, but, in addition, some scattering of the radiation occurs. If the frequency content of the scattered radiation is analysed, there will be observed to be present, the wavenumber $\tilde{\nu}_0$ associated with the incident radiation and also, in general, pairs of new wavenumbers of the type $\tilde{\nu}' = \tilde{\nu}_0 \pm \tilde{\nu}_M$. In molecular systems, the wavenumbers $\tilde{\nu}_M$ are found to lie principally in the ranges associated with transitions between rotational, vibrational, and electronic levels. The scattered radiation usually has polarization characteristics

W. Kiefer and D. A. Long (eds.), Non-Linear Raman Spectroscopy and Its Chemical Applications, 99–130.

different from those of the incident radiation, and both the intensity and the polarization of the scattered radiation depend on the direction of observation.

The radiation scattered without change of wavenumber is known as Rayleigh scattering when the scattering centres are very much smaller than the wavelength of the incident radiation as is the case for molecules. Scattering of radiation with modified wavenumbers of the type $\tilde{\nu}_0 \pm \tilde{\nu}_M$ is called Raman scattering. The Raman bands at wavenumbers less than the incident wavenumber (i.e. of the type $\tilde{\nu}_0 - \tilde{\nu}_M$) are referred to as Stokes bands, and those at wavenumbers greater than the incident wavenumber (i.e. of the type $\tilde{\nu}_0 + \tilde{\nu}_M$) as anti-Stokes bands. Rayleigh scattering is generally several orders of magnitude more intense than Raman scattering. A typical Rayleigh and Stokes Raman spectrum is shown in Fig. 1(a).

The origin of the modified frequencies (or wavenumbers) found in Raman scattering may be explained in terms of energy transfer between the scattering system and the incident radiation. When a system interacts with radiation of wavenumber $\tilde{\nu}_0$, it may make an upward transition from a lower energy level E_1 to an upper energy level E_2. It must then acquire the necessary energy, $\Delta E = E_2 - E_1$, from the incident radiation. The energy ΔE may be expressed in terms of a wavenumber $\tilde{\nu}_M$ associated with the two levels involved, where $\Delta E = hc\tilde{\nu}_M$. This energy requirement may be regarded as being provided by the annihilation of one photon of the incident radiation of energy $hc\tilde{\nu}_0$ and the simultaneous creation of a photon of smaller energy $hc(\tilde{\nu}_0 - \tilde{\nu}_M)$, so that scattering of radiation of lower wavenumber, $\tilde{\nu}_0 - \tilde{\nu}_M$, occurs. Alternatively, the interaction of the radiation with the system may cause a downward transition from a higher energy level E_2, if the system happens already to be in that excited level, to a lower energy level E_1, in which case it makes available energy $E_2 - E_1 = hc\tilde{\nu}_M$. Again a photon of the incident radiation of energy $hc\tilde{\nu}_0$ is annihilated, but in this instance there is simultaneously created a photon of higher energy $hc(\tilde{\nu}_0 + \tilde{\nu}_M)$, so that scattering of radiation of higher wavenumber, $\tilde{\nu}_0 + \tilde{\nu}_M$, occurs.

In the case of Rayleigh scattering, although there is no resultant change in the energy state of the system, the system still participates directly in the scattering act, causing one photon of incident radiation $hc\tilde{\nu}_0$ to be annihilated and a photon of the same energy to be created simultaneously, so that scattering of radiation of unchanged wavenumber, $\tilde{\nu}_0$, occurs. Although the involvement of the system is not apparent in the wavenumber of the scattered radiation, other properties of the scattered radiation are characteristic of the scattering system and confirm its participation in the scattering act.

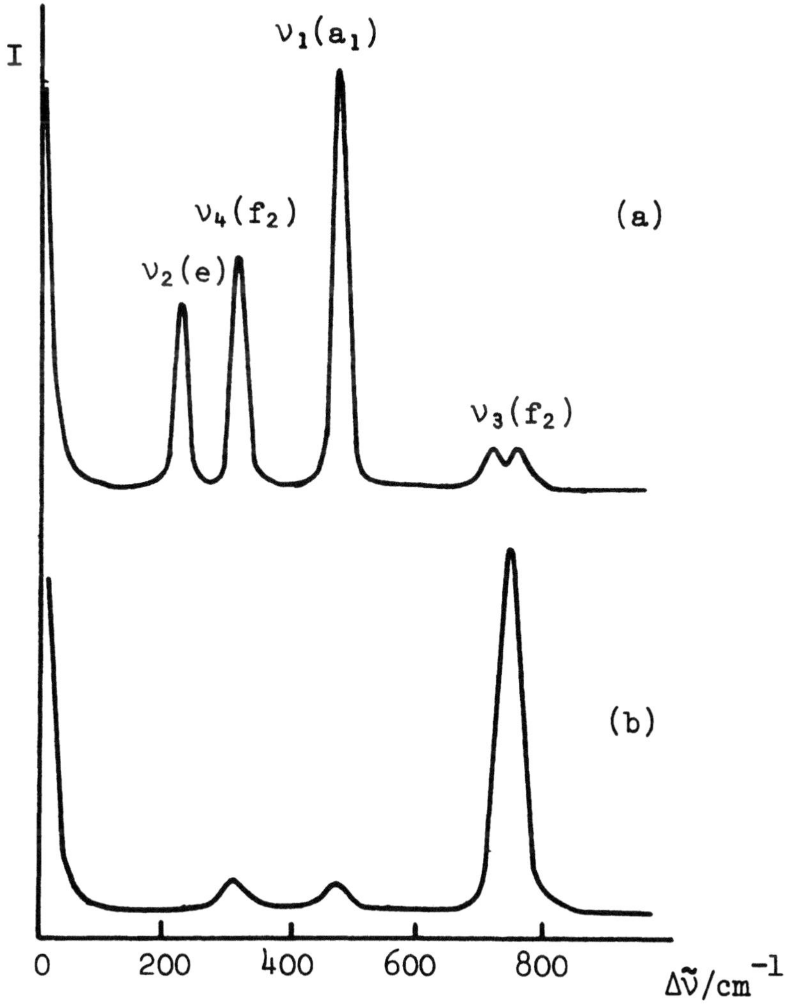

Fig. 1. (a) Rayleigh and Raman spectra; (b) hyper Rayleigh and hyper Raman spectra of CCl_4 (liquid).

3. HYPER RAYLEIGH AND HYPER RAMAN SCATTERING

When a system is illuminated with radiation of wavenumber $\tilde{\nu}_0$ of adequately large irradiance, the scattered radiation is found to include wavenumbers of the type $2\tilde{\nu}_0$ and $2\tilde{\nu}_0 \pm \tilde{\nu}_M$, where $\tilde{\nu}_M$ is a wavenumber associated with a transition between two levels of the scattering molecules. The scattering at $2\tilde{\nu}_0$ is called hyper Rayleigh scattering, that at $2\tilde{\nu}_0 - \tilde{\nu}_M$, Stokes hyper Raman scattering and that at $2\tilde{\nu}_0 + \tilde{\nu}_M$ anti-Stokes hyper Raman scattering. Such scattering also has polarization characteristics different from those of the incident radiation; and both the intensity and polarization of this scattered radiation depend on the direction of observation. Hyper Rayleigh and hyper Raman scattering is many orders of magnitude less intense than Rayleigh and Raman scattering. However the wavenumbers $\tilde{\nu}_M$ which are observed in hyper Raman scattering can be different from those observed in Raman scattering and infra-red absorption and even when they are not, the intensities, polarization characteristics and band contours contain new information. Thus interest in hyper Rayleigh and hyper Raman scattering lies in their potential as sources of new spectroscopic information.

The hyper Rayleigh and Stokes hyper Raman spectra of CCl_4 are shown in Fig. 1(b). Comparison with the Rayleigh and Stokes Raman spectrum in Fig. 1(a) illustrates some of the differences between the two types of scattering.

Hyper Rayleigh and hyper Raman scattering can also be given formal explanations in terms of energy transfer schemes. Each scattering act is associated with the simultaneous annihilation of *two* photons of the incident radiation energy $hc\tilde{\nu}_0$ and the creation of *one* photon of energy which is scattered. The energy of the scattered photon is thus $hc(2\tilde{\nu}_0 - \tilde{\nu}_M)$ for Stokes hyper Raman scattering in which the energy of the scattering system increases by $hc\tilde{\nu}_M$, $hc(2\tilde{\nu}_0 + \tilde{\nu}_M)$ for anti-Stokes hyper Raman scattering in which the energy of the scattering system decreases by $hc\tilde{\nu}_M$ and $hc2\tilde{\nu}_0$ for hyper Rayleigh scattering in which there is no resultant change in the energy of the scattering system.

4. THE INDUCED ELECTRIC DIPOLE

The energy exchange models relate the observed wavenumbers in the scattered radiation to the energy levels of the scattering system but afford no explanation of the mechanism of the interaction and in consequence no information about selection rules, intensities, band contours, directional properties, and states of polarization.

Such detailed knowledge can only be obtained from quantum mechanical treatments. However it is useful to consider initially certain aspects of a classical treatment which can be carried over into a quantum mechanical treatment.

According to the classical theory of electromagnetic radiation, electric and magnetic multipoles oscillating with a given frequency radiate electromagnetic radiation of that frequency. Light scattering phenomena may be given a classical explanation in terms of the electromagnetic radiation produced by such multipoles *induced* in a scattering system by the electric and magnetic fields of the incident radiation.

In what follows we shall confine ourselves to *electric dipoles* induced by the electric field of the incident radiation.

The induced electric dipole moment P is related to the electric field E of the radiation by the power series

$$P = P^{(1)} + P^{(2)} + P^{(3)} + \dots \tag{1}$$

where

$$P^{(1)} = \alpha \cdot E \tag{2}$$

$$P^{(2)} = \tfrac{1}{2}\beta : EE \tag{3}$$

and

$$P^{(3)} = \tfrac{1}{6}\gamma \vdots EEE \tag{4}$$

In these equations, P and E are vectors with units of C m and V m^{-1}, respectively, and α, β, and γ, which relate P and E, are tensors since, in general, the direction of the induced dipole is not the same as that of the electric field producing it: i.e., in general, each component of $P^{(1)}$ is a different linear combination of the components of E, each component of $P^{(2)}$ is a different linear combination of the components of E, and so on. The full significance of these tensor relationships will emerge as we proceed.

α is termed the polarizability tensor and is a second-rank tensor whose components have units of CV^{-1} m^2; β is the hyperpolarizability tensor and is a third-rank tensor whose components have units of CV^{-2} m^3; γ is termed the second hyperpolarizability tensor and is a fourth-rank tensor whose components have units of CV^{-3} m^4. The polarizabilities, α β, and γ can be regarded as measures of the ease with which electrons can be displaced to produce an electric dipole under the action of an electric field.

Typical orders of magnitude for components of α , β , and γ are as follows: α, 10^{-40} CV^{-1} m^2; β, 10^{-50} CV^{-2} m^3; and γ, 10^{-61} CV^{-3} m^4. For such values, if the contribution to the induced dipole from $P^{(2)}$ is to reach one per cent of that from $P^{(1)}$ the electric field intensity must be of the order of 10^9 V m^{-1}. This requires radiation with an irradiance of the order of 10^{15} Wm^{-2} which is normally reached only in focused giant pulse laser beams.

The first-order term $P^{(1)}$ has the correct frequency dependence for Rayleigh and Raman scattering. The second-order term $P^{(2)}$ does not include any induced dipoles whose frequencies correspond to normal Rayleigh and Raman scattering, but gives rise to hyper Rayleigh and hyper Raman scattering, and other non-linear phenomena. Although the third-order term $P^{(3)}$ does include induced dipoles with the correct frequency dependence for Rayleigh and Raman scattering, the contributions from such dipoles is quite negligible. $P^{(3)}$ also gives rise to a variety of non-linear phenomena including second hyper Rayleigh and Raman scattering associated with frequencies of the type $3\tilde{\nu}_O$ and $3\tilde{\nu}_O \pm \tilde{\nu}_M$ respectively. The amplitudes of these induced dipoles will determine the intensities of the scattering processes with which they are associated according to classical radiation theory.

5. THE TRANSITION ELECTRIC DIPOLE

A full quantum theory of spectroscopic processes should treat the radiation and the molecule as a complete system in which both have quantised energies, and explore how energy may be transferred between the radiation and the molecule as a result of their interaction. Correct results for Rayleigh, Raman, hyper Rayleigh and hyper Raman scattering may be obtained by a compromise procedure in which the radiation is treated classically and is regarded as the source of a perturbation of the molecular system which is treated quantum mechanically.

In the quantum mechanical picture the oscillating electric dipole P of classical theory is replaced by a transition electric dipole $[P]_{fi}$ associated with a transition between an initial state i and a final state f. The amplitude of the transition electric dipole determines the intensity, and its time dependence, the frequency of the radiation associated with the transition.

For an unperturbed system in the state i the time-dependent wave function, $\Psi_i^{(o)}$ is given by

$$\Psi_i^{(o)} = \psi_i \exp(-i\omega_i t) \tag{5}$$

where ψ_i is the corresponding time-independent wave function and $\hbar\omega_i$ is the energy of the state i. We note that ψ_i is associated with an arbitrary phase factor $\exp(i\delta_i)$.

When the system is perturbed the time-dependent wave function Ψ_i' may be expressed as

$$\Psi_i' = \Psi_i^{(o)} + \Psi_i^{(1)} + \Psi_i^{(2)} + \ldots. \tag{6}$$

where $\Psi_i^{(1)}$ is the first-order perturbation term, $\Psi_i^{(2)}$ is the second-order perturbation term, and so on.

Hence $[\mathcal{P}]_{fi}$, the electric dipole transition moment for the transition f ← i, when the system is perturbed, is given by

$$[\mathcal{P}]_{fi} = \langle\Psi_f' |\mathcal{P}|\Psi_i'\rangle \tag{7}$$

$\mathcal{P}$ is the electric dipole moment operator for the system defined by

$$\mathcal{P} = \sum_j e_j \boldsymbol{r}_j \tag{8}$$

where e_j is the charge, $\boldsymbol{r}_j$ the position vector of the j-th particle, and the summation is over all particles.

On introducing into eq. (7) the expansions of the perturbed wave functions Ψ_f' and Ψ_i' of the form given by eq. (6), we find that

$$[\mathcal{P}]_{fi} = [\mathcal{P}^{(o)}]_{fi} + [\mathcal{P}^{(1)}]_{fi} + [\mathcal{P}^{(2)}]_{fi} + \ldots. \tag{9}$$

where

$$[\mathcal{P}^{(o)}]_{fi} = \langle\Psi_f^{(o)}|\mathcal{P}|\Psi_i^{(o)}\rangle \tag{10}$$

$$[\mathcal{P}^{(1)}]_{fi} = \langle\Psi_f^{(1)}|\mathcal{P}|\Psi_i^{(o)}\rangle + \langle\Psi_f^{(o)}|\mathcal{P}|\Psi_i^{(1)}\rangle \tag{11}$$

$$[\mathcal{P}^{(2)}]_{fi} = \langle\Psi_f^{(1)}|\mathcal{P}|\Psi_i^{(1)}\rangle + \langle\Psi_f^{(2)}|\mathcal{P}|\Psi_i^{(o)}\rangle + \langle\Psi_f^{(o)}|\mathcal{P}|\Psi_i^{(2)}\rangle \tag{12}$$

The transition moment $[\mathcal{P}^{(o)}]_{fi}$ relates to a direct transition between the unperturbed states f and i, and will not be considered further here. We shall find that the first-order term $[P^{(1)}]_{fi}$ includes terms that relate to normal Rayleigh and Raman scattering, and that the second-order term contains terms relating to hyper Rayleigh and Raman scattering, and so on. We see, therefore, that eq. (9) with the direct transition moment $[P^{(0)}]_{fi}$ omitted, is closely analogous to eq. (1).

When the time-dependent perturbation theory is followed through it is found that the first-order transition electric dipole associated with Rayleigh and Raman scattering can be written in the form

$$[P^{(1)}]_{fi} = [\alpha]_{fi} \cdot E \tag{13}$$

which is the quantum mechanical analogue of eq. (2) and the second-order transition electric dipole associated with hyper Rayleigh and hyper Raman scattering in the form

$$[P^{(2)}]_{fi} = \frac{1}{2}[\beta]_{fi} : EE \tag{14}$$

which is the quantum mechanical analogue of eq. (3).

$[\alpha]_{fi}$ is termed the transition polarizability tensor and $[\beta]_{fi}$ the transition hyperpolarizability tensor. It is now clear that these transition tensors play the determining roles in the scattering processes we are considering and we now turn our attention to a detailed survey of their nature and properties.

6. SOME GENERAL PROPERTIES OF A SECOND RANK TENSOR

A. Introduction

The classical polarizability tensor α and the transition polarizability tensor $[\alpha]_{fi}$ are second rank tensors and it is convenient to summarize the general properties of a second rank tensor before proceeding to the specific case of the transition polarizability tensor. For economy in notation we shall relate these general properties to the polarizability tensor α. The equation

$$P = \alpha \cdot E \tag{15}$$

implies that the magnitudes of the components of P are related to the magnitudes of the components of E, by the following three linear equations:

$$P_x = \alpha_{xx}E_x + \alpha_{xy}E_y + \alpha_{xz}E_z$$
$$P_y = \alpha_{yx}E_x + \alpha_{yy}E_y + \alpha_{yz}E_z \qquad (16)$$
$$P_z = \alpha_{zx}E_x + \alpha_{zy}E_y + \alpha_{zz}E_z$$

The nine coefficients α_{ij} constitute the components of the polarizability tensor $\boldsymbol{\alpha}$. We may write these nine components in an ordered array as follows:

$$\begin{matrix} \alpha_{xx} & \alpha_{xy} & \alpha_{xz} \\ \alpha_{yx} & \alpha_{yy} & \alpha_{yz} \\ \alpha_{zx} & \alpha_{zy} & \alpha_{zz} \end{matrix} \qquad (17)$$

and this constitutes a representation of the tensor $\boldsymbol{\alpha}$. If $\alpha_{xy} = \alpha_{yx}$, $\alpha_{xz} = \alpha_{zx}$, and $\alpha_{yz} = \alpha_{zy}$ the tensor is symmetric and has a maximum of six distinct components.

The nature of the tensor components and their inter-relation depends on the system. For example the transition polarizability tensor may be real or complex and symmetric or non-symmetric, depending on the molecular system involved in the scattering and the relationship between its energy levels and the photon energy of the incident electromagnetic radiation. To maintain generality we shall present the second rank tensor properties for the general case of nine distinct components and for the special case of the real symmetric tensor.

B. Rotation of axes

The relationship between the tensor components in one Cartesian system x, y, z and another Cartesian system x', y', z' with the same origin but a different orientation in space is given by

$$\alpha_{xy} = \sum_{x'y'} \alpha_{x'y'} \cos(xx')\cos(yy') \qquad (18)$$

where the direction cosine, cos(xx'), is the cosine of the angle between the x-axis and x'-axis, and so on, and the summation is over all possible pairs of Cartesian axes in the x'y'z' system, *viz.*, x'x', y'y', z'z', x'y', y'x', x'z', z'x', y'z', and z'y'.

C. Invariants

Although the values of individual components of the polarizability tensor change on rotation of the axis system, certain combinations of them remain invariant. For the general case these are the mean polarizability a defined by

$$a = \frac{1}{3}(\alpha_{xx}+\alpha_{yy}+\alpha_{zz}) \; ; \tag{19}$$

the anisotropy γ defined by

$$\gamma^2 = \frac{1}{2}\Big[(\alpha_{xx}-\alpha_{yy})^2+(\alpha_{yy}-\alpha_{zz})^2+(\alpha_{zz}-\alpha_{xx})^2+\frac{3}{2}\{(\alpha_{xy}+\alpha_{yx})^2+ (\alpha_{yz}+\alpha_{zy})^2+(\alpha_{zx}+\alpha_{xz})^2\}\Big] \tag{20}$$

and the second anisotropy, δ defined by

$$\delta^2 = \frac{3}{4}\{(\alpha_{xy}-\alpha_{yx})^2+(\alpha_{yz}-\alpha_{zy})^2+(\alpha_{zx}-\alpha_{xz})^2\} \tag{21}$$

For the symmetric tensor the mean polarizability a remains unchanged, γ^2 reduces to

$$\gamma^2 = \frac{1}{2}\{(\alpha_{xx}-\alpha_{yy})^2+(\alpha_{yy}-\alpha_{zz})^2+(\alpha_{zz}-\alpha_{xx})^2+ 6(\alpha_{xy}^2+\alpha_{yz}^2+\alpha_{zx}^2)\} \tag{22}$$

and δ^2 becomes zero.

D. Space averages

When considering scattering from a system of molecules freely rotating it is necessary to average over all orientations taken up by the system with respect to the electric field. Since the power of the scattered radiation from a freely rotating induced dipole will be determined by the space-averaged value of the square of the induced dipole, we shall require the average of the squares of the tensor components over all orientations of the system-fixed axes relative to a set of space-fixed axes which we shall here denote as x, y, z. The necessary results may be obtained by starting with eq. (18) where x', y', z' are regarded as the system-fixed axes, squaring and averaging over the appropriate cosine terms. From such calculations we find that for the symmetric tensor

$$\overline{\alpha^2_{xx}} = \overline{\alpha^2_{yy}} = \overline{\alpha^2_{zz}} = \frac{45a^2+4\gamma^2}{45} \qquad (23)$$

$$\overline{\alpha^2_{yx}} = \overline{\alpha^2_{yz}} = \overline{\alpha^2_{zx}} = \frac{\gamma^2}{15} \qquad (24)$$

$$\overline{\alpha_{xx}\alpha_{yy}} = \overline{\alpha_{yy}\alpha_{zz}} = \overline{\alpha_{zz}\alpha_{xx}} = \frac{45a^2-2\gamma^2}{45} \qquad (25)$$

and all other terms, i.e., those which involve any subscript once only, like $\overline{\alpha_{xx}\alpha_{xy}}$, and so on, are zero. It should not be surprising that these space averages are expressed in terms of the invariants of the polarizability tensor.

When the tensor is not symmetric the following additional space averages are required

$$\overline{\left(\frac{\alpha_{xy}-\alpha_{yx}}{2}\right)^2} = \overline{\left(\frac{\alpha_{xz}-\alpha_{zx}}{2}\right)^2} = \overline{\left(\frac{\alpha_{yz}-\alpha_{zy}}{2}\right)^2} = \frac{\delta^2}{9} \qquad (26)$$

The classical space averaging is equivalent to summing and averaging over all permitted rotational transitions. Thus if the rotational structure is not of interest the intensities and polarization properties arising from vibrational transitions in freely rotating molecules can be calculated using the classical space averages of the squares of components of the vibrational transition polarizability.

E. Tensor decomposition

The second rank tensor α can always be written as the sum of three other tensors: an isotropic tensor α_{iso}, an anisotropic tensor α_{aniso} and an anti-symmetric tensor α_{anti}. These tensors are defined as follows:

$$\alpha_{iso}: \quad \begin{matrix} a & 0 & 0 \\ 0 & a & 0 \\ 0 & 0 & a \end{matrix} \qquad (27)$$

$$\alpha_{aniso}: \quad \begin{matrix} \alpha_{xx} - a & \frac{\alpha_{xy} + \alpha_{yx}}{2} & \frac{\alpha_{xz} + \alpha_{zx}}{2} \\ \frac{\alpha_{yx} + \alpha_{xy}}{2} & \alpha_{yy} - a & \frac{\alpha_{yz} + \alpha_{zy}}{2} \\ \frac{\alpha_{zx} + \alpha_{xz}}{2} & \frac{\alpha_{zy} + \alpha_{yz}}{2} & \alpha_{zz} - a \end{matrix} \tag{28}$$

$$\alpha_{anti}: \quad \begin{matrix} 0 & \frac{\alpha_{xy} - \alpha_{yx}}{2} & \frac{\alpha_{xz} - \alpha_{zx}}{2} \\ -\left(\frac{\alpha_{xy} - \alpha_{yx}}{2}\right) & 0 & \frac{\alpha_{yz} - \alpha_{zy}}{2} \\ -\left(\frac{\alpha_{xz} - \alpha_{zx}}{2}\right) & -\left(\frac{\alpha_{yz} - \alpha_{zy}}{2}\right) & 0 \end{matrix} \tag{29}$$

It is clear that

$$\alpha = \alpha_{iso} + \alpha_{aniso} + \alpha_{anti} \tag{30}$$

and that α_{iso} and α_{aniso} are both symmetric with $\alpha_{ij} = \alpha_{ji}$ and α_{anti} is antisymmetric with $\alpha_{ij} = -\alpha_{ji}$. Thus we can rewrite eq. (30) as

$$\alpha = \alpha_{sym} + \alpha_{anti} \tag{31}$$

where

$$\alpha_{sym} = \alpha_{iso} + \alpha_{aniso} \tag{32}$$

If α is symmetric then clearly the definition of α_{iso} remains unchanged, α_{aniso} becomes

$$\alpha_{aniso}: \quad \begin{matrix} \alpha_{xx} - a & \alpha_{xy} & \alpha_{xz} \\ \alpha_{yx} & \alpha_{yy} - a & \alpha_{yz} \\ \alpha_{zx} & \alpha_{zy} & \alpha_{zz} - a \end{matrix} \tag{33}$$

(with $\alpha_{xy} = \alpha_{yx}$, $\alpha_{yz} = \alpha_{zy}$ and $\alpha_{zx} = \alpha_{xz}$) and α_{anti} is zero.

The traces of the products of these tensors with themselves are readily shown to be as follows:

$$\mathrm{Trace}\{\boldsymbol{\alpha}_{iso} : \boldsymbol{\alpha}_{iso}\} = 3a^2 \tag{34}$$

$$\mathrm{Trace}\{\boldsymbol{\alpha}_{aniso} : \boldsymbol{\alpha}_{aniso}\} = \frac{2}{3}\gamma^2 \tag{35}$$

where γ^2 is given by eq. (20) if the tensor is not symmetric and by eq. (22) if the tensor is symmetric.

$$\mathrm{Trace}\{\boldsymbol{\alpha}_{anti} : \boldsymbol{\alpha}_{anti}\} = \frac{2}{3}\delta^2 \tag{36}$$

Traces of cross products of the type $\mathrm{Trace}\{\boldsymbol{\alpha}_{iso} : \boldsymbol{\alpha}_{aniso}\}$ etc. are all zero. This has important consequences because it means that the intensity of Rayleigh and Raman scattering from freely rotating molecules can always be expressed as the sum of an isotropic part (involving only a^2), an anisotropic part (involving only γ^2) and an antisymmetric part (involving only δ^2). For a real symmetric tensor there will be no antisymmetric contribution to the scattering.

F. Irreducible components

The Cartesian irreducible components of $\boldsymbol{\alpha}$ are given by

$$\alpha_{xy} = a^{(0)}\delta_{xy} + \alpha_{xy}^{(1)} + \alpha_{xy}^{(2)} \tag{37}$$

where $a^{(0)} = a$ which is defined by eq. (19),

$$\alpha_{xy}^{(1)} = \frac{1}{2}(\alpha_{xy} - \alpha_{yx}) \tag{38}$$

and

$$\alpha_{xy}^{(2)} = \frac{1}{2}(\alpha_{xy} + \alpha_{yx}) - \delta_{xy}a \tag{39}$$

$a^{(0)}$ is the isotropic part, $\alpha_{xy}^{(1)}$ the antisymmetric part and $\alpha_{xy}^{(2)}$ the anisotropic part and constitute the components of $\boldsymbol{\alpha}_{iso}$, $\boldsymbol{\alpha}_{anti}$ and $\boldsymbol{\alpha}_{aniso}$ respectively. The components of each of the three tensors transform among themselves under the operations of the rotation group (that is they are defined to reduce the Cartesian $\boldsymbol{\alpha}$ tensor into additive parts which do not mix under rotations) and form the bases for the representations of a scalar, a vector and a symmetric traceless second rank tensor. Because of the connection with angular momentum theory they are said to have weights $j = 0$, 1 and 2 respectively (denoted by the superscripts on the irreducible components) each having

(2j + 1) components, thus accounting for the 9 independent components of a second rank tensor.

7. THE TRANSITION POLARIZABILITY TENSOR

A. Quantum mechanical formula

The transition polarizability tensor $[\alpha]_{fi}$ introduced in eq. (13) can be related to the energy levels of a scattering system using perturbation theory, the most convenient approach being to use diagrammatic perturbation techniques. The relevant perturbation diagrams are shown in Fig. 2 and from these the following expression for $[\alpha_{xy}]_{fi}$ is quite readily obtained

$$[\alpha_{xy}]_{fi} = \frac{1}{\hbar}\sum_{r}\left\{\frac{[P_y]_{fr}[P_x]_{ri}}{\omega_{rf}+\omega_0} + \frac{[P_x]_{fr}[P_y]_{ri}}{\omega_{ri}-\omega_0}\right\} \tag{40}$$

where the summation is over all states r of the system except i and f. ω_0 is the circular frequency of the incident radiation. $\hbar\omega_r$ is the energy of the state r and

$$\omega_{ri} = \omega_r - \omega_i \quad \text{etc.} \tag{41}$$

$$[P_y]_{fr} = \langle\Psi_f|P_y|\Psi_i\rangle \quad \text{etc.} \tag{42}$$

where P_y is the y component of the electric dipole moment operator defined by eq. (8).

Eq. (40) applies in non-resonant conditions, that is provided $\omega_0 \neq \omega_{ri}$. In resonant conditions damping factors must be introduced into the frequency denominators and these will be discussed in a later section.

We now examine the implications of eq. (40) for Rayleigh and Raman scattering.

B. The role of the states r

We see that the numerators of the transition polarizability components always contain products of components of transition moments of the types $[P_x]_{ri}$, involving the initial state and the state r, and $[P_y]_{fr}$, involving the final state and the state r. Since these transition moments always occur as products, if the scattering tensor component is to be non-zero, there must exist in the system at least one state r which has a non-zero dipole transition moment to both the initial state and the final state. It must be stressed that this requirement does not mean that the transition from state i to state f, which is associated with Raman scattering, occurs in two distinguishable stages via a transition to an 'intermediate' state r. We note also that incident radiation of energy $\hbar\omega_0$ is not absorbed, since there is no requirement that the energy $\hbar\omega_0$ corresponds to an energy difference between two discrete states of the system, i.e., $\hbar\omega_0$ does not have to be equal to $\hbar\omega_r$. Further there is no restriction upon the states r relative to ω_0. Indeed, the state r can lie above the final state f, below the initial state i, or between i and f.

C. Coherence properties of the scattered radiation

When the phase factors in the time-independent wave functions ψ_i, ψ_r, and ψ_f are taken into account, the Raman transition moment $[P^{(1)}]_{fi}$ of eq. (13) will have the phase factor $\exp\{-i(\delta_f - \delta_i)\}$ which will vary arbitrarily from one scattering molecule to the next, and so Raman scattering will be incoherent. In the case of the Rayleigh transition moment $[P^{(1)}]_{ii}$, the phase factor is unity since $\exp\{-i(\delta_i - \delta_i)\} = 1$, and so Rayleigh scattering is coherent, provided the state i is non-degenerate. If the state i is degenerate, then in addition to coherent Rayleigh scattering involving the same state i there is an incoherent contribution involving transitions between the various degenerate states i.

D. Selection rules

We have already seen that the existence of at least one level r which has a non-zero dipole transition moment with both the initial *and* final states of the system is a necessary condition for the transition polarizability to be non-zero. The selection rules can, however, be cast in a form which involves only the properties of the initial and final states and not the properties of the states r.

It can be shown that $[\alpha_{xy}]_{fi}$ has the same transformation properties as

$$\langle f|xy|i\rangle \tag{43}$$

Thus, the general condition for $[\alpha_{xy}]_{fi}$ to be non-zero is that the product $\psi_f xy \psi_i$ belongs to a representation which contains the totally symmetric species; and the selection rules follow from this.

It should be noted that the general condition we have stated, while a necessary one, is not a sufficient one.

Although a transition polarizability satisfies the symmetry requirements, it may still have a near-zero value.

E. Frequency dependence

Because the denominators of the transition polarizability components contain frequency terms of the type $\omega_{rf} + \omega_o$ and $\omega_{ri} - \omega_o$, the magnitude of the transition dipole moment is, in general, dependent on the excitation frequency ω_o. Only if the excitation frequency ω_o is much smaller than the transition frequencies ω_{rf} and ω_{ri} can the transition polarizability be regarded as independent of ω_o, so that the intensity of scattering follows a fourth power law. As ω_o approaches a transition frequency, the intensity of scattering will depart from the fourth power law. In particular, if ω_o is reasonably close to ω_{ri}, the frequency of an absorption band of the system, the transition polarizability components will have values which are greatly enhanced compared with the values when ω_o is remote from ω_{ri}. Such anomalously intense Raman scattering is termed resonance Raman scattering.

Resonance Raman scattering can exhibit selection rules and polarization phenomena which are different from ordinary Raman scattering. This is because, as we shall see subsequently, when ω_o is much smaller than any transition frequency the transition polarizability tensor acquires symmetry properties that are not possessed by the general tensor. The lower symmetry of the general tensor can mean, for example, that transitions inactive in normal Raman scattering become active in resonance Raman scattering.

When $\omega_o \approx \omega_{ri}$, it would appear that the transition polarizability will tend to infinity. However, the formulae we have derived here do not apply in this limiting case. In a more general treatment, the lifetimes of the states r have to be taken into account. The states r are assumed to decay exponentially with time so that the stationary state r, defined by a form of eq. (5), becomes a quasi-stationary state given by

$$\Psi_r = \psi_r \exp\{-i(\omega_r - \frac{1}{2} i\Gamma_r)t\} \tag{44}$$

where Γ_r is inversely proportional to the lifetime of the state r. The finite lifetimes of the states r can therefore be incorporated into our formulae simply by changing to complex energies, i.e., replacing $\hbar\omega_r$ by $\hbar\omega_r - \frac{1}{2}i\hbar\Gamma_r$. Thus, the frequency denominators in the transition polarizability will now be of the type $(\omega_{ri} - \omega_o - i\Gamma_{ri})$ where $\Gamma_{ri} = \frac{1}{2}(\Gamma_r - \Gamma_i)$ and is related to the width of the r ← i transition. Clearly the new frequency denominators do not become zero when $\omega_o \approx \omega_{ri}$.

F. Symmetry properties

We consider, first, the symmetry properties of the complex Rayleigh scattering tensor with components $[\alpha_{xy}]_{ii}$. Using the Hermitian property of the transition moment, namely, that

$$[\tilde{P}_x]_{ri} = [\tilde{P}_x^*]_{ir} \tag{45}$$

it is readily shown that

$$[\tilde{\alpha}_{xy}]_{ii} = [\tilde{\alpha}_{yx}]_{ii} \tag{46}$$

and thus the tensor itself is Hermitian. If the components of the tensor are real then the tensor is symmetric, since

$$[\alpha_{xy}]_{ii} = [\alpha_{yx}]_{ii} \tag{47}$$

If the components of the tensor are complex we may write

$$[\tilde{\alpha}_{xy}]_{ii} = [\alpha_{xy}]_{ii} - i[\alpha'_{xy}]_{ii} \tag{48}$$

where $[\alpha_{xy}]_{ii}$ is real and $i[\alpha'_{xy}]_{ii}$ imaginary. Now, using eq. (47), we may write

$$[\tilde{\alpha}^*_{yx}]_{ii} = [\alpha_{yx}]_{ii} + i[\alpha'_{yx}]_{ii} \tag{49}$$

It follows from eq. (46) that $[\alpha'_{xy}]_{ii} = -[\alpha'_{yx}]_{ii}$. Thus, the imaginary part of the complex tensor is always antisymmetric, whereas the real part is always symmetric. For Rayleigh scattering, there can be no antisymmetric real part and no symmetric imaginary part of the scattering tensor. Now a complex tensor requires complex wave functions and these can only exist for systems in the presence of external magnetic fields or where there are internal magnetic perturbations, as in the case of spin-orbit interactions. Thus, in the absence of magnetic perturbation, the Rayleigh scattering tensor is real and symmetric. It should be noted that there is no restriction on the frequency denominators; the symmetry properties of the Rayleigh scattering tensor are the same for transparent and absorbing samples.

For the complex Raman scattering tensor with components $[\tilde{\alpha}_{xy}]_{fi}$, it is easily shown that

$$[\tilde{\alpha}_{xy}]_{fi} \neq [\tilde{\alpha}^*_{yx}]_{fi} \tag{50}$$

and thus, in general, the scattering tensor is not Hermitian and so is not symmetric, even if the tensor components are real.

We shall now develop the Placzek polarizability theory which will reveal under what conditions the tensor can be first Hermitian and then symmetric.

G. The Placzek polarizability theory

The wave functions ψ_i, ψ_r, and ψ_f involved in the transition polarizabilities are the time-independent total wave functions of the system. We now separate the wave functions into their electronic, vibrational, and rotational parts. We write, for ψ_i, for example

$$\psi_i = \psi_{e^i}\psi_{v^iR^i} \tag{51}$$

In eq. (51), ψ_{e^i} is the electronic part of the wave function with the set of electronic quantum numbers e^i, and is a function of the electronic and nuclear coordinates; and $\psi_{v^iR^i}$ is the vibrational and rotational part of the wave function with the set of vibrational quantum numbers v^i and the set of rotational quantum numbers R^i. The wave function $\psi_{v^iR^i}$ can be expanded as

$$\psi_{v^iR^i} = \phi_{v^i}\Theta_{R^i} \tag{52}$$

where ϕ_{v^i} is the vibrational wave function which is a function of the normal coordinates of vibration and Θ_{R^i} is the rotational wave function which is a function of the Euler angles which define the orientation of the molecule with respect to space-fixed axes. In what follows we shall be concerned with the separation of the electronic part from the vibrational and rotational parts, and it will thus not be necessary to introduce eq. (52) into eq. (51).

If the expansions of ψ_i, ψ_r, and ψ_f are introduced into eq. (40), the resulting expression for the transition polarizability is rather complicated, and the electronic and nuclear contributions appear to be inextricably mixed. However, if certain special conditions are satisfied, the scattering tensor associated with a transition in which only the vibrational and/or rotational quantum numbers change and the system starts and finishes in the ground electronic state is given by

$$[\tilde{\alpha}_{xy}]_{fi} = \langle v^f R^f | [\tilde{\alpha}_{xy}]_{e^o e^o} | v^i R^i \rangle \tag{53}$$

where

$$[\tilde{\alpha}_{xy}]_{e^o e^o} = \frac{1}{\hbar} \sum_r \left\{ \frac{[\tilde{P}_y]_{e^o e^r} [\hat{P}_x]_{e^r e^o}}{\bar{\omega}_{e^r e^o} + \omega_o} + \frac{[\tilde{P}_x]_{e^o e^r} [\hat{P}_y]_{e^r e^o}}{\bar{\omega}_{e^r e^o} - \omega_o} \right\} \tag{54}$$

In eq. (54), e^o and e^r refer, respectively, to the *electronic* quantum numbers of the ground state and an intermediate state r, and the summation is over all electronic states; $\omega_{e^r e^o}$ is the average frequency separation of the ground electronic state and the electronic r; and the transition moment components in eq. (54) relate to transitions between *electronic* states, in equilibrium nuclear geometry so that

$$[\tilde{P}_y]_{e^r e^o} = \langle e^r | P_{y_e} | e^o \rangle, \quad \text{etc.} \tag{55}$$

where P_{y_e} is the y-component of the *electronic* part of the electric dipole operator (i.e., in the definition of the dipole moment operator given by eq. (8) the summation is now confined to the electrons).

The rather extensive manipulations that are necessary to arrive at eqs. (53) and (54) will not be given here. We shall content ourselves with noting the special conditions that have to be introduced in the course of the manipulations. These are:

(a) The excitation frequency ω_o must be much larger than the frequency associated with any vibration or rotation transition of the system.
(b) The excitation frequency ω_o must be much less than any electronic transition frequency $\bar{\omega}_{e^r e^o}$ of the system.
(c) The ground electronic state must not normally be degenerate (although there may be special cases where this is not a necessary condition).

The special form of the transition polarizability given by eq. (53) is Hermitian, since $[\tilde{\alpha}]_{e^o e^o}$ is Hermitian. Thus, when the conditions in (a), (b), and (c) are satisfied, the Raman transition polarizability has the same symmetry properties as the Rayleigh scattering tensor, i.e., it has only a real symmetric part and an antisymmetric imaginary part.

In the absence of magnetic perturbations, the electronic wave functions of the ground state are real, $[\alpha]_{e^oe^o}$ is real and hence symmetric and so $[\alpha_{xy}]_{fi}$ is real and symmetric. We have then in place of eqs. (53) and (54)

$$[\alpha_{xy}]_{fi} = \langle v^f R^f | [\alpha_{xy}]_{e^oe^o} | v^i R^i \rangle \qquad (56)$$

and

$$[\alpha_{xy}]_{e^oe^o} = \frac{1}{\hbar} \sum_r \left\{ \frac{[P_y]_{e^oe^r}[P_x]_{e^re^o}}{\bar{\omega}_{e^re^o} + \omega_0} + \frac{[P_x]_{e^oe^r}[P_y]_{e^re^o}}{\bar{\omega}_{e^re^o} - \omega_0} \right\} \qquad (57)$$

$[\alpha_{xy}]_{e^oe^o}$ is the electronic polarizability in the ground electronic state.

In addition to the symmetry properties with which $[\alpha_{xy}]_{e^oe^o}$ is endowed, it is a function of the nuclear coordinates only, and not of the electron coordinates.

Thus, it is only when these several conditions are met that the polarizability can be regarded as a frequency-independent, symmetric real tensor which is a function simply of the nuclear coordinates.

In resonance Raman scattering the simplifications of the Placzek theory no longer apply, and we have to deal in general with a Raman scattering tensor which is both unsymmetric and complex. However, in the absence of magnetic perturbation, the tensor becomes real although remaining unsymmetric.

8. THE TRANSITION HYPERPOLARIZABILITY TENSOR

A. Introduction

In general a third rank tensor has 27 components. In the case of the transition hyperpolarizability tensor $[\beta]_{fi}$, if the two incident photons have different energies then the tensor has no permutational symmetry of its indices x, y, z whatsoever and so the general case of 27 components of the type β_{xyz} has to be considered. However for the special case of hyper Rayleigh and hyper Raman scattering where the two incident photons have the same energy the transition hyperpolarizability tensor $[\boldsymbol{\beta}]_{fi}$ can be shown to be symmetric in its last two indices, i.e., $\beta_{xyz} = \beta_{xzy}$ etc. Such a third rank tensor has only 18 components.

Because we are only interested in this special case here, it is convenient to consider first the quantum mechanical formula for the transition hyperpolarizability $[\boldsymbol{\beta}]_{fi}$ associated with hyper Rayleigh and hyper Raman scattering as this enables us to establish the special symmetry properties of this particular third rank tensor. Other relevant properties of this tensor will then be treated. Some of these are of general applicability to any third rank tensor.

B. Quantum mechanical formula

The transition hyperpolarizability tensor $[\boldsymbol{\beta}]_{fi}$ introduced in eq. (14) can be related to the energy levels of a scattering system using diagrammatic perturbation theory. The relevant diagrams for the special case when the energies of the two incident photons are the same and each equal to $\hbar\omega_0$ are shown in Fig. 3 and lead to the following expressions for $[\beta_{xyz}]_{fi}$

$$[\beta_{xyz}]_{fi} = \frac{1}{\hbar^2}\sum_{r,s}\left\{ \frac{[P_x]_{fs}[P_z]_{sr}[P_y]_{ri}}{(\omega_{ri}-\omega_0)(\omega_{si}-2\omega_0)} + \frac{[P_z]_{fs}[P_x]_{sr}[P_y]_{ri}}{(\omega_{ri}-\omega_0)(\omega_{sf}+\omega_0)} \right.$$

$$+ \frac{[P_x]_{fs}[P_y]_{sr}[P_z]_{ri}}{(\omega_{ri}-\omega_0)(\omega_{si}-2\omega_0)} + \frac{[P_y]_{fs}[P_x]_{sr}[P_z]_{ri}}{(\omega_{ri}-\omega_0)(\omega_{sf}+\omega_0)}$$

$$\left. + \frac{[P_z]_{fs}[P_y]_{sr}[P_x]_{ri}}{(\omega_{rf}+2\omega_0)(\omega_{sf}+\omega_0)} + \frac{[P_y]_{fs}[P_z]_{sr}[P_x]_{ri}}{(\omega_{rf}+2\omega_0)(\omega_{sf}+\omega_0)} \right\} \tag{58}$$

where the summation is over all states r and s of the system except i and f. The definitions of the frequency terms ω_{ri} etc., and the electric dipole transition amplitudes $[P_y]_{fs}$ etc., follow from eqs. (41) and (42) respectively.

Eq. (58) applies in non-resonant conditions, that is provided $\omega_o \neq \omega_{ri}$ and $\omega_o \neq \omega_{si}$. In resonant conditions damping factors must be introduced into the frequency denominators as discussed in section 7E for the transition polarizability tensor.

C. Index symmetry properties

It is readily seen that $[\beta_{xyz}]_{fi}$ as defined in eq. (58) is symmetric in the last two indices so that $[\beta_{xyz}]_{fi} = [\beta_{xzy}]_{fi}$. It is important to note that there is not full permutation symmetry of the indices.

D. The role of the states r and s

For $[\beta_{xyz}]_{fi}$ to be non-zero there must exist in the system at least two states r and s for which there are non-zero electric dipole transition moments involving the states i and r, r and s, and s and f. The involvement of two states r *and* s is to be contrasted with the involvement of only one state r in the case of $[\alpha_{xy}]_{fi}$. The two situations are compared in Figs. 4a and b.

E. Placzek approximation

If the Placzek conditions discussed in section 7G are applied to the transition hyperpolarizability tensor then for vibrational and rotational transitions within the ground electronic state the tensor with components $[\beta_{xyz}]_{fi}$ may be expressed in the form

$$[\beta_{xyz}]_{fi} = \langle v^f R^f | [\beta_{xyz}]_{e^oe^o} | v^i R^i \rangle \qquad (59)$$

where $[\beta_{xyz}]_{e^oe^o}$ is analogous to $[\alpha_{xy}]_{e^oe^o}$ and is a function of the nuclear coordinates only. However unlike $[\alpha_{xy}]_{e^oe^o}$, $[\beta_{xyz}]_{e^oe^o}$ does not acquire any additional index symmetry; it is still only symmetric in the last two indices. However if the following conditions are satisfied

$$\frac{\omega_o}{\omega_{to}} \ll 1 \text{ and } \frac{2\omega_o + \omega_{fi}}{\omega_{to}} \ll 1 \qquad (60)$$

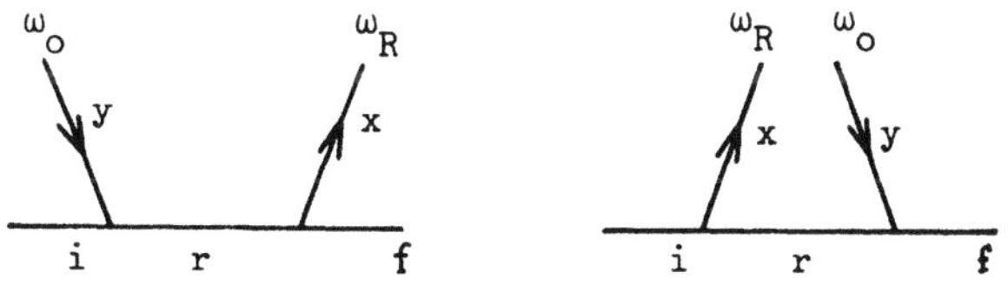

Fig. 2. Perturbation diagrams for transition polarizability $[\alpha_{xy}]_{fi}$ (ω_o incident; $\omega_R = \omega_o + \omega_{if}$ scattered).

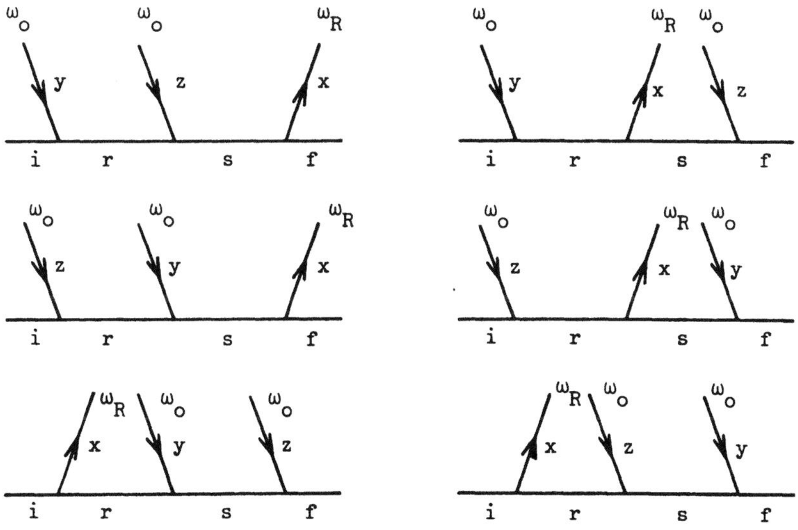

Fig. 3. Perturbation diagrams for transition hyperpolarizability $[\beta_{xyz}]_{fi}$ (ω_o incident, $\omega_R = 2\omega_o + \omega_{if}$ scattered).

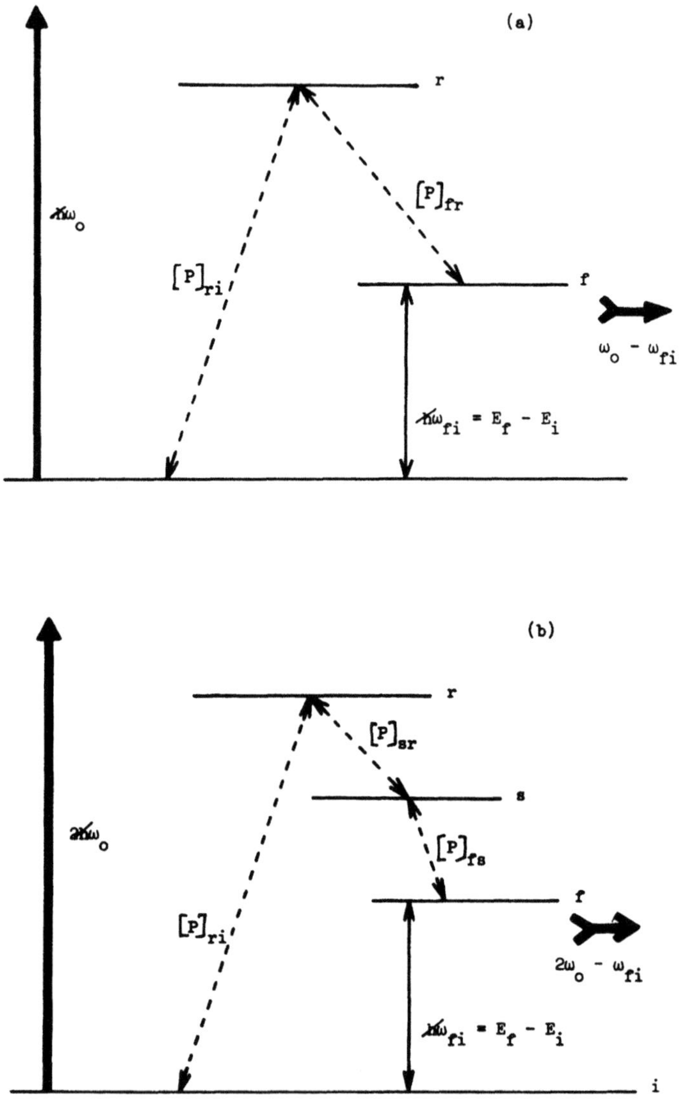

Fig. 4. (a) the states i, r and f and $[\alpha]_{fi}$; (b) the states i, r, s and f and $[\beta]_{fi}$.

where t = r or s the tensor becomes symmetric. The non-symmetric part of the tensor thus becomes less significant as the photon energies become smaller. However under normal conditions the frequency ratios in eq. (60) can lie in the range 0.25 to 0.50 and so it is not a good approximation to ignore the asymmetric part of the tensor.

F. Frequency dependence

Resonance enhancement can result if $\omega_{ri} \approx \omega_0$ or $\omega_{si} \approx 2\omega_0$. If both conditions can be satisfied simultaneously, intensity enhancements of the order of 10^{10} could be possible.

G. Rotation of axes

The relationship between the hyperpolarizability tensor components in one Cartesian system x, y, z and another Cartesian system x', y', z' with the same origin but a different orientation in space is given by

$$\beta_{xyz} = \sum_{x'y'z'} \beta_{x'y'z'} \cos(xx')\cos(yy')\cos(zz') \tag{61}$$

where cos(xx') is the direction cosine between the x and x' axes and so on and the summation is over all possible triples of axes in the x'y'z' system. This is a general relationship for any third rank tensor.

H. Space averages

In this section we shall use X, Y, Z to denote a space-fixed Cartesian axis system and i, j, k to denote a molecule-fixed Cartesian system. For the hyperpolarizability tensor with symmetry in the last two indices, i.e., $\beta_{ijk} = \beta_{ikj}$ etc., the space averages of quadratic functions of the hyperpolarizability tensor components referred to the space-fixed axes X, Y, Z are given in Table 1 in terms of quadratic functions of hyperpolarizability tensor components referred to molecule-fixed axes which are listed in Table 2 (1).

It should be noted that the existence of these six quadratic functions does not imply that there are six independent variables which can be measured in appropriate light scattering experiments. Because an assembly of freely orienting molecules (e.g. a liquid or a gas) is isotropic whatever the structure of the individual molecules, there are only five observables. The relationship of the six quadratic space averages to these five observables (denoted by a, b, c, d and e) is included in Table 1.

Table 1. Expectation values of quadratic functions of β in space-fixed coordinates expressed as coefficients of quadratic functions of β in molecule-fixed coordinates (see Table 2).

	α	β_1	β_2	β_3	β_4	β_5	γ_1	γ_2	γ_3	δ_1	δ_2
$\langle\beta_{ZZZ}^2\rangle = 2a+c+4(b+d+e)$	$\frac{1}{7}$	$\frac{2}{35}$	$\frac{2}{35}$	$\frac{1}{35}$	$\frac{2}{35}$	$\frac{1}{35}$	$\frac{1}{105}$	$\frac{1}{105}$	$\frac{2}{105}$	$\frac{1}{210}$	$\frac{1}{105}$
$\langle\beta_{XZZ}^2\rangle = 2a+c$	$\frac{1}{35}$	$\frac{4}{105}$	$\frac{-1}{35}$	$\frac{3}{35}$	$\frac{-1}{35}$	$\frac{2}{105}$	$\frac{1}{35}$	$\frac{-1}{210}$	$\frac{-1}{105}$	$\frac{1}{70}$	$\frac{-1}{210}$
$\langle\beta_{XZZ}\beta_{XYY}\rangle = c$	$\frac{1}{105}$	$\frac{2}{35}$	$\frac{-1}{105}$	$\frac{1}{35}$	$\frac{-1}{105}$	$\frac{-1}{210}$	$\frac{8}{105}$	$\frac{1}{105}$	$\frac{-1}{21}$	$\frac{-1}{84}$	$\frac{1}{105}$
$\langle\beta_{ZZZ}\beta_{ZYY}\rangle = c+2e$	$\frac{1}{35}$	$\frac{11}{105}$	$\frac{1}{210}$	$\frac{2}{105}$	$\frac{1}{210}$	$\frac{-1}{70}$	$\frac{1}{35}$	$\frac{-1}{210}$	$\frac{1}{42}$	$\frac{-1}{420}$	$\frac{-1}{210}$
$\langle(\beta_{ZYZ}+\beta_{ZZY})^2\rangle = 4(a+b+d)$	$\frac{4}{35}$	$\frac{-4}{35}$	$\frac{2}{105}$	$\frac{8}{105}$	$\frac{2}{105}$	$\frac{8}{105}$	$\frac{-2}{105}$	$\frac{1}{70}$	$\frac{-4}{105}$	$\frac{1}{42}$	$\frac{1}{70}$
$\langle(\beta_{XYZ}+\beta_{XZY})^2\rangle = 4a$	$\frac{4}{105}$	$\frac{-4}{105}$	$\frac{-4}{105}$	$\frac{4}{35}$	$\frac{-4}{105}$	$\frac{1}{21}$	$\frac{-2}{21}$	$\frac{-1}{35}$	$\frac{8}{105}$	$\frac{11}{210}$	$\frac{-1}{35}$

Table 2. Quadratic functions of β in molecule-fixed axes.

$$\alpha = \sum_{i} \beta_{iii}^2,$$

$$\beta_1 = \sum_{i,j} \beta_{iii}\beta_{ijj},$$

$$\beta_2 = \sum_{i,j} \beta_{iii}(\beta_{jij} + \beta_{jji}),$$

$$\beta_3 = \sum_{i,j} \beta_{ijj}^2,$$

$$\beta_4 = \sum_{i,j} \beta_{ijj}(\beta_{jij} + \beta_{jji}),$$

$$\beta_5 = \sum_{i,j} (\beta_{jij} + \beta_{jji})^2,$$

$$\gamma_1 = \sum_{i,j,k} \beta_{ijj}\beta_{ikk},$$

$$\gamma_2 = \sum_{i,j,k} (\beta_{jij} + \beta_{jji})(\beta_{kik} + \beta_{kki}),$$

$$\gamma_3 = \sum_{i,j,k} \beta_{ijj}(\beta_{kik} + \beta_{kki}),$$

$$\delta_1 = \sum_{i,j,k} (\beta_{ijk} + \beta_{ikj})^2,$$

$$\delta_2 = \sum_{i,j,k} (\beta_{ijk} + \beta_{ikj})(\beta_{jik} + \beta_{jki}).$$

Table 3. Space averages of quadratic β functions for full permutation symmetry.

$$\overline{\beta_{XXX}^2} = \tfrac{1}{7}\sum_{i}\beta_{iii}^2 + \tfrac{6}{35}\sum_{i\neq j}\beta_{iii}\beta_{ijj} + \tfrac{9}{35}\sum_{i\neq j}\beta_{iij}^2 + \tfrac{6}{35}\sum_{ijk,\ \text{cyclic}}\beta_{iij}\beta_{jkk} + \tfrac{12}{35}\beta_{ijk}^2,$$

$$\overline{\beta_{XYY}^2} = \tfrac{1}{35}\sum_{i}\beta_{iii}^2 - \tfrac{2}{105}\sum_{i\neq j}\beta_{iii}\beta_{ijj} + \tfrac{11}{105}\sum_{i\neq j}\beta_{iij}^2 - \tfrac{2}{105}\sum_{ijk,\ \text{cyclic}}\beta_{iij}\beta_{jkk} + \tfrac{8}{35}\beta_{ijk}^2,$$

$$\overline{\beta_{XYZ}^2} = \tfrac{1}{105}\sum_{i}\beta_{iii}^2 - \tfrac{1}{35}\sum_{i\neq j}\beta_{iii}\beta_{ijj} + \tfrac{2}{35}\sum_{i\neq j}\beta_{iij}^2 - \tfrac{1}{35}\sum_{ijk,\ \text{cyclic}}\beta_{iij}\beta_{jkk} + \tfrac{1}{7}\beta_{ijk}^2.$$

Table 5. Irreducible components of the β tensor.[†]

β_1	$\beta_{xxx}+\beta_{yyx}+\beta_{zzx}$	set A	weight-1
β_2	$\beta_{xxy}+\beta_{yyy}+\beta_{zzy}$		
β_3	$\beta_{xxz}+\beta_{yyz}+\beta_{zzz}$		
β_4	$\beta_{xxx}+\beta_{xyy}+\beta_{xzz}$	set B	
β_5	$\beta_{yxx}+\beta_{yyy}+\beta_{yzz}$		
β_6	$\beta_{zxx}+\beta_{zyy}+\beta_{zzz}$		
β_7	$\beta_{xyz}-\beta_{yzx}$		weight-2
β_8	$\sqrt{\frac{1}{3}}\,(\beta_{yzx}-2\,\beta_{zxy}+\beta_{xyz})$		
β_9	$\sqrt{\frac{1}{3}}\,(\beta_{yyz}-\beta_{zyy}+\beta_{zxx}-\beta_{xxz})$		
β_{10}	$\sqrt{\frac{1}{3}}\,(\beta_{yzz}-\beta_{zzy}+\beta_{xxy}-\beta_{yxx})$		
β_{11}	$\sqrt{\frac{1}{3}}\,(\beta_{zzx}-\beta_{xzz}+\beta_{xyy}-\beta_{yyx})$		
β_{12}	$\sqrt{\frac{2}{3}}\,(\beta_{xyz}+\beta_{yzx}+\beta_{zxy})$		weight-3
β_{13}	$\frac{1}{2}\,(\beta_{xxx}-2\,\beta_{yyx}-\beta_{xyy})$		
β_{14}	$\frac{1}{2}\,(\beta_{yyy}-2\,\beta_{xxy}-\beta_{yxx})$		
β_{15}	$\sqrt{\frac{1}{6}}\,(2\,\beta_{xxz}+\beta_{zxx}-2\,\beta_{yyz}-\beta_{zyy})$		
β_{16}	$\sqrt{\frac{1}{10}}\,(2\,\beta_{zzz}-2\,\beta_{xxz}-2\,\beta_{yyz}-\beta_{zxx}-\beta_{zyy})$		
β_{17}	$\sqrt{\frac{1}{60}}\,(8\,\beta_{zzx}+4\,\beta_{xzz}-3\,\beta_{xxx}-2\,\beta_{yyx}-\beta_{xyy})$		
β_{18}	$\sqrt{\frac{1}{60}}\,(8\,\beta_{zzy}+4\,\beta_{yzz}-3\,\beta_{yyy}-2\,\beta_{xxy}-\beta_{yxx})$		

[†] after Andrews and Thirunamachandran (5).

In the special case where the hyperpolarizability tensor has full permutation symmetry of its indices only three space averaged quadratic functions are non-zero. These three functions have the special forms given in Table 3. These special forms may be obtained from Tables 1 and 2 by introducing full permutation symmetry of the indices i, j, k. With full permutation symmetry there are only two observables (a = b, c = d = e).

I. Irreducible components

A third rank tensor without any index symmetry has 27 independent components. In the case of symmetry in the last two indices the number of independent components is 18 and for full permutation symmetry of the indices there are only 10. The distributions of components amongst terms of weights, 0, 1, 2 and 3 is given in Table 4 for the three index symmetry cases. A set of irreducible components for the hyperpolarizability tensor with symmetry in the last two indices is given in Table 5.

Table 4. Numbers of irreducible components and their weights for third rank tensors of various index symmetries

Index Symmetry	Total Number of Irreducible Components	Number of terms of			
		Weight 0 (2j+1=1)	Weight 1 (2j+1=3)	Weight 2 (2j+1=5)	Weight 3 (2j+1=7)
none	27	1	3	2	1
ijk=ikj	18	0	2	1	1
full permutation symmetry	10	0	1	0	1

J. Selection rules

It can be shown that $[\beta_{xyz}]_{fi}$ transforms as

$$\langle f|\beta_{xyz}|i\rangle \tag{62}$$

and β_{xyz} as the product xyz. Thus, the general selection rule for hyper Raman activity is that the triple direct product of the species of the wave functions for the states i and f, and xyz must contain the totally symmetric species.

Thus to predict hyper Raman activity requires a knowledge for each point group of the symmetry classes to which the products xyz belong.

The first detailed tabulation of the transformation properties of β_{ijk} was given by Cyvin, Rauch and Decius (2) but this related only to the case where β_{ijk} has full permutation symmetry. Subsequently Christie and Lockwood (3) gave the transformation properties for the case where β_{ijk} has index symmetry only in the last two indices. They tabulated separately the transformation properties for the symmetric and non-symmetric parts of β_{ijk}. Long (4) has given convenient tables of these results correcting some misprints in the original paper of Cyvin, Rauch and Decius which have been carried over into other publications.

However as Andrews and Thirunamachandran (5) have pointed out Christie and Lockwood used an over-complete set of tensor components (20 symmetric and 16 non-symmetric components). Andrews and Thirunamachandran have proposed a set of 18 irreducible components (Table 5) upon which to base transformation properties and selection rules. They also give the relationships between five independent quadratic functions of the hyperpolarizability and the 18 irreducible components of Table 5. These five independent quadratic functions which correspond to a, b, c, d, e of Table 1 involve only the following five types of products of the irreducible components ${}^1A_\beta {}^1B_\beta$; ${}^1A_\beta {}^1A_\beta$; ${}^1B_\beta {}^1B_\beta$; ${}^2\beta {}^2\beta$; ${}^3\beta {}^3\beta$ where the superscripts refer to the classification by weights in Table 5.

Andrews and Thirunamachandran have given the irreducible representations spanned by the 18 irreducible components for molecules of point group symmetry D_{5h}, D_{6h}, D_{4d}, D_{6d}, O_h, I_h and $D_{\infty h}$. These are reproduced in Table 6. The corresponding representations for other point groups of lower symmetry can be obtained using correlation tables.

9. FURTHER READING

The rotational structure of vibrational bands in hyper Raman spectra of gases has been thoroughly reviewed by Altmann and Strey (6). The theory of hyper Raman scattering in crystals has been treated by Zaborobnev and Ovander (7) and by Andrews (8). The theory of Raman scattering is covered in some detail by Long (4). A general review of non-linear light

Table 6. Infra-red, Raman, and hyper-Raman activity: Irreducible representations of dipole μ, polarizability α, and hyperpolarizability β.†

Group	Infrared	Raman		Hyper-Raman		
	μ (weight 1)	α (weight 0)	α (weight 2)	β (weight 1)	β (weight 2)	β (weight 3)
D_{5h}	$A_2'' + E_1'$	A_1'	$A_1' + E_2' + E_1''$	$2A_2'' + 2E_1'$	$A_1'' + E_1' + E_2''$	$A_2'' + E_1' + E_2' + E_2''$
D_{6h}	$A_{2u} + E_{1u}$	A_{1g}	$A_{1g} + E_{1g} + E_{2g}$	$2A_{2u} + 2E_{1u}$	$A_{1u} + E_{1u} + E_{2u}$	$A_{2u} + B_{1u} + B_{2u} + E_{1u} + E_{2u}$
D_{4d}	$B_2 + E_1$	A_1	$A_1 + E_2 + E_3$	$2B_2 + 2E_1$	$B_1 + E_1 + E_2$	$B_2 + E_1 + E_2 + E_3$
D_{6d}	$B_2 + E_1$	A_1	$A_1 + E_2 + E_5$	$2B_2 + 2E_1$	$B_1 + E_1 + E_4$	$B_2 + E_1 + E_3 + E_4$
O_h	T_{1u}	A_{1g}	$E_g + T_{2g}$	$2T_{1u}$	$T_{2u} + E_u$	$A_{2u} + T_{1u} + T_{2u}$
I_h	T_{1u}	A_g	H_g	$2T_{1u}$	H_u	$T_{2u} + G_u$
$D_{\infty h}$	$\Sigma_u^+ + \Pi_u$	Σ_g^+	$\Sigma_g^+ + \Pi_g + \Delta_g$	$2\Sigma_u^+ + 2\Pi_u$	$\Sigma_u^- + \Pi_u + \Delta_u$	$\Sigma_u^+ + \Pi_u + \Delta_u + \Phi_u$

†after Andrews and Thirunamachandran (5).

scattering has been given by French and Long (9).

ACKNOWLEDGMENTS

I gratefully acknowledge permission to reproduce Tables 1 and 2 from reference 1 and Tables 5 and 6 from reference 5.

REFERENCES

1. Behrsohn, R., Pao, Y-H. and Frisch, H.L., J. Chem. Phys., 45, 3184 (1966).

2. Cyvin, S.J., Rauch, J.E. and Decius, J.C., J. Chem. Phys., 43, 4083 (1965).

3. Christie, J.H. and Lockwood, D.J., J. Chem. Phys., 54, 1141 (1971).

4. Long, D.A., Raman Spectroscopy, McGraw-Hill, London, 1977.

5. Andrews, D.L. and Thirunamachandran, T., J. Chem. Phys., 68, 2941 (1978).

6. Altmann, K. and Strey, G., J. Raman Spectrosc., 12, 1 (1982).

7. Zavorobnev, Y.D. and Ovander, L.N., Phys. Status Solidi (b), 68, 443 (1975).

8. Andrews, D.L., Mol. Phys., 37, 325 (1979).

9. French, M.J. and Long, D.A., in Molecular Spectroscopy (Barrow, R.F., Long, D.A. and Millen, D.J., eds.), Specialist Periodical Reports, The Chemical Society, London, 1973, 4, 225.

CLASSICAL APPROACH TO THIRD-ORDER NON-LINEAR SUSCEPTIBILITIES

J. W. Fleming and A.B. Harvey

Chemistry Division
Naval Research Laboratory
Washington, D.C. 20375 USA

1. INTRODUCTION

The important quantitiy in any nonlinear light scattering phenomenon is the nonlinear molecular susceptibility

$$\chi \rightarrow \chi^{(N)}_{ijkl} \; (\omega_{N+1}, \omega_1, \omega_2, \omega_3, \; \ldots) \qquad (1)$$

where N indicates Nth order dependence on the inducing electric field and the subscripts i, j, k, l, ... denote directions in a Cartesian coordinate system indicating polarization directions of the generated and incident radiation. In general, ~ denotes a vector, ≈ denotes a tensor of rank two and ≋ will identify tensors of rank greater than two. The susceptibility $\chi^{(N)}$ is a tensor of rank N+1. Third order effects include Coherent Stokes and Anti-Stokes Raman Scattering, the a.c. and d.c. Kerr effects, third harmonic generation, self-focussing, self-phase modulation, stimulated Brillouin scattering, stimulated Raman scattering, Raman induced Kerr effect and inverse Raman.

The common quantity in each of these is the third-order nonlinear molecular susceptibility, a tensor of rank four. The third-order susceptibility $\chi^{(3)}$ is actually a property of the microscopic sample. To separate bulk properties from molecular contribution, χ can be redefined in the following way:

$$\chi = N\mathcal{L}(\alpha + \beta + \gamma + \ldots) \qquad (2)$$

Here, α is recognized as the linear polarizability. Higher order terms, denoted by β, γ, ... are referred to as the first, second, ..., hyperpolarizabilities, respectively. In general

$$\chi = \chi_L + \chi_{NL} \qquad (3)$$

W. Kiefer and D. A. Long (eds.), Non-Linear Raman Spectroscopy and Its Chemical Applications, 131–140.

where L and NL indicate linear and nonlinear, respectively. The bulk third order susceptibility $\chi^{(3)}$ is proportional to the second hyperpolarizability, γ, where no superscript indicating the order is necessary. $\chi^{(3)}$ then is a function of the total number of molecules per unit volume N, the second hyperpolarizability γ, and a local field correction factor $\mathcal{L}$. The local field correction factor relates the electric field experienced by the particle due to the presence of surrounding medium to the inducing field. Its form will depend on the type of interaction. To describe the propagation of the fields in the medium one must employ Maxwell's equations.

2. MAXWELL'S EQUATIONS AND NONLINEAR POLARIZATION

In order to treat the classical development of the electric susceptibility (see references [1]-[4] for general review works) one must start with Maxwell's equations. The system we wish to characterize is non-magnetic and non-conducting: ie. no free charges or free currents. In order to show explicit orientational relationships between electric field polarization and the susceptibility, the discussion here will be limited to isotropic materials such as molecular liquids and gases. In general these media would be in thermodynamic equilibrium except for the presence of electromagnetic radiation propagating through them. For this system Maxwell's equations can be written in the following form [5]:

$$\nabla \cdot \mathbf{D}(\mathbf{r},t) = 0 \tag{4}$$

$$\nabla \cdot \mathbf{H}(\mathbf{r},t) = 0 \tag{5}$$

$$\nabla \times \mathbf{E}(\mathbf{r},t) + \mu_0 \dot{\mathbf{H}}(\mathbf{r},t) = 0 \tag{6}$$

$$\nabla \times \mathbf{H}(\mathbf{r},t) - \dot{\mathbf{D}}(\mathbf{r},t) = 0 \tag{7}$$

where the dot indicates time derivative. The constitutive relationship important to this discussion is

$$\mathbf{D}(\mathbf{r},t) = \varepsilon_0 \mathbf{E}(\mathbf{r},t) + \mathbf{P}(\mathbf{r},t) \tag{8}$$

Included in the above equations are the permittivity of free space ε_0, and the permeability of free space μ_0. By definition $c^2 = 1/\mu_0\varepsilon_0$. The fields in the above equations are the electric field $\mathbf{E}(\mathbf{r},t)$, the magnetic field $\mathbf{H}(\mathbf{r},t)$, the electric dipole per unit volume $\mathbf{P}(\mathbf{r},t)$ and the electric displacement $\mathbf{D}(\mathbf{r},t)$. As is indicated, each of the fields is a function of position and time. In general, each of these fields is composed of many frequencies and it is convenient to use the Fourier transform to obtain equations for their frequency components. Thus for the electric polarization and electric field

$$P(t) = \int_{-\infty}^{\infty} P(\omega) e^{-i\omega t} d\omega \quad (9)$$

$$E(t) = \int_{-\infty}^{\infty} E(\omega) e^{-i\omega t} d\omega \quad (10)$$

with corresponding inverse transforms

$$P(\omega) = \frac{1}{2\pi} \int_{-\infty}^{\infty} P(t) e^{i\omega t} dt \quad (11)$$

$$E(\omega) = \frac{1}{2\pi} \int_{-\infty}^{\infty} E(t) e^{i\omega t} dt \quad (12)$$

For $\underset{\sim}{E}(\underset{\sim}{r},t)$ and $\underset{\sim}{P}(\underset{\sim}{r},t)$ to be real they must obey the time invariance principle. This may be shown by taking the complex conjugate of eq. (10)

$$E^*(\omega) = \frac{1}{2\pi} \int_{-\infty}^{\infty} E^*(t) e^{-i\omega t} dt \quad (13)$$

as one relationship and letting $\omega \to -\omega$ as another

$$E(-\omega) = \frac{1}{2\pi} \int_{-\infty}^{\infty} E(t) e^{-i\omega t} dt \quad (14)$$

If E(t) is real, ie. $E(t) = E^*(t)$
then

$$E(-\omega) = E^*(\omega) \quad (15)$$

The same may be shown for P(t).

Maxwell's equations can be arranged to show how the electric field and the induced polarization are coupled together. Taking the curl of both sides of eq. (6)

$$\underset{\sim}{\nabla} x \underset{\sim}{\nabla} x \underset{\sim}{E}(\underset{\sim}{r},t) + \underset{\sim}{\nabla} x \mu_o \dot{\underset{\sim}{H}}(\underset{\sim}{r},t) = 0 \quad (16)$$

But since ∇ and $\partial/\partial t$ commute

$$\underset{\sim}{\nabla} x \frac{\partial}{\partial t} \underset{\sim}{H}(\underset{\sim}{r},t) = \frac{\partial}{\partial t} \underset{\sim}{\nabla} x \underset{\sim}{H}(\underset{\sim}{r},t) \quad (17)$$

then combining eqs. (7), (8) and (16)

$$\underset{\sim}{\nabla} x \underset{\sim}{\nabla} x \underset{\sim}{E}(\underset{\sim}{r},t) + \mu_o \varepsilon_o \ddot{\underset{\sim}{E}}(\underset{\sim}{r},t) = -\mu_o \ddot{\underset{\sim}{P}}(\underset{\sim}{r},t) \quad (18)$$

The wave equation for each frequency component may be obtained using eq. (10) in eq. (18) and evaluating the second time derivative

$$\underset{\sim}{\nabla} x \underset{\sim}{\nabla} x \underset{\sim}{E}(\underset{\sim}{r},\omega_i) - \frac{\omega_i^2}{c^2} \underset{\sim}{E}(\underset{\sim}{r},\omega_i) = \omega_i^2 \mu_o \underset{\sim}{P}(\underset{\sim}{r},\omega_i) \quad (19)$$

For weak fields $\underset{\sim}{P}(\underset{\sim}{r},\omega)$ will be linear in $\underset{\sim}{E}(\underset{\sim}{r},\omega)$ or

$$\underset{\sim}{P}_L(\underset{\sim}{r},\omega) = \varepsilon_o \underset{\approx}{\chi}_L(\omega)\underset{\sim}{E}(\underset{\sim}{r},\omega) \qquad (20)$$

$\chi_L(\omega)$ will be recognized as the linear susceptibility (see eq. (3)). For intense fields the polarization can be induced by fields at different frequencies. This may be represented by

$$\underset{\sim}{P}(\underset{\sim}{r},\omega) = \underset{\sim}{P}_L(\underset{\sim}{r},\omega) + \underset{\sim}{P}_{NL}(\underset{\sim}{r},\omega) \qquad (21)$$

where NL indicates the nonlinear contributions. Substitution into the wave equation yields

$$\underset{\sim}{\nabla}x\underset{\sim}{\nabla}x\underset{\sim}{E}(\underset{\sim}{r},\omega_i)-\frac{\omega_i^2}{c^2}(1+\underset{\approx}{\chi}_L)\underset{\sim}{E}(\underset{\sim}{r},\omega_i)$$

$$= \mu_o\omega_i^2\ \underset{\sim}{P}_{NL}(\underset{\sim}{r},\omega_i) \qquad (22)$$

The complex dielectric constant $\varepsilon(\omega)$ is defined as

$$\varepsilon(\omega) = 1+\underset{\approx}{\chi}_L(\omega) \qquad (23)$$

so that eq. (22) may be rewritten as

$$\underset{\sim}{\nabla}x\underset{\sim}{\nabla}x\underset{\sim}{E}(\underset{\sim}{r},\omega_i)-\frac{\omega_i^2}{c^2}\varepsilon(\omega)\underset{\sim}{E}(\underset{\sim}{r},\omega_i) = \mu_o\omega_i^2\underset{\sim}{P}_{NL}(\underset{\sim}{r},\omega_i) \qquad (24)$$

The nonlinear polarization $\underset{\sim}{P}_{NL}(\underset{\sim}{r},\omega)$ in eq. (23) is the induced polarization from all fields interacting with the medium which give rise to an ω_i frequency dependence. The combination of frequency components and the associated equations form a set of coupled equations. These equations cannot be solved in closed form. However, for our discussion we have assumed a non-magnetic medium with no free charges or currents. We will further assume that for $\underset{\sim}{P}_{NL}(\underset{\sim}{r},\omega)$, all terms higher than third-order in the electric field are negligible. With these assumptions the above equations can be used as a starting point for a description of the relationship between $\underset{\sim}{P}_{NL}$ and the applied fields.

3. CLASSICAL MODELS OF NONLINEARITY

A. General

When electromagnetic radiation interacts with a molecular system, the incident radiation induces a polarization in the medium which can be related to the incident radiation through the molecular susceptibility $\underset{\approx}{\chi}$.

$$\underset{\sim}{P} = \varepsilon_o\underset{\approx}{\chi}\underset{\sim}{E} \qquad (25)$$

If the molecular system is composed of N polarizable particles per unit volume then the induced polarization may be expressed as

$$\underset{\sim}{P} = N\mathscr{L}\underset{\sim}{d} \qquad (26)$$

where $\underset{\sim}{d}$ is the induced dipole moment resulting from the interaction. In this development we assume that the Born-Oppenheimer approximation is valid [6]. This approximation relies on the fact that for optical frequencies, electrons adiabatically follow both the nuclear motions and the changing optical field. The induced dipole moment can be related to the applied field through the polarizability so that

$$\underset{\sim}{d} = \underset{\approx}{\alpha}\underset{\sim}{E} \tag{27}$$

This results in an expression for $\underset{\approx}{\chi}$ in terms of the polarizability

$$\underset{\approx}{\chi} = N\mathscr{L}\underset{\approx}{\alpha}/\varepsilon_o \tag{28}$$

For the development of the source of nonlinear terms in a classical picture two approaches can be used. One can introduce the nonlinearity into the driving force (ie. a system of harmonically bound particles in a nonlinear field) or one can consider an anharmonic oscillator. We will consider both of these models.

B. HARMONIC OSCILLATOR MODEL

The motion of the electrons can be described as that of a harmonic oscillator. According to Placzek's polarizability theory [7] the induced polarizability α, due to an external field can be related to the distribution of a bond coordinate q, in the oscillator. Expanding $\alpha(q)$ about its equilibrium value ($q_{equil}=0$) gives

$$\alpha(q) = \alpha_o + \left(\frac{\partial\alpha}{\partial q}\right)_o q + \ldots \tag{29}$$

where only terms to first order in q are considered. The force experienced by the oscillator due to this induced polarizability is

$$F = \frac{1}{2}\left(\frac{\partial\alpha}{\partial q}\right)_o E^2 \tag{30}$$

so that the equation of motion described by the harmonic oscillator equation becomes

$$\ddot{q} + \Gamma_v\dot{q} + \omega_v^2 q = \frac{F}{\mu} \tag{31}$$

where Γ_v is a damping constant, ω_v is the resonant frequency of vibration and μ is the reduced mass of the oscillator. Thus,

$$\ddot{q} + \Gamma_v\dot{q} + \omega_v^2 q = \frac{1}{2\mu}\left(\frac{\partial\alpha}{\partial q}\right)_o E^2 \tag{32}$$

The total incident field in general can be a single frequency, two frequencies or the completely general case of three distinct frequencies. Since we are concerned with nonlinear Raman scattering we will develop the case for CARS. The case for other nonlinear Raman scattering techniques follow a similar derivation. If the total incident field $\underset{\sim}{E}_T$, is the sum of two fields, one can write

$$\underset{\sim}{E}_T = \frac{1}{2}\left[\underset{\sim}{E}_1^o e^{i(\underset{\sim}{k}_1\cdot\underset{\sim}{r}-\omega_1 t)} + \underset{\sim}{E}_2^o e^{i(\underset{\sim}{k}_2\cdot\underset{\sim}{r}-\omega_2 t)} + c.c.\right] \tag{33}$$

Assuming a solution to the equation of motion of the form

$$q = \frac{1}{2} q_o\, e^{-i[(\underset{\sim}{k}_1-\underset{\sim}{k}_2)\cdot\underset{\sim}{r}-(\omega_1-\omega_2)t]} + c.c. \tag{34}$$

gives the following, upon substitution in eq. (32)

$$q = \frac{1}{\mu}\left(\frac{\partial\alpha}{\partial q}\right)_o \frac{E_1^o E_2^{o*} e^{-i[(\underset{\sim}{k}_1-\underset{\sim}{k}_2)\cdot\underset{\sim}{r}-(\omega_1-\omega_2)t]}}{\omega_v^2-(\omega_1-\omega_2)^2+i(\omega_1-\omega_2)\Gamma_v} \tag{35}$$

For convenience

$$D(\omega) \equiv \omega_v^2-\omega^2+i\omega\Gamma_v = D^*(-\omega) \tag{36}$$

The relationship of q to χ can be seen by substitution of eq. (29) in (28)

$$\underset{\approx}{\chi} = N\mathcal{L}\underset{\approx}{\alpha}_o/\varepsilon_o + \frac{N\mathcal{L}}{\varepsilon_o}\left(\frac{\partial\alpha}{\partial q}\right)_o q \tag{37}$$

The first term can be related to the linear susceptibility

$$\underset{\approx}{\chi}_L = N\mathcal{L}\underset{\approx}{\alpha}/\varepsilon_o \tag{38}$$

and the second term to the nonlinear susceptibility.

$$\chi_{NL} = \frac{N\mathcal{L}}{\varepsilon_o}\left(\frac{\partial\alpha}{\partial q}\right)_o q \tag{39}$$

Substitution of eq. (35) into (39) results in several terms, if the total solution to the equation of motion is considered. Again we consider the case where $\omega_3 = 2\omega_1-\omega_2$. Frequency combinations except $\pm(2\omega_1-\omega_2)$ will not be significantly enhanced at $\omega_v = \omega_1-\omega_2$ and can be combined and collectively called χ^{NR}. Thus, keeping only those terms which have a $\pm(2\omega_1-\omega_2)$ frequency dependence yields a resonant third order nonlinear susceptibility of the form

$$\chi^{(3)}(\omega_3,\omega_1,\omega_1,-\omega_2) = \frac{N}{3\mu}\left(\frac{\partial\alpha}{\partial q}\right)_o^2 \frac{\Delta N_v}{D(\omega_1-\omega_2)} \tag{40}$$

ΔN_v has been added to account for the difference in population between the lower state (usually the ground state) and the state connected by the transition. Roughly it is proportional to $N_{initial}-N_{final}$ and approaches unity for low temperatures and zero for infinite temperatures. The result comes directly from a quantum mechanical derivation and for this work we will assume $\Delta N_v \approx 1$. The frequency designation follows that of N. Bloembergen [1]. The first frequency in parentheses ω_3, is that of the induced polarization and is equal to the algebraic sum of the remaining frequencies, ie. for CARS

$$(\omega_3,\ \omega_1,\ \omega_1,\ -\omega_2) \rightarrow \omega_3 = \omega_1+\omega_1-\omega_2 \tag{41}$$

The factor of 3 in eq. (40) is due to the fact that there are 3!/2! ways to achieve a frequency combination taking three frequencies where two are the same. It should be noted that reported values of $\chi^{(3)}$ may differ by a factor of three (or four) and the reader is reminded to check this before using values from the literature. By definition, the induced polarization for CARS then is

$$P^{(3)}(\omega_3) = 3\varepsilon_o\chi^{(3)}(\omega_3,\omega_1,\omega_1,-\omega_2)E_1^2\,E_2^* \tag{42}$$

C. ANHARMONIC OSCILLATOR MODEL

Another derivation for the development of the classical nonlinear susceptibility is to introduce an anharmonic term into the oscillator equation for a bound system. The equation of motion may be written as [8]

$$\ddot{q} + \Gamma_v\dot{q} + \omega_v^2 q + \lambda q^2 = -\frac{eE}{M} \tag{43}$$

where the force is that experienced by a charged particle in an electric field. This is a second order nonlinear, nonhomogeneous differential equation and approximation methods must be used. A perturbation solution [1] can be found by assuming a solution of the form

$$q = q_1 + \lambda q_2 + \lambda^2 q_3 \tag{44}$$

with

$$q^2 = q_1^2 + 2\lambda q_1 q_2 + \lambda^2(q_2^2 + 2q_1q_3) + \ldots \tag{45}$$

Substituting eq. (44) into (43) and equating terms independent of λ gives

$$\ddot{q}_1 + \Gamma_v\dot{q}_1 + \omega_v^2 q_1 = -\frac{eE}{M} \tag{46}$$

Again the total applied field will be the sum of two fields (see eqs. (33) and (34)). Substituting eqs. (33) and (34) into eq. (46) gives

$$q = \frac{-e}{2M}\left[\underset{\sim}{E}_1^o\,\frac{e^{i(\underset{\sim}{k}_1\cdot\underset{\sim}{r}-\omega_1 t)}}{D(\omega_1)} + \underset{\sim}{E}_2^o\,\frac{e^{i(\underset{\sim}{k}_2\cdot\underset{\sim}{r}-\omega_2 t)}}{D(\omega_2)} + \text{c.c.}\right] \tag{47}$$

where $D(\omega)$ was defined in eq. (36). Equating terms in λ gives

$$\ddot{q}_2 + \Gamma_v\dot{q}_2 + \omega_v^2 q_2 = -q_1^2 \tag{48}$$

The q_1^2 term will lead to exponentials depending on $2\omega_1$, $2\omega_2$, and $\omega_1\pm\omega_2$. Assuming a solution of the form

$$q_2 = \frac{q_2^o}{2}\,e^{i(\underset{\sim}{k}_1-\underset{\sim}{k}_2)\cdot\underset{\sim}{r}-i(\omega_1-\omega_2)t} + \text{c.c.} \tag{49}$$

gives

$$q_2 = \frac{-e^2}{2M^2}\left\{\underset{\sim}{E}_1^o \underset{\sim}{E}_2^{o*} \frac{e^{i[\underset{\sim}{k}_1-\underset{\sim}{k}_2)\cdot\underset{\sim}{r}-(\omega_1-\omega_2)t]}}{D(\omega_1-\omega_2)D(\omega_1)D^*(\omega_2)} + c.c.\right\} \tag{50}$$

Equating terms in λ^2 gives

$$\ddot{q}_3 + \Gamma_v \dot{q}_3 + \omega_v^2 q_3 = -2q_1 q_3 \tag{51}$$

Here again the term on the right will lead to a mixing of frequencies. Assuming a solution of the form

$$q_3 = \frac{q_3^o}{2} e^{i(\underset{\sim}{k}_3\cdot\underset{\sim}{r}-\omega_3 t)} + c.c. \tag{52}$$

where

$$\underset{\sim}{k}_3 = 2\underset{\sim}{k}_1-\underset{\sim}{k}_2 \; ; \; \omega_3 = 2\omega_1-\omega_2 \tag{53}$$

gives

$$q_3 = \frac{-e^3}{2M^3}\left[E_1^{o2} E_2^{o*} \frac{e^{i(\underset{\sim}{k}_3\cdot\underset{\sim}{r}-\omega_3 t)}}{D(\omega_3)D^2(\omega_1)D(\omega_1-\omega_2)D^*(\omega_2)} + c.c.\right] \tag{54}$$

This term will give a dipole moment and a polarization which when compared to eq. (42) gives the third order susceptibility

$$\chi^{(3)}(\omega_3) = \frac{N\lambda^2 e^4/M^3}{D(\omega_3)D^2(\omega_1)D(\omega_1-\omega_2)D^*(\omega_2)} \tag{55}$$

Equation (42) along with eq. (54) are to be substituted into eq. (24) in order to describe the polarization of waves with the desired frequency dependence.

4. NONLINEAR SUSCEPTIBILITY AND THE RAMAN CROSS-SECTION

For a sense of the relationship of the nonlinear susceptibility to the Raman contribution in the scattering process let us return to the polarizability results and develop χ in terms of the Raman cross-section.

The polarizability derivative can be related to the polarizability matrix elements $\alpha_{ij,v}$ [9]. For the i→j transition

$$\alpha_{ij,v} \equiv \left(\frac{\hbar}{2M\omega_v}\right)^{1/2}\left(\frac{\partial\alpha_{ij}}{\partial q_v}\right) \tag{56}$$

The square of the polarizability matrix is proportional to the normal Raman cross-section. Since α^R is a tensor, squaring adds additional indices to the individual elements. This development deals with an isotropic medium so that indices must be permuted to account for indistinguishability. The following orientational average results [8]

$$\langle(\alpha_{ij,v}^R)^2\rangle \equiv \frac{\langle\alpha_{ij}\alpha_{kl}+\alpha_{ik}\alpha_{jl}\rangle_v}{8} \tag{57}$$

For our purpose here, and the specific case of CARS, k and j will be chosen equivalent. The resulting expression is

$$<(\alpha^{R}_{ij,v})^2> \equiv \frac{<\alpha_{ij}\alpha_{kl}>_v}{4} \tag{58}$$

Electromagnetic scattering theory can be used to derive the relationship of the polarizability derivative matrix to the Raman differential scattering cross-section [10]. The result is

$$\frac{d\sigma}{d\Omega} = (\alpha^{R})^2 \left(\frac{\omega_s}{c}\right)^4 \tag{59}$$

where ω_s is the frequency of the Stokes scattered Raman radiation and c is the speed of light. Using the results from eq. (55) and (53)

$$\left(\frac{\partial\alpha_{ij}}{\partial q_v}\right) = \frac{\omega_v <\alpha_{ij}\alpha_{kl}>_v}{2\hbar} \tag{60}$$

so that

$$\chi^{(3)}_{ijkl}(\omega_3,\omega_1,\omega_1,-\omega_2) = \frac{N\mathcal{L}}{6\hbar}\frac{\omega_v <\alpha_{ij}\alpha_{kl}>_v}{D(\omega_1-\omega_2)} \tag{61}$$

or in terms of the Raman cross-section, the right hand side of eq. (61) becomes

$$= \frac{c^4}{\omega_v^4}\frac{N\mathcal{L}}{6\hbar}\frac{d\sigma_v}{d\Omega}\frac{\omega_v}{D(\omega_1-\omega_2)} \tag{62}$$

The total susceptibility is then

$$\chi^{(3)T}_{ijkl}(\omega_3) = \chi^{(3)NR}_{ijkl}(\omega_3) + \frac{N\mathcal{L}}{6\hbar}\frac{\omega_v <\alpha_{ij}\alpha_{kl}>}{\omega_v^2-(\omega_1-\omega_2)^2+i(\omega_1-\omega_2)\Gamma_v} \tag{63}$$

5. SUMMARY

The development of the relationship of an induced nonlinear polarization in a medium to the electric field or fields present has been outlined using classical methods. These derivations allow us to write explicit expressions for the third-order nonlinear susceptibility. Although the classical approach to $\chi^{(3)}$ is extremely useful, a quantum mechanical derivation is needed for the explicit dependence of $\chi^{(3)}$ on the incident and generated frequencies, explicit polarization information and contributions from electronic resonance enhancement. Nonetheless, the expressions derived and subsequent assumptions are valuable tools in understanding problems involved in nonlinear Raman scattering processes.

REFERENCES

1. Bloembergen, N., Nonlinear Optics, W.A. Benjamin, New York, 1965.

2. Dewitt, R.N., Harvey, A.B., and Tolles, W.M., Theoretical Development of Third-Order Susceptibility as Related to Coherent Anti-Stokes Raman Spectroscopy, Naval Research Laboratory Memorandum Report 3260, April 1976.

3. Andersen, H.C., and Hudson, B.S., in Molecular Spectroscopy, Vol. 5, p. 142, (Barrow, R.F., Long, D.A., Sheridan, J., eds.) Chemical Society Specialist Periodical Reports, London, 1977.

4. Butcher, P.N., Nonlinear Optical Phenomena, Engineering Experiment Station, Ohio State University Bulletin 100, 1965.

5. Jackson, J.D., Classical Electrodynamic, John Wiley and Sons, New York, 1962

6. Hellwarth, R.W., in Progress in Quantum Electronics, Vol. 5, p.1, 1977.

7. Placzek, G., in Handbuch der Radiologie, Vol. 6, Part II, Ch. 3, pp. 205-374, (E. Marx, ed.), Akad. Verlagsges. Leipzig, 1934.

8. Loudon, R., The Quantum Theory of Light, Oxford Univ. Press, Oxford, 1973.

9. Levenson, M.D., and Bloembergen, N., Phys. Rev. B10, 4447 (1974).

10. Maier, M., Kaiser, W. and Giordmaine, J.A., Phys. Rev. 177, 580 (1969).

HYPER RAYLEIGH AND HYPER RAMAN SCATTERING

GROUP THEORY FOR VARIOUS RAMAN SCATTERING PROCESSES

H.W. Schrötter

Sektion Physik
der LMU München
München, West Germany

1. INTRODUCTION

It is well known that the selection rules for the activity of vibrations of molecules in the Raman effect are different from those for infrared absorption. For molecules having a center of symmetry the principle of mutual exclusion holds: vibrations that are active in infrared absorption are inactive in the Raman effect and vice versa. Highly symmetric molecules may have quite a number of vibrations that are inactive in both effects, the so-called silent modes. Of the 20 vibrations of the benzene molecule, for instance, 4 are infrared active, 7 Raman active, and 9 silent modes.

The selection rules for the hyper Raman effect are governed by the transformation properties of the third rank tensor of the hyperpolarizability as opposed to the second rank tensor of the linear polarizability for the linear Raman effect and the vector of the dipole moment for infrared absorption. Therefore the selection rules are more relaxed and some of the "silent" modes become active in the hyper Raman effect and may be directly observed, as pointed out in the preceding and the following paper by D.A. Long.

Further silent modes may become observable through the second hyper Raman effect based on the fourth rank tensor of the second hyperpolarizability, therefore also the selection rules for this effect are considered here.

There exists an extended literature on group theoretical methods in molecular spectroscopy. Only the classical textbook

W. Kiefer and D. A. Long (eds.), Non-Linear Raman Spectroscopy and Its Chemical Applications, 143–163.

of Wilson, Decius, and Cross (1) and the corresponding chapters in the books on Raman spectroscopy by Brandmüller and Moser (2), Sushchinskij (3), and Long (4) shall be mentioned here. The selection rules for the hyper Raman effect were derived by Cyvin, Rauch, and Decius (5) and for the second hyper Raman effect by Christie and Lockwood (6). A quantum electrodynamical derivation of the hyper Raman selection rules, also for chiral molecules, can be found in papers by Andrews and Thirunamachandran (7).

Here a very simple explanation of group theoretical methods to calculate the number of vibrations of a molecule in the various irreducible representations (symmetry species) and their activity in infrared absorption, the linear Raman effect, and the first and second hyper Raman effect will be given without any mathematical proof. The Schoenflies notation common in molecular spectroscopy and the Hermann-Mauguin or international notation known from crystallography will be used in parallel.

Only vibrational selection rules and only symmetric tensors are considered here for simplicity. The rotational selection rules of the hyper Raman effect are treated in a review article by Altmann and Strey (8). Antisymmetric contributions to the polarizability occur in resonance Raman spectra (9). The antisymmetric components of the polarizability and first hyperpolarizability (6, 10) tensors are included in the tables of symmetry classes in Long's book (4). The selection rules for the first and second hyper Raman effect in crystals have been calculated by Winter (11) for the 32 crystallographic point groups using the method of the projection operator (12, 13). The resulting tables can be obtained by writing to Sektion Physik der Universität München, Lehrstuhl J. Brandmüller, Schellingstrasse 4/III, D-8000 München 40.

2. MOLECULAR SYMMETRY

Rigid molecules, and only these are considered here, can be transformed into identical configurations by symmetry operations. In every symmetry operation a point, a straight line (axis), a plane, or the whole space remain fixed; these geometrical elements are called symmetry elements. The symmetry operations are characterized by the symmetry elements of a molecule and form a group; because at least one point remains fixed in the operations, the group is called the point group of the molecule. In a molecular vibration the atoms are displaced from their equilibrium positions and the symmetry of the molecule is changed. According to their symmetry or antisymmetry with respect to the different symmetry elements the vibrations are grouped into the different symmetry species of a point group. The terminology will be explained using the examples of the molecules benzene C_6H_6 and cyclohexane C_6H_{12}.

2.1. Symmetry Operations and Symmetry Elements

2.1.1. Rotations or proper symmetry operations. The equilibrium structure of the benzene molecule is shown in Fig. 1.

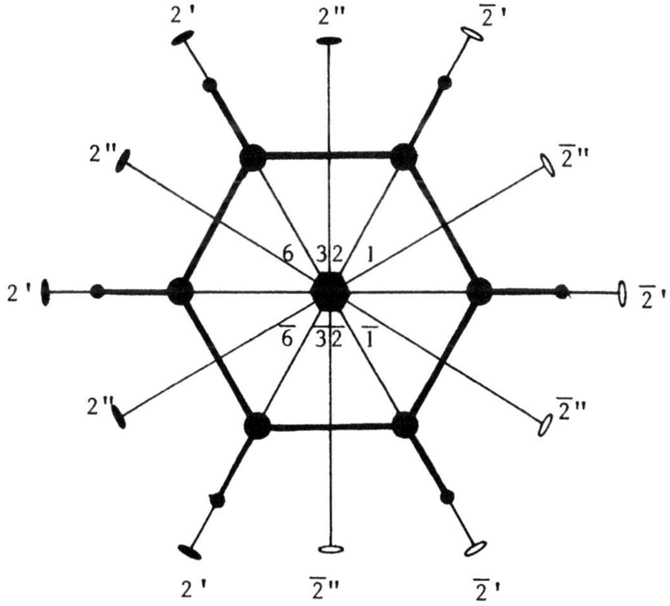

Fig. 1 Equilibrium structure of benzene C_6H_6 and its symmetry elements. Instead of 2', 2" and $\bar{2}'$, $\bar{2}''$ it may be better to write 2_a, 2_b and $\bar{2}_a$, $\bar{2}_b$, respectively.

It can be transformed into an identical configuration by rotation through angles of $\phi = \frac{2\pi}{p} \cdot k$ with p=6 and k=0,1,2,...,p-1=5 about an axis perpendicular to the plane of the molecule. These symmetry operations are represented by the following symmetry elements: identity $C_6^0 = E \equiv 1$ for k=0, six-fold axis $C_6^1 = C_6 \equiv 6$ for k=1, three-fold axis $C_6^2 = C_3 \equiv 3$ for k=2, two-fold axis $C_6^3 = C_2 \equiv 2$ for k=3. A rotation through 240° for k=4 is equivalent to a rotation through -120°, the corresponding symmetry element is therefore denoted as $C_6^4 = C_3^2 = -C_3 \equiv -3$; for k=5 one obtains $C_6^5 = -C_6 \equiv -6$. The equivalent symmetry elements are said to belong to one class.

The benzene molecule of Fig. 1 can also be brought into identical configurations by rotations through 180° about the three two-fold axes $C_2' \equiv 2'$ lying in the plane of the molecule and likewise the three two-fold axes $C_2''\equiv 2''$. Altogether we have so far found the symmetry elements $E\equiv 1$, $2\ C_6\equiv 6$, $2\ C_3\equiv 3$, $C_2\equiv 2$, $3\ C_2'\equiv 2'$, $3\ C_2''\equiv 2''$ of the benzene molecule.

The structure of the non-planar cyclohexane molecule is shown in Fig. 2. The dominating symmetry operation is now a rotation through 120° about a three-fold axis $C_3\equiv 3$. Further we

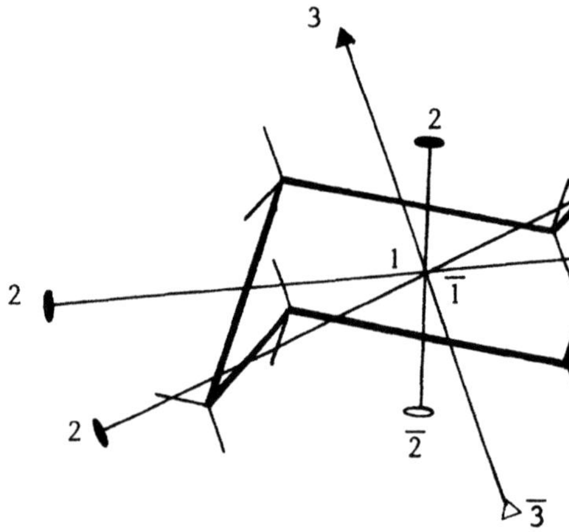

Fig. 2 Equilibrium structure of cyclohexane C_6H_{12} and its symmetry elements.

find $C_3^2=-C_3\equiv -3$, and $C_3^0=E\equiv 1$ must always be present. Three two-fold symmetry axes intersect the C-C bonds in the middle. Altogether the proper symmetry elements of cyclohexane are $E\equiv 1$, $2\ C_3\equiv 3$, $3\ C_2\equiv 2$.

2.1.2. Improper symmetry operations. In addition to the symmetry operations that are performed by pure rotations we can find others, namely reflections at a mirror plane or at a point (the center of symmetry). When the Schoenflies notation is used, the improper symmetry operations are described as rotations through angles of $\phi = \frac{2\pi}{p}$ followed by a reflection at a mirror plane perpendicular to the rotation axis. The corresponding symmetry elements are called rotary-reflection axes or alternating axes and denoted as S_p. In the international system the rotation is followed by an inversion through a point in the center. The improper axes are then called inversion axes and denoted as $\bar{1}$, $\bar{2}$, $\bar{3}$ and so on.

In our example of the benzene molecule we find a center of symmetry $\overline{1}$ which is equivalent to $S_2=i$. Two-fold inversion axes $\overline{2}$ are equivalent to mirror planes $S_1=\sigma$. One of these is perpendicular to the six-fold symmetry axis of the benzene molecule and coincides with the plane of the molecule; it is denoted as σ_h. Three others are vertical to the molecular plane and contain four atoms ($\overline{2}'' \equiv S'' = \sigma_v$) and further three are diagnonal planes and do not contain any atoms ($\overline{2}' \equiv S_1' = \sigma_d$). A three-fold inversion axis $\overline{3}$ coincides with the six-fold rotation axis, is equivalent to a six-fold rotary-reflection axis S_6 and occurs twice in this class ($S_6^5 = -S_6$). Finally the six-fold inversion axis $\overline{6}$ is found to be equivalent to $S_3^2 = -S_3$ and also occurs twice ($S_3 \equiv -\overline{6}$). The improper symmetry elements of the benzene molecule are therefore $i \equiv \overline{1}$, $2\ S_3 \equiv \overline{6}$, $2\ S_6 \equiv \overline{3}$, $\sigma_h \equiv \overline{2}$, $3\sigma_d \equiv \overline{2}'$, $3\sigma_v \equiv \overline{2}''$.

In the cyclohexane molecule we find also a center of symmetry $i \equiv \overline{1}$, two three-fold inversion axes $\overline{3} \equiv S_6$ in one class and three diagonal mirror planes $3\sigma_d \equiv \overline{2}$.

2.1.3. The transformation matrices. The symmetry operations can be performed by linear transformations of the coordinates of the atoms in the molecule. The corresponding transformation matrices are (for active transformations)

$$C(\phi) = \begin{pmatrix} \cos\phi & -\sin\phi & 0 \\ +\sin\phi & \cos\phi & 0 \\ 0 & 0 & 1 \end{pmatrix} \tag{1}$$

for proper symmetry operations and

$$S(\phi) = \begin{pmatrix} \cos\phi & -\sin\phi & 0 \\ +\sin\phi & \cos\phi & 0 \\ 0 & 0 & (-1)^k \end{pmatrix} \tag{2}$$

for improper symmetry operations, with $\phi = \dfrac{2\pi k}{p}$.

The trace of these matrices represents the contribution of one atom, which does not change its position in the symmetry operation, to the character of this symmetry operation S_j, namely

$$\chi_0(S_j) = 2 \cdot \cos \frac{2\pi k}{p} + 1 \tag{3}$$

Table 1. Contributions to Character of Symmetry Operations per Unshifted Atom χ_0, Character of the Representation for Polarizability Components χ_α, for First Hyperpolarizability Components χ_β, and for Second Hyperpolarizability Components χ_γ.

	χ_0	χ_α	χ_β	χ_γ
$1\equiv C_1=E$	3	6	10	15
$2\equiv C_2$	-1	2	-2	3
$3\equiv C_3,\ C_3^2$	0	0	1	0
$4\equiv C_4,\ C_4^3$	1	0	0	1
$5,5^4\equiv C_5,\ C_5^4$	$\frac{1}{2}(1+\sqrt{5})$	1	0	0
$5^2,5^3\equiv C_5^2,C_5^3$	$\frac{1}{2}(1-\sqrt{5})$	1	0	0
$6\equiv C_6,\ C_6^5$	2	2	1	0
$\bar{1}\equiv S_2=i$	-3	6	-10	15
$\bar{2}\equiv S_1=\sigma$	1	2	2	3
$\bar{3}\equiv S_6,\ S_6^5$	0	0	-1	0
$\bar{4}\equiv S_4,\ S_4^3$	-1	0	0	1
$\overline{10},\ \overline{10}^9\equiv S_5^3,\ S_5^7$	$-\frac{1}{2}(3+\sqrt{5})$	$2+\sqrt{5}$	$-(3+\sqrt{5})$	$3+\sqrt{5}$
$\overline{10}^7,\ \overline{10}^3\equiv S_5,\ S_5^9$	$-\frac{1}{2}(3-\sqrt{5})$	$2-\sqrt{5}$	$-(3-\sqrt{5})$	$3-\sqrt{5}$
$\bar{6}\equiv S_3,\ S_3^5$	-2	2	-1	0

or

$$\chi_0(S_j) = 2\cos\frac{2\pi k}{p} + (-1)^k . \qquad (4)$$

These contributions of one atom to the character of the symmetry operation are given in Table 1 for further use.

2.2 Point Groups

The symmetry elements of a molecule form a group in the mathematical sense, and because at least one point remains unshifted in the symmetry operation, it is called the point group of the molecule. The elements of the group have to fulfil the following properties: a) the product of two elements is also an element of the group, b) the associative law of multiplication is valid, c) the set includes an identity element, d) to every element exists the inverse element within the set. These postulates are fulfilled for all molecular point groups. Table 2 gives a number of examples for point groups. Instead of $\bar{2}$ the letter m is frequently used for a mirror plane in the international notation.

Table 2. Examples of Point Groups

International notation		Schoenflies notation	
1 =	{1}	C_1 =	{E}
2 =	{1,2}	C_2 =	$\{E, C_2\}$
3 =	$\{1,3,3^2\}$	C_3 =	$\{E, C_3, C_3^2\}$
6 =	$\{1,6,3,2,3^2,6^5\}$	C_6 =	$\{E, C_6, C_3, C_2, C_3^2, C_6^5\}$
$\bar{1}$ =	$\{1,\bar{1}\}$	C_i =	{E,i}
m =	{1,m}	C_s =	$\{E, \sigma\}$
2mm =	{1,2,m,m'}	C_{2v} =	$\{E, C_2, \sigma_v, \sigma_v'\}$
mmm =	$\{1,2,2',2'',\bar{1},m,m',m''\}$	D_{2h} =	$\{E, C_2, C_2', C_2'', i, \sigma, \sigma', \sigma''\}$
3m =	$\{1,3,3^2,3m\}$	C_{3v} =	$\{E, 2C_3, 3\sigma_v\}$
$\bar{3}$m =	$\{1, 2\cdot 3, 3\cdot 2, \bar{1}, 2\cdot\bar{3}, 3m\}$	D_{3d} =	$\{E, 2C_3, 3C_2, i, 2S_6, 3\sigma_d\}$
6/mmm =	$\{1, 2\cdot 6, 2\cdot 3, 2, 3\cdot 2', 3\cdot 2'',$ $\bar{1}, 2\cdot\bar{6}, 2\cdot\bar{3}, m, 3m', 3m''\}$	D_{6h} =	$\{E, 2C_6, 2C_3, C_2, 3C_2', 3C_2'', i,$ $2S_3, 2S_6, \sigma_h, 3\sigma_d, 3\sigma_v\}$
$\overline{10}$m2 =	$\{1, 2\cdot 5, 2\cdot 5^2, 5\cdot 2,$ $m, 2\cdot\overline{10}, 2\cdot\overline{10}^7, 5m'\}$	D_{5h} =	$\{E, 2C_5, 2C_5^2, 5C_2, \sigma_h,$ $2S_5, 2S_5^3, 5\sigma_v\}$

The point group of a molecule can be found in an unambiguous way by following the scheme given by Zeldin (14) which is reproduced in Ref. (15).

2.3. Irreducible Representations of Point Groups

So far we have considered the atoms in the molecules to be in their equilibrium positions. When the molecule vibrates, however, the structure and herewith the symmetry of the molecule is a periodic function of time, and at the maximum of the displacement due to a non-degenerate normal vibration the structure is either symmetric or anti-symmetric to a symmetry element of the molecule in its vibrational ground state.

According to their behavior against the various symmetry operations of a point group the normal vibrations of a molecule are classified into various symmetry species. These are identical with the irreducible representations of the point group. The notation for the irreducible representations has been introduced by Placzek (16).

A denotes symmetry with respect to the main rotation axis or, if there is none, to three perpendicular two-fold axes or simply serves as basis for indices.

B denotes anti-symmetry with respect to the main rotation axis or to two of the three two-fold axes just mentioned.

E denotes two-fold degeneracy.

F or T denotes three-fold degeneracy.

g or u as lower index denotes symmetry or anti-symmetry with respect to a center of inversion.

' or " as upper index denotes symmetry or anti-symmetry with respect to a mirror plane.

1,2,3... as lower index distinguish the remaining symmetry species.

The characters of the irreducible representations have been compiled in the character tables of the point groups which can be found in many books (1, 2, 12, 15). In Table 3 the character table of the point group $\bar{3}m \equiv \mathcal{D}_{3d}$ is reproduced. The cyclohexane molecule belongs to this point group and shall serve as example to demonstrate the use of the character table to calculate the number of normal vibrations in a symmetry species and its activity in the various Raman effects.

Table 3. Character table of the point group $\bar{3}m \equiv \mathcal{D}_{3d}$ with application to the cyclohexane molecule.

$\bar{3}m$	1	2·3	3·2	$\bar{1}$	2·$\bar{3}$	3m	$\bar{n}^{(\gamma)}$	$n_M^{(\gamma)}$	$n_\alpha^{(\gamma)}$	$n_\beta^{(\gamma)}$	$n_\gamma^{(\gamma)}$
$\mathcal{D}_{3d}$	E	$2C_3$	$3C_2$	i	$2S_6$	$3\sigma_d$	Vib.	IR	Ra	HR	2.HR
A_{1g}	1	1	1	1	1	1	6	0	2	0	4
A_{2g}	1	1	-1	1	1	-1	2	0	0	0	1
E_g	2	-1	0	2	-1	0	2·8	0	4	0	10
A_{1u}	1	1	1	-1	-1	-1	3	0	0	1	0
A_{2u}	1	1	-1	-1	-1	1	5	1	0	3	0
E_u	2	-1	0	-2	1	0	2·8	2	0	6	0
$\chi_0(S_j)$	3	0	-1	-3	0	1	48	3	6	10	15
a_j	(18)	(0)	(0)	0	0	6					
a_j-2	16	-2	-2	-	-	-					
$\bar{\chi}(S_j)$	48	0	2	0	0	6					
$\chi_M(S_j)$	3	0	-1	-3	0	1					
$\chi_\alpha(S_j)$	6	0	2	6	0	2					
$\chi_\beta(S_j)$	10	1	-2	-10	-1	2					
$\chi_\gamma(S_j)$	15	0	3	15	0	3					

The character table forms the left upper part of Table 3. In the first two lines the symmetry elements are given in international and Schoenflies notation. The numbers in front of the symbols for the symmetry elements are the number of elements in each class which will be denoted by g_j. In the first column we find the symbols for the symmetry species. We see that there are as many symmetry species as classes of symmetry elements, so that the characters of the irreducible representations form a square matrix. The character +1 means symmetry and -1 anti-symmetry with respect to the symmetry operation, in the degenerate species E_g and E_u the degree of degeneracy d=2 is found in the second column under $1 \equiv E$.

2.4. Calculation of the Number of Normal Vibrations of a Molecule in the Symmetry Species of the Point Group

A non-linear molecule consisting of N atoms has 3N-6 vibrational degrees of freedom. The distribution of these proper normal vibrations over the various symmetry species can be calculated by the formula

$$\bar{n}^{(\gamma)} = \frac{d^{(\gamma)}}{g} \sum_j g_j \cdot \chi_j^{(\gamma)*} \cdot \bar{\chi}(S_j) \ , \qquad (5)$$

where $\bar{n}^{(\gamma)}$ is the number of normal vibrations in the symmetry species γ, $d^{(\gamma)}$ the degree of degeneracy of the symmetry species γ, g the order of the point group, which is equal to the total number of symmetry elements in the group, g_j the number of symmetry elements in one class, $\chi_j^{(\gamma)*}$ the complex conjugate of the character of the irreducible representation, and $\bar{\chi}(S_j)$ the character of the symmetry operation.

$$\bar{\chi}(S_j) = (a_j - 2)\chi_0(S_j) \qquad (6)$$

for proper symmetry operations, and

$$\bar{\chi}(S_j) = a_j \cdot \chi_0(S_j) \qquad (7)$$

for improper symmetry operations, where a_j is the number of atoms on a symmetry element, which remain unshifted by the symmetry operation, and $\chi_0(S_j)$ the contribution of one atom to the character of the symmetry operation (see eqns. (3) and (4) and Table 1). The properties of the individual molecule enter only through a_j into the consideration, apart from the determination of the point group. The translational and rotational degrees of freedom are split off by subtracting 2 from a_j for proper symmetry operations in eq. (6) and by writing $\bar{\chi}(S_j)$ and $\bar{n}^{(\gamma)}$ instead of $\chi(S_j)$ and $n^{(\gamma)}$.

In order to calculate the number of normal vibrations in the various symmetry species for the cyclohexane molecule, we first look up $\chi_0(S_j)$ for the symmetry elements in Table 1 and write the numbers into the ninth line of Table 3. Then we look at Fig. 2 and try to find out, how many atoms there are on the symmetry elements. Of course all 18 atoms remain unshifted when the identity operation $1\equiv E$ is applied. Two C- and four H-atoms are located on the diagonal mirror planes $m\equiv\sigma_d$, all other a_j are zero. These numbers are written into line 10 of Table 3 and in line 11 we have to subtract 2 for the proper symmetry operations (also from 0). Now line 9 is multiplied with line 11 or 10 and the result for $\bar{\chi}(S_j)$ is written into line 12.

After these preparations we can proceed to the calculation of $\bar{n}^{(\gamma)}$ by eq. (5). g=12 for our point group $\bar{3}m\equiv\mathcal{D}_{3d}$. To obtain $\bar{n}^{(A_{1g})}$, we have to multiply g_j (line 1 or 2 of Table 3) with $\chi_j^{(A_{1g})}$ (line 3) and $\bar{\chi}(S_j)$ (line 12), sum up the products for all classes of symmetry elements, and divide by g.

$$\bar{n}^{(A_{1g})} = \frac{1}{12} \quad (1\cdot1\cdot48+2\cdot1\cdot0+3\cdot1\cdot2+1\cdot1\cdot0+2\cdot1\cdot0+3\cdot1\cdot6)=$$

$$= \frac{1}{12} \; (48+6+18) = \frac{72}{12} = 6 .$$

The result must always be an integer and is written into column 8, line 3 of Table 3.

As an example for a degenerate symmetry species we calculate $\bar{n}^{(E_u)}$ with $d^{(E_u)}=2$.

$$\bar{n}^{(E_u)} = \frac{2}{12}\,(1\cdot2\cdot48+0+0+0+0+0) = 2\cdot\frac{96}{12} = 2\cdot8 .$$

For clarity it is convenient to write 2·8 instead of 16 into column 8, line 8 of Table 3, because the result means that 8 doubly degenerate vibrations of species E_u, which contribute 16 degrees of freedom, are to be expected for the cyclohexane molecule.

The remaining results for $\bar{n}^{(\gamma)}$ are given in column 8 of Table 3. The sum of this column must always be 3N-6, in our case 3·18-6=48, the number of all vibrational degrees of freedom.

3. SELECTION RULES

Now we are ready to calculate the selection rules for infrared absorption and for the different Raman scattering processes.

3.1. Selection Rules for Infrared Absorption

Electromagnetic radiation can only be absorbed by a molecule, when its frequency is in resonance with a molecular eigen frequency and the molecular dipole moment is changing with this frequency. This holds in the dipole approximation, where magnetic dipoles, electric quadrupoles and higher multipoles are neglected. When the molecular energy level corresponds to a vibration, the electromagnetic radiation has a frequency in the infrared region of the spectrum.

The electric dipole moment $\vec{\mu}$ is a vector with the components μ_ρ. The transformation properties of these components are the same as that of the cartesian coordinates. Therefore the character of the transformation matrix of the dipole moment is equal to the contribution of the coordinates of one atom to the character of the symmetry operation

$$\chi_M(S_j) = \chi_O(S_j) = 2\cos\frac{2\pi k}{p} + (\pm 1)^k \tag{8}$$

and can be found in Table 1. For completeness we write the numbers into line 13 of Table 3 for our example.

Now one can calculate in which symmetry species the vibration results in a change of one or more dipole moment components from the formula

$$n_M^{(\gamma)} = \frac{d^{(\gamma)}}{g} \sum_j g_j \cdot \chi_j^{(\gamma)*} \cdot \chi_M(S_j) \ , \tag{9}$$

where $n_M^{(\gamma)}$ is the number of dipole moment components in species γ.

In our example we obtain for instance for the species A_{2u}

$$n_M^{(A_{2u})} = \frac{1}{12} \ (1\cdot 1\cdot 3+0+3\cdot(-1)\cdot(-1)+1\cdot(-1)\cdot(-3)+0+3\cdot 1\cdot 1) =$$

$$= \frac{1}{12} \ (3+3+3+3) = \frac{12}{12} = 1 \ ,$$

so that the 5 vibrations of species A_{2u} are found to be infrared active. The other $n_M^{(\gamma)}$ are given in column 9 of Table 3 and we see that the other two dipole moment components occur in species E_u, whereas the vibrations of the gerade species are infrared inactive as required by the principle of mutual exclusion. The sum of all $n_M^{(\gamma)}$ must always be 3, because there are 3 components of a dipole moment.

3.2. Selection Rules for the Linear Raman Effect

Raman scattering is based on the temporal change of the polarizability tensor with the components $\alpha_{\rho\sigma}$. The components of a second rank tensor are transformed as the products of the cartesian coordinates $\rho\sigma$. The character of its representation is

$$\chi_\alpha(S_j) = 2\cos\frac{2\pi k}{p}\left(2\cos\frac{2\pi k}{p} + (\pm 1)^k\right) \tag{10}$$

and the resulting numbers are also given in Table 1 and are transferred to line 14 in Table 3.

The selection rules for the components of a symmetric polarizability tensor can be obtained from the formula

$$n_\alpha^{(\gamma)} = \frac{d^{(\gamma)}}{g} \sum_j g_j \cdot \chi_j^{(\gamma)*} \cdot \chi_\alpha(S_j) \tag{11}$$

where $n_\alpha^{(\gamma)}$ is the number of polarizability components that changes with a vibration of species γ.

For our example cyclohexane the calculation is performed for species E_g.

$$n_\alpha^{(E_g)} = \frac{2}{12}\left(1\cdot 2\cdot 6+0+0+1\cdot 2\cdot 6+0+0\right) =$$

$$= \frac{2}{12}\,(12+12) = 2\cdot\frac{24}{12} = 2\cdot 2 = 4$$

and the result is transferred to column 10 of Table 3. For the totally symmetric representation A_{1g} the calculation yields $n_\alpha^{(A_{1g})} = 2$, all other $n_\alpha^{(\gamma)}$ are zero, the sum must always be 6, the number of independent components of a symmetric second rank tensor.

If the polarizability tensor can also be partly anti-symmetric, a separate calculation must be performed with the character of the representation of an anti-symmetric tensor,

$$\chi_{\bar{\alpha}}(S_j) = \pm\, 2\cos\frac{2\pi k}{p} + 1 \tag{12}$$

where $\bar{\alpha}$ denotes the anti-symmetric part of the polarizability tensor. We see, that

$$\chi_{\bar{\alpha}}(S_j) = \chi_0(S_j) \quad \text{for proper and}$$

$$\chi_{\bar{\alpha}}(S_j) = -\chi_0(S_j) \quad \text{for improper symmetry}$$

operations. The result of the calculation is $n_{\alpha}^{(A_{2g})} = 1$ and $n_{\bar{\alpha}}^{(E_g)} = 2$ with the other $n_{\bar{\alpha}}^{(\gamma)} = 0$. This means, that the two silent A_{2g} vibrations of cyclohexane may become Raman active with resonance excitation.

With the number of non-zero polarizability components in a given symmetry species we do not yet know, which components these are. But with more detailed calculations also these results can be obtained and are included in the character tables (1, 2, 4, 15). As example the polarizability components for point group $\bar{3}m \equiv \mathcal{D}_{3d}$ are given in Table 4.

In general the totally symmetric irreducible representation is always Raman active and yields polarized Raman lines. For molecules with a center of inversion the principle of mutual exclusion results, gerade vibrations being Raman active and ungerade vibrations infrared active.

But our example cyclohexane shows, that there are also silent modes (2 A_{2g} and 3 A_{1u}) which can not be observed directly with either of the two effects. For these silent modes the hyper Raman effects become important.

3.3. Selection Rules for Hyper Raman Scattering

The intensity of hyper Raman scattering is dependent on the temporal change of the hyperpolarizability with the normal vibrations of the molecule. The hyperpolarizability is a third rank tensor with the components $\beta_{\rho\sigma\tau}$ which transform as the triple products of the coordinates $\rho\sigma\tau$. The character of the representation of a symmetric third rank tensor is

Table 4. Components of dipole moment vector, polarizability and hyperpolarizability tensor derivatives for point group $\bar{3}m \equiv \mathcal{D}_{3d}$.

Species	Vector	Symmetric tensor	Anti-symmetric tensor
A_{1g}		$\alpha_{xx}+\alpha_{yy}, \alpha_{zz}$	
A_{2g}			$\bar{\alpha}_{xy}$
E_g		$(\alpha_{xx}-\alpha_{yy}, \alpha_{xy})$ $(\alpha_{yz}, \alpha_{zx})$	 $(\bar{\alpha}_{yz}, \bar{\alpha}_{zx})$
A_{1u}		$\beta_{xxx}-3\beta_{xyy}$	$\bar{\beta}_{xyz}-\bar{\beta}_{yzx}$
A_{2u}	μ_z	$\beta_{zzz}, \beta_{zxx}+\beta_{yyz},$ $\beta_{yyy}-3\beta_{xxy}$	$\bar{\beta}_{xxz}+\bar{\beta}_{yyz}$
E_u	(μ_x, μ_y)	$(\beta_{zzx}, \beta_{yzz})$ $(\beta_{xyz}, \beta_{zxx}-\beta_{yyz})$ $(\beta_{xxx}+\beta_{xyy}, \beta_{yyy}+\beta_{xxy})$	$(\bar{\beta}_{xxy}, \bar{\beta}_{yyx})$ $(\bar{\beta}_{zzx}, \bar{\beta}_{zzy})$ $(\bar{\beta}_{xxz}-\bar{\beta}_{yyz}, \bar{\beta}_{xyz}+\bar{\beta}_{yzx})$

$$\chi_\beta(S_j) = 2\cos\frac{2\pi k}{p}\left(4\cos^2\frac{2\pi k}{p} \pm 2\cos\frac{2\pi k}{p} - 1\right) \quad (13)$$

and the resulting numbers for the important symmetry operations are given in Table 1. For cyclohexane the results are written into line 15 in Table 3.

The formula for the selection rules is again very similar to the previous ones, the number of hyperpolarizability components changing in symmetry species γ is

$$n_\beta^{(\gamma)} = \frac{d^{(\gamma)}}{g} \sum_j g_j\, \chi_j^{(\gamma)*} \cdot \chi_\beta(S_j) \ . \quad (14)$$

For cyclohexane the silent A_{1u} modes become hyper Raman active,

$$n_\beta^{(A_{1u})} = \frac{1}{12}(1\cdot1\cdot10+2\cdot1\cdot1+3\cdot1\cdot(-2)+1\cdot(-1)\cdot(-10)+(-10)+2\cdot(-1)\cdot(-1)+$$

$$+3\cdot(-1)\cdot2) = \frac{1}{12}(10+2-6+10+2-6) = \frac{12}{12} = 1 \ ,$$

so that one hyperpolarizability component derivative is different from zero. The other components occur in the infrared active species A_{2u} and E_u. This reflects the general result that all infrared active vibrations are also hyper Raman active.

The distribution of the 10 components of the symmetric hyperpolarizability tensor derivative and their linear combinations in the three hyper Raman active species is given in Table 4. In addition the 8 components of the antisymmetric part of the tensor are given, however, they do not lead to additional hyper Raman activity in this point group.

The hyper Raman spectrum of cyclohexane has been investigated by Hiraya, Udagawa, and Ito (17). In addition to the established infrared active fundamentals (18) they found a strong hyper Raman band at 1107 cm^{-1} which they assign to either the $\nu_8(A_{1u})$ or $\nu_9(A_{1u})$ fundamental vibration. An investigation of the hyper Raman spectrum with better resolution might show, whether only one or perhaps both fundamental vibrations occur in this wavenumber region. It may be mentioned at this point that Falge and Schuller (19) found two weak infrared bands also in the vapor phase at 1095 cm^{-1} and 1113 cm^{-1} which they assigned to triple and quadruple combinations, because a silent mode should not show up in the spectrum of the free molecule.

3.4. Selection Rules for Second Hyper Raman Scattering

From Table 3 we see that the two A_{2g} vibrations are silent also in the first hyper Raman effect. In order to observe them directly the second hyper Raman effect would have to be used, the scattered intensity of which is caused by a temporal change of the components $\gamma_{\rho\sigma\tau\upsilon}$ of the fourth rank tensor of the second hyperpolarizability. This tensor transforms as the quadruple product of the coordinates $\rho\sigma\tau\upsilon$. The character of the representation is now

$$\chi_{\gamma}(S_j) = 2\cos\frac{2\pi k}{p}\left(8\cos^3\frac{2\pi k}{p} \pm 4\cos^2\frac{2\pi k}{p} - 4\cos\frac{2\pi k}{p} \mp 1\right) + 1, \qquad (15)$$

these numbers are given in Table 1 in the last column, and for cyclohexane in the last line of Table 3.

The number of second hyperpolarizability components changing with vibrations in symmetry species γ is calculated with the formula

$$n_{\gamma}^{(\gamma)} = \frac{d^{(\gamma)}}{g} \sum_j g_j \cdot \chi_j^{(\gamma)*} \cdot \chi_{\gamma}(S_j) \quad . \qquad (16)$$

For species A_{2g} in the point group $\bar{3}m \equiv \mathcal{D}_{3d}$ this becomes

$$n_{\gamma}^{(A_{2g})} = \frac{1}{12} \quad (1\cdot1\cdot15+0+3\cdot(-1)\cdot3+1\cdot1\cdot15+0+3\cdot(-1)\cdot3) \quad =$$

$$= \frac{1}{12} \quad (15-9+15-9) \quad = \frac{12}{12} = 1$$

and this result is transferred to the last column of Table 3. The other components show up in species A_{1g} and E_g, their total number is 15 for a symmetric second hyperpolarizability tensor. The components itself were tabulated by Christie and Lockwood (6). All Raman active vibrations are also active in the second hyper Raman effect.

Table 5. Number of vibrations, infrared activity (M), Raman (α), hyper-Raman (β), and second hyper Raman (γ) activity for linear molecules with N atoms

---

a) Point group $\mathcal{C}_{\infty v} \equiv \infty m$

$(A_1) = \Sigma^+(N-1, M, \alpha, \beta, \gamma)$

$(A_2) = \Sigma^-(0,-,-,-,-)$

$(E_1) = \Pi(2\cdot(N-2), M, \alpha, \beta, \gamma)$

$(E_2) = \Delta(0,-,\alpha,\beta,\gamma)$

$(E_3) = \Phi(0,-,-,\beta,\gamma)$

b) Point group $\mathcal{D}_{\infty h} \equiv {}^{\infty}/m\ m$

N even or [N odd], respectively

$(A_{1g}) = \Sigma_g^+ \left(\frac{N}{2} \left[\frac{N-1}{2}\right], -, \alpha, -, \gamma\right)$

$(A_{2g}) = \Sigma_g^- (0,-,-,-,-)$

$(E_{1g}) = \Pi_g \left(2\cdot\left(\frac{N}{2} - 1\right) \left[2\cdot\frac{N-3}{2}\right], -, \alpha, -, \gamma\right)$

$(E_{2g}) = \Delta_g (0,-,\alpha,-,\gamma)$

....

$(A_{1u}) = \Sigma_u^- (0,-,-,-,-)$

$(A_{2u}) = \Sigma_u^+ \left(\frac{N}{2} - 1 \left[\frac{N-1}{2}\right], M, -, \beta, -\right)$

$(E_{1u}) = \Pi_u \left(2\cdot\left(\frac{N}{2} - 1\right) \left[2\cdot\frac{N-1}{2}\right], M, -, \beta, -\right)$

$(E_{2u}) = \Delta_u (0,-,-,\beta,-)$

....

No experimental results on the second hyper Raman effect are known until now.

The various techniques of active or coherent Raman spectroscopy, that are based on the nonlinear third order susceptibility $\chi^{(3)}$, have the same selection rules as the linear Raman effect, because $\chi^{(3)}$ is linearly proportional to the Raman scattering cross sections (see e.g. Ref. (20)).

4. LINEAR MOLECULES

Linear molecules have 3N-5 vibrational degrees of freedom and belong to the continuous point groups $\infty m \equiv \mathcal{C}_{\infty v}$ or $\infty/mm \equiv \mathcal{D}_{\infty h}$. For completeness, in Table 5 the number of vibrations in the various symmetry species is given for a linear molecule with N atoms, together with the selection rules for infrared absorption and the Raman effects.

5. CONCLUSION

The scope of this article is to give the reader a very simple introduction into the use of group theory in molecular spectroscopy and its application to the derivation of the selection rules for the various Raman scattering processes. For a more rigorous treatment and a deeper understanding the specialized literature has to be consulted.

In freely oriented molecules in the gaseous or liquid state only the invariants of the hyperpolarizability tensor determine the depolarization ratios of hyper Raman bands (7, 8) by which an experimental distinction of the symmetry species is possible. In crystals the relative magnitudes of the hyper Raman tensor components itself can be determined (11), in analogy to the determination of Raman tensor components (21-25).

REFERENCES

1. Wilson, E.B. Jr., Decius, J.C., and Cross, P.C., Molecular Vibrations, McGraw Hill, New York, 1955.

2. Brandmüller, J., and Moser, H., Einführung in die Ramanspektroskopie, Steinkopff, Darmstadt, 1962.

3. Sushchinskij, M.M., Spektry Kombinatsionnogo Rasseyaniya Molekul i Kristallov, Izdatelstvo Nauka, Moskau 1969; Raman Spectra of Molecules and Crystals, Israel Program for Scientific Translations, New York 1972; Ramanspektren von Molekülen und Kristallen, Heyden, Rheine 1974.

4. Long, D.A., Raman Spectroscopy, McGraw Hill, New York, 1977.

5. Cyvin, S.J., Rauch, J.E., and Decius, J.C., J. Chem. Phys. 43, 4083 (1965).

6. Christie, J.H., and Lockwood D.J., J. Chem. Phys. 54, 1141 (1971).

7. Andrews, D.L., and Thirunamachandran, J. Chem. Phys. 68, 2941 (1978); 70, 1027 (1979).

8. Altmann, K., and Strey, G., J. Raman Spectrosc. 12, 1 (1982).

9. Spiro, T.G., and Strekas, T.C., Proc. Natl. Acad. Sci. USA 69, 2622 (1972).

10. Altmann, K., and Strey, G., Z. Naturforsch. 32a, 307 (1977).

11. Winter, F.X., Dissertation LMU München 1977.

12. Tinkham, M., Group Theory and Quantum Mechanics, McGraw Hill, New York, 1964.

13. Birman, J.L., Theory of Crystal Space Groups and Infra-Red and Raman Lattice Processes of Insulating Crystals, in: Handbuch der Physik XXV, 2b (Flügge S., ed.), Springer, Berlin, 1974.

14. Zeldin, M., J. Chem. Education 43, 17 (1966).

15. Salthouse, J.A., and Ware, M.J., Point Group Character Tables and Related Data, Cambridge University Press, Cambridge, 1972.

16. Placzek, G., Rayleigh-Streuung und Ramaneffekt, in: Handbuch der Radiologie VI, part 2 (Marx, E., ed.), Akadem. Verlagsges., Leipzig 1934, p. 205.

17. Hiraya, A., Udagawa, Y., and Ito, M., Chemistry Letters (Japan) 1979, p. 433.

18. Wiberg, K.B., and Shrake, A., Spectrochim. Acta 27A, 1139 (1971) and 29A, 583 (1973).

19. Falge, H.J., and Schuller, E., Proc. 5th Int. Conf. Raman Spectrosc. (Schmid, E.D., et al., eds.), H.F. Schulz, Freiburg 1976, p. 50.

20. Eesley, G.L., Coherent Raman Spectroscopy, Pergamon Press, Oxford 1981.

21. Loudon, R., Adv. in Physics 13, 423 (1964).

22. Poulet, H., Mathieu, J.P., Spectres de vibration et symétrie des cristaux, Gordon & Breach, Paris 1970.

23. Behringer, J., Springer Tracts in Mod. Phys. 68, 161 (1973).

24. Claus, R., Merten, L., Brandmüller, J., Springer Tracts in Mod. Phys. 75, 1 (1975).

25. Rousseau, D.L., Bauman, R.P., Porto, S.P.S., J. Raman Spectrosc. 10, 253 (1981).

THEORY OF HYPER RAYLEIGH AND HYPER RAMAN SCATTERING

D. A. Long

School of Chemistry
University of Bradford
Bradford, West Yorkshire
United Kingdom

1. INTRODUCTION

It is not possible to consider in detail all aspects of the theory of hyper Rayleigh and hyper Raman scattering here. The majority of this chapter will be devoted to consideration of the theory of vibrational scattering from an assembly of freely rotating non-interacting molecules.

We shall first consider briefly a classical treatment of hyper Rayleigh and hyper Raman scattering to establish the frequency dependence of the scattering. We shall then introduce the transition hyperpolarizability tensor and proceed to a detailed consideration of the selection rules and polarization properties associated with vibrational hyper Raman scattering. Some specific cases will then be examined.

2. CLASSICAL THEORY

To avoid over-generalisation (which would encumber us with subscripts and in any case does not relate to real experimental situations) we shall consider the following special case.

The scattering system is located at the origin of a set of Cartesian space-fixed axes X, Y, Z. The radiation of circular frequency ω_0 (and adequate irradiance!) is incident along the Z axis and is plane polarized such that $E_Y \neq 0$ and $E_X = E_Z = 0$. The scattering system is considered to be a single *non-rotating* diatomic molecule with one molecular vibration of circular frequency ω_M and the scattering is observed along X. We thus

W. Kiefer and D. A. Long (eds.), Non-Linear Raman Spectroscopy and Its Chemical Applications, 165–179.

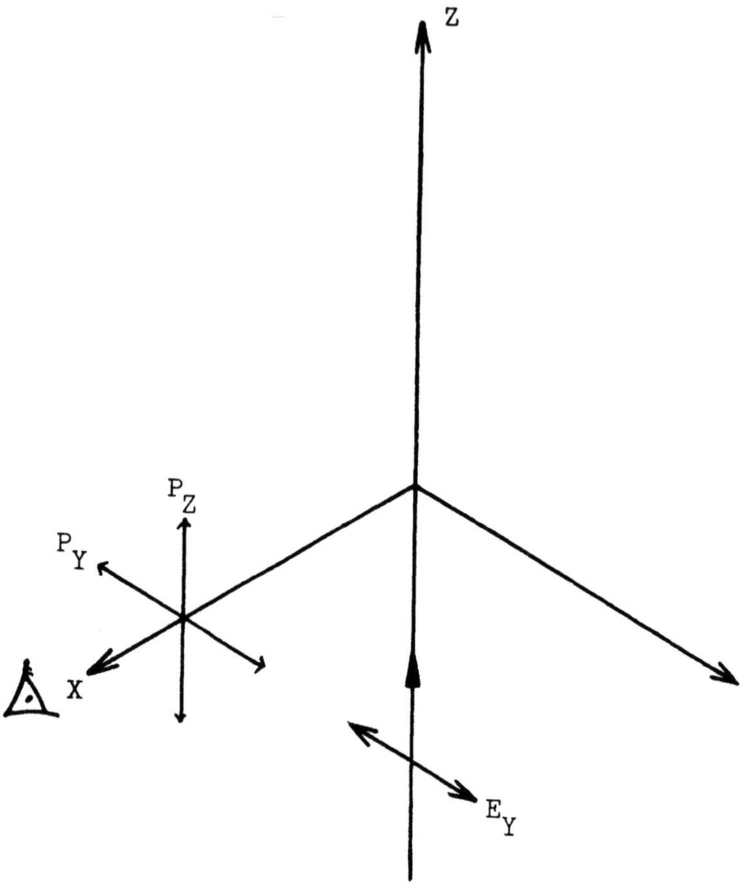

Fig. 1. Illumination-observation geometry.

Note: following Long (1) the symbol ${}^{\perp}I_{\perp}(\pi/2)$ is to be read sequentially as incident radiation polarized *perpendicular* ($\perp$) to the scatter plane, *radiant intensity* (I) of scattered radiation polarized *perpendicular* ($\perp$) to the scatter plane and propagating along a direction in the scatter plane making an *angle* ($\pi/2$) to the direction of the incident radiation.

need to calculate the Y and Z components of the second order induced electric dipole, $P_Y^{(2)}$ and $P_Z^{(2)}$ (see Fig. 1)

It follows from eq. (3) in the previous chapter that

$$P_Y^{(2)} = \frac{1}{2}\beta_{YYY}E_Y^{\ 2} \tag{1}$$

and

$$P_Z^{(2)} = \frac{1}{2}\beta_{ZYY}E_Y^{\ 2} \tag{2}$$

Since in what follows we are interested only in the time dependence of $P^{(2)}$ we can for convenience in the ensuing mathematics temporarily drop the subscripts and write as a generalisation of eqs. (1) and (2)

$$P^{(2)} = \frac{1}{2}\beta E^2 \tag{3}$$

We now introduce the time dependence of the quantities in eq. (3). We write for radiation of circular frequency ω_o

$$E = E_o \cos \omega_o t \ ; \tag{4}$$

for the normal coordinate Q with molecular frequency ω_M, in the simple harmonic approximation

$$Q = Q_o \cos \omega_M t \tag{5}$$

where Q_o is the amplitude; and for the dependence of β on Q to a first approximation

$$\beta = \beta_o + \left(\frac{\partial\beta}{\partial Q}\right)_o Q \tag{6}$$

where β_o is the equilibrium value of the hyperpolarizability. Introducing eqs. (4), (5), and (6) into eq. (3), we obtain, after a little rearrangement and the use of trigonometric identities

$$2 \cos A \cos B = \cos (A + B) + \cos (A - B) \tag{7}$$

$$2 \cos^2 A = 1 + \cos 2A \tag{8}$$

the following expression for $P^{(2)}$

$$P^{(2)} = P^{(2)}(\omega{=}0) + P^{(2)}(2\omega_o) + P^{(2)}(2\omega_o \pm \omega_M) + P^{(2)}(\omega_M) \tag{9}$$

where

$$P^{(2)}(\omega = 0) = \frac{1}{4}\beta_o E_o^2 \tag{10}$$

$$P^{(2)}(2\omega_o) = \frac{1}{4}\beta_o E_o^2 \cos 2\omega_o t \tag{11}$$

$$P^{(2)}(2\omega_o \pm \omega_M) = \frac{1}{8}\left(\frac{\partial\beta}{\partial Q}\right)_o Q_o E_o^2\{\cos(2\omega_o + \omega_M)t + \cos(2\omega_o - \omega_M)t\} \tag{12}$$

$$P^{(2)}(\omega_M) = \frac{1}{4}\left(\frac{\partial\beta}{\partial Q}\right)_o Q_o E_o^2 \cos\omega_M t \tag{13}$$

All these second-order induced dipoles depend, as expected, on the square of the amplitude of the electric field intensity.

We see that through the hyperpolarizability, second order (or non-linear) induced dipoles with five different frequency dependences can result. Two will not concern us further here: $P^{(2)}(\omega = 0)$ which gives radiation at zero frequency, i.e., a static electric field, and $P^{(2)}(\omega_M)$ which produces radiation at the molecular frequency ω_M. Of interest to us are $P^{(2)}(2\omega_o)$ which produces radiation at $2\omega_o$, i.e., hyper Rayleigh scattering, and $P^{(2)}(2\omega_o \pm \omega_M)$ which produce radiation at $2\omega_o \pm \omega_M$, i.e., Stokes and anti-Stokes hyper Raman scattering.

We have thus accounted for the observed wavenumbers or frequencies of non-linear origin which are termed hyper Rayleigh and Raman scattering. In addition, we see that for hyper Rayleigh scattering to occur at least one of the components of the equilibrium hyperpolarizability tensor β_o must be non-zero; and for hyper Raman scattering at least one of the components of the derived hyperpolarizability tensor must be non-zero. For further insight into the selection rules and polarization properties, we must turn to the transition hyperpolarizability.

3. THE TRANSITION HYPERPOLARIZABILITY

For the special illumination and observation case defined in Section 2 above and Fig. 1, the time-independent quantum mechanical analogues of eqs. (1) and (2) are

$$\left[P_{Y_o}^{(2)}\right]_{fi} = \frac{1}{2}\left[\beta_{YYY}\right]_{fi} E_{Y_o}^2 \tag{14}$$

and

$$\left[P_{Z_o}^{(2)}\right]_{fi} = \frac{1}{2}\left[\beta_{ZYY}\right]_{fi} E_{Y_o}^2 \qquad (15)$$

where the subscript 'o' denotes an amplitude. Eqs. (14) and (15) relate to Stokes hyper Raman scattering of frequency dependence $2\omega_o-\omega_M$, when the subscripts 'fi' refer to a fundamental vibrational transition $(v_f = 1) \leftarrow (v_i = 0)$ associated with a given normal mode Q_M of circular frequency ω_M; and to hyper Rayleigh scattering of frequency dependence $2\omega_o$ when $f = i$.

In the Placzek approximation for a fundamental vibrational transition associated with Q_M

$$\left[\beta_{YYY}\right]_{1,0} = \langle\phi_1(Q_M)|\beta_{YYY}|\phi_o(Q_M)\rangle \qquad (16)$$

with β_{YYY} a function of nuclear coordinates only. In the approximation of electrical and mechanical harmonicity eq. (16) becomes

$$\left[\beta_{YYY}\right]_{1,0} = \left(\frac{\partial\beta_{YYY}}{\partial Q_M}\right)_o b_M \qquad (17)$$

where

$$b_M^2 = \frac{h}{4\pi\omega_M} \qquad (18)$$

4. INTENSITIES AND DEPOLARIZATION RATIOS

The intensity of scattering (power per steradian) at 90° to a dipole is given by (1)

$$I = \frac{\omega^4 P_o^2}{32\pi^2\varepsilon_o c^3} \qquad (19)$$

where ω is the circular frequency of the radiation, P_o is the amplitude of the induced electric dipole, ε_o is the permittivity of free space and c the velocity of light.

Thus the intensities of scattering along the X direction with electric vectors perpendicular and parallel to the scatter plane XZ are given by

$$^{\perp}I_{\perp}\left(\frac{\pi}{2}\right) = \frac{\omega^4 \left[P_{Y_0}^{(2)}\right]_{fi}^2}{32\pi^2 \varepsilon_0 c^3} \tag{20}$$

and

$$^{\perp}I_{\parallel}\left(\frac{\pi}{2}\right) = \frac{\omega^4 \left[P_{Z_0}^{(2)}\right]_{fi}^2}{32\pi^2 \varepsilon_0 c^3} \tag{21}$$

where ω is the circular frequency of the scattered radiation, that is $2\omega_0-\omega_M$ for Stokes hyper Raman and $2\omega_0$ for hyper Rayleigh scattering. Using eqs. (14) and (15) and introducing the irradiance $\mathcal{I}$ defined (1) by

$$\mathcal{I} = \frac{1}{2} c\varepsilon_0 g E_0^2 \tag{22}$$

where g is the coherence factor

$$^{\perp}I_{\perp}\left(\frac{\pi}{2}\right) = \frac{\omega^4 \left[\beta_{YYY}\right]_{fi}^2 \mathcal{I}^2}{32\pi^2 \varepsilon_0^3 c^5} \tag{23}$$

$$^{\perp}I_{\parallel}\left(\frac{\pi}{2}\right) = \frac{\omega^4 \left[\beta_{ZYY}\right]_{fi}^2 \mathcal{I}^2}{32\pi^2 \varepsilon_0^3 c^5} \tag{24}$$

For $\mathcal{N}$ freely rotating molecules we then have for incoherent scattering

$$^{\perp}I_{\perp}\left(\frac{\pi}{2}\right) = \frac{\omega^4 \mathcal{N} \overline{\left[\beta_{YYY}\right]_{fi}^2} \mathcal{I}^2}{32\pi^2 \varepsilon_0^3 c^5} \tag{25}$$

$$^{\perp}I_{\parallel}\left(\frac{\pi}{2}\right) = \frac{\omega^4 \mathcal{N} \overline{\left[\beta_{ZYY}\right]_{fi}^2} \mathcal{I}^2}{32\pi^2 \varepsilon_0^3 c^5} \tag{26}$$

and

$$\rho_{\perp}\left(\frac{\pi}{2}\right) = \frac{{}^{\perp}I_{\parallel}\left(\frac{\pi}{2}\right)}{{}^{\perp}I_{\perp}\left(\frac{\pi}{2}\right)} \tag{27}$$

$$= \frac{\overline{\left[\beta_{ZYY}\right]^2_{fi}}}{\overline{\left[\beta_{YYY}\right]^2_{fi}}} \tag{28}$$

In terms of the five types of β products introduced by Andrews and Thirunamachandran and discussed in the previous chapter $\rho_{\perp}\left(\frac{\pi}{2}\right)$ has the form

$$\rho_{\perp}\left(\frac{\pi}{2}\right) = \frac{-28\left({}^{1A}\beta\,{}^{1B}\beta\right) + 7\left({}^{1A}\beta\,{}^{1A}\beta\right) + 28\left({}^{1B}\beta\,{}^{1B}\beta\right) + 35\left({}^{2}\beta\,{}^{2}\beta\right) + 20\left({}^{3}\beta\,{}^{3}\beta\right)}{28\left({}^{1A}\beta\,{}^{1B}\beta\right) + 28\left({}^{1A}\beta\,{}^{1A}\beta\right) + 7\left({}^{1B}\beta\,{}^{1B}\beta\right) + 30\left({}^{3}\beta\,{}^{3}\beta\right)} \tag{29}$$

It follows from eq. (27) that the numerator of eq. (29) gives the dependence of ${}^{\perp}I_{\parallel}\left(\frac{\pi}{2}\right)$ and the denominator the dependence of ${}^{\perp}I_{\perp}\left(\frac{\pi}{2}\right)$ on the five types of β products.

For the special case of complete permutation symmetry of the indices the β products in eq. (29) of weight 2 are zero and the weight 1 products become equal with a coefficient of 63.

For complete permutation symmetry of the indices $\rho_{\perp}\left(\frac{\pi}{2}\right)$ can be readily expressed in terms of quadratic functions of hyperpolarizability components referred to molecule-fixed axes using Table 3 of the previous chapter.

5. GENERALISATIONS

The classification of β tensor components according to their weights 1, 2 and 3 affords a very useful basis for discussion of selection rules, and intensity and polarization characteristics in vibrational hyper Raman scattering.(2).

Two important generalisations can be made regarding selection rules. First, all ir active modes are also hyper-Raman active since the weight 1 components of the β tensor transform like the electric dipole moment. Secondly, for centrosymmetric molecules since only *ungerade* modes are hyper-Raman allowed, such modes are Raman inactive. However, for noncentrosymmetric systems vibrational modes with the same symmetry as a weight 2 component of the β tensor can be Raman active. Although this is a necessary requirement, it is not sufficient because the weight 2 components of the Raman and hyper-Raman tensors, in general, transform differently under reflection. For example, for D_{4d} (previous chapter) only E_2 is *both* Raman and hyper-Raman active. However, in the pure rotation (chiral) groups C_n, D_n, T, O and I any vibration belonging to the same representation as a weight 2 component of the β tensor is also Raman active.

Hyper Raman active modes can be classified into six types.

Class A: For modes belonging to this class the β tensor has contributions of weights 1, 2, and 3. These modes are also ir active. In view of the remarks made above, no general rule can be made about Raman activity; each case has to be analysed separately. Examples of Class A are the $\nu_3(E)$ antisymmetric stretching and $\nu_4(E)$ bending modes of ammonia which are Raman active and the $\nu_7(B_{1u})$ molecular bending mode of ethylene which is Raman inactive.

Class B: For these modes the β tensor has contributions of weights 2 and 3. They are clearly ir inactive. Examples are the E'' modes (such as the ν_{13} methylene rocking mode) of cyclopropane and the Raman inactive $\nu_4(A_u)$ torsional mode of ethylene.

Class C: These modes have both weights 1 and 3 contributions. Consequently, they are always ir active and Raman inactive. An example of this type is the $\nu_3(A_{2u})$ out-of-plane bending mode in xenon tetrafluoride of D_{4h} symmetry.

Class D: For the class D modes the contributions are of weight 3 alone. Hence, these modes are both ir and Raman inactive. An example of this type is the $\nu_6(B_{1u})$ hydrogen stretch in benzene

Class E: The β tensor for these modes has weight 2 contributions only. Hence, these modes are ir inactive, but can be Raman active. For example, the Raman-active $\nu_2(E)$ deformation modes of methane belong to this class, as does the Raman-inactive $\nu_4(A_1'')$ methylene twisting mode of cyclopropane.

Class F: The final class contains those modes which have only weight 1 contributions. These modes are ir active but Raman inactive. Molecules belonging to the common molecular point groups cannot have modes of this type; such modes are found only in molecules of icosahedral symmetry.

Certain generalisations regarding $\rho_\perp\left(\frac{\pi}{2}\right)$ values can also be made. It is clear from eq. (29) that for class D modes which have weight 3 tensor products only, $\rho_\perp\left(\frac{\pi}{2}\right) = \frac{2}{3}$; and for class E modes which have weight 2 contributions only the ratio is ∞. This latter behaviour is unknown in Raman scattering except under resonance conditions. The class D and class E results are of special interest since they correspond to modes which are active in hyper-Raman scattering but not in ir absorption; the other class of ir inactive modes, namely, class B, should exhibit depolarization ratios intermediate between the class D and class E values. Thus in general all ir-inactive, hyper-Raman-active modes do not have the same depolarization ratios.

For the special case of complete permutation symmetry of the indices weight 2 contributions are zero. Thus would have an upper limit of $\frac{2}{3}$ and cannot go to infinity; class E modes would be inactive; and class B and class D modes, that is ir inactive, hyper-Raman-active modes, would have the same $\rho_\perp\left(\frac{\pi}{2}\right)$ values.

6. EXAMPLES

The application of the foregoing theory to several molecules will now be considered.

A. Benzene, C_6H_6 (point group D_{6h})

This molecule has 30 fundamental modes of vibration and 20 distinct frequencies. We see from Table 1 that 4 of these are infra-red active ($1\ A_{2u} + 3\ E_{1u}$) and another 7 are Raman active ($2\ A_{1g} + 1\ E_{1g} + 4\ E_{2g}$). Thus 9 modes are spectroscopically inactive (as fundamentals) using normal spectroscopic techniques. In the hyper Raman effect 10 modes are active, the 4 infra-red active modes and 6 otherwise inactive modes ($2\ B_{1u}$, $2\ B_{2u}$ and $2\ E_{2u}$). The B_{1u} and B_{2u} modes belong to class D (weight 3 only) and so have $\rho_\perp\left(\frac{\pi}{2}\right) = 2/3$.

It is interesting to note that the antisymmetric (weight 1) α tensor component belongs to class A_{2g}, so that for conditions under which the α tensor becomes non-symmetric this A_{2g} mode which is otherwise inaccessible would be active. The modes of

Table 1. Symmetry classes for μ, α, and β for point group D_{6h} and relation to benzene fundamental vibrations.

Symmetry species	μ	α	weights	β	weights	Number of frequencies
A_{1g}		α	0, 2			2
A_{2g}		α	1			1
B_{1g}						0
B_{2g}						2
E_{1g}		α	1, 2			1
E_{2g}		α	2			4
A_{1u}				β	2	0
A_{2u}	μ			β	1, 3	1
B_{1u}				β	3	2
B_{2u}				β	3	2
E_{1u}	μ			β	1, 2, 3	3
E_{2u}				β	2, 3	2

class B_{2g} which are infra-red, Raman and hyper Raman inactive are active in the second hyper Raman effect which is controlled by the fourth rank second hyperpolarizability tensor γ.

B. Ethylene, C_2H_4 (point group D_{2h})

Ethylene has 12 vibrational modes of symmetry species $A_g(3)$, $B_{1g}(2)$, $B_{2g}(1)$, $A_u(1)$, $B_{1u}(1)$, $B_{2u}(2)$, and $B_{3u}(2)$. Of these only the six with *ungerade* character are hyper-Raman active; the B_u modes are also ir active. The representations spanned by the components of the β tensor are

Weight 1: $2B_{1u} + 2B_{2u} + 2B_{3u}$,

Weight 2: $2A_u + B_{1u} + B_{2u} + B_{3u}$;

Weight 3: $A_u + 2B_{1u} + 2B_{2u} + 2B_{3u}$.

Thus, the B_u vibrational modes belong to class A (containing weights 1, 2, and 3) and the A_u modes belong to class B (with weights 2 and 3 only).

The $\nu_4(A_u)$ torsional mode of ethylene, which is both ir and Raman inactive, provides another example of the usefulness of hyper-Raman spectra in the study of molecular vibrations.

An example of class A modes is the molecular bending mode ν_1 of B_{1u} symmetry.

C. Cyclopropane, C_3H_6 (point group D_{3h})

Cyclopropane has 21 fundamental modes of vibration: $3\ A_1' + 1\ A_2' + 4\ E' + 1\ A_1' + 2\ A_2'' + 3E''$. The components of the β tensor associated with the $\nu_4(A_1'')$ methylene twisting mode are entirely of weight 2 character; this mode, which is ir and Raman inactive, belongs to class E. On the assumption of full index symmetry for the β tensor this mode would not be hyper Raman active. As noted earlier the depolarization ratio for this type of mode is infinite.

D. Xenon tetrafluoride, XeF_4 (point group D_{4h})

The molecule XeF_4 has a $\nu_5(B_{2u})$ out-of-plane bending mode belonging to class B. This is the only vibrational mode of XeF_4 in this class. The $\nu_3(A_{2u})$ out-of-plane bending vibration is an example of a class C mode (with weights 1 and 3).

E. Methane, CH_4 (point group T_d)

The $\nu_2(E)$ mode of methane belongs to class E (weight 2 only). With full permutation symmetry this mode would not be hyper Raman active.

7. EXPERIMENTAL TECHNIQUES

Hyper Rayleigh and Raman scattering are many orders of magnitude less intense than their linear counterparts. For example, a typical nanosecond pulse from a giant-pulse laser would deliver about 3×10^{17} photons to the sample at frequency ω_0, but the number of photons of frequency $2\omega_0 - \omega_M$ which the detector would receive, ranges from a few photons per laser pulse down to one photon every hundred laser pulses, depending on the sample. There are two methods of tackling the problem of observing scattered radiation of such low intensities: single-channel detection or multichannel detection. In single-channel detection, the number of photons in a small wavenumber range $\Delta\tilde{\nu}$, averaged over a large number of laser pulses, is determined using photon counting equipment, and the procedure repeated over successive $\Delta\tilde{\nu}$ until the desired wavenumber range is covered. In multichannel detection, an image intensifier is used to intensify the complete spectrum to a level at which it can be scanned by a television camera. In this way, all the spectral information can be recorded for each laser pulse and the observation time much reduced. For most samples, it is necessary to keep the electric field intensity low enough to avoid dielectric breakdown. The consequent low level of signal usually means it is necessary to average over some thousands of laser pulses. The repetition rate of a ruby laser is restricted to about one pulse per second, and thus 6000 laser pulses require 100 minutes. Clearly, with single-channel detection, the total time involved would be prohibitively long except for strong signals and low resolution, and ruby lasers are usually used with multichannel detection. Nd-YAG lasers have much higher repetition rates and can be used with single-channel detection. A description of a typical multichannel system has been given by French and Long (3); and of single-channel systems by Vogt and Neumann (4) and by Schmid and Schrötter (5).

8. SOME EXPERIMENTAL RESULTS

The first hyper Rayleigh and hyper Raman spectra were reported by Terhune, Maker and Savage (6). The following account of subsequent experimental work is intended to be representative rather than exhaustive.

A. Gases

Only a few gases have been studied. These include ethane, ethene and methane (7). Only vibrational spectra at modest resolution have so far been obtained. No resolved rotational structure has yet been reported.

B. Liquids

Maker (8) has analysed hyper Rayleigh scattering from a number of liquids. A number of group IV tetrahalides have been studied by Long and co-workers (9). Semi-quantitative intensities and depolarization ratios were measured and discussed. Other liquids whose hyper Raman spectra have been reported include water (3, 6) and tetrachloroethene (5).

C. Crystals

A number of crystal systems have been investigated. These include NH_4Cl (10); NH_4Br (10); calcite (11, 12); $NaNO_2$ (13, 14); $NaNO_3$ (10, 15, 16); $LiNbO_3$ (17); $SrTiO_3$ (18, 19, 20, 21, 22, 23); caesium and rubidium halides

(24, 25, 26); rutile (27, 28, 29); PbI_2 (30); CuBr (31); diamond (32); quartz (33, 34).

D. Special effects

Stimulated hyper Raman scattering has been observed from sodium vapour (35); resonance hyper Raman scattering from CdS (36); and surface enhanced hyper Raman scattering from SO_3^{2-} ions adsorbed on silver powder (37).

REFERENCES

1. Long, D.A., Raman Spectroscopy, McGraw-Hill, London, 1977.

2. Andrews, D.L. and Thirunamachandran, T., J. Chem. Phys., 68, 2941 (1978).

3. French, M.J. and Long, D.A., J. Raman Spectrosc., 3, 391 (1975).

4. Vogt, H. and Neumann, G., Phys. Status Solidi (b), 92, 57 (1979).

5. Schmid, W.J. and Schrötter, H.W., Chem. Phys. Lett., 45, 502 (1977).

6. Terhune, R.W., Maker, P.D. and Savage, C.M., Phys. Rev. Lett., 14, 681 (1965).

7. Verdieck, J.F., Peterson, S.H., Savage, C.M. and Maker, P.D., Chem. Phys. Lett., 7, 219 (1970).

8. Maker, P.D., Phys. Rev. A, 1, 923 (1970).

9. Dines, T.J., French, M.J., Hall, R.J.B. and Long, D.A., Proc. Fifth Int. Conf. Raman Spectrosc., (Schmid, E.D. et al., eds) H.F. Schulz Verlag, Freiburg 1976, p. 707.

10. Long, D.A. in Nonlinear Behaviour of Molecules, Atoms & Ions in Electric, Magnetic or Electromagnetic Fields, (Neel, L. ed.) Elsevier, Amsterdam, 1979, p. 153.

11. Polivanov, Y.N. and Sayakhov, R.S., Sov. Phys. Solid State, 20, 1564 (1978).

12. Polivanov, Y.N. and Sayakhov, R.S., Phys. Status Solidi (b), 103, 89 (1981).

13. Weinmann, D. and Vogt, H., Phys. Status Solidi (a), 23, 463 (1974).

14. Vogt, H., Phys. Status Solidi (b), 58, 705 (1973).

15. Vogt, H. and Neumann, G., Phys. Status Solidi (b), 86, 615 (1978).

16. Ortmann, L. and Vogt, H., Optics Communications, 16, 234 (1976).

17. Denisov, V.N., Mavrin, B.N., Podobedov, V.B. and Sterin, K.E., Optics Communications, 26, 372 (1978).

18. Vogt, H. and Rossbroich, G., Phys. Rev. B, 24, 3086 (1981).

19. Denisov, V.N., Mavrin, B.N., Podobedov, V.B. and Sterin, K.E., Sov. Phys. Solid State, 22, 1657 (1980).

20. Inoue, K., Asai, N. and Sameshima, T., J. Phys. Soc. Japan, 50, 1291 (1981).

21. Vogt, H. and Neumann, G., J. Phys. Soc. Japan, 49, 2417 (1980).

22. Inoue, K. and Sameshima, T., J. Phys. Soc. Japan, 47, 2037 (1979).

23. Vogt, H. and Rossbroich, G., Ferroelectrics, 37, 713 (1981).

24. Vogt, H. and Neumann, G., Phys. Status Solidi (b), 92, 57 (1979).

25. Vogt, H., Proc. Fifth Int. Conf. Raman Spectrosc., (Schmid, E.D., et al., eds.) H.F. Schulz-Verlag, Freiburg 1976, p. 742.

26. Vogt, H. and Neumann, G., Optics Communications, 19, 108 (1976).

27. Denisov, V.N., Mavrin, B.N., Podobedov, V.B. and Sterin, K.E., Optics Communications, 34, 357 (1980).

28. Denisov, V.N., Mavrin, B.N., Podobedov, B.N., Sterin, K.E., and Varshal, B.G., Opt. Spectrosc. (USSR) 49, 221 (1980).

29. Varshal, B.G., Denisov, V.N., Mavrin, B.N., Parlova, G.A., Podobedov, V.B. and Sterin, K.E., Opt. Spectrosc. (USSR), 47, 344 (1979).

30. Yu, Q.H. and Levy, R., J. Lumin., 24/25, 417 (1981).

31. Honerlage, B., Rossler, U., Phach, V.D., Bivas, A. and Grun, J.B., Phys. Rev. B, 22, 797 (1980).

32. Yu, W. and Alfano, R.R., Phys. Rev. A, 11, 188 (1975).

33. Denisov, V.N., Mavrin, B.N., Podobedov, V.B. and Sterin, K.E., J. E. T. P. Lett., 32, 316 (1980).

34. Denisov, V.N., Mavrin, B.N., Podobedov, V.B. and Sterin, K.E., Sov. Phys. Solid State, 20, 2016 (1978).

35. Cotter, D., Hanna, D.C., Tuttlebee, W.H.W. and Yuratich, M.A., Optics Communications, 22, 190 (1977).

36. Polivanov, Y.N. and Sayakhov, R.S., J. E. T. P. Lett., 30, 580 (1979).

37. Murphy, D.V., von Raben, K.U., Chang, R.K. and Dorain, P.B., Chem. Phys. Lett., 85, 43 (1982).

STIMULATED RAMAN EFFECT

STIMULATED RAMAN SCATTERING

A. Laubereau

Physikalisches Institut
Universität Bayreuth
Bayreuth, W. Germany

I. INTRODUCTION

Inelastic light scattering has proved to be highly valuable for the study of elementary excitations of gases, liquids and solids. The parameters of the scattered radiation, e.g. frequency position, spectral band shape, degree of polarization and scattering efficiency provide detailed information on the material under consideration. Using conventional light sources, i.e. working at low intensity the scattering intensity is very weak and is proportional to the intensity of the incident light beam. At elevated intensity level the situation may change significantly as already predicted in the 1930's. |1| Nevertheless the first discovery of stimulated Raman scattering was accidental using a nitrobenzene Kerr shutter as Q-switch of a ruby laser. |2| With intense laser sources a variety of new scattering phenomena has been observed during the past years that are discussed in detail in the various chapters of this book. |3| The present article intends to provide a basic understanding of stimulated scattering and to reveal the close relationship with spontaneous scattering, coherent scattering and inverse Raman effect.

II. A SIMPLE QUANTUM MECHANICAL TREATMENT

We first outline a quantum mechanical picture which describes both spontaneous and stimulated scattering. |4,5| The inelastic light-scattering process is shown in Fig. 1a. An incident photon $\hbar\omega_L$ is destroyed, a scattered photon $\hbar\omega_S$ generated, while the difference in energy $\hbar(\omega_L-\omega_S)=\omega_0$ is absorbed by the scattering medium. The excitation $\hbar\omega_0$ of the material system is often a

W. Kiefer and D. A. Long (eds.), Non-Linear Raman Spectroscopy and Its Chemical Applications, 183–209.

vibrational or rotational transition of a molecule. The frequency ω_S is called Stokes frequency and is smaller than the incident light frequency. The stimulated process is schematically represented in Fig. 1b. In this case a light wave at frequency ω_S is incident on the material system simultaneously with a light wave at ω_L. A photon $\hbar\omega_S$ is added to the wave at ω_S which consequently becomes amplified while the incident light beam loses a quantum $\hbar\omega_L$.

Comparison of Figs. 1a and b shows that the elementary process in both cases is identical, the difference being the probability of the scattering event.

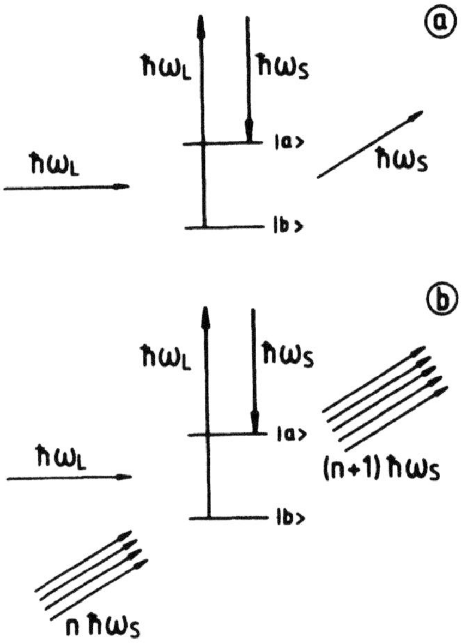

Fig. 1: Spontaneous and stimulated Stokes scattering

For simplicity we describe the physical system (e.g. molecules) interacting with the light field as an ensemble of harmonic oscillators j with transition frequency ω_o. The elongation operators q_j of the oscillators are expressed by the corresponding creation and annihilation operators, and the Hamiltonian of the unperturbed oscillators reads

$$H_v = \sum_{k_v} \hbar\omega_o \left(b_{k_v}^{+} \, b_{k_v} + \frac{1}{2} \right) \qquad (1)$$

Here the sum over the individual molecules j in the interaction volume is replaced by the sum over the wave vectors k_v in Fourier space. The local operator q_j of molecule j is connected with the operators b_{k_v}, $b_{k_v}^+$ according to

$$q_j = \left(\frac{\hbar}{2\omega_o mN}\right)^{1/2} \sum_{k_v} \exp(ik_v R_j)(b_{k_v} + b^+_{-k_v}) \tag{2}$$

N_o is the number of molecules (oscillators) in the interaction volume V. R_j and m respectively denote the position and reduced mass of the individual oscillator.

The linearly polarized electromagnetic field is described by

$$H_f = \sum_{k_f} \hbar\omega_{k_f} (a^+_{k_f} a_{k_f} + \frac{1}{2}) \tag{3}$$

The a_k's denote the annihilation operators of the electromagnetic modes labeled by the wave vector k_f. ω_{k_f} represents the mode frequency which is determined by k_f via the known photon dispersion. The operator E of the electromagnetic field amplitude is connected to the operators a_{k_f}, $a_{k_f}^+$ by

$$E = i\left(\frac{2\pi\hbar\omega_f}{\mu^2 V}\right)^{1/2} \sum_f \left[a^+_{k_f} \exp(-ik_f R_j) - a_{k_f} \exp(ik_f R_j)\right] \tag{4}$$

μ denotes the optical refractive index. Eq. 4 applies for electrostatic units which will be used in the following.

The total unperturbed system is represented by the sum of the Hamiltonians, $H_v + H_f$. The interaction responsible for Raman scattering is treated in the electric dipole approximation, where the energy of the induced dipoles is given by $-\int \alpha E dE$. We are interested here only in the coupling which originates from the elongation q of the molecules. Consequently we expand the molecular polarizability α as a function of elongation, $\alpha \approx \alpha_o + \partial\alpha/\partial q$, and retain only the term $\partial\alpha/\partial q$. |6| The tensor character of α is ignored here (leading to isotropic scattering) yielding the interaction Hamiltonian

$$H_{int} = -\frac{1}{2}\frac{\partial\alpha}{\partial q} \sum_j q_j E(R_j) E(R_j) \tag{5}$$

Energy and momentum conservation greatly reduce the number of electro-magnetic modes, which interact in the scattering process. The electric field in Eq. 5 consists of the incident laser field, and of scattered Stokes and anti-Stokes components. Here, we are interested in Stokes scattering only and restrict ourselves to the laser and the first order Stokes field, $E = E_L + E_S$. For this situation only terms proportional to $E_L(R_j) \times E_S(R_j)$ in Eq. 5 fulfil energy conservation. The total set of electromagnetic modes k_f is split up in the group of laser modes k_L, and Stokes modes k_S (and additional modes which do not show up explicitly). Using Eqs. 2, 4 and 5 it can be shown that the interaction Hamiltonian takes the form

$$H_{int} \propto \sum_{k_L, k_S} \left[b_{k_v}^{+} a_{k_S}^{+} a_{k_L} + b_{k_v} a_{k_S} a_{k_L}^{+} \right] \delta(k_L - k_S - k_v) \qquad (6)$$

It is easily seen that the first term in Eq. 6 represents the creation of a Stokes photon, as illustrated in Fig. 1. The inverse process is introduced by the second term.

Under steady-state conditions and for weak intensities with small scattering rates we may apply stationary perturbation theory. The rate of scattered photons is proportional to

$$\frac{dn_S}{dt} \propto |\langle f|H_{int}|i\rangle|^2 \qquad (7)$$

where the final states $|f\rangle$ are connected to the initial states $|i\rangle$ by energy conservation. Substitution of Eq. (6) shows that matrix elements of the form $\langle f|b_{k_v}^{+} a_{k_S}^{+} a_{k_L}|i\rangle$ and $\langle f|b_{k_v} a_{k_S} a_{k_L}^{+}|i\rangle$ have to be evaluated.

It is convenient to use the occupation number representation and write the initial state $|i\rangle$ in terms of the initial occupation numbers n_S, n_v, and n_L of the Stokes modes, molecular system, and laser modes: $|i\rangle = |n_S, n_v, n_L\rangle$. Due to the well-known selection rules of the creation and annihilation operators of harmonic oscillators, the matrix elements are non-vanishing for the two final states $|n_S + 1, n_v + 1, n_L - 1\rangle$ and $|n_S - 1, n_v - 1, n_L + 1\rangle$:

$$\langle n_S+1, n_v+1, n_L-1|H_{int}|n_S, n_v, n_L\rangle \propto \left[(n_S+1)(n_v+1)n_L\right]^{1/2}$$

and (8)

$$\langle n_S-1, n_v-1, n_L+1|H_{int}|n_S, n_v, n_L\rangle \propto \left[n_S n_v (n_L+1)\right]^{1/2}$$

For the simple case of a single laser, Stokes and oscillator mode we obtain

$$\frac{dn_S}{dt} = \frac{N_o}{V^2}\left(\frac{\partial\alpha}{\partial q}\right)^2 \frac{4\pi^3\hbar\omega_L\omega_S}{\mu_L^2\mu_S^2 m\omega_o}\left[n_L+n_Ln_S+n_Ln_v-n_Sn_v\right] \tag{9a}$$

$$\omega_S = \omega_L-\omega_o \tag{9b}$$

$$k_S = k_L-k_v \tag{9c}$$

The first term of the r.h.s. of Eq. 9a represents spontaneous Raman scattering. It describes the scattering of laser light off the quantum fluctuations, when the population of Stokes modes is negligible, $n_S \simeq 0$. The second term (depending on n_S) describes the rate of Stokes light generation resulting from the amount of Stokes photons present. For a large Stokes population, $n_S \gg 1$, the scattering rate increases rapidly. This term is responsible for the stimulated Raman effect. The third term proportional to $n_L \times n_v$ corresponds to parametric coupling between the Stokes light field and the material excitation. In general, n_v is small and approximately equal to the thermal equilibrium value and may be neglected in many cases. Even in the stimulated process, where a large number of Stokes photons and a corresponding number of material transitions (phonons) are generated, n_v is negligible due to the stronger damping of the material excitation as compared to photons, i.e. $n_v \ll n_S$. The last term on the r.h.s. of Eq. 9a is negligible for the same reason.

Spontaneous scattering ($n_S \ll 1$)

We generalize the result of Eq. 9 to the multiple mode case. Summation over the number of interacting Stokes modes in the solid angle $\Delta\Omega$ and introducing the laser intensity $I_L = (\hbar\omega_L c/\mu_L V)\Sigma n_L$ yields the scattered Stokes power

$$P_S = \hbar\omega_S \frac{dn_S}{dt} = V\,N\,\frac{\partial\sigma}{\partial\Omega} I_L\,\Delta\Omega \tag{10a}$$

where we have introduced the integrated scattering cross section

$$\frac{\partial\sigma}{\partial\Omega} = \left(\frac{\partial\alpha}{\partial q}\right)^2 \frac{\omega_S^4\,\mu_S\hbar}{2c^4\mu_L m\omega_o} \tag{10b}$$

The Raman polarizability $\partial\alpha/\partial q$ may be evaluated by a semi-classical calculation using time-dependent perturbation theory. |5|

Stimulated scattering ($n_S >> 1$)

We retain only the second term on the r.h.s. of Eq. 9a, substitute the photon numbers by the corresponding intensity values and transform $dI/dt \rightarrow cdI/dx\mu$ to obtain

$$\frac{dI_S}{dx} = g_S \; I_L \; I_S \tag{11a}$$

where the gain factor g_S collects the various material parameters. x denotes the propagation coordinate of the Stokes field. A solution of (11a) is

$$I_S(x) = I_S(x=0) \; \exp(g_S I_L x) \tag{11b}$$

Equation (11b) predicts exponential growth of the scattered Stokes light. The result holds in the stationary case and for small conversion efficiency so that the incident laser light is not notably effected by the scattering process.

Apart from geometrical factors, the scattering is isotropic in the plane perpendicular to the electric field vector of the incident light field. This fact is well known for spontaneous scattering and also holds for stimulated scattering. On the other hand, the large effect of the scattering geometry should be noted. Eq. 11b shows that the interaction length enters the scattering process exponentially. For high gain systems, $g_S I_L x_{max} >> 1$, Stokes radiation in forward or backward direction with respect to the incident laser beam may find a significant larger interaction length than in transverse direction; as a result the exponential amplifications parallel to the laser beam may be drastically larger by several orders of magnitude.

The close relationship between spontaneous and stimulated scattering is demonstrated by an experiment, where the stimulated Raman effect was measured over eleven orders of magnitude in liquid N_2. |7| In Fig. 2 the observed Stokes output power in forward direction is plotted versus imput laser power. Starting at weak Stokes emission, one sees the linear range of spontaneous Raman scattering (a), (note the logarithmic ordinate scale). At higher input power exponential build-up of Stokes power occurs (curve b). At large Stokes amplification saturation occurs due to depletion of incident laser light. The rapid rise of Stokes power (c) is explained by optical feed-back which increases the effective amplification length in the sample cell.

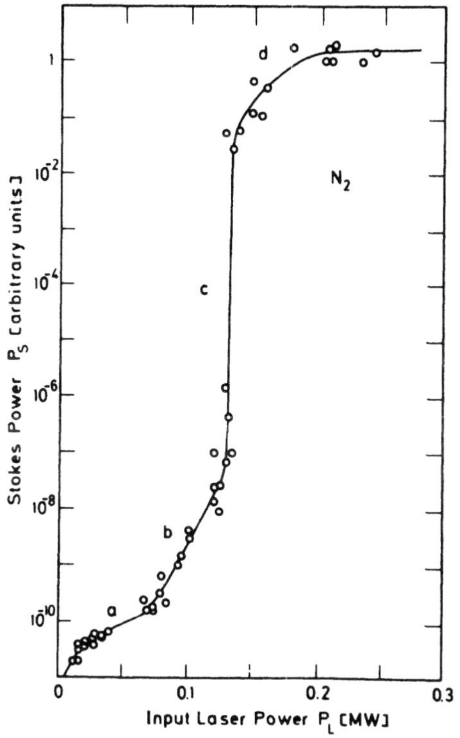

Fig. 2: Spontaneous (a) and stimulated (b-d) Stokes generation in liquid nitrogen (after Ref. 7).

Equivalent Noise Input

In many high gain experiments the stimulated scattering builds up from quantum noise. While the exponential amplification process is well accounted for by semiclassical theory (see next section) the starting condition supplied by spontaneous scattering has to be evaluated from a full quantum theory. Including the spontaneous scattering term (note Eq. 9a) in Eq. 11a, we have to take into account the number density of Stokes modes in the solid angle $\Delta\Omega$ and in the Stokes frequency band $\Delta\omega$($\sim$ spontaneous linewidth), $\omega_S^2\ \mu_S^3 \Delta\Omega\Delta\omega/8\pi^3 c^3$. One finds

$$\frac{dI_S}{dx} = g_S\ I_L (I_S + \frac{\hbar\omega_S^3\ \mu_S^3}{8\pi^3 c^2}\ \Delta\Omega\Delta\omega) \tag{12}$$

Integration yields

$$I_S(x) = I_{SN} \left| \exp(g_S I_L x) - 1 \right| \tag{13}$$

where

$$I_{SN} = \frac{\hbar \omega_S^{\,3} \; \mu_S^{\,3}}{8\pi^3 c^2} \Delta\Omega\Delta\omega$$

The equivalent noise input I_{SN} represents the initial condition for stimulated amplification. A value of 10^{-3} W/cm^2 is estimated for $\omega_S/2\pi c \simeq 10^4$ cm^{-1}, $\Delta\omega/2\pi c \simeq 5$ cm^{-1} and $\Delta\Omega \simeq 1\times10^{-4}$ sr.

Inverse Raman scattering; laser depletion

Instead of relating the scattering rate to the creation of Stokes photons (Eq. 7) we may consider the annihilation rate of laser photons, since for every created Stokes photon a laser photon has to be annihilated. This means conservation of the total number of photons, $dn_L/dt + dn_S/dt = 0$. For spontaneous scattering this effect is not measurable because of the small $d\sigma/d\Omega$. For the stimulated process the scattering efficiency may be larger by many orders of magnitude and deserves some consideration. Eq. 11a is supplemented by

$$\frac{dI_L}{d_x} = - \frac{\omega_S}{\omega_L} g_S I_S \cdot I_L \tag{14}$$

indicating rapid loss of laser intensity when the scattered Stokes light approaches this intensity level. For an intense beam at the frequency ω_S, Eq. 14 predicts exponential loss of incident radiation at the high frequency $\omega_S+\omega_o$ $(=\omega_L)$. This is called inverse Raman scattering. |8| Conventionally in these experiments the lower frequency component is called "laser" (instead of "Stokes") and the higher frequency component "anti-Stokes" (instead of "laser").

III. SEMICLASSICAL THEORY

The simple quantum theory discussed above is valid in the steady state case, only. It gives no information on the coherence properties of the scattered radiation and of the material excitation. The semiclassical theory we wish to derive now will give a more detailed picture of the physical situation. |9| It has been shown above that stimulated Raman scattering occurs for large occupation numbers of the Stokes modes, so that the light field may be treated classically. The development from incoherent, spontaneously scattered radiation to a coherent, intense Stokes field by high gain stimulated amplification is not contained in the present treatment.

The physical system is now described by a two-level model, which is felt to be more realistic than the harmonic oscillator picture on account of vibrational anharmonicities etc. Transitions to other energy levels are assumed to be far off-resonance and may be included via a nonresonant nonlinear susceptibility.

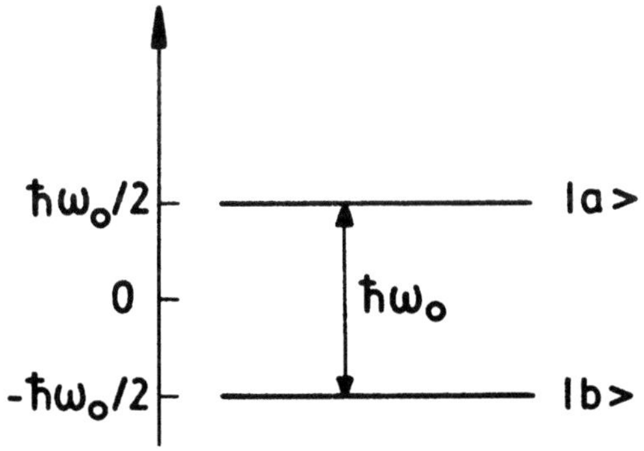

Fig. 3: Two-level system

The two-level system is depicted in Fig. 3. The unperturbed molecule possesses the upper state $|a\rangle$ and the lower state $|b\rangle$. The energy separation is $\hbar\omega_o$. The Hamiltonian is written as the sum of the free system and the Raman interaction; the latter has been discussed in Eq. (5):

$$H = H_o + H_{int} = H - \frac{1}{2} \sum_j \frac{\partial\alpha}{\partial q} q_j \, E_L(j) \, E_S(j) \qquad (15)$$

The wave function of an individual molecule j is given by the linear combination $|\Psi_j\rangle = a_j(t)|a\rangle + b_j(t)|b\rangle$. The density operator of the molecular system is defined by

$$\rho = \frac{1}{N} \sum_j^N |\Psi_j\rangle\langle\Psi_j| \qquad (16)$$

This operator is valuable for the evaluation of the ensemble-averaged expectation value of an arbitrary observable A: $\langle A\rangle = Tr(A\rho)$. The time evolution of the system is approximately governed by the Redfield equation |10|

$$\frac{d}{dt}\rho_{kl} = \frac{i}{\hbar}[\rho,H]_{kl} + \sum_{m,n} R_{klmn}(\rho_{mn}-\overline{\rho_{mn}}) \quad (17)$$

(where k, l, m, n = a, b). The first term on the r.h.s. of Eq. 17 describes the molecular dynamics in the absence of damping. Relaxation processes of the system via fluctuating forces are included by the second term, which is valid for weak damping and for time intervals long compared to the time scale of the damping processes, e.g. time between collisions in a gas etc. Eq. 17 has been derived first for magnetic dipole transitions of spin systems. It may be shown by the help of symmetry arguments that the tensor R has only the following non-vanishing elements

$$\begin{aligned} R_{aa,bb} &= R_{bbaa} = -1/T_1 \\ R_{ab,ab} &= R_{baba} = -1/T_2 \end{aligned} \quad (18)$$

The time constants T_1 and T_2 are the relaxation times of the ensemble of two-level systems. $\overline{\rho_{mn}}$ in Eq. 17 denotes the equilibrium value of ρ_{mn}.

The four matrix elements ρ_{aa}, ρ_{ab}, ρ_{ba} and ρ_{bb} describe the physical system. On account of the normalization condition $\rho_{aa} + \rho_{bb} = 1$, three independent variables govern the time dependence. In this article we introduce the occupation probability $n = \rho_{aa}$ of the upper state and the collective amplitude

$$\langle q\rangle = Tr(\rho q) \simeq q_{ab}(\rho_{ab}+\rho_{ba}) \quad (19)$$

The closure relation $|a\rangle\langle a|+|b\rangle\langle b|=1$ and the approximation $q_{aa} \simeq q_{bb} \simeq 0$ have been used for the derivation of Eq. (19). q_{ab} denotes the matrix element $\langle a|q|b\rangle$. Inserting $\langle q\rangle$ and n in Eq. 17 yields the differential equations |11|:

$$\frac{\partial^2\langle q\rangle}{\partial t^2} + \frac{2}{T_2}\frac{\partial\langle q\rangle}{\partial t} + \omega_o^2\langle q\rangle = \frac{\omega_o q_{ab}^2}{\hbar}\frac{\partial\alpha}{\partial q}E_L E_S(1-2n) \quad (20a)$$

$$\frac{\partial n}{\partial t} + \frac{1}{T_1}(n-\bar{n}) = \frac{1}{2\hbar\omega_o}\frac{\partial\alpha}{\partial q}E_L E_S\frac{\partial\langle q\rangle}{\partial t} \quad (20b)$$

On the left hand side of equation (20a), a term $\langle q\rangle T_2^2/2$ was omitted assuming small damping, $1/T_2 \ll \omega_o$. In equation (20b), a term proportional to $\langle q\rangle/T_2$ has been neglected in comparison with $\partial\langle q\rangle/\partial t$ for the same reason. The dephasing time T_2 and the population lifetime T_1 predict exponential decay of the collective amplitude $\langle q\rangle$ and the excited state population n, respectively. |9|

It is interesting to see from equations (20) that the collective amplitude of the molecular ensemble obeys the equation of motion of a classical damped oscillator. The driving term on the right hand side depends upon the population n of the upper state. A change of sign occurs for population inversion, $n>1/2$. The coupling between the light field and the molecular transition is determined by $\partial\alpha/\partial q$ and the matrix element q_{ab}. Having in mind an anharmonic vibration, we use the result $q_{ab} \simeq (\hbar/2m\omega_o)^{1/2}$.

The propagation of the light field is governed by the classical electromagnetic theory. The following wave equation is obtained from Maxwell's equations for the Stokes field |12|:

$$\nabla^2 E_S - \frac{\mu_S^2}{c^2}\frac{\partial^2}{\partial t^2} E_S = \frac{4\pi}{c^2}\frac{\partial^2}{\partial t^2} \tilde{P}_S^{NL} \tag{21a}$$

The analogous expression holds for the laser field E_L. The nonlinear contribution $\tilde{P}_S^{NL}$ acts as a source term on the right hand side and is responsible for the stimulated Raman process. It has two contributions: A first term due to the interaction of the light field with the two-level systems, and a second part, $\tilde{P}_{NR}$, resulting from all the other off-resonance transitions of the molecules. The latter part is important only for diluted systems and/or for weak Raman lines (see below).

$$\tilde{P}_S^{NL} = N \frac{\partial\alpha}{\partial q} \langle q\rangle E_L + \tilde{P}_{NR} \tag{21b}$$

$\tilde{P}_{NR}$ will be ignored in the following section for simplicity. N is the number density of molecules. Since $\partial\alpha/\partial q$ is treated as a scalar quantity, Eq. 21b refers to highly polarized Raman lines. Scattered light polarized perpendicular to the incident laser light is not included in the calculation.

Stimulated Scattering in Forward Direction

Solutions for the stimulated scattering in the forward direction are obtained using the ansatz of plane waves for the vibrational amplitude $\langle q\rangle$, the electromagnetic field and the nonlinear polarization:

$$\langle q\rangle = \frac{1}{2} Q \exp(ik_v x - i\omega_v t) + c.c.$$

$$E_j = \frac{1}{2} A_j \exp(ik_j x - i\omega_j t) + c.c. \tag{22}$$

$$\tilde{P}_j^{NL} = \frac{1}{2} P_j^{NL} \exp(ik_j x - i\omega_j t) + c.c.$$

where j = L,S. ω_v and k_v respectively denote the frequency and k-vector of the material excitation which may not exactly coincide with the values ω_o and k_o at resonance. A_L and A_S are the amplitudes of the laser and Stokes field propagating in the x-direction.

Inserting Eq. 22 into Eqs. 20-21, equating terms with equal frequency, $\omega_v = \omega_L - \omega_S$, and equal wave vectors, $k_v = k_K - k_S$, gives a new set of differential equations. For pulse durations t_p and for dephasing times T_2 in the picosecond range or longer, various terms may be neglected, since $\partial/\partial t \ll \omega_j$, $\partial^2/\partial t^2 \ll \omega_j^2$, and $1/T_2 \ll \omega_v$ (j = L,S,v). With these approximations we find:

$$\frac{\partial Q}{\partial t} + \frac{i}{2m\omega_j}\left(\omega_o^2 - \omega_j^2 - i\frac{2\omega_v}{T_2}\right) Q = \frac{i}{4m\omega_j}\frac{\partial\alpha}{\partial q} A_L A_S^* (1-2n) \tag{23a}$$

$$\frac{\partial A_S}{\partial x} + \frac{\mu_S}{c}\frac{\partial A_S}{\partial t} = \frac{i\pi\omega_S}{c\mu_S} N \frac{\partial\alpha}{\partial q} A_L Q^* \tag{23b}$$

$$\frac{\partial A_L}{\partial x} + \frac{\mu_L}{c}\frac{\partial A_L}{\partial t} = \frac{i\pi\omega_L}{c\mu_L} N \frac{\partial\alpha}{\partial q} A_S Q \tag{23c}$$

$$\frac{\partial n}{\partial t} + \frac{1}{T_1}(n - \bar{n}) = \frac{i\omega_v}{8\hbar\omega_o}\frac{\partial\alpha}{\partial q}\left(A_L A_S^* Q^* - A_L^* A_S Q\right) \tag{23d}$$

Equations (23) represent a complete set of differential equations which are applicable to a variety of experimental situations. Scattering in backward direction is governed by a similar set of equations |3|.

IV. STIMULATED RAMAN SCATTERING UNDER QUASI-STATIONARY CONDITIONS

Most stimulated Raman experiments are performed with pulsed lasers under quasi-stationary conditions. When the dephasing time T_2 is short compared to the pulse duration, the first term may be dropped in Eq. 23a. We also neglect the small population changes generated by stimulated Raman scattering, and use the thermal equilibrium value, $n \simeq \bar{n}$. Now Eq. 23d is completely separated and will not be furtherly considered. Substitution of Q (Eq. 23a) in

Eqs. 23b, c and transformation to a retarded time frame $t' = t - x/v$, $x' = x$ leads to

$$\frac{\partial A_S}{\partial x'} = \frac{i\pi\omega_S}{2\mu_S cm} \left(\frac{\partial\alpha}{\partial q}\right)^2 N \frac{1-2\bar{n}}{\omega_o^2 - (\omega_L-\omega_S)^2 + 2i(\omega_L-\omega_S)T_2} |A_L|^2 A_S \tag{24}$$

and a similar expression for A_L. v denotes the group velocity. Integrating Eq. 24 for small Stokes conversion and converting to intensities ($I_{L,S} = c\mu_{L,S}|A_{L,S}|^2/8\pi$) reproduces Eq. 11a:

$$I_S(x') = I_S(0) \exp(gI_L x') \tag{25a}$$

Here, we have introduced a frequency dependent gain factor

$$g(\omega_S) = \frac{16\pi^2\omega_S N}{\mu_S\mu_L c^2 m} \left(\frac{\partial\alpha}{\partial q}\right)^2 \frac{(1-2\bar{n})\omega_v/T_2}{(\omega_o^2-\omega_v^2)^2 + 4\omega_v^2/T_2^2} \tag{25b}$$

($\omega_v = \omega_L - \omega_S$). It is interesting to see that the gain factor has a Lorentzian frequency dependence with linewidth (FWHH) $\Delta\omega = 2/T_2$. Maximum amplification occurs at resonance $\omega_L - \omega_S = \omega_o$ with

$$g_S = \frac{4\pi^2\omega_S^2(1-2\bar{n})}{\mu_S\mu_L c^2 m\omega_o} \left(\frac{\partial\alpha}{\partial q}\right)^2 N\, T_2 \tag{26a}$$

or using Eq. 10b,

$$g_S = \frac{8\pi^2 c^2(1-2\bar{n})}{\mu_S\mu_L \hbar\, \omega_S^3} \frac{\partial\sigma}{\partial\Omega} N\, T_2 \tag{26b}$$

For a strong Raman line we typically have $\partial\sigma/\partial\Omega \simeq 5 \times 10^{-30}$ cm^2. For $N \simeq 10^{22}$ cm^2 (liquid), $\omega_S/2\pi c \simeq 10^4$, and $T_2 \simeq 10^{-12}$ s we estimate $g_S \simeq 10^{-9}$ cm/W. A value of the product pump intensity I_L times interaction length ℓ, $I_L \times \ell \simeq 10^9$ W/cm is necessary for amplification of the Stokes wave by a factor of e; this corresponds, for example, to 100 MW/cm^2 over a cell length of 10 cm. For the Raman generator where the Stokes emission builds up in a travelling wave process from quantum noise (see Eq. 13) one estimates an amplification factor of $\sim$exp(25) to reach a conversion efficiency of $I_S(\ell)/I_L \simeq 10^{-2}$. In other words, even larger values of $I_L \times \ell$ of several 10^{10} W/cm are required in this high gain situation. As a consequence competing nonlinear effects may occur in many experi-

TABLE 1. Gain factors for the stimulated Raman process in various materials (after Ref. 3)

Substance	Pressure P atm	Temp. T K	Pump wavelength λ_L Å	Frequency shift $\omega_0/2\pi c$ cm^{-1}	Linewidth $\Delta\tilde{\nu}_v$ cm^{-1}	Gain factor g_S cm/W
Measurements with Raman generator:						
H_2 (gas)	> 10	77	6943	4155		10^{-9}
		300	6943	4155		1.5×10^{-9}
O_2		90	6943	1552	0.117	1.6×10^{-8}
N_2		77	6943	2326.5	0.067	1.6×10^{-8}
CS_2		300	6943	655.6	0.50	1×10^{-8}
C_6H_6		300	5300	992.2	2.15	5×10^{-9}
H_2O		300	6943	~3290	~200	1.4×10^{-10}
Calcite		300	5300	1086	1.2	5.5×10^{-9}
Measurements with Raman amplifiers:						
H_2 (gas)	> 40	300	6943	4151		forward: 1.5×10^{-9}
						backward: 1.5×10^{-9}
N_2 (gas)	1	300	2330.72	0.075		3.5×10^{-11}
CH_3OH		300	6943	2837	18	4×10^{-10}
CCl_4		300	5976	458		1.3×10^{-9}
C_2Cl_4		300	5976	447		1.7×10^{-9}
fused quartz		300	4954	420		1.7×10^{-11}
Values calculated from spontaneous Raman data:						
Nitrobenzene		300		1345	6.6	2.1×10^{-9}
Bromobenzene		300		1000	1.9	1.5×10^{-9}
Chlorobenzene		300		1002	1.6	1.9×10^{-9}
Toluene		300		1003	1.94	1.2×10^{-9}
Ethyl alcohol		300	5300	2928	17.4	5.11×10^{-9}
Isopropyl alcohol		300	5300	2882	26.7	9.2×10^{-10}
Acetone		300	5300	2925	17.4	1.17×10^{-9}
CCl_3CH_3		300	5300	2939	5.2	5.11×10^{-9}
$LiNbO_3$		80	6943	258	7	2.87×10^{-8}
$Ba_2NaNb_5I_{15}$		80	6943	655		1.89×10^{-8}
$LiTaO_3$		80	6943	215	12	1×10^{-8}
SiO_2		300	6943	467		8×10^{-10}
Diamond		300	6943	1332	2.04	6.9×10^{-9}

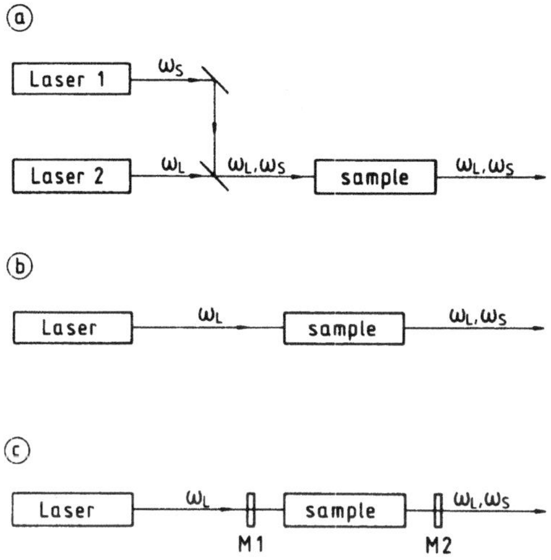

Fig. 4: Schematic of experimental setups:
a) Raman amplifier in forward direction;
b) Raman generator in forward direction;
c) Raman oscillator with longitudinal pump geometry.

ments, e.g. stimulated Brillouin scattering and/or optical self-focussing. For a quantitative study of the stimulated Raman effect the experimental conditions have to be carefully adjusted to avoid disturbing other nonlinearities.

In Table I data on a number of vibrational transitions with efficient stimulated scattering are compiled. It is interesting to note that there is good agreement between the experimental and calculated g_S-values, substantiating the theory presented here.

The stimulated amplification process in forward direction has been applied in three kinds of experiments: a) Raman amplifier with well-defined input Stokes beam; b) Raman generator; and c) Raman oscillator with optical feedback for the Stokes wave. The experimental setups are schematically shown in Fig. 4. The Raman amplifier is often used at small intensity level ($g_S I_L x \ll 1$) so that the exponential in Eq. 25a may be expanded to first order, yielding $I_S \propto 1 + g_S I_L x'$. This is the situation relevant for CAR-, RIKE- and Stokes gain spectroscopy. |3| The Raman oscillator displays a threshold behaviour. If R denotes the relative feedback of Stokes light per cavity transit (e.g. combined mirror reflectivity) net amplification occurs for

$$R \exp(2\, g_S\, I_L\, \ell) > 1 \tag{27}$$

Efficient Stokes production may occur in the oscillator set-up slightly above threshold at an intensity level significantly smaller than for the generator system.

Eq. 25 predicts a frequency dependence of the exponential gain factor which follows the shape of the spontaneous Raman band. This fact has been verified experimentally. |13| An example is given in Fig. 5 where the isotopic substructure of the totally symmetric vibration of C_2Cl_4 is studied. |14| The different combination of the ^{35}Cl and ^{37}Cl isotopes give rise to small variations of the vibrational frequencies on account of the different masses.

The liquid was contained in a hollow glass fibre to obtain high intensity over a long interaction length. The gain profile was measured in a Raman amplifier using a first dye laser as pump laser and a second, frequency-tunable dye laser as input Stokes beam. The measured frequency dependence of the gain factor is depicted in Fig. 5. The spectral distribution is in excellent agreement with the shape of the differential Raman cross section which is measured by spontaneous Raman scattering.

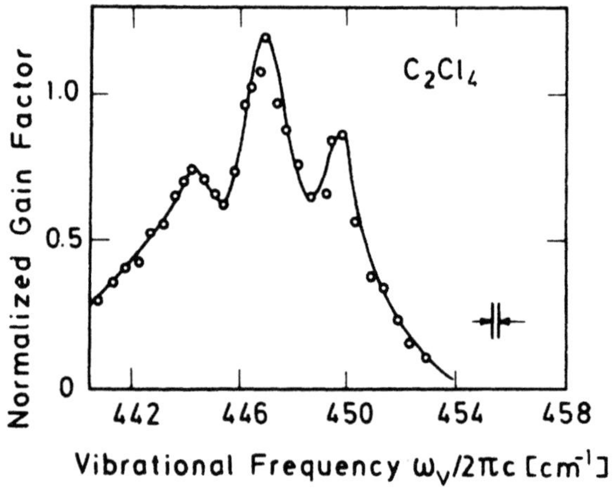

Fig. 5: Stokes gain versus frequency in C_2Cl_4 (after Ref. 14).

The spectral intensity distribution of the Stokes emission of the generator and oscillator system is governed by the frequency dependence of $g(\omega_S)$. From Eq. 25b we conclude that the Stokes line will appear exactly at the band center of spontaneous Stokes scattering (if competing nonlinear processes are absent, e.g. self-focussing). Since the amplification is more efficient close to the resonance ($\omega_S=\omega_L-\omega_o$), gain narrowing occurs; i.e. bandwidth of the stimulated emission is notably smaller than the spontaneous linewidth under high gain conditions |15|.

$\chi^{(3)}$ Formalism

For the theoretical description of various nonlinear optical phenomena, a power expansion of the induced macroscopic polarization as a function of the applied electromagnetic field is generally used. |11,16| This approach is convenient for parametric processes and scattering effects. For stimulated scattering the term of interest is proportional to E^3:

$$\tilde{P}^{NL}(t) = \tilde{\chi}^{(3)}(t)\, EEE \tag{28a}$$

The proportionality factor in Eq. 28a is the third-order nonlinear susceptibility $\tilde{\chi}^{(3)}$ in the time representation.

Under stationary and quasi-stationary conditions $P^{NL}(t)$ may be described by the electric fields at time t, the $\chi^{(3)}$ being time-independent:

$$\tilde{P}_\ell^{NL}(t) = \sum_{m,n,o}^{3} \tilde{\chi}^{(3)}_{lmno} E_m(t) E_n(t) E_o(t) \tag{28b}$$

The subscripts l, m, n, o denote the vector components in the x, y and z directions. Eqs. 28 emphasize the point of view of a parametric light-light interaction. Certain microscopic aspects, e.g. the excited-state population, are not evident. It is convenient to describe the nonlinear polarization and the electric field by their frequency components. For a set of discrete frequencies ω_p we write:

$$E_\ell(t) = \frac{1}{2}\sum_p A_\ell(\omega_p)\exp(-i\omega_p t) + c.c.$$
$$\tilde{P}_\ell^{NL}(t) = \frac{1}{2}\sum_p P_\ell^{NL}(\omega_p)\exp(-i\omega_p t) + c.c. \tag{29}$$

Using the relation $\omega_\delta=\omega_\alpha+\omega_\beta+\omega_\gamma$ we find for the frequency component $P_\ell^{NL}(\omega_\delta)$ from Eqs. 28 and 29

$$P_\ell^{NL}(\omega_\delta) = \sum_{m,n,o} \chi^{(3)}_{lmno}(-\omega_\delta;\omega_\alpha,\omega_\beta,\omega_\gamma)E_m(\omega_\alpha)E_n(\omega_\beta)E_o(\omega_\gamma) \tag{28c}$$

The ordering of the electric field components $E(\omega_\alpha)$ to $E(\omega_\gamma)$ in Eq. 28c is arbitrary; therefore the frequencies may be interchanged in the argument of $\chi^{(3)}$. The frequency values of the photons which are annihilated in the nonlinear process conventionally enter the argument of $\chi^{(3)}$ with positive signs; the generated frequencies appear with negative signs. The Stokes Raman scattering is described in the formalism by $\chi^{(3)}(-\omega_S; \omega_L, -\omega_L, \omega_S)$.

The frequency dependence of $\chi^{(3)}$ is complicated in general and has been discussed extensively in the literature. |16,17| The magnitude of $\chi^{(3)}$ for Stokes scattering is resonantly enhanced if the frequency difference $\omega_L-\omega_S$ coincides with an eigenfrequency of the physical system. In the resonance region, $\chi^{(3)}$ may be split into two components: a resonant contribution χ_{res} which is strongly frequency dependent and a non-resonant contribution χ_{NR} due to all other transition frequencies which are far off-resonance. We therefore write

$$\chi^{(3)} = \chi_{res} + \chi_{NR} \tag{30}$$

and accordingly

$$\tilde{P}_S^{NL} = \tilde{P}_{res} + \tilde{P}_{NR} \tag{31a}$$

where

$$\tilde{P}_{res} = \chi_{res}\, EEE; \quad \tilde{P}_{NR} = \chi_{NR}\, EEE \tag{31b}$$

In the following we neglect the tensor character of $\chi^{(3)}$. This is equivalent to treating $\chi^{(3)}$ as a scalar. Comparison with Eq. 21b yields

$$\tilde{P}_{res} = N \frac{\partial\alpha}{\partial q} \langle q \rangle\, E = \chi_{res}\, EEE \tag{31c}$$

Substitution of Eqs. 22 and 23a for the stationary case considered here $(\partial Q/\partial t \approx 0)$ leads to | 3 |:

$$\chi^{(3)}(-\omega_S; \omega_L, -\omega_L, \omega_S) =$$
$$= \frac{1}{4m}\left(\frac{\partial\alpha}{\partial q}\right)^2 \frac{N(1-\bar{n})}{\omega_o^2-(\omega_L-\omega_S)^2+2i(\omega_L-\omega_S)/T_2} + \chi_{NR} \tag{32}$$

The first part on the right hand side of Eq. 32 denotes the resonant part. $\chi^{(3)}$ is complex in general, $\chi^{(3)} = \chi' - i\chi''$. The frequency dependence is depicted in Fig. 6. The nonresonant susceptibility χ_{NR} is real in most practical cases and only slightly frequency dependent. At resonance, $\omega_L, -\omega_S = \omega_o$, the imaginary part χ''_{res} has its extremum while $\chi'_{res} = 0$.

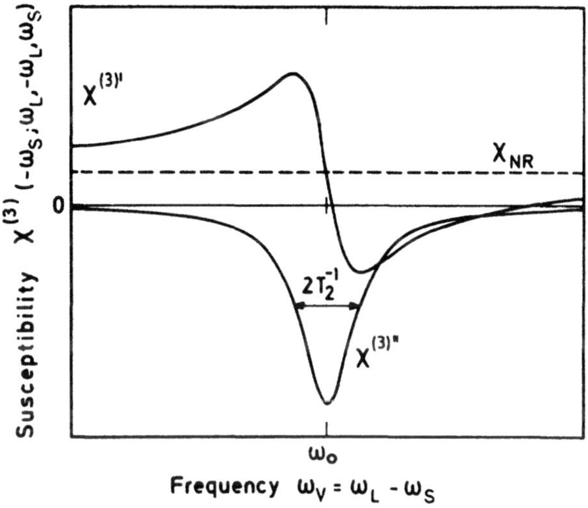

Fig. 6: Resonance structure of the nonlinear susceptibility for stimulated Stokes generation.

Only the imaginary part of $\chi^{(3)}$ leads to Stokes amplification. The real part represents intensity-dependent changes of refractive index which are responsible for frequency modulation and self-focussing effects. |3| Inserting Eqs. 31 in Eq. 21a leads to an expression analogous to Eq. 24. The result is again exponential Stokes amplification (note Eqs. 25) with gain factor

$$g(\omega_S) = - \frac{32 \pi^2 \omega_S}{\mu_S \mu_L c^2} \chi''(-\omega_S; \omega_L, -\omega_L, \omega_S) \qquad (33)$$

Eq. 33 demonstrates that stimulated scattering may be discussed in terms of $\chi^{(3)}$ without introducing a microscopic picture of the light-matter interaction.

V. TRANSIENT STIMULATED SCATTERING

The transient situation deserves special attention since it enables us to determine directly important material parameters of the molecular transition under investigation. |9,18-20| In particular, experimental techniques have been devised to study the dephasing time T_2 and the energy relaxation time T_1 in the liquid state. |21,22| We will restrict the discussion to forward scattering in the resonant case ($\omega_S = \omega_L - \omega_o$) with small conversion. Transforming Eqs. 23 to the retarded time frame, t' = t-x/v, x' = x and neglecting group dispersion we obtain

$$\left(\frac{\partial}{\partial t} + \frac{1}{T_2}\right) Q^*(x',t') = -i\kappa_1 A_L^* A_S$$

$$\frac{\partial}{\partial x'} A_S(x',t') = i\kappa_2 A_L Q^* \tag{34}$$

$$\kappa_1 = \frac{1-2\bar{n}}{4m(\omega_L-\omega_S)} \frac{\partial \alpha}{\partial q}; \; \kappa_2 = \frac{\pi \omega_S}{c\mu_S} N \frac{\partial \alpha}{\partial q}$$

In Eqs. 34 we have neglected the small population changes setting $n \simeq \bar{n}$. Since pump depletion (and also linear losses) are neglected we have $E_L = E_L(t')$. Eqs. 34 may be integrated using Riemann's method. Solutions have been published containing modified Bessel functions I_0 and I_1 of zeroth and first order. |19,20| Qualitatively speaking, transient scattering has similar properties as stationary scattering in many respects. Important points are:

(i) Time is required for the material to respond to the light fields making the scattering less efficient; the latter fact may be compensated for by higher pump intensities.

(ii) Material excitation still survives when the external light field has been switched off.

An example for the increasing intensity requirements of transient scattering is shown in Fig. 7. A (high gain) Raman generator with constant Stokes efficiency η_S is considered for Gaussian laser pulses of duration t_p. The required gain product $g_S I_L x$ for η_S = 0.1 % and 1 % is plotted as a function of the ratio t_p/T_2. For decreasing values of t_p/T_2 the transient character of the stimulated Raman process becomes more dominant and larger values of $g_S I_L x$ are necessary to achieve the same efficiency η_S.

The generated Stokes pulse is shortened and slightly delayed with respect to the incident laser pulse. |19| An example for the time evolution of the material excitation is shown in Fig. 8. The coherent excitation Q^2 is plotted as a function of time for three parameters t_p/T_2. The ordinate is in units of $\hbar/m\omega_o$ and refers to a typical experimental situation with x = 1 cm, η_S = 1 %, and Gaussian laser pulses. The coherent excitation grows very fast on account of the rapid onset of stimulated scattering. When the pumping field has terminated, the vibrational excitation relaxes freely. The exponential decay of the Q^2-curves directly gives the relaxation time $T_2/2$. The transient material excitation may be directly observed by coherent Raman scattering of delayed probing pulses. |21,22| A similar time dependence is evaluated (Eq. 23d)

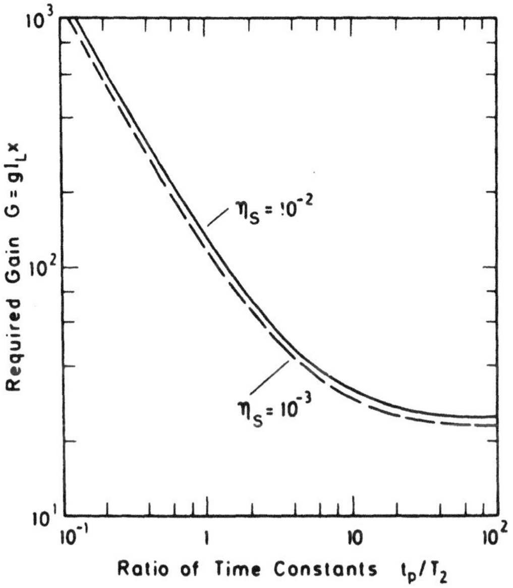

Fig. 7: Required gain product $g_S I_L x$ versus pulse duration t_p for constant energy conversion efficiency.

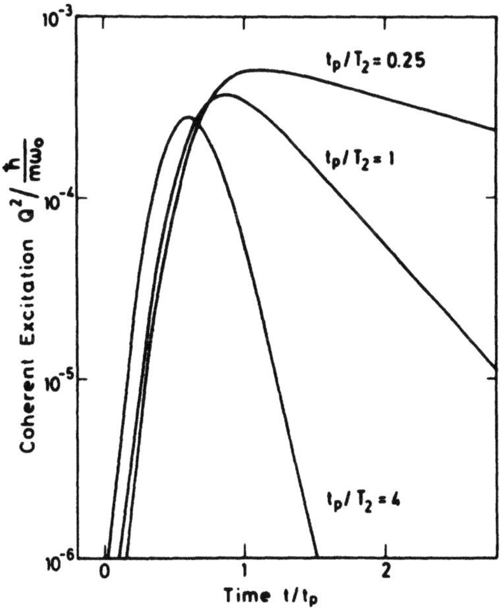

Fig. 8: Coherent vibrational excitation versus time for three values of the parameter t_p/T_2.

for the excess population $n-\bar{n}$ which decays with population lifetime T_1. |9|

The decay of the material excitation in Fig. 8 represents the Raman analogue to free induction decay for an electric (or magnetic) dipole transition. In the terminology of coherent effects the Stokes pulse propagation discussed here represents small area pulse propagation. Additional coherent phenomena occur in the transient case at high conversion efficiencies and for large population changes. |23,24| A major difficulty is the weakness of the Raman coupling. The observation of a Raman echo was recently reported for a spin-flip transition in CdS at 1.6 K. |25|

VI. SIMULTANEOUS ANTI-STOKES AND HIGHER ORDER STOKES GENERATION

The discussion given above has been restricted to the presence of two frequency components of the electromagnetic field. At high scattering efficiencies additional frequency components show up in the spectrum. Only a few remarks shall be given here on these more complicated situations. The generation of new Stokes components shifted by multiples of the resonance frequency ω_o to lower frequencies is easily understood as a multiple step process. |3,26| After the generation of intense first-Stokes light, this component in turn serves as a pump source generating 2nd Stokes emission and so on. These processes are facilitated by the presence of the coherent excitation <q> which provides an efficient initial condition for stimulated amplification by coherent Stokes Raman scattering. In other words, higher order Stokes production need not develop from quantum noise.

Of special interest is the anti-Stokes generation in the frequency region $\omega_a = \omega_L + \omega_o$. It has been mentioned above (Section II) that laser-anti-Stokes coupling leads to stimulated losses of the ω_a component (provided that the molecular transition is not population-inverted). So, anti-Stokes generation in the general case is possible only via Stokes-anti-Stokes coupling. In fact, producing a coherent excitation <q> by the Stokes process allows anti-Stokes generation via coherent Raman scattering off the material excitation |11| represented by

$$\frac{\partial A_a}{\partial x} + \frac{\mu_a}{c}\frac{\partial A_a}{\partial t} = i\,\frac{\pi\,\omega_a}{\mu_a c}\,N\,\frac{\partial \alpha}{\partial q}\,A_L Q\,\exp(i\Delta k_a x) \tag{35}$$

Eq. 35 is the anti-Stokes analogue to Eq. 23b and is derived from the nonlinear wave equation in the same way as for the Stokes process. Since the anti-Stokes process is the inverse process, the molecular transition is not pumped but de-excited by the generation of ω_a-photons. As a consequence, Eq. 23a for the amplitude Q is modified by an additional term on the r.h.s. introducing loss via anti-Stokes scattering |3|

$$\frac{\partial Q}{\partial t} + \frac{i}{2\omega_v}\left(\omega_o^2 - \omega_v^2 - i\,\frac{2\omega_v}{T_2}\right) Q = \tag{36}$$

$$= \frac{i}{4m\omega_v}\,\frac{\partial\alpha}{\partial q}\,(1-2n)\left[A_L A_S^* + A_a A_L^* \exp(i\Delta k_a x)\right]$$

The factor Δk_a in Eqs. 35 and 36 denotes the k-mismatch between Stokes and anti-Stokes scattering, $\Delta k_a = k_S + k_a - 2k_L$, for a collinear geometry. For $k_a = 0$, the anti-Stokes process totally inhibits the stimulated build-up of the coherent amplitude Q and no amplification of Stokes and anti-Stokes light occurs; i.e. the two wave-picture considered above does not apply, Stokes-anti-Stokes coupling must not be omitted. For real systems, we have $\Delta k_a \neq 0$ due to dispersion of the optical refractive index. For off-axis geometries, the value of Δk_a also depends on emission angles (particularly important for gases). The situation is illustrated by Fig. 9 where the Raman amplification for Stokes and anti-Stokes light is plotted versus mismatch Δk_a. |27|

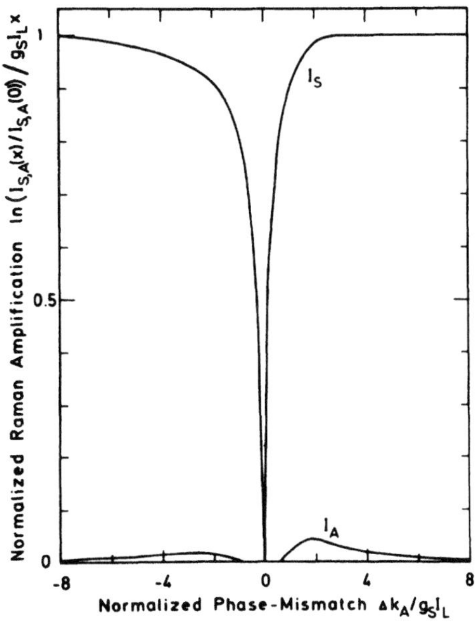

Fig. 9: Stimulated Stokes and anti-Stokes amplification versus k-mismatch Δk_a (after Ref. 27).

The Fig. shows that no amplification occurs for $\Delta k_a = 0$ while for $|\Delta k_a \ell| > 2g_S I_L$ the deviations from the two-wave-picture of Sections II-V are negligible. For the Raman generator ($g_S I_L \ell \simeq 25$) the required mismatch $\Delta k_a > 50/\ell$ is readily fulfilled in liquids and solids. One has $\Delta k_a > 10^2\ cm^2$ in many practical cases and a cell length of several cm to tens of cm, so that the theory of Sections II-V applies.

In gases the situation has to be carefully chequed. A non-collinear geometry may be used to suppress anti-Stokes coupling (important for Raman gain spectroscopy). For backward scattering the mismatch Δk_a is very large in all cases and the effect of anti-Stokes coupling negligible.

VII. CONCLUDING REMARKS

In conclusion we briefly point to several important applications of stimulated scattering. The stationary stimulated process is used in various nonlinear Raman techniques, e.g. CARS, RIKES, gain spectroscopy, inverse Raman scattering, which are discussed in detail by several chapters of this volume. Transient stimulated scattering is highly valuable as an excitation process for time-resolved investigations of elementary excitations. |28|

The high conversion efficiency of several 10 per cent makes stimulated Raman scattering a versatile tool for the generation of intense radiation at new light frequencies. For example, a cw Raman oscillator was devised with a 100 m long fused silica fibre pumped by an argon laser which generated intense light tunable over 300 cm^{-1}. |29| A quasi-cw Raman oscillator with high pressure H_2-gas pumped by a tunable dye laser is a versatile light source for coherent light in visible to the far infrared; the long-wavelength ($\sim 100\mu m$) range is reached by efficient 4th order Stokes generation. |30,31| Tunable IR radiation may be also produced by spin-flip Raman scattering in InSb pumped by a CO- or CO_2-laser. |32,33|

The high nonlinearity of the process allows the generation of shorter pulses. Using backward scattering in a Raman generator with nanosecond pump pulses, picosecond pulses were generated. |34| Sub-picosecond pulses have been produced recently in a synchronously coupled Raman oscillator excited by a train of picosecond pulses. |35|

This short list of applications demonstrates the impressive progress made in the field of stimulated Raman scattering during the past two decades.

REFERENCES

1. Göppert-Mayer, M., Ann. Physik 9, 273 (1931).

2. Woodbury, E.J., and Ng,W.K., Proc. IRE 50, 2367 (1962).

3. For a review, see Penzkofer, A., Laubereau,A. and Kaiser, W., Prog. Quant. Electr. 6, 55 (1979).

4. Hellwarth, R.W., Phys. Rev. 130, 1850 (1963); Grob, K., Z. Physik, 184, 395 (1965); Bloembergen, N., Ann. J. Phys. 35, 989 (1967).

5. Wang, C.S., in"Quantum Electronics", Vol. 1, edited by H. Rabin and C.L. Tang, Academic Press, New York, 1975.

6. Placzek, G., in "Handbuch der Radiologie", edited by E. Marx, Akademische Verlagsgesellschaft, Leipzig, 1934.

7. Grun, J.B.,McQuillan, A.K.,and Stoicheff,B.P., Phys. Rev. 180, 61 (1969).

8. Jones, W.J. and Stoicheff,B.P., Phys. Rev. Lett. 13, 657 (1964).

9. For a review, see Laubereau, A., and Kaiser, W., Rev. mod. Phys. 50, 3607 (1978).

10. Redfield, A.G., IBM J. Res. Dev. 1, 19 (1957); Slichter, C.P., "Principles of Magnetic Resonance" (Harper and Row, New York, 1963).

11. Giordmaine, J.A., and Kaiser,W., Phys. Rev. 144, 676 (1966); Maier, M., Kaiser,W., Giordmaine, J.A., Phys. Rev. 177, 380 (1969).

12. Bloembergen, N., in "Nonlinear Optics" (Benjamin, New York, 1965).

13. Lallemand, P., Simova, P., Bret, G. , Phys. Rev. Lett. 17, 1239 (1966).

14. Görner, H., Maier, M., Kaiser, W., J. Raman Spectr. 2, 363 (1974).

15. McQuillan, A.K., Clements, W.R.L., and Stoicheff, B.P., Phys. Rev. A1, 628 (1970).

16. Armstrong, J.A., Bloembergen, N., Ducuing, J., Pershan, P.S., Phys. Rev. 127, 1918 (1962);
Flytzanis, C., in "Quantum Electronics", Vol. 1, edited by H. Rabin and C.L. Tang, Academic Press, New York, 1975.

17. Ducuing, J., in "Quantum Optics", edited by R.J. Glauber (Academic Press, New York, 1969).

18. Wang, C.S., Phys. Rev. 182, 482 (1969).

19. Carman, R.L., Shimizu, F., Wang, C.S., and Bloembergen, N., Phys. Rev. A2, 60 (1970).

20. Akhmanov, S.A., Chirkin, A.S., Brabovich, K.N., Kovrigin, A.I., Khokhlov, R.V., and Sukhorukov, A.P., IEEE J. Quant. Electr. QE-4, 598 (1968);
Akhmanov, S.A., Mat. Res. Bull. 4, 455 (1969);
Akhmanov, S.A., Drabovich, K.N., Sukhorukov, A.P., and Chirkin,A.S., Soviet Phys. JETP 32, 266 (1971).

21. von der Linde, D., Laubereau, A., and Kaiser, W., Phys. Rev. Lett. 26, 954 (1971);
Alfano, R.R., and Shapiro,S.L., Phys. Rev. Lett. 26, 1247 (1971).

22. Laubereau, A., von der Linde, D., and Kaiser, W., Phys. Rev. Lett. 28, 1162 (1972).

23. Lowdermilk, W.H., and Kachen,G.I., Opt. Commun. 18, 68 (1976).

24. Tanno, N., Shirahata, T., Jokota, K., and Inaka, H., Phys. Rev. A12, 189 (1975);
Hoshimiya, T., Inaba, H., and Tanno, N., Opt. and Quant. Electr. 9, 448 (1977).

25. Hu, P., Geschwind, S., and Jedju, T.M., Phys. Rev. Lett. 20, 1357 (1976).

26. Chiao, R., and Stoicheff, B.P., Phys. Rev. Lett. 12, 290 (1964).

27. Shen, Y.R., and Bloembergen, N., Phys. Rev. 137A, 1787 (1965).

28. See Chapter by Kaiser, W., in this volume.

29. Stolen, R.H., Ippen, E.P., and Tynes, A.R., Appl. Phys. Lett. 20, 62 (1972).

30. Schmidt, W., and Appt, W., IEEE J. Quant. Electr. QE-10, 792 (1974);
Kuhl, J., and Schmidt, W., Appl. Phys. 3, 251 (1974).

31. Ambartsumyan, R.V., Apatin, V.M., and Letokhov, V.S., Soviet Phys. JETP. Lett. 15, 237 (1972);
Grasyuk, A.Z., Zubarev, I.G., Kotov, A.V., Mikhailov, S.I., and Smirnov, V.G., Opt. Commun. 18, 211 (1976).

32. Patel, C.K.N., and Shaw, E.D., Phys. Rev. Lett. 24, 451 (1970);
Patel, C.K.N., Shaw, E.D., and Kerl, R.J., Phys. Rev. Lett. 25, 8 (1970).

33. For a review, see Smith, S.D., Dennis, R.B., and Harrison, R.G., Prog. Quant. Electr. 5, 205 (1977).

34. Maier, M., Kaiser, W., and Giordmaine, J.A., Phys. Rev. Lett. 17, 1275 (1966);
Culver, W.H., Vanderslice, J.T.A., and Townsend, V.W.T., Appl. Phys. Lett. 12, 189 (1968).

35. Colles, M.J., Appl. Phys. Lett. 19, 23 (1971).

RAMAN INVESTIGATIONS OF POLYATOMIC MOLECULES IN THE LIQUID STATE USING PICOSECOND LIGHT PULSES
I. DIRECT OBSERVATION OF THE POPULATION LIFETIME T_1 AND THE DEPHASING TIME T_2 OF VIBRATIONAL MODES

W. Kaiser and W. Zinth

Physik Department der
Technischen Universität München
München, West Germany

1. INTRODUCTION

With modern Laser-Raman spectrometers it is readily possible to measure Raman transitions with high accuracy. In spite of substantial progress in recent years basic problems remain. First of all, it is difficult to separate the Raman band of interest from the background made up of other neighboring or weak underlying Raman transitions. In polyatomic molecules there exist numerous overtones and combination modes which contribute - possibly enhanced by Fermi resonance - to the observed scattered intensity. Additional weaker transitions may be burried under the main Raman bands giving rise to distortion of the band profile.

Assuming the line shape being well established serious difficulties are encountered when one tries to interprete the physical origin of the Raman band-width. In general, there are several mechanisms of line broadening which may or may not contribute to the observed Raman band. We point to the collision induced dephasing, to the energy relaxation of the excited vibrational state, to the possible isotopic multiplicity of the transition, and finally, to inhomogeneous broadening due to varying molecular surrounding.

In recent years we succeeded in measuring directly the population life-time T_1 and the dephasing time T_2 of various vibrational modes in the time domain without resort to the transition band-width /1-5/. For polyatomic molecules in the liquid state at 300 K both time constants T_1 and T_2 are very short, frequently of the order of picoseconds. As a consequence, we had to

W. Kiefer and D. A. Long (eds.), Non-Linear Raman Spectroscopy and Its Chemical Applications, 211–225.

design experimental systems which enable us to measure vibrational time constants on the time scale of 10^{-12} s. In this summary we introduce two experimental techniques and present some examples to illustrate the state of our knowledge. The interested reader is referred to the original literature cited at the end of this article.

2. THE POPULATION LIFE-TIME T_1

The population life-time T_1 of a specific vibrational state is measured as follows. A first ultrashort infrared pulse tuned to the resonance frequency of the investigated transition excites molecules by direct IR absorption and a second visible pulse of variable delay time monitors the instantaneous population of the excited molecules. The spontaneous anti-Stokes Raman signal is used as a direct measure of the excess population. This measuring technique has the great advantage of providing very direct information of the populated vibrational mode. The observation of various anti-Stokes frequencies tells us accurately which vibrational mode is populated at a certain time. In this way we are able to observe the time dependence of the primary excitation and follow the vibrational energy through different decay channels. On the other hand, the T_1 technique has one major shortcoming. The Raman scattering cross-sections notoriously are extremely small and the population densities achieved with the short excitation pulse are only of the order of $\leq 1\%$. As a result, the number of scattered photons is very limited, approximately 10^2 to 10^3 per pulse, making the data acquisition a non-trivial task /4/.

A. Experimental System to Measure T_1

In Fig. 1, we show schematically an experimental system. Two pulses varying widely in frequency are generated. The tunable infrared pulse at ν_1 is made via an optical three-photon process in three $LiNbO_3$ crystals /9/. Typical values of the IR pulses are 10^{15} photons per pulse, spectral band width $\Delta\tilde{\nu}_{IR} \sim 20$ cm^{-1}, and beam divergence of $\sim 10^{-3}$ mrad. The monitoring green pulse is produced at the second harmonic frequency $\tilde{\nu}_2 = 18{,}990$ cm^{-1} in a nonlinear KDP crystal. The two pulses are focussed - with a proper delay time - onto a small sample which is located at the focus of an ellipsoidal mirror. Approximately 50% of the anti-Stokes scattered photons are collected by this mirror and imaged at the entrance slit of a spectrometer which is connected to a photomultiplier. We point to a thin $LiNbO_3$ crystal (middle, r.h.s.) in Fig. 1, where the two input pulses of frequencies ν_1 and ν_2 are optically mixed to form the sum frequency at the different delay settings. In this way we observe the cross-correlation function in each experimental run. The peak of the cross-correlation function marks the zero point of our time scale (where the maxima of both pulses overlap) and the width of the cross-correlation curve gives

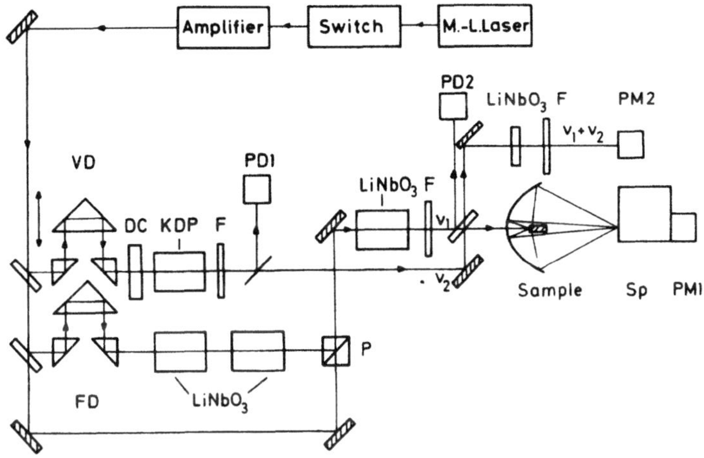

Fig. 1 Experimental system to measure the population life-time T_1 of vibrational modes. The exciting infrared pulse ν_1 is generated by the optical parametric process in three $LiNbO_3$ crystals. The probe pulse at the second harmonic frequency ν_2 is prepared with the help of a KDP crystal. An elipsoidal mirror collects the scattered anti-Stokes photons. The small $LiNbO_3$ crystal at the r.h.s. serves to measure the cross-correlation curve of the two pulses at the frequencies ν_1 and ν_2.

us valuable information on the time resolution of the individual experiment.

B. Results and Discussion of T_1 Data

We have investigated numerous molecules and found widely varying values of the population life-times between 1 ps (the time resolution of our system) and 200 ps in polyatomic molecules at room temperature /1,2,4/. Special attention was paid to the CH-stretching modes in the frequency range of 3000 ± 100 cm^{-1} where direct pumping by IR pulses is readily possible. At the beginning of our investigations the large spread in T_1 values was quite puzzling. The T_1 results appeared to be very sensitive to the specific level structure of the individual molecule. During the past years we have made considerable progress in our understanding of the physical phenomena which determine the population life-time in polyatomic molecules. Two processes should be distinguished. (i) Very rapid exchange of vibrational energy occurs between similar fundamental vibrations of near energies. For example, we have experimental evidence that collision-induced interaction leads to rapid energy exchange ($\leq$ 1 ps) between the symmetric and the asymmetric stretching modes of the CH_2 group. As a result, during the excitation process with pulses of 5 ps duration we populate both vibrational levels. Similarly, we see very fast energy exchange ($\leq$ 1 ps) between infrared and Raman active modes within the CH fundamentals. This observation has consequences for the decay process which starts - after the pump pulse has left the medium - from a reservoir of two or more energy states. (ii) Vibrational energy is transferred from the CH-stretching modes ($\sim$ 3000 cm^{-1}) via overtones and combination modes to lower energy states. Intramolecular anharmonic coupling, the Fermi resonance, between neighboring energy states is of major importance in this respect. The degree of Fermi resonance in a certain frequency range manifests itself in the infrared and Raman spectra. Overtones and higher order combination modes borrow intensity from CH-stretching modes. We define as a measure of Fermi resonance mixing the intensity ratio, R, between the final and initial state taken from the infrared or Raman spectrum. In a recent publication a formula was derived which allows to estimate the life time T_1 of vibrational states /1/:

$$T_1 = N(1+R)^2 R^{-1} \exp(\omega/\Omega)^{2/3} T_2(f) \qquad (1)$$

N corresponds to the number of states initially excited, R is a measure of the Fermi resonance, and $T_2(f)$ stands for the dephasing time of the final state. $T_2(f)$ may be estimated from the Raman line-width $\Delta\tilde{\nu}$ as $T_2(f) = (2\pi c\Delta\tilde{\nu})^{-1}$ ($T_2(f)$ is equal to $T_2/2$ measured in coherent Raman experiments) /3/. The frequency ω represents the energy difference between the initial and the final state. Ω has a value close to 100 cm^{-1}. In some cases it is difficult to determine the $T_2(f)$ value on account of a complex band structure. Another difficulty arises for very strong vibrational coupling. Their Fermi-resonance interaction may become so strong that energy transfer occurs very fast, i.e., within the time of the excitation pulse.

As an example we present data of the two molecules CH_2CCl_2 and trans CHClCHCl, the IR-absorption spectra and Raman spectra of which are shown in Fig. 2 and 3 /4/. At first we are concerned with the frequency range from 2950 to 3250 cm^{-1}.

In Fig. 2 we see the symmetric (ν_1) and asymmetric (ν_7) CH_2-stretching vibrations of $CH_2{=}CCl_2$ in strong Fermi resonance with combination modes. In particular, there is strong Fermi resonance between ν_1 and $\nu_2+\nu_3$, both of A_1 symmetry, and between ν_1 and $\nu_2+\nu_6+\nu_{11}$, both of B_1 symmetry /4/. This observation suggests that we have to consider two decay channels for the CH_2-stretching modes. For the decay $\nu_1 \rightarrow \nu_2+\nu_3$ we estimate the intensity ratio $R = 0.2\pm0.05$ and calculate $T_2(f) = 0.3$ ps from the Raman line-width of $\Delta\tilde{\nu} = 17$ cm^{-1}. With N = 2 (two CH_2 modes) and $\omega = 40$ cm^{-1} we obtain from Eq. 1 a value of $T_1 = 7.8\pm3$ ps.

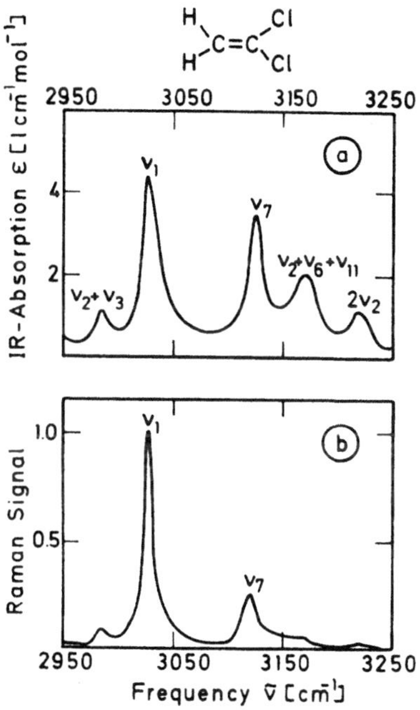

Fig. 2 Infrared absorption (a) and Raman (b) spectra of CH_2CCl_2 between 3950 and 3250 cm^{-1}. Two combination tones are in strong Fermi resonance with the two CH-stretching modes at ν_1 and ν_7.

For the second decay channel $\nu_7 \rightarrow \nu_2+\nu_6+\nu_{11}$ we take the numbers $R = 0.6\pm0.1$, $T_2(f) = 0.3$ ps, $N = 2$, and $\omega = 45$ cm^{-1} and calculate $T_1 = 4.5\pm1$ ps. Adding the two decay rates gives an estimate for the relaxation time for the two CH_2-stretching modes of $T_1 = 2.9\pm0.6$ ps.

In Fig. 4a we present experimental data of the direct determination of the T_1 value of the CH_2-stretching modes. The ν_1 mode at 3036 cm^{-1} is resonantly excited by an infrared pulse and the anti-Stokes scattering of the same ν_1 mode is observed. The scattered Raman signal rises to a slightly delayed ($\sim$ 1.8 ps) maximum during the excitation process and decays with a relaxation time of $T_1 = 3\pm1$ ps. This experimental finding is in excellent agreement with the calculated value for T_1 discussed in the preceding paragraph. The broken curves in Fig. 4 are cross-correlation curves. As pointed out above, they determine the zero point on the time axis and give a good indication of the time resolution of the experiment.

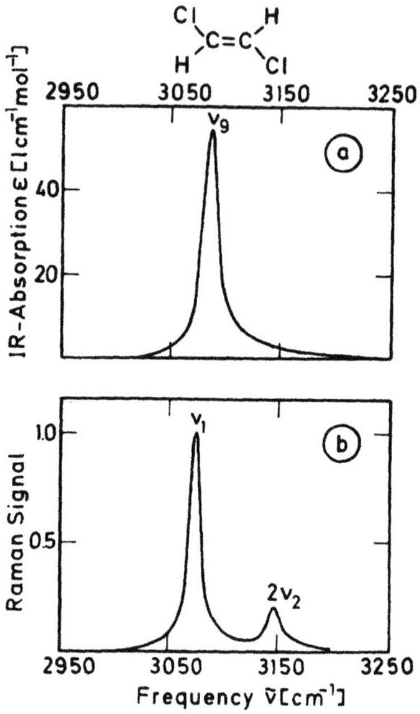

Fig. 3 Infrared absoption (a) and Raman (b) spectra of trans CHClCHCl. There is less Fermi resonance mixing than in Fig. 2.

In Fig. 3a we see the infrared active CH-stretching mode ν_9 and in Fig. 3b the Raman active symmetric ν_1 mode of trans CHCl=CHCl. Here we find a considerably smaller Fermi resonance. The Raman spectrum of Fig. 3b suggests some anharmonic coupling between ν_1 and $2\nu_2$, both of A_g symmetry. With the values $R = 0.15 \pm 0.02$, $T_2 = 0.3$ ps, $N = 2$, and $\omega = 80\ cm^{-1}$ we calculate $T_1 = 13$ ps. It should be noted that there might be additional weak Fermi resonance between the ν_9 mode and higher combination modes (e.g., $\nu_2+\nu_5+\nu_{10}$) burried under the high frequency tail of the ν_9 fundamental. These additional decay channels may reduce somewhat the estimated T_1 value.

The time dependence of the CH-stretching modes of trans CHCl=CHCl is depicted in Fig. 4b. The molecule is excited via the ν_9 mode at 3084 cm^{-1} and the population of the ν_1 mode at 3073 cm^{-1} is monitored by anti-Stokes Raman scattering. The rapid rise of the Raman signal, i.e., the fast population of the ν_1 mode, gives clear evidence of the quick energy exchange between the two CH fundamentals ν_1 and ν_9. The delayed maximum (3.7 ps) and the slow decay of the signal curve suggest a long lifetime of the two CH-stretching modes of $T_1 = 10 \pm 2$ ps. This number is in good agreement

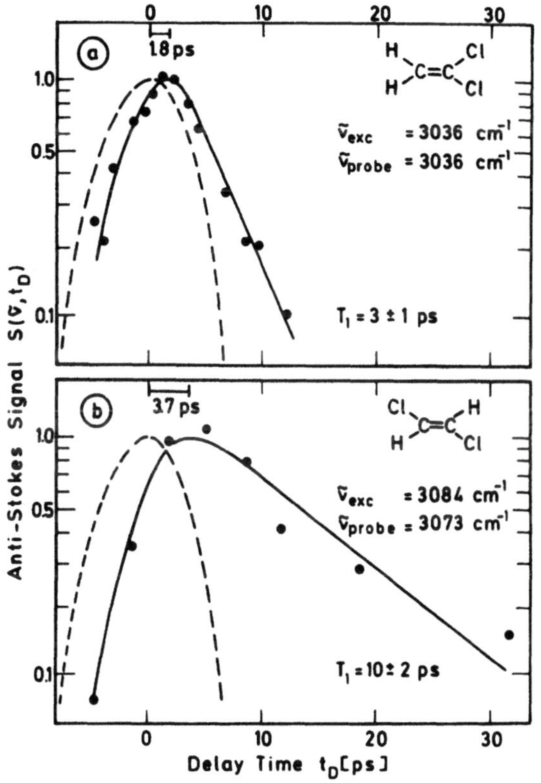

Fig. 4 Anti-Stokes scattering signal versus delay time of the probing pulse. (a) CH_2CCl_2 in CCl_4 (c = 0.35 m.f.). The decay of the CH_2-stretching mode at 3036 cm^{-1} is shown. (b) trans CHClCHCl in CCl_4 (c = 0.35 m.f.). The CH-stretching mode at 3084 cm^{-1} is excited and the mode at 3073 cm^{-1} is monitored. The broken curves are the cross-correlation functions of the IR exciting and green probing pulses.

with the value estimated above. The small intramolecular coupling gives rise to the longer vibrational life time.

In the preceding sections we have given strong arguments that the initially excited CH-stretching modes interact with well defined overtones or combination modes. The question now arises whether we are able to see the population and depopulation of the subsequent energy states. The answer is positive. Under favorable conditions, i.e., for subsequent states with sufficiently high Raman scattering cross-sections and with long life times, we indeed see the delayed population of a lower vibrational level.

As an example we discuss our observations of the molecule $CH_2{=}CCl_2$. We pointed out above that the two CH_2-stretching modes decay via the combination mode $\nu_2+\nu_3$ and $\nu_2+\nu_6+\nu_{11}$. Both combination modes contain the infrared and Raman active CH-bending mode at $\nu_2 = 1616\ cm^{-1}$. We have searched for the anti-Stokes Raman signal of this bending mode and, indeed, found evidence of its population. The scattered ν_2 signal shifted at $1616\ cm^{-1}$ has a maximum delayed by roughly 1 ps relative to the maximum of the ν_1 curve of Fig. 4a and the decay time was approximately equal to the ν_1 data. Our observations suggest a short life time of the ν_2 level of $T_1 < 2$ ps. This estimate is quite consistent with the infrared and Raman spectra of the ν_2 mode where strong coupling with a nearby ($\omega \simeq 50\ cm^{-1}$) overtone $2\nu_9$ is very apparent. As an additional decay route one has to consider the coupling to the ν_3 fundamental which is only $220\ cm^{-1}$ apart from the ν_2 mode.

The results discussed in the preceding paragraph give interesting information of the decay of the ν_2 mode. We do not know, however, whether we probe the combination state or the ν_2 fundamental. In the first case, the total vibrational energy of the molecule is approximately $3000\ cm^{-1}$, in the second case, the total energy is $1616\ cm^{-1}$ and the energy difference is resonantly transferred to neighboring molecules.

3. DEPHASING TIME T_2

In our experiments vibrational modes are coherently excited via the stimulated Raman effect and the time dependence of the coherent vibrational excitation Q is monitored by a delayed and properly matched probe pulse.

A. Experimental System to Measure T_2

There are two ways to generate a short coherent excitation. (i) An ultrashort intensive and bandwidth limited laser pulse traverses the sample and builds up a coherent excitation via transient stimulated Raman scattering /6,8/. In this case only one vibration, the transition with the highest Stokes gain, becomes excited. In Fig. 5a an experimental system is depicted schematically. The laser pulse (e.g., a frequency doubled pulse at $\tilde{\nu} = 18{,}990\ cm^{-1}$) passes through the polarizer P1 and the sample, but the generated (parallel polarized) Stokes light is blocked by P2. The beamsplitter BS produces the time delayed probe pulse which is rotated by 90°. The probe pulse travels collinearly with the exciting pulse through the sample. The coherently scattered Stokes light of the probe pulse passes the polarizer P2 and is measured with the photomultiplier PM. The system of Fig. 5a has the advantage of simplicity but lacks the flexibility to study

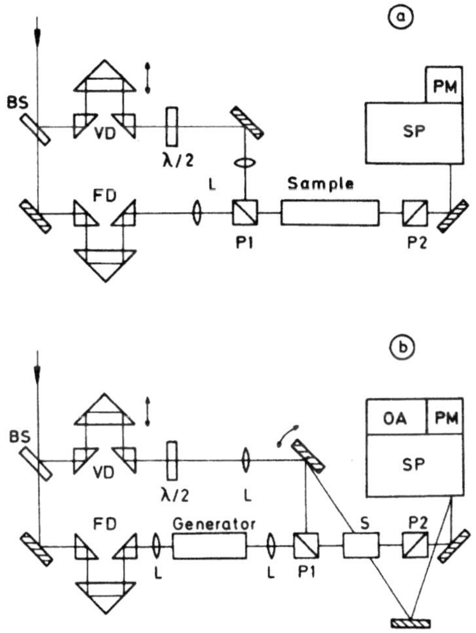

Fig. 5 a and b Schematics of two experimental systems used to study coherent probe scattering in the time and frequency domain. Beam splitter BS, variable and fixed delay VD and FD, wave plate λ/2, lenses L, polarizers P1 and P2 (crossed), samples S, spectrometers SP, photomultipliers PM, and optical multichannel analyser OA.

different vibrational modes of the same molecule. In addition, high light intensities of the laser pulse are required over a substantial interaction length in order to generate a sufficient coherent vibrational excitation. It has been shown that the intense excitation pulse is able to cause a substantial phase modulation and spectral broadening of the probe pulse via the nonresonant susceptibility $\chi_{NR}^{(3)}$ /5/.

(ii) Of advantage is the coherent excitation of the sample by two pulses of different frequencies, ν_L and ν_S. Considerably lower light intensities are needed and different modes may be investigated. In addition, the frequency dependence of the coherent excitation may be studied and new spectral features are observed. Two experimental systems have been used successfully for the coherent study of molecular vibrations. A first experimental set-up is depicted in Fig. 5b. The two ultrashort pulses at ν_L and ν_S are produced in an extra generator cell via stimulated Raman scattering and the material excitation is studied in the sample cell S. For interrogation at the Stokes frequency the probe pulse is introduced via the polarizer P1; it travels through the sample collinearly with the two exciting pulses. When coherent anti-Stokes scattering is observed the probe pulse crosses the coherently excited volume under a well defined phase-matching angle. The coherent scattering signal is measured as a function of delay time with the help of a photomultiplier at the exit slit of a spectrometer. If one is interested in the spectrum of the scattered signal, a high resolution spectrometer (2 m grating, 0.5 cm^{-1} resolution) in conjunction with an optical multichannel analyser proved to be most informative. More recently /10/, a second experimental system was used to measure dephasing times T_2 of various vibrational modes and molecules. The two ultrashort light pulses were generated by two synchronously pumped tunable dye lasers with a small jitter between the two pulses of less than 2 ps. While the intensities of the individual pulses are very small, the high repetition rate of CW mode-locked laser systems allows the application of sensitive photon counting devices.

B. Results and Discussion of T_2 Data

In this section we report on three physical situations. First, we consider the simplest case where coherent Raman scattering is observed on a single homogeneously broadened vibrational transition. Next, we discuss data obtained on multiple homogeneously broadened transitions and finally, we make some comments on inhomogeneously broadened Raman lines.

1. Single Homogeneously Broadened Transitions. With experimental systems described in Fig. 5a or b it is readily possible to coherently excite individual vibrational transitions of polyatomic molecules in the liquid state. On a semilog plot of the scattered probe signal versus delay time of the probe pulse one obtains an exponential slope which allows to deduce directly the dephasing time T_2. A typical example is presented in Fig. 6 where the symmetric CH_3-stretching mode at 2939 cm^{-1} is investigated. The open circles in the Figure correspond to coherent scattering signals. The pump pulse builds up a coherent excitation around $t_D = 0$ with a slightly delayed maximum. The free relaxation of the coherent amplitude Q may be followed experimentally over several orders of magnitude (four orders in Fig. 6). For the specific vibration of liquid CH_3CCl_3 we find $T_2 = 2.5 \pm 0.3$ ps. This important material parameter should be compared with the time constant deduced from the spontaneous Raman line-width $\Delta\tilde{\nu}_R = 4.3$ cm^{-1}. From $T_2 = 1/\pi c \Delta\tilde{\nu}_R$ we calculate a time value of 2.5 ps which fully agrees with our experimental time measurements. This finding tells us that the spontaneous Raman line-width is fully accounted for by the dephasing process of the vibrational transition. Obviously, other line-broadening factors, i.e., inhomogeneous broadening, appear to be small for this vibrational mode and for this molecular surrounding.

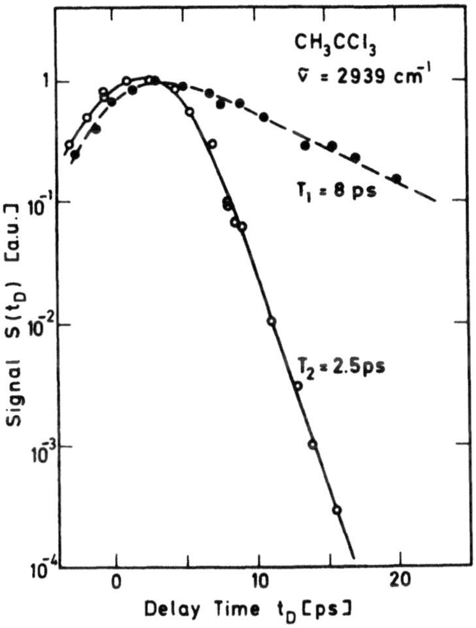

Fig. 6 The symmetric CH_3-stretching mode at ν = 2939 cm^{-1} of CH_3CCl_3 is investigated. Open circles: The coherently scattered signal of the probe pulse versus delay time gives a short dephasing time T_2. Full circles: The incoherently scattered anti-Stokes signal of the interrogating pulse provides the longer population lifetime T_1.

In this connection it is interesting to briefly return to the population life-time T_1 discussed above. In Fig. 6 we show experimental points (full circles) which correspond to a T_1 measurement. The data of the same symmetric CH_3 mode of CH_3CCl_3 suggest a life time of T_1 = 8 ps, much longer than the dephasing time of T_2 = 2.5 ps. These findings tell us that the phase memory is lost much faster than the vibrational energy in this vibrational mode and molecule.

2. Multiple Homogeneously Broadened Transitions. With very short laser pulses several neighboring vibrational transitions may be excited simultaneously via transient stimulated Raman scattering /5/. In the presence of the pump and Stokes pulse all molecules vibrate at the frequency of the driving force $\omega_Q = \omega_L - \omega_S$ and the coherent excitation has the wave vector $k_Q = k_L - k_S$. During the following free relaxation, however, the situation changes substantially. The molecules return to their resonance frequencies ω_j and the different vibrational components vibrate with individual wave vectors k_j. Since the molecules were coherently excited,

they start their free relaxation with fixed phases. Interference occurs between the different components which influences the observation of the decay of the total coherent amplitude.

In the special case of equidistant transition frequencies one finds destructive and constructive interference with an occurrance time of the maxima $t_M = 2\pi/\Delta\omega$, where $\Delta\omega$ is the frequency difference between two neighboring components /5/. From the envelope of the decay curve we may deduce the molecular dephasing time T_2. Liquid CCl_4 represents an ideal medium for the experimental investigation of coherent Raman probe scattering from multiple transitions. The spontaneous Raman spectrum of CCl_4 around 460 cm^{-1} (insert of Fig.7) consists of four peaks with a distance of 3 cm^{-1}. The different transition frequencies are caused by the chlorine isotopes ^{37}Cl and ^{35}Cl in CCl_4. The experiments were performed with the generator set-up of Fig. 1a. We carefully controlled the duration of the laser pulses measuring the autocorrelation function before and after each measurement. The laser intensity was adjusted making the conversion of laser energy into Stokes energy small, approximately 5%. Two representative results are shown in Fig. 7. The coherent scattering signal is plotted as a function of the delay time t_D. Time zero was determined by the peak of the autocorrelation curve with an accuracy of better than 1 ps. The experimental points are averaged over more than 15 individual measurements. The liquid cell has a length of 10 cm. Fig. 7a shows data taken with pulses of a mean duration of 8 ps. The signal curve rises to its maximum at $t_D = 0$ and decays subsequently approximately exponentially with a time constant of close to 3 ps. The small modulation superimposed on the exponential decay is only seen in very careful measurements. Using the same geometry but shorter light pulses with a mean duration of $t_p = 3$ ps, a completely different pattern of the signal curve is found. A pronounced modulation with distinct minima at $t_p = 5$ ps and $t_p = 17$ ps is found in Fig. 7b. The solid curves in Fig. 7a and b are calculated with parameters relevant for CCl_4. We see from Fig. 7 that both calculated curves fit the experimental points very well for $t_D > 10$, i.e., for the time range of free relaxation of the molecular ensemble. The experimental data provide the dephasing time $T_2 = 6.0\pm0.4$ ps and the beating interval gives the frequency difference of two neighboring components, $\Delta\omega/2\pi c = 3.1\pm0.2$ cm^{-1}, in agreement with spontaneous Raman measurements.

Fig. 7 shows for delay times $t_D \lesssim 0$ a considerably larger scattering signal than predicted from the theory of stimulated Raman scattering. Obviously, the additional scattering component is of different physical origin. A clue for the interpretation comes from the following two experimental observations. First, the extra signal is directly related to the laser pulse which peaks at $t_D = 0$. Adding to the solid curve a term proportional to the laser pulse one obtains the broken lines which fit the experimental data

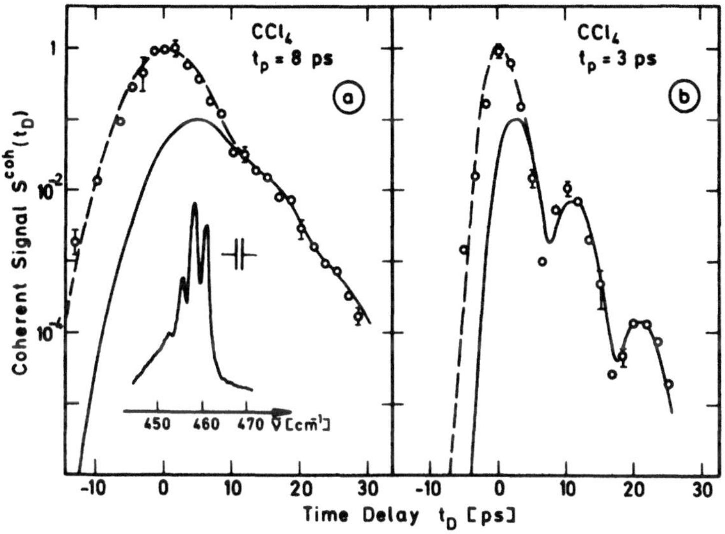

Fig. 7 a and b Experimental results of time resolved coherent Raman probe scattering on the symmetric tetrahedral mode of CCl_4 around 460 cm^{-1}; insert: Raman spectrum. Two different durations of the exciting and probing laser pulses are used: t_p = 8 ps (a) and t_p = 3 ps (b). Both curves allow the determination of the dephasing time T_2. The additional signal near t_D = 0 (broken curve) is attributed to nonresonant light scattering.

perfectly. Second, the spectrum of the scattered signal was measured at t_D = 0 and found quite broad, approximately 30 cm^{-1}. This spectrum differes strongly from the narrow spectra obtained from the delayed scattered pulses with t_D > 5 ps. We strongly feel that the additional scattering signal is generated by parametric four-photon interaction. This process is fast, follows the laser pulse at saturation, and its spectrum is not sharply peaked at the Raman frequency.

3. Inhomogeneously Broadened Raman Lines. In liquids inhomogeneously broadened vibrational transitions are frequently encountered. The interaction of a molecule with various surroundings leads to slight changes of the resonance frequencies. The spontaneus Raman band has no longer a Lorentzian shape and the dephasing time T_2 may not be deduced from the half width of the spontaneous Raman band.

Vibrational transitions with inhomogeneously broadened Raman bands are readily excited with short laser pulses. The stimulated Stokes pulse grows rapidly with the input pulse and the Stokes frequency ω_S corresponding to the maximum of the inhomogeneous distribution. The coherent vibrations are driven during the excitation process at the center frequency $\omega_o = \omega_L - \omega_S$. Neighboring frequencies are excited within a bandwidth $d\Omega$ determined by the band width of the driving force (proportional to $E_L E_S$). In general, $d\Omega$ is determined by the short duration t_{pS} of the Stokes pulse; i.e., $d\Omega \simeq 1/t_{pS}$. When the exciting force vanishes the vibrating molecules return to their resonance frequencies and the coherent amplitudes at each frequency $\omega_o + \Delta\omega$ decay with a dephasing time T_2. Calculating the total coherent amplitude we find a destructive interference between the coherent amplitudes at different frequency positions. The total coherent amplitude drops with a time of approximately $1/d\Omega$; i.e., the coherent amplitude exists for a very short time (approximately for the duration of the short stimulated Stokes pulse). The time dependence of the coherent probe signal reproduces the shape of the probing laser pulse and gives no information on the material excitation.

We note that selective phase matching of a molecular subgroup (suggested previously) cannot be realized in coherent probing experiments /3,5/.

4. CONCLUSIONS

We have shown in this paper that different picosecond techniques allow the direct determination of important molecular parameters. The investigation of the population lifetimes T_1 give valuable information on the dissipation of vibrational energy in the electronic ground state of polyatomic molecules. The importance of anharmonic coupling between vibrational modes is well documented. The coherent excitation of vibrational modes proves to be interesting in various ways. We can directly measure the dephasing time T_2, we find vibrational beating between neighboring states, and we are able to obtain new spectroscopic information by prolonged probing of the coherently molecular ensemble /5,7/. The latter technique will be discussed in the following article.

REFERENCES

1. Fendt, A., Fischer, S.F., and Kaiser, W., Chem. Phys. 57, 55 (1981).

2. Fendt, A., Fischer, S.F., and Kaiser, W., Chem. Phys. Lett. 82, 350 (1981).

3. Laubereau, A. and Kaiser, W., Rev. Mod. Phys. 50, 607 (1978).

4. Kolmeder, C., Zinth, W., and Kaiser W., to be published in Chem. Phys. (1982).

5. Zinth, W., Polland, H.J., Laubereau, A., and Kaiser, W., Appl. Phys. B26, 77 (1981).

6. Zinth, W. and Kaiser, W., Optics Commun. 32, 507 (1980).

7. Zinth, W., Nuss, M.C., and Kaiser, W., to be published in Chem. Phys. Lett. (1982).

8. Zinth, W., Laubereau, A., and Kaiser, W., Optics Commun. 22, 161 (1977).

9. Seilmeier, A. and Kaiser, W., Appl. Phys. 23, 113 (1980).

10. Kuhl, J., von der Linde, D., to be published.

RAMAN INVESTIGATIONS OF POLYATOMIC MOLECULES IN THE LIQUID STATE USING PICOSECOND LIGHT PULSES
II. A NEW RAMAN SPECTROSCOPY WITH SHORT EXCITATION AND PROLONGED INTERROGATION

W. Zinth and W. Kaiser

Physik Department der
Technischen Universität München
München, West Germany

1. INTRODUCTION

Modern Raman spectroscopic equipment allows a spectral resolution of approximately 0.1 cm^{-1}. A number which is quite sufficient for the investigation of liquids, where the Raman transitions have line widths of several wave numbers. Difficulties arise in congested spectral regions where broad spectral features suggest the existence of several transitions. The best instrumental resolution does not permit the classification of band intensities and frequency positions.

We report here on a Raman technique which provides, in a novel way, spectral information on homogeneously broadened transitions. New Raman lines are detected in congested spectral regions and peak positions of Raman lines in liquids are measured with an unprecedented accuracy.

2. SPECTRAL NARROWING OF HOMOGENEOUSLY BROADENED TRANSITIONS

Homogeneously broadened transitions have a Lorentzian band shape produced by the exponential decay of a time dependent quantity. When the population of excited molecules decays with the life time T_1 one obtains a line width $\Delta\nu_{pop} = 1/2\pi T_1$. Collisions between molecules cause an exponential decay of the coherency among vibrating molecules. The pure dephasing time T_{ph} determines the width $\Delta\nu_{coll} = 1/\pi T_{ph}$. We recall that the time constant T_1 is related to an intensity decay while T_{ph} is defined as the decay time

W. Kiefer and D. A. Long (eds.), Non-Linear Raman Spectroscopy and Its Chemical Applications, 227–237.

of the coherent amplitude. In liquids both broadening mechanisms are frequently of the same order of magnitude and must be added. Experimentally we observe: $\Delta\nu = \Delta\nu_{pop} + \Delta\nu_{coll} = 1/\pi T_2$. A steady-state experiment provides for homogeneously broadened lines the spontaneous width $\Delta\nu$ which is determined by the correlation function of the vibrational amplitudes /1/. In a coherent transient experiment the observation of a longer part of the exponential decay is possible. The increased measuring time improves the spectral resolution. Our technique is based on short excitation and prolonged interrogation (SEPI) of molecular states /2,3,4/.

During the short transient excitation process the molecules are driven by two light pulses at the difference frequency ν_D. Raman transitions at and close to ν_D become excited with coherent amplitudes Q_j. The coherent excitation persists even when the two pumping pulses have left the sample. The molecules regain their resonance frequencies ν_j and the amplitudes Q_j decay with the dephasing times T_{2j}. During the free relaxation process a third delayed probe pulse of long duration interacts with the coherently vibrating molecules. It generates coherently scattered, Raman shifted, light containing valuable information on the molecules. For a suitable shape of the probing pulse (e.g. Gaussian) only the molecules oscillating in phase contribute to the coherent signal; molecules which have suffered dephasing collisions are out of step and are not observed. The time of observation - i.e., the spectral resolution - is determined by the third, the probing pulse.

3. THEORY

During the excitation process the two incoming light pulses with amplitudes E_L, E_S, frequencies ω_L, ω_S and wave vectors k_L, k_S generate the coherent amplitude Q with frequency ω_Q and wave vector k_Q /1,2/.

$$\frac{\partial Q}{\partial t} + \frac{1}{T_2} Q \propto E_L E_S^* \tag{1}$$

During the excitation ω_Q and k_Q are determined by the driving fields: $\omega_Q = \omega_L - \omega_S$; $k_Q = k_L - k_S$. After the excitation the coherent amplitude decays exponentially with the time constant T_2 and oscillates at its resonance frequency ω_o with the wave vector $k_o = \omega_o n/c$. In the probing process a third pulse E_{L2} with (ω_{L2}, k_{L2}) interacts with Q and generates coherently scattered Stokes light:

$$\frac{\partial E_{S2}}{\partial x} \propto E_{L2}\, Q^* \exp\left[-i(\omega_{L2} - \omega_o - \omega_{S2})t + i(k_{L2} - k_o - k_{S2})x\right] \tag{2}$$

E_{S2} is emitted under phase-matching conditions $\vec{k}_{S2} = \vec{k}_{L2} - \vec{k}_o$ with

$\omega_{S2} = \omega_{L2} - \omega_o$. We recall that anti-Stokes scattering is treated by a similar equation. When several Raman lines are excited simultaneously, Eq.(2) may be extended to give:

$$\frac{\partial E_{S2}}{\partial x} \propto E_{L2} \times \sum_j Q_j^* \exp\left[-i(\omega_{L2}-\omega_j-\omega_{S2j})t + i(k_{L2}-k_j-k_{S2j})x\right] \qquad (3)$$

The contribution of nonresonant susceptibility $\chi_{NR}^{(3)}$ may be ommitted when the probing pulse E_{L2} is sufficiently delayed.

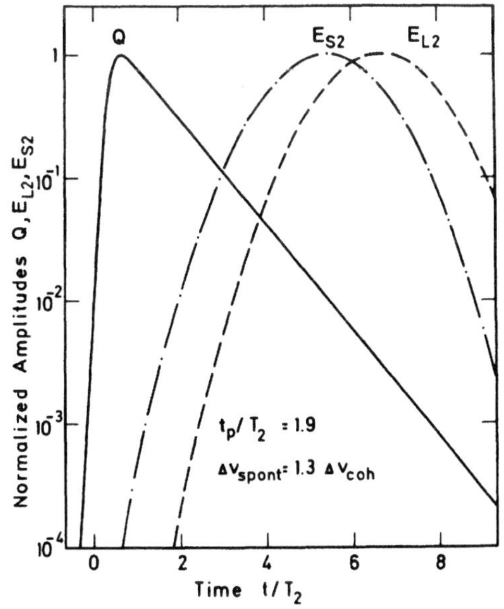

Fig. 1 Calculated coherent excitation and probing versus time. The coherently scattered field E_{S2} has the same duration and shape as the Gaussian probing pulse E_{L2}.

In Fig.1 the probing process is shown as a function of time. Short pulses prepare a coherent excitation Q at time zero and the amplitude Q is probed by the pulse E_{L2}. In our example we use Gaussian shaped probing pulses. According to Fig. 1 the coherently scattered pulse has the same shape and duration as the probing pulse (when influence of the non-exponential dependence of Q near time zero is neglected). Only the peak position of the scattered pulse is shifted to earlier times. Its frequency width $\Delta\nu_{coh}$ is the same as the width of the probing pulse of duration t_p, $\Delta\nu_{coh} = 2\ell n2/\pi t_p = 0.44/t_p$. Using sufficiently long Gaussian shaped probing pulses, $t_p > 1.4\ T_2$, the width of the coherent spectrum becomes smaller than the spontaneous Raman line, i.e., $\Delta\nu_{coh} < \Delta\nu_{spont} = 1/\pi T_2$.

In the case of several molecular transitions the spectrum of the SEPI signal is calculated from Eq.(3). Using a Gaussian shape of the probing pulse and disregarding a possible overlap of the spectra we obtain:

$$I(\omega) \propto \sum_j I_{oj} \exp\left[- 2t_D/T_{2j} - \frac{(\omega - \omega_{S2j})^2}{(\pi\Delta\nu_{L2}\sqrt{\ln 2})^2}\right] \tag{4}$$

where I_{oj} accounts for the degree of excitation of the jth component at time zero and t_D is the time delay between excitation and probing. According to Eq.(4) the SEPI spectrum consists of a sum of bands with Gaussian spectral shape and with a half width determined by the interrogating pulse. The peak positions of the different bands provide the frequencies of the vibrational modes: $\omega_j = \omega_{L2} - \omega_{S2j}$. Modes with shorter dephasing times or with smaller excitation have smaller peak intensities in the SEPI spectrum.

A. Optimum Resolution of SEPI Spectra

We have seen above that improved spectral resolution requires long probing pulses. These long probing pulses should interact with

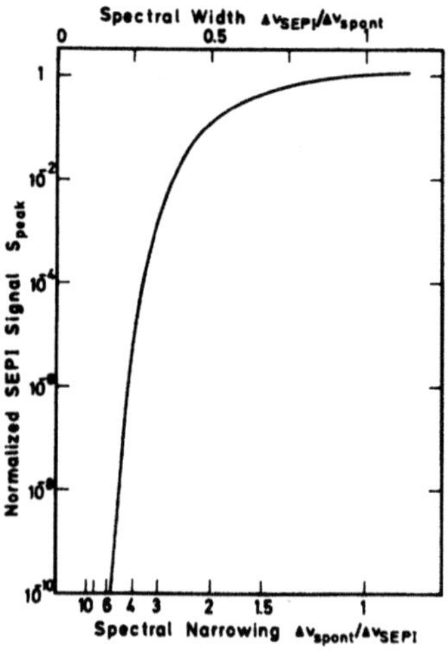

Fig. 2 Peak intensity of delayed probing, the SEPI spectrum as a function of spectral narrowing. In order to enjure only probing of freely relaxing molecules longer delay times t_D have to be used for longer probing pulses. A spectral narrowing by a factor of four gives rise to a signal reduction by a factor of 10^6.

the freely relaxing molecules and should not be substantially affected by the excitation process. In order to avoid the influence of the excitation the delay time t_D must be sufficiently long. Unfortunately, the scattered signal decreases with t_D according to Eq.(4). In Fig. 2 the normalized peak of a SEPI band, S_{peak}, is plotted versus the spectral narrowing $\Delta\nu_{spont}/\Delta\nu_{SEPI}$. For calculating the curves of Fig. 2 the delay times t_D are chosen in such a way that 90% of the scattered energy is generated from the freely relaxing molecules. In practical coherent experiments we start (at $t_D = 0$) with signal to background ratios of 10^6. According to Fig.2 a narrowing by a factor of four should be possible.

4. EXPERIMENTAL SYSTEM

The short dephasing times in molecular liquids require picosecond pulses in order to measure SEPI spectra. It is convenient

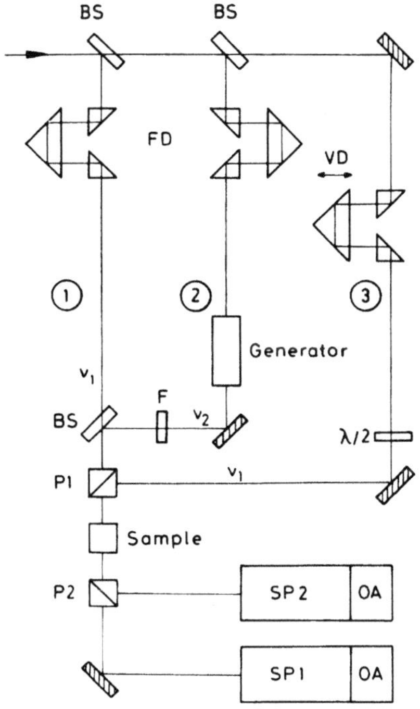

Fig. 3 Schematic of the experimental system used for the study of SEPI spectra of liquids. Beam splitters BS, variable and fixed delay VD and FD, polarizers P1 and P2, blocking filter F, spectrographs SP1 and SP2, used in conjunction with optical spectrum analysers OA.

to use exciting pulses with a band width of $\approx 10\ cm^{-1}$ tuned in steps over a larger frequency range. For each excitation band the coherent spectrum should be recorded by a spectrometer or spectrograph with sufficient resolution. Fig. 3 shows the schematic of our experimental system. At the top, l.h.s., a single, frequency doubled pulse from a mode-locked Nd-glass laser system enters the Figure /5/. This pulse of frequency $\nu_1 = 18{,}990\ cm^{-1}$ and 8 ps duration is split in three parts by two consecutive beam splitters. The pulse in beam 1 passes through the polarizer P1 and the sample but is blocked by the polarizer P2 in its straight path. In the center beam 2 of Fig. 3 a new frequency ν_2 is produced in the generator by a stimulated Raman process. Changing the medium of the generator one readily obtains pulses of different frequencies ν_2. These pulses are blocked by the polarizer P2 and are spectrally monitored by spectrograph SP2. On account of the transient generation process the pulses of frequency ν_2 are shorter in duration than the incident pulses ν_1 by a factor of approximately three /6/. The two pulses of the beams 1 and 2 enter simultaneously the sample coherently exciting molecular vibrations via transient stimulated Raman scattering at a frequency $\nu_D = \nu_1 - \nu_2$. In the optical path 3 a delayed pulse with polarisation perpendicular to the pulses of 1 and 2 is produced. This third pulse interacts with the coherently excited molecules of the sample producing a Raman shifted signal pulse. Using Stokes scattering the three pulses travel collinearly through the sample. When anti-Stokes scattering is used, the probing pulse crosses the beam direction of the exciting pulses at the phase matching angle. The spectrum of the coherently scattered light is studied by a 2 m spectrograph SP1 and a cooled optical spectrum analyser OA. The experimental system has a resolution of $0.2\ cm^{-1}$ per channel and an absolute accuracy of the frequency scale of $0.4\ cm^{-1}$.

5. EXPERIMENTAL RESULTS

We have performed SEPI measurements on a number of organic and anorganic liquids. First we present results on liquid CH_3CCl_3 where a single Lorentzian shaped Raman line exists at 2939 cm^{-1}. This transition allows to see the influence of different delay times t_D. As a second example we show the results on liquid C_6H_{12} where broad and overlapping lines occur between 2850 and 2940 cm^{-1}, in the common spontaneous Raman spectrum.

A. Liquid 1,1,1-Trichloroethane

SEPI spectra of the symmetrical CH_3-stretching modes of liquid CH_3CCl_3 are shown in Fig. 4. A duration of the probing pulse of 6.5 ps was used. The peak of the coherent excitation occurs near t_D = 5 ps. The two upper SEPI spectra in Fig. 4a and b taken at t_D = 1.5 and 8.5 ps are strongly affected by the excitation process

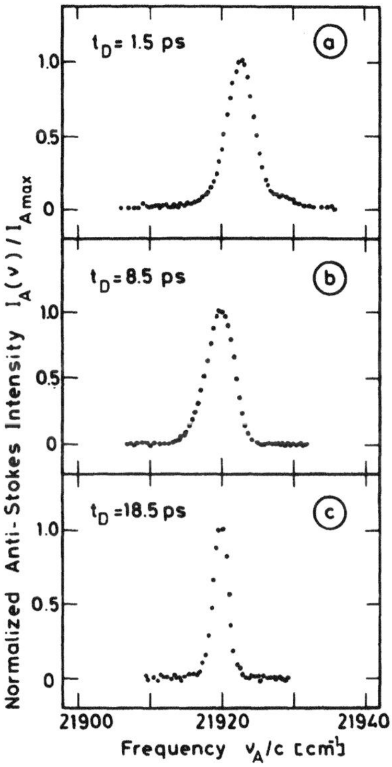

Fig. 4 Coherent probing (SEPI) spectra of liquid CH_3CCl_3 taken as a function of the delay time. During the excitation (Fig. 4a and b) the spectra are broader than the spectrum obtained at t_D = 18.5 ps (Fig.4c). The duration of the exciting Stokes and probing laser pulses is 3 ps and 6.5 ps, respectively.

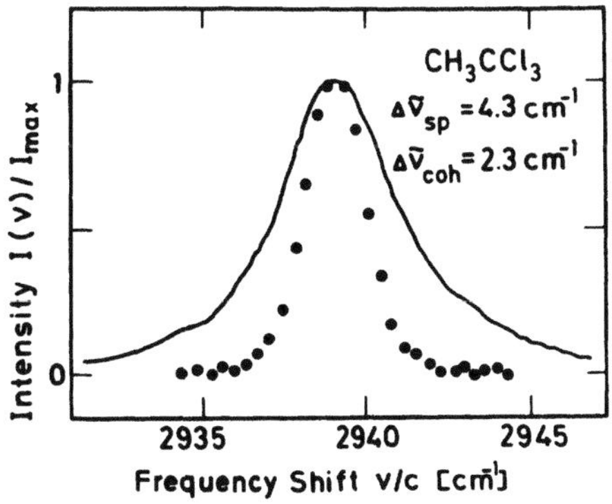

Fig. 5 Spontaneous Raman spectrum (solid curve) and coherent probing (SEPI) spectrum taken at t_D = 18.5 ps. The CH_3-stretching mode of CH_3CCl_3 is investigated. Note the narrow SEPI spectrum.

and by $\chi_{NR}^{(3)}$. The spectra are broad, 5 cm^{-1}, and the positions of the maxima are shifted. For longer delay times t_D, e.g. for t_D = 18.5 ps, the spectrum becomes narrower and indicates a Raman shift of 2938.2 cm^{-1} in full agreement with the spontaneous shift of 2939 cm^{-1}. In Fig. 5 we compare the spontaneous Raman spectrum with the SEPI spectrum. The peak position of the two spectra coincide. But the SEPI spectrum is narrower by a factor of two. It has the same width as the (simultaneously measured) spectrum of the probing pulse.

B. Liquid Cyclohexane

As another example for the short excitation and prolonged interrogation (SEPI) technique we present Raman data of cyclohexane in the small frequency range between 2850 cm^{-1} and 2940 cm^{-1}. In Fig. 6b the polarized spontaneous Raman spectrum is depicted. This spectrum was taken with an Ar^+ laser and a Raman spectrometer with a resolution better than 1 cm^{-1}. The three strong Raman bands correspond to CH-stretching modes and the diffuse spectrum between 2860 cm^{-1} and 2920 cm^{-1} is considered to be due to overlapping overtones and combination modes which are enhanced by Fermi resonance with the fundamentals /7,8/.

In Fig. 6c we show three SEPI spectra on an expanded scale (factor 3.7). Each spectrum was obtained by a single laser shot. On the r.h.s. we present the sharp SEPI band corresponding to the CH-stretching mode at 2923 cm^{-1}. The small line width of 2.3 cm^{-1} allows to determine the peak position accurately to 2922.0±0.7 cm^{-1}. We note that the SEPI band is considerably smaller than the corresponding band in the spontaneous Raman spectrum of Fig. 6b, the latter being asymmetric on account of other smaller Raman transitions. The Raman transition at 2923 cm^{-1} was excited using ethylene glycol, $(CH_2OH)_2$, in the generator cell of Fig. 3.

The SEPI spectrum of Fig. 6c, middle, shows four Raman transitions. Lines as close as 2.5 cm^{-1} are clearly resolved The four transitions are hidden under the wing of the strong Raman band at 2923 cm^{-1}; they cannot be detected in the conventional Raman spectrum of Fig. 6b. The SEPI spectrum is obtained by using an exciting pulse ν_2 with a frequency band extending from 2900 cm^{-1} to 2920 cm^{-1} (dimethyl sulfide, C_2H_6S, in the generator).

In Fig. 6c, l.h.s., we depict a SEPI spectrum obtained after excitation by a ν_2 pulse with a spectral band width extending from 2875 cm^{-1} to 2890 cm^{-1} (propylene oxide, C_3H_6O). We find two distinct Raman bands at 2877.5 cm^{-1} and 2887 cm^{-1}. The band at 2877.5 cm^{-1} has never been reported previously. It is buried in the diffuse part of the conventional Raman spectrum (see Fig. 6b).

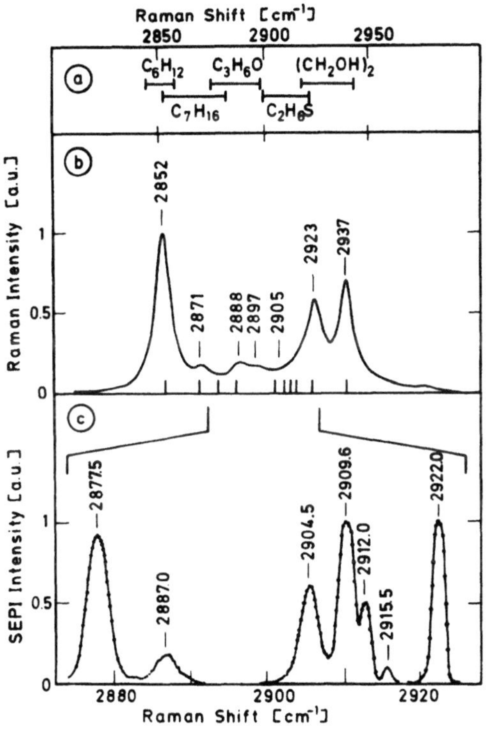

Fig. 6 a, b, c Experimental results of short excitation and prolonged interrogation (SEPI) spectroscopy of C_6H_{12}. (a) Frequency range of the various generator liquids used in the experiment. (b) Polarized spontaneous Raman spectrum of C_6H_{12} recorded with a resolution of 1 cm^{-1}. The frequency positions of the resonances found in SEPI spectra are marked by vertical lines. (c) Three SEPI spectra taken with different generator liquids. New Raman lines are detected and the spectral resolution is improved. (Note, the frequency scale of c is 3.7 times larger than the one of b).

A final assignment of the new Raman lines between 2870 cm^{-1} and 2920 cm^{-1} has not yet been made. Inspection of the lower fundamental modes suggest overtones and combination modes in this frequency range. Of special interest is the new Raman line at 2912 cm^{-1} which coincides precisely with an infrared active mode of the molecule. It appears that we observe here a Raman forbidden mode.

6. ADDITIONAL OBSERVATIONS AND COMMENTS

The following points are relevant for the application of the SEPI technique: (i) The frequency positions of the observed Raman lines are independent of the excitation conditions since we observe freely relaxing molecules. We have tested this notion by exciting our sample with narrow or with broad pulses of similar central frequency ν_2. This experiment is readily performed using different media in the generator cell. The advantage of a broad frequency spectrum of the incident pulse is to provide initial conditions for several Raman transitions in congested frequency regions. One can observe several Raman lines with one shot (see Fig. 6c, middle). (ii) In SEPI experiments the exciting and interrogating pulses should not overlap temporarily in order to avoid the generation of a coherent signal via the nonresonant four-photon parametric process. For this reason, the delay time of the third probing pulse has to be sufficiently large. One roughly estimates delay times of t_D = 20 to 25 ps for dephasing times $T_{2j} \simeq 1$ ps and Gaussian pulses of 8 ps duration. The SEPI spectra are observed with good accuracy, approximately five orders of magnitude below the peak value at $t_D = 0$. (iii) The maxima of the SEPI spectra are not proportional to the Raman scattering cross-section, since the initial conditions of the exciting pulses and the T_{2j} times are important parameters for the observed magnitude of the generated signal (Eq.(4)). SEPI spectra taken for different delay times allow an estimate of the dephasing times T_{2j}. (iv) The frequency precision of the generated Stokes spectrum depends upon the frequency stability of the interrogating pulse. For highest accuracy the frequency ν_1 has to be measured simultaneously with the SEPI spectrum. Interrogating pulses with a chirped frequency spectrum give unwanted shifts of the SEPI spectra and thus should be avoided. (v) The scattering process may also be performed on the anti-Stokes side of the spectrum. The disturbing interference found in stationary CARS spectroscopy does not occur for the delayed probing used with the SEPI spectroscopy /2,9/. (vi) A simultaneous measurement of the coherent Stokes and anti-Stokes SEPI spectra allows to eliminate the effect of a chirped probing pulse. In this way the absolute frequency position is obtained with high accuracy /2/.

7. CONCLUDING REMARKS

The data presented in this article convincingly show the usefulness of the short excitation and prolonged probing technique. It is possible to obtain molecular information which are not found by other existing spectroscopic techniques.

REFERENCES

1. Laubereau, A., Kaiser, W., Rev. Mod. Phys. 50, 607 (1978); Penzkofer, A., Laubereau, A., and Kaiser, W., Prog. Quant. Electron. 6, 55 (1979).

2. Zinth, W., Optics Commun. 34, 479 (1980).

3. Zinth, W., Polland, H.J., Laubereau, A., and Kaiser, W., Appl. Phys. B26, 77 (1981).

4 Zinth, W., Nuss, M.C., and Kaiser, W., to be published in Chem. Phys. Lett. (1982).

5. Zinth, W., Laubereau, A., and Kaiser, W., Optics Commun. 22, 161 (1977).

6. Carman, R.L., Shimizu, F., Wang, C.S., and Bloembergen, N., Phys. Rev. A2, 60 (1970); Akhmanov, S.A., Brabovich, K.N., Sukhorukov, A.P., and Chirikin, A.S., Sov. Phys. JETP 32, 266 (1971).

7. Kohlrausch, K.W.F., Wittek, W., Z. phys. Chem. B48, 177 (1941).

8. Wiberg, K.B., Shrake, A., Spectrochim. Acta 27A, 1139 (1971).

9. Zinth, W., Laubereau, A., and Kaiser, W., Optics Commun. 26 457 (1978).

CARS AND APPLICATIONS

APPLICATION OF CLASSICAL THEORY OF CARS TO DIATOMIC MOLECULES IN THE GAS PHASE

W. Kiefer

Physikalisches Institut der Universität Bayreuth,
D-8580 Bayreuth, F.R. Germany

1 INTRODUCTION

Linear Raman scattering can be described classically when excitation frequencies are used which are far from resonance with electronic absorption frequencies. If this is not the case, the scattering amplitude must be derived by quantum-mechanical calculation of the linear polarizability [1]. Similarly, CARS intensities can be obtained classically by combining the classical theory of the linear Raman effect with a simple damped, field-driven, harmonic oscillator model [2,3], whereas for the case of electronic resonances one has to use again quantum-mechanics for the determination of the CARS amplitude. The latter is given through the third-order non-linear CARS susceptibility [4].

In the present article we will show how the simple classical CARS theory is able to well explain experimentally ovserved CARS spectra of diatomic nonabsorbing molecules in the gas phase. After a brief review of the classical theory of linear Raman scattering in diatomics we will outline the corresponding theory for CARS. Special emphasis will be given to interference effects between neighbouring spectral lines such as the rotational lines within a Q-branch of a diatomic molecule. Applying this theory, simulated CARS spectra for various different bandwidths will be given, which significantly influence the magnitude of the interactions. Specifically we will examine the interference between two individual lines as a function of the bandwidth to frequency separation (Γ/δ) ratio. We will then proceed to calculate the Q-branch spectra of the diatomic molecules O_2, N_2 and NO at various spectral bandwidths. These three molecules were chosen because they represent situations where the frequency spacings

W. Kiefer and D. A. Long (eds.), Non-Linear Raman Spectroscopy and Its Chemical Applications, 241–260.

between neighbouring Raman-active lines are relatively large (for O_2), moderate (for N_2) and relatively small (for NO). Fianally we will compare calculated CARS spectra with experimentally obtained one's.

2. BRIEF REVIEW OF THEORY ON LINEAR RAMAN SCATTERING FROM DIATOMIC MOLECULES IN THE GAS PHASE

Placzek has shown in his famous polarizability theory [5] that under special conditions the Raman intensity can be determined classically. For Raman scattering into the full solid angle 4π one uses the total scattering cross section σ. The differential scattering cross section $d\sigma/d\Omega$ is used for observation with a limited acceptance angle into a certain direction. Both, σ and $d\sigma/d\Omega$ contain the invariants of the polarizability tensor derived with respect to the molecular normal vibration q, namely the linear average a and the anisotropy γ^2 [6]. For the most important case of linearly polarized incident light and observation at 90° to both the polarization vector and the direction of incidence, one obtains [6] for the differential Raman scattering cross section of a molecule which vibrates with frequency ν:

$$\frac{d\sigma}{d\Omega} = \frac{2^4\pi^4(\nu_o-\nu)^4 b^2}{1-\exp(-hc\nu/kT)} g\ (a^2 + \frac{7}{45}\gamma^2), \qquad (1)$$

where ν is the exciting frequency, $b^2 = h/8\pi^2 c\nu$ is the square of the zero point amplitude of the normal vibration q, h is the Planck constant, c is the velocity of light in vacuum, k is the Boltzmann constant, T is the temperature and g is the degeneracy of the normal vibration.

The total intensity for ν, scattered by a macroscopic number of molecules into the solid angle Ω_m is given by [7]:

$$I = I_\perp + I_\parallel = I_o nL \int_{\Omega_m} (\frac{d\sigma}{d\Omega}) d\Omega \qquad (2)$$

where the integration is over Ω_m, n is the density of scattering molecules, L is the length from which scattered photons are collected and which corresponds to the ratio of the scattering volume and the effective cross-sectional area of the laser beam, and I_o is the intensity of the incident light. If an analyzer is used for the scattered light, one can separate $I_\parallel$ and $I_\perp$. For this case, one obtains:

$$I_\perp \sim (\frac{3}{45}\gamma^2) \qquad (3)$$

and

$$I_\parallel \sim (a^2 + \frac{4}{45}\gamma^2) \qquad (4)$$

For room temperature and vibrational frequencies $\nu > 1000\ cm^{-1}$, the exponential function in the denominator of Equation 1 is negligible (< 1%).

Now we allow the molecule to simultaneously rotate and restrict ourselves to vibrational-rotational transitions with $\Delta v = 1$ and $\Delta J = 0$ (Q-branch) in a diatomic molecule. Then, the space average square of the transition polarizability tensor component is given by [8]:

$$\overline{[\alpha_{ii}]^2_{v',J' \leftarrow v'',J''}} = a^2 + b_{J,J} \cdot \frac{4}{45} \gamma^2 \tag{5}$$

where

$$b_{J,J} = \frac{J(J+1)}{(2J-1)(2J+3)} \tag{6}$$

is the appropriate rotational intensity factor. Note, that the suscripts ii in Equation 5 denote $I_{\parallel}$ observation. For $I_{\perp}$ observation we would obtain:

$$\overline{[\alpha_{ij}]^2_{v',J' \leftarrow v'',J''}} = b_{J,J} \cdot \frac{3}{45} \gamma^2 \tag{7}$$

Using Equations 1,5 and 6 (or 7 instead of 5) one is therefore able to calculate the differential Raman scattering cross section for the rotational lines within a Q-branch of a diatomic molecule. Of course, the (2J+1) degeneracy and the statistical weight factor g_I due to the nuclear spin have to be taken into account additionally for this purpose.

2. CLASSICAL THEORY OF CARS

A. Outline of theory

It is well known, that the intensity of a CARS signal is proportional to the square of the absolute magnitude of the third-order non-linear susceptibility:

$$I_{CARS} \sim |\chi^{(3)}_{CARS}|^2 \tag{8}$$

As has been shown elsewhere [2,3], the third-order non-linear susceptibility at a particular CARS frequency difference $\omega_L - \omega_S$ (ω_L, ω_S are the angular frequencies of the two lasers which generate the non-linear polarization at frequency $\omega_{AS} = 2\omega_L - \omega_S$) can be expressed for a single isolated spectral line of frequency ω_R (e.g. a rotational-vibrational Raman line) by

$$\chi^{(3)}_{CARS} = \frac{2Nc^4}{\hbar \omega_S^4} \left(\frac{d\sigma}{d\Omega}\right) \frac{\omega_R}{\omega_R^2 - (\omega_L - \omega_S)^2 - i(\omega_L - \omega_S)\Gamma} + \chi'_{NR} \tag{9}$$

where $N = n\Delta_j$ is the population difference between the lower and upper states for a particular transition j, which accounts for statistical distribution [4,9], that is, $\Delta_j = 1$ for low temperature, $\Delta_j = 0$ for infinite temperature. The factor Δ_j is an integral part of the quantum mechanical approach by density matrix methods [2,4]. $d\sigma/d\Omega$ is the differential Raman scattering cross section as discussed above, Γ is the spectral band-width for the corresponding Raman transition (FWHM), and χ'_{NR} is the non-resonant susceptibility which arises from background electronic susceptibility contributions and remote Raman resonances. As can be seen from Equation 9, the total susceptibility is a sum of a frequency-dependent resonant part and a nearly frequency-independent non-resonant part. Since χ'_{NR} is in many cases very small (see however Ref. 1) and additionally only a slowly varying function over the spectral region of interest, it may be neglected for the gas phase studies discussed here.

Note, that the resonant third-order susceptibility is a complex quantity which can be separated into a real and an imaginary part (from now on we set $\chi_{NR} = 0$):

$$\chi^{(3)} = \chi' + i\chi'' \tag{10}$$

In order to explicitely derive the real (χ') and the imaginary (χ'') parts of $\chi^{(3)}$ we rewrite Equation 9 according to [10-12]:

$$\chi^{(3)} = AN \frac{\omega_R/(\omega_R-\Delta)}{\omega_R^2/(\omega_R-\Delta)-(\omega_R-\Delta)-i\Gamma} \tag{11}$$

where

$$A = \frac{2c^4}{\hbar\omega_S^4}\left(\frac{d\sigma}{d\Omega}\right) \tag{12}$$

and where

$$\Delta = \omega_R-(\omega_L-\omega_S) \tag{13}$$

If we assume that $\Delta/\omega_R \ll 1$, which is usually a very valid approximation in the vicinity of a resonance, and if we recognize that then

$$\frac{1}{(\omega_R-\Delta)} = \frac{1}{\omega_R} + \frac{\Delta}{\omega_R{}^2} + \frac{\Delta^2}{\omega_R{}^3} + \ldots\ldots$$

$$\approx \frac{1}{\omega_R} + \frac{\Delta}{\omega_R{}^2}\ , \tag{14}$$

Equation 11 may be rewritten as

$$\chi^{(3)} = \frac{AN}{2\Delta-i\Gamma} \tag{15}$$

This expression is easy to separate into real (χ') and imaginary (χ'') parts:

$$\chi' = \frac{2AN\Delta}{4\Delta^2+\Gamma^2} \tag{16a}$$

$$\chi'' = \frac{AN\Gamma}{4\Delta^2+\Gamma^2}\ . \tag{16b}$$

The CARS intensity is then proportional to the square of the absolute third-order non-linear CARS susceptibility:

$$I \sim |\chi^{(3)}_{CARS}|^2 = (\chi')^2 + (\chi'')^2$$
$$= \frac{4A^2N^2\Delta^2}{(4\Delta^2+\Gamma^2)^2} + \frac{A^2N^2\Gamma^2}{(4\Delta^2+\Gamma^2)^2} \qquad (17)$$

B. Single isolated Raman resonance

We first apply the expressions derived above for a single isolated Raman resonance at frequency ω_R.

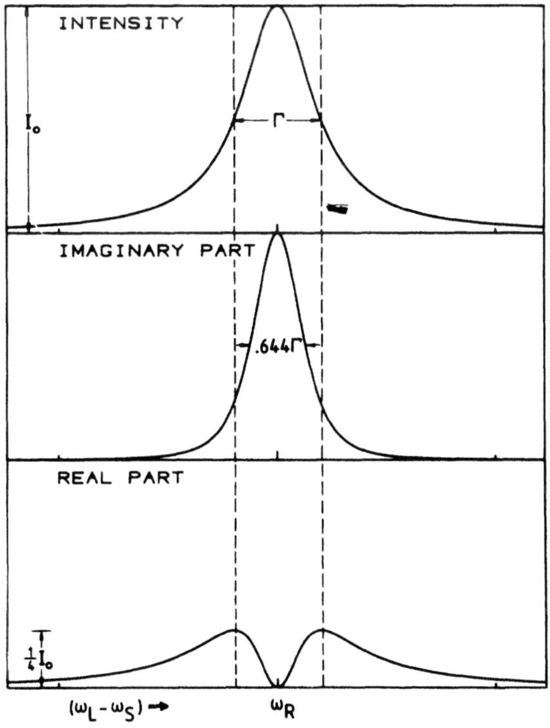

Fig.1 CARS intensity (upper field), contribution from imaginary part ($(\chi'')^2$, middle field) and contribution from real part ($(\chi')^2$, lower field) to the CARS signal for a single isolated Raman resonance at frequency ω_R.

Figure 1 shows in the lower and middle field the dispersion of real $(\chi')^2$ and imaginary $(\chi'')^2$ contributions to $|\chi^{(3)}_{CARS}|^2$, as well as in the upper field their sum ($|\chi^{(3)}_{CARS}|^2$) or equivalently the intensity of the CARS signal, for a single isolated Raman resonance at frequency ω_R. While the contribution from the real part of the susceptibility does not add intensity to the maximum of the band ($\Delta = 0$), it does broaden the bandwidth of the resultant line over that of the imaginary part by a factor of 1.55. The plot of $|\chi^{(3)}_{CARS}|^2$ in the upper field of Figure 1 has a Lorentzian line shape similar to a Raman line in the linear Raman effect.

It may be stated that the real part of the susceptibility is associated to the nonlinear refractive index, and the imaginary part is related to the normal Raman transition [3]. This picture is very similar to the interpretation of the linear (complex) refractive index, where the real and imaginary parts correspond to the dispersion and the absorption, respectively. The simple Lorentz model of the oscillating electron in atoms driven by the incident electromagnetic wave, which leads to the complex quantity of the linear refractive index has its analogue in CARS: the driving force is due to the electric field with difference frequency $(\omega_L-\omega_S)$ and the oscillating frequency is ω_R, which is the molecular (rotational-) vibrational frequency.

The CARS signal of a single Raman resonance as shown in Figure 1 does not reveal any unusual spectral feature. However, as soon as there are other resonances situated nearby, the band shapes may be distorted considerably as will be shown next.

C. Closely spaced Raman resonances

We now consider a series of closely spaced spectral lines with frequencies ω_k. Each Raman resonance which is spaced by

$$\Delta_k = \omega_k-(\omega_L-\omega_S) \tag{18}$$

from the frequency of the driving field $(\omega_L-\omega_S)$ contributes to both the real and imaginary part, leading to a total real and a total imaginary part given by

$$\chi' = \sum_k \frac{2A_k N_k \Delta_k}{4\Delta_k^2+\Gamma_k^2} \tag{19a}$$

$$\chi'' = \sum_k \frac{A_k N_k \Gamma_k}{4\Delta_k^2+\Gamma_k^2} \tag{19b}$$

Because the expressions in Equation 19 are summations and because each Δ_k may be positive or negative depending on the actual value of $\omega_L-\omega_S$ (which is tuned during a CARS experiment), considerable interference in the real part of the susceptibility between closely lying Raman resonances may be expected. These interference effects are discussed in the following where we consider two adjacent Raman resonances.

D. Interference between two adjacent Raman resonances

Let us now examine the simple case of two Raman resonances with same differential scattering cross section ($A_1=A_2=A$), equal bandwidth ($\Gamma_1=\Gamma_2=\Gamma$) and we also assume equal population difference ($N_1=N_2=N$). The two lines are separated by $\delta = \omega_1-\omega_2$. Figure 2 shows the calculated CARS spectra of such a situation as well as the real (χ'^2) and imaginary (χ''^2) contributions to the CARS intensity ($\chi'^2+\chi''^2$). It can be seen (Figure 2a) that at $\Gamma/\delta = 0.02$ very little interference between the lines is present and only a small amount of asymmetry is present in the real contribution to each spectral band.

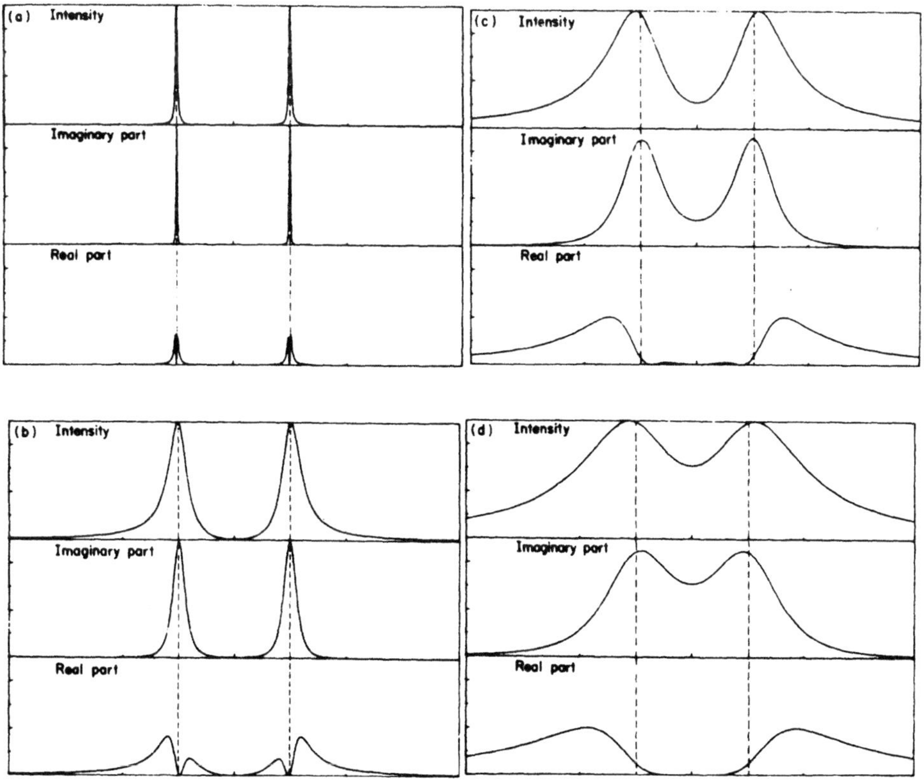

Fig. 2 Calculated CARS spectra (upper fields) of two Raman bands of equal intensity as a function of Γ/δ (= Raman band width/frequency separation between the two Raman lines) a) $\Gamma/\delta = 0.02$; b) $\Gamma/\delta = 0.2$; c) $\Gamma/\delta = 0.6$; d) $\Gamma/\delta = 1.0$. Displayed are also the imaginary ((χ''^2), middle fields) and real ((χ'^2), lower fields) contributions to the CARS intensity ($\chi'^2+\chi'^2$) [12].

For larger Γ (and Γ/δ) values, however, the asymmetry becomes significant as the interference between the two lines causes considerable cancellation of χ' in the region between them. This interference is easily understood by considering Equations 18 and 19 for k = 1,2. Since Δ_1 and Δ_2 will have different signs when $(\omega_L-\omega_S)$ is between ω_1 and ω_2, the contributions from the two lines to the real susceptibility will tend to cancel. When $(\omega_L-\omega_S)$ is outside this region (i.e. less than both ω_1 and ω_2 or greater than both of these), Δ_1 and Δ_2 will have the same sign and their contributions to χ' will be additive.

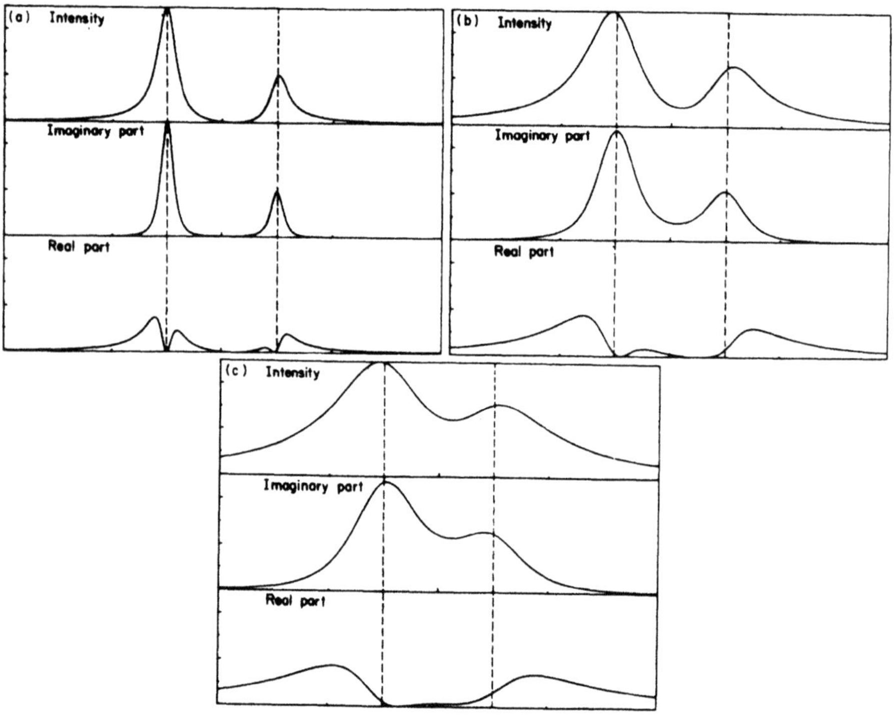

Fig. 3 Calculated CARS spectra(upper fields) of two Raman bands of unequal intensity (I_2/I_1 = 0.4) as a function of Γ/δ: a) Γ/δ = 0.2; b) Γ/δ = 0.6; c) Γ/δ = 1.0; [12].

Thus at the larger values of Γ/δ, at which interference increases, increased cancellation results in a decreased χ' in the region between the two lines. However, χ' increases in the regions outside the spectral lines due to the additive effects. This results in the shift of the maxima of the composite CARS spectrum away from the actual positions of the two resonances, thereby increasing the apparent separation between the lines. Furthermore, the maximum line intensity is somewhat enhanced. For Γ/δ = 0.6 the band maxima in Figure 2c) are separated by $1.84 \times \Gamma$ even though the resonance frequencies are only $1.67 \times \Gamma$ apart. These synthetically drawn curves illustrate the slightly better "resolving" power inherent in CARS.

In Figure 3 we show for three different values of Γ/δ similar curves for the case where the second line has an intensity of o.40 of the first (i.e. $A_2^2N_2^2 = 0.4\ A_1^2N_1^2$). Here, of course, the interferences inbetween the two Raman resonances lead to asymmetries for the real part contribution.

The examples given in Figures 2 and 3 represent only the interactions between two spectral lines. In actual spectra, such as the Q-branch CARS spectra of diatomic molecules, the interference between numerous rotational-vibrational lines will be present. Examples of such systems will be considered next.

3. APPLICATION TO Q-BRANCHES OF DIATOMIC MOLECULES

A. Q-branch of a diatomic molecule

• For the specific examination of the Q-branch spectra of diatomic molecules, the summation in Equation 19 must be carried out over the various rotational levels designated by J = 0,1,2,... With negligible error we may assume that the differential Raman scattering cross section $d\sigma/d\Omega$ for each of the transitions starting at the various J-levels is identical. This is justified by the fact that the rotational intensity factor given in Equation 6 very rapidly approaches the constant value 1/4 for increasing J. For example, the $b_{5,5}$ value differs from the $b_{\infty,\infty}$ value only by 2.5 %. However, for exact calculations the $b_{J,J}$ factor can, of course, be easily retained.

Setting $b_{J,J}$ = constant leads to $A_k = A$ = constant for closely spaced Q-branch transitions ($\omega_S \approx$ constant, see Equation 12). Similarly, we may take the bandwidths to be independent of rotational level, namely $\Gamma_k = \Gamma$ = constant. The latter approximation can certainly be made in the pressure-broadening regime.

With these approximations Equation 19 then becomes:

$$\chi' = 2A \sum_J \frac{N_J \Delta_J}{4\Delta_J^2 + \Gamma^2} \tag{20a}$$

$$\chi'' = A \sum_J \frac{N_J \Gamma}{4\Delta_J^2 + \Gamma^2} \tag{20b}$$

where A is given in Equation 12 and where

$$N_J = (2J+1) g_I N_o \exp(-E_J/kT) \tag{21}$$

and

$$\Delta_J = \omega_J - (\omega_L - \omega_S) \quad (22)$$

with

$$\omega_J = \omega_o - \alpha J(J+1) + \beta J^2(J+1)^2 \quad (23)$$

Usually β is very small and the third term in Equation 23 can therefore be neglected.

The (2J+1) term in Equation 21 reflects the magnetic degeneracy of each J level and g_I represents the nuclear statistical weight; N_o is the molecular population in the J = 0 state and E_J is the energy of the rotational level, i.e. hcBJ(J+1). The molecular populations of the excited vibrational states are negligible for the normal range of temperatures T and for light molecules. ω_o,α,β and B are the usual molecular constants.

For the calculations of the CARS spectra Equations 8 and 20 - 23 were programmed into a mini-computer and the spectra were plotted with an X-Y-plotter.

B. Q-branch CARS spectra of O_2

The first example we choose is O_2 since the linespacing is relatively large in comparison to other molecules. This is because the nuclear statistical weight g_I is zero for J = even and therefore half the Q-branches are missing. The line frequencies were calculated from the expression [13]

$$\omega_J = 1556.384 - 0.01587\ J\ (J+1)\ cm^{-1} \quad (24)$$

and a value of B = 1.4376478 cm^{-1} was used to determine the Boltzmann distribution.

Figure 4 shows the calculated CARS spectra ($|\chi^{(3)}|^2 = \chi'^2 + \chi''^2$) together with the real (χ'^2) and imaginary (χ''^2) contribution for six different Γ values. For Γ = 0.01 cm^{-1} the interference between the various lines is minor (the centre gap in the real contribution is present but difficult to make out on the scale shown), but it increases steadily with increasing Γ. It can be seen that the interference has the effect of cancelling the higher frequency lobe in the real part of most lines except that an increasingly larger contribution develops on the high frequency side of the highest frequency (J=1) line. This broad contribution, which can have a substantial effect on the composite spectrum even at Γ = 0.08 cm^{-1} is due to the fact that all of the Δ_J in Equation 20 in this region have the same sign and their contributions to χ' are additive. Because χ' is squared to obtain its contribution to the CARS intensity, this effect is further enhanced.

It is interesting to compare the outlined calssical theory

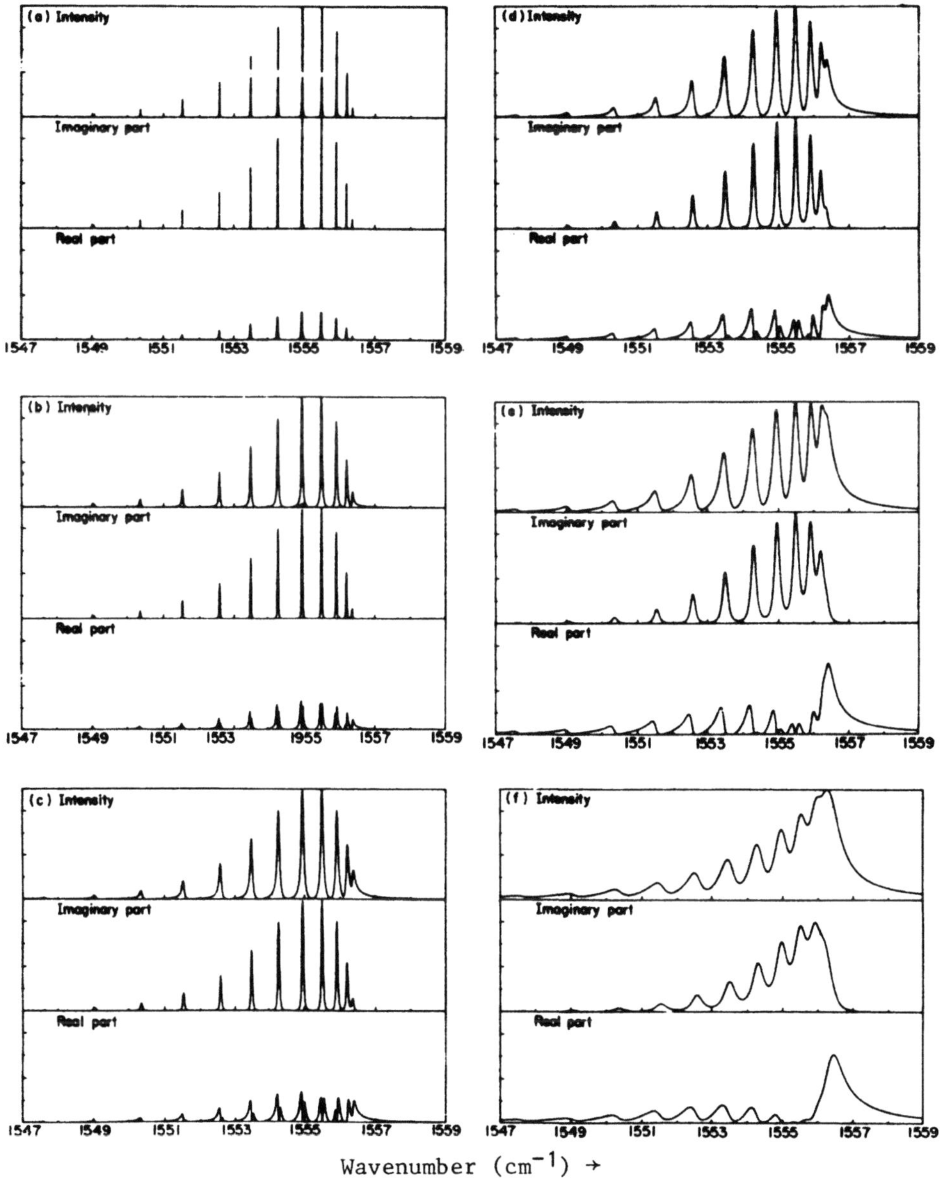

Fig. 4 Calculated CARS spectra (labelled by "Intensity") for the Q-branch of O_2 gas as a function of Γ: a) $\Gamma = 0.01\ cm^{-1}$, max = 1.02×10^6; b) $\Gamma = 0.04\ cm^{-1}$, max = 6.43×10^4; c) $\Gamma = 0.08\ cm^{-1}$, max = 1.64×10^4; d) $\Gamma = 0.15$, max = 5.11×10^3; e) $\Gamma = 0.25\ cm^{-1}$, max = 2.22×10^3; f) $\Gamma = 0.50$, max = 1.30×10^3.
The spectra were calculated for T = 293 K [12].

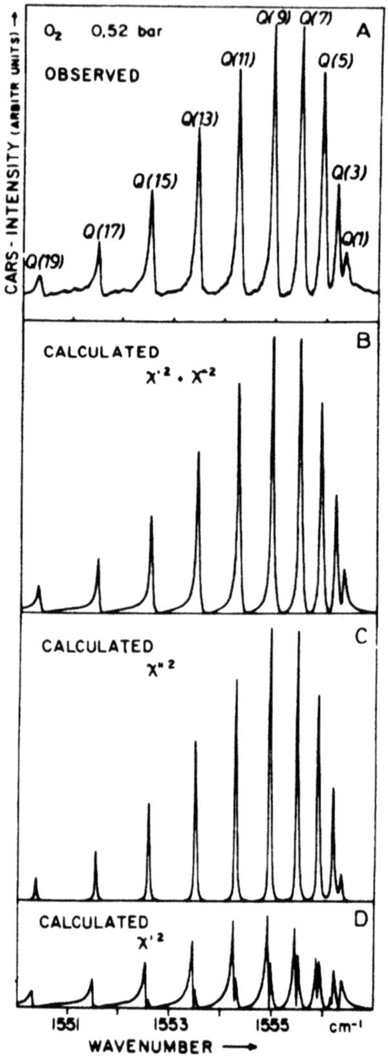

Fig. 5 Experimentally observed high resolution CARS spectrum of O_2 at a pressure of 0.52 bar (A); B: Calculated CARS spectrum; C: Contribution from imaginary part (χ''^2); D: Contribution from real part (χ'^2). Calculation was performed for Γ = 0.08 cm^{-1} [10].

with experimentally observed data. Field A of Figure 5 shows the observed high resolution CARS spectrum of oxygen obtained at 0.52 bar gas gressure [10]. The Q-branches are clearly resolved even for low J numbers (Q(1),Q(3)). Fields D and C of Figure 5 give the spectral intensity contributions of the real (χ'^2) and imaginary (χ''^2) part of the susceptibility, respectively, whereas field B displays the sum of χ'^2 and χ''^2. The calculated spectrum, where the intensity is proportional to $\chi'^2 + \chi''^2$ is in very good agreement with the observed spectrum. Even the asymmetries on the high frequency side for J = 1 as discussed above can be seen. Also the asymmetries on the low frequency side for higher J transitions agree nicely in the observed and calculated spectra. A Γ value of 0.08 cm^{-1} was used for the calculations used for comparison with experimental results. This value was derived from the data reported by Fletcher and Rayside [13] who reported $\Gamma = 0.14\ cm^{-1}$ for atmospheric pressure.

C. Q-branch CARS spectra of N_2

For the nitrogen Q-branch we have computed [12] the frequencies from:

$$\omega_J = 2329.92 - 0.017292\ J(J+1)\ cm^{-1} \qquad (25)$$

and used B = 1.989574 cm^{-1} [14] to determine the population distribution. The nuclear statistical weights for N_2 are $g_I = 2$ for even and $g_I = 1$ for odd J.

The calculated CARS spectra for N_2 for several Γ values are shown in Figure 6. The effects of changing Γ are similar to those found for oxygen except that the closer spacing of the lines and the alternating intensity due to the nuclear statistical weights do have pronounced influence.

In Figure 7 we show an experimentally observed CARS spectrum of the Q-branch of N_2 in air in the upper field, while the lower field displays the calculated spectrum. Again, high resolution and good agreement between experiment and theory were achieved. From Figures 6 and 7 it is clearly evident that in both the experimental as well as the calculated spectrum there is large contribution of CARS intensity at the high frequency wing for $\Gamma = 0.1\ cm^{-1}$, which again is due to the constructive interference of the contribution of the real part of the susceptibility.

In Figures 6 and 7 one also can nicely see that the CARS intensity is proportional to the square of what one would observe in a linear Raman spectrum. In the latter the intensity ratio between lines with odd and even J values could be $g_I(\text{even}):g_I(\text{odd}) = 2:1$. In the CARS spectrum this ratio turns out to be 4:1.

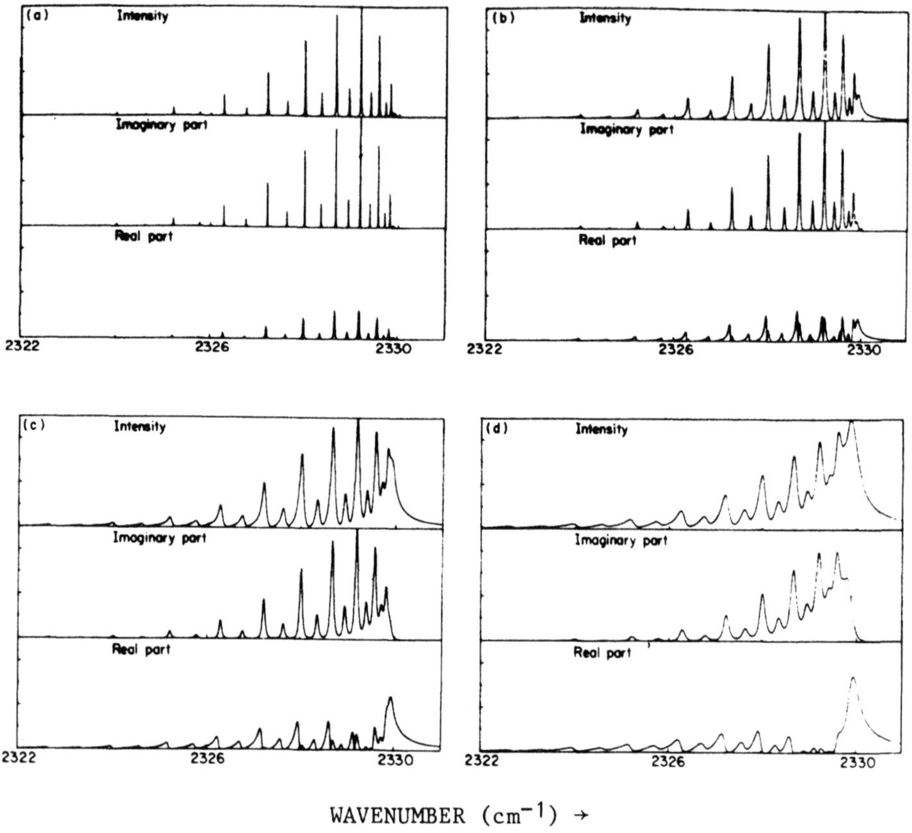

Fig. 6 Calculated CARS spectra for the Q-branch of N_2 gas as a function of Γ: a) Γ = 0.01 cm^{-1}, max = $2.96x10^6$; b) Γ = 0.05 cm^{-1}, max = $1.22x10^4$; c) Γ = 0.10 cm^{-1}, max = $3.47x10^4$; d) Γ = 0.20 cm^{-1}, max = $1.59x10^4$. The spectra were calculated for T = 293 K [12].

D. Q-branch CARS spectra of NO

In the last example we discuss a molecule which shows relative complexity due to the overlap of two Q-branch series resulting from two closely spaced electronic states. In Figure 8 we first show part of an experimental CARS spectrum of NO for higher J-levels in order to demonstrate this. This spectrum was observed by Beckmann et al. [15] from a sample at $1.33x10^4$ Pa. The two series originating from the $^2\Pi_{1/2}$ and $^2\Pi_{3/2}$ electronic states are assigned to $\Delta J = 0$ transitions starting at given J levels. The stronger bands originate from molecules in the $^2\Pi_{1/2}$ ground electronic state, while the weaker ones result from the $^2\Pi_{3/2}$ state which lies 124 cm^{-1} above the ground state. At higher wavenumbers

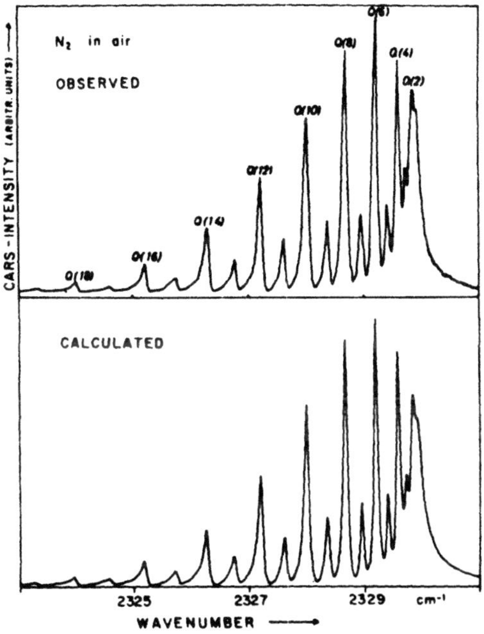

Fig. 7 Experimentally observed and calculated ($\Gamma = 0.1\ cm^{-1}$) high resolution CARS spectrum of nitrogen in air [10].

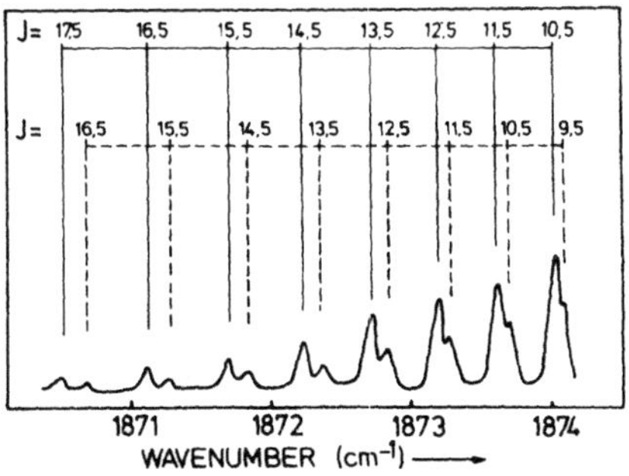

Fig. 8 Part of the observed medium high-resolution CARS spectrum of the Q-branch of NO (low frequency side), indicating the rotational structure due to two electronic states ($^2\Pi_{1/2}$ and $^2\Pi_{3/2}$) [15].

virtual overlap of the Q-branches from each of the two series occurs (see Figure 9).

Similar calculations as for O_2 and N_2 were also performed for this slightly more complex molecule [12,15]. The frequencies of the line positions were calculated from recent IR values [16,17] and by utilizing a correction matrix to determine the Λ doubling perturbation [18]. The Q-branch frequencies, before application for the correction matrix, for the two series are given by

$$\omega_J(^2\Pi_{1/2}) = 1876.08959 - 0.017115\, J\,(J+1) + 1.332\times10^{-7}\, J^2(J+1)^2\ \mathrm{cm}^{-1} \qquad (26)$$

and

$$\omega_J(^2\Pi_{3/2}) = 1875.88082 - 0.018019\, J\,(J+1) - 0.119\cdot10^{-7}\, J^2\,(J+1)^2\,\mathrm{cm}^{-1} \qquad (27)$$

The rotational constants for the vibrational ground state in the two electronic states are $B(^2\Pi_{1/2}) = 1.6720433\ \mathrm{cm}^{-1}$ and $B(^2\Pi_{3/2}) = 1.7201377\ \mathrm{cm}^{-1}$. For the calculation of the spectra, the differential Raman scattering cross section $(d\sigma/d\Omega)$ was assumed to be identical for both electronic states. This assumption was confirmed by the experiment (see below).

In Figure 9 experimentally observed CARS spectra [15] of the Q branch of NO are displayed in the upper fields of each quarter part for various gas pressures.In the second fields from above we show in each quarter part calculated CARS spectra which are the composite spectra of real and imaginary contributions $(\chi'^2 + \chi''^2)$. Again we display separately the pure imaginary (χ''^2) and real (χ'^2) part as indicated in each of the 4 quarter parts of Figure 9.

Comparing the observed with the corresponding calculated spectra one notices fairly good agreement between both. The only fitting parameters for the calculated spectra have been the maximum peak intensity and the damping term Γ whose value has been chosen as indicated in each diagram. Because of the good agreement between the calculated total CARS intensity and the observed spectra, these studies allow one consequently to determine separately the pressure dependence of the contributions from real and imaginary part of $\chi^{(3)}$. Owing to pressure broadening, the bands in the imaginary part spectrum which simply reflect the square of the normal Raman spectrum, broaden with increasing gas pressure (compare the third fields of the inserts of Figure 9 to each other).Again, considerable interference between adjacent Raman lines occurs for the contribution of the real part of $\chi^{(3)}$ which causes asymmetry in the band profiles. The magnitude of the interference itself is dependent on Γ, but this interference can already be recognized for small Γ. In the quarter part labeled by pressure 1.33×10^4Pa, one can observe destructive interference

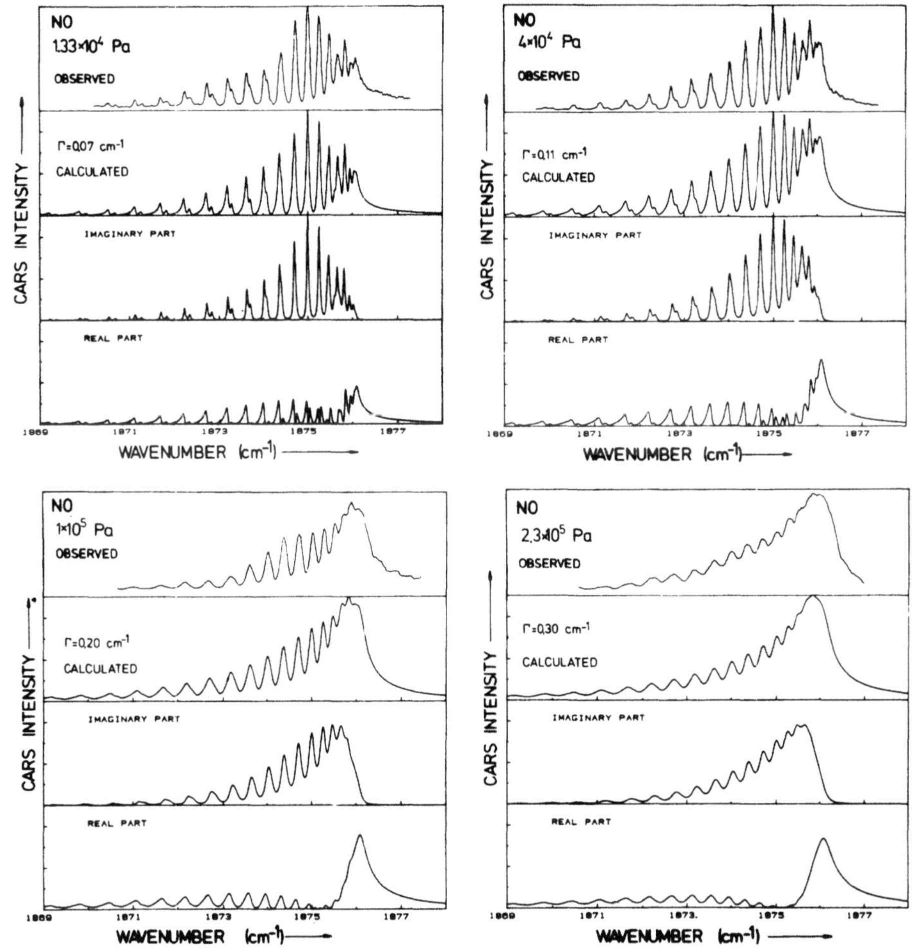

Fig. 9 Experimentally observed and calculated medium high resolution CARS spectra of nictric oxide as a function of pressure. Pressure range is 1.33×10^4 - 2.3×10^5 Pa. Displayed are also the contributions from real (χ'^2) and imaginary (χ''^2) parts of the susceptibility [15].

between about 1874 and 1876 cm^{-1}, whereas for wavenumbers higher than the band origin (1876.08 cm^{-1}), constructive interference takes place leading to a broad continuous band with decreasing intensity for increasing wave numbers. This is again consequence of the fact that the (χ')'s for each transition have the same sign in this region. This broad contribution which has a substantial effect on the composite spectrum increases relative to the contribution of the imaginary part for increasing gas pressure.

This can be readily recognized if one compares the corresponding spectral regions in the real part spectra of Figure 9. Simultaneously also, the effect of destructive interference between the Raman lines increases with increasing pressure. Consequently, for higher pressure ($\sim 10^5$ Pa), there is only a relatively small contribution of the real part of $\chi^{(3)}$ to the composite spectrum within the spectral region close to the center of the Q branch (~ 1875 cm^{-1}).

The good agreement between experimental and calculated CARS spectra allows one to separately study the pressure dependence of real and imaginary parts of the third-order non-linear susceptibility for molecules in the gaseous phase. As far as we know, this is the first time [15] that CARS studies of such a type have been made. However, it should be mentioned, that Owyoung [19] has obtained spectra of both the real and imaginary parts of $\chi^{(3)}$, for H_2 gas in the motionally narrowed pressure regime, by means of cw stimulated Raman gain spectroscopy.

The studies reported in Reference 15 allowed furthermore to derive a ratio for differential Raman scattering cross sections of a vibrational transition originating from two different electronic states. Assuming thermal equilibrium, and taking into account that the CARS signal is proportional to $(d\sigma/d\Omega)^2$ the authors of Reference 15 could conclude by comparision of observed and calculated CARS spectra, that the differential Raman scattering cross section is identical for both electronic states; a ratio of $(d\sigma/d\Omega)\Pi_{1/2}$: $(d\sigma/d\Omega)\Pi_{3/2} = 1.01 \pm 0.02$ was found. It should be noted that the calculated spectra displayed in Figure 9 were obtained by applying this experimental result.

4. CONCLUSION

In this article we have outlined a classical theory for the calculation of $\chi^{(3)}$ for CARS excitation. The theory was then applied to the Q-branches of the diatomic molecules O_2, N_2 and NO. Excellent agreement was obtained between this theory and experimentally obtained CARS data. Special emphasis has been put on the discussion of interference effects between adjacent Raman lines. It has also been demonstrated that these interferences between various lines are a function of bandwidth.

It should be stated, however, that the simple theory is valid only for the (electronic) non-resonance case. For the latter, as mentioned already above, the third-order non-linear susceptibility has to be treated quantum-mechanically [1,4]. Furthermore, the theory has to be expanded when studies are performed in the pressure range where Dicke narrowing may occur [20] or when high electromagnetic fields are applied such that Stark effects change the CARS spectrum [21,22].

Although most of the content reported here, has been published previously in detail [10-12,15], the author of this article has felt, that it may be advantageous to have it presented comprehensively in a tutorial way - a particular aim of an Advanced Study Institute.

ACKNOWLEDGEMENTS

The author is pleased to acknowledge here the collaboration of various friends. Dr. W. Nitsch has measured and calculated during his Ph.D. work at the University of Munich the CARS spectra of oxygen and nitrogen. Based on Dr. Nitsch's work, Prof. J. Laane has performed many computer calculations during his one year stay as Humboldt U.S. Senior Awardee in my laboratory at the University of Bayreuth. The experimental CARS spectra of nitric oxide were measured by Mr. A. Beckmann and Mr. H. Fietz. To all of them I would like to express my sincere gratitude for the very stimulating collaboration and for fruitful discussions. Finally, financial support from the Deutsche Forschungsgemeinschaft as well as the technical assistance of W. Griessl, R. Haagen, P. Hurych and J. Zeller is gratefully acknowledged.

REFERENCES

1. see for example the article by Beckmann, A., Baierl, P., and Kiefer, W. in this book, page 393.

2. DeWitt, R.N., Harvey, A.B., and Tolles, W.M., NRL Memorandum Report 3260, 1976.

3. Tolles, W.M., Nibler, J.W., McDonald, J.R., and Harvey, A.B., Applied Spectrosc. 31, 253 (1977).

4. Druet, S.A.J., Attal, B., Gustafson, T.K., and Taran, J.P.E., Phys. Rev. A 18, 1529 (1978).

5. Placzek, G. in Handbuch der Radiologie, Marx, G., ed., Vol. 6, Pt. 2, Akademische Verlagsgesellschaft, Leipzig, 1934, p. 205.

6. for a_j and γ_j^2 see for example: Schrötter, H.W. and Klöckner, H.W., in Raman Spectroscopy of Gases and Liquids, Weber, A., ed., Topics in Current Physics, Vol. 11, Springer, Berlin, 1979, p. 123.

7. Fenner, W.R., Hyatt, H.A., Kellam, J.M., and Porto, S.P.S., J. Opt. Soc. Am. 63, 73 (1973).

8. Long, A.D., Raman Spectroscopy, McGraw Hill, New York, 1977, p. CRS 13.

9. Moya, F., Druet, S.A.J., and Taran, J.P.E., Opt. Commun., 13, 169 (1975).

10. Nitsch, W., and Kiefer, W., Opt. Commun. 23, 240 (1977).

11. Nitsch, W., and Kiefer, W., J. Mol. Struct. 45, 343 (1978).

12. Laane, J., and Kiefer, W., J. Raman Spectrosc. 9, 353 (1980).

13. Fletcher, U.H., and Rayside, J.S., J. Raman Spectrosc. 2, 3 (1974).

14. Bendtsen, J., J. Raman Spectrosc. 70, 560 (1974).

15. Beckmann, A., Fietz, H., Kiefer, W., and Laane, J., Phys. Rev. A24, 2518 (1981).

16. Valentin, A., Boissy, J.P., Cardinet, Ph., Henry, A., Chen, D.W., and Rao, K.N., C.R. Acad. Sci. B 283, 233 (1976).

17. Valentin, A., Henry, A., Cardinet, Ph., LeMoal, M.F., Chen, D.W., and Rao, K.N., J. Mol. Spectrosc. 70, 9 (1978).

18. Saenger, K.L., Zare, R.N., and Mathews, C.W., J. Mol. Spectrosc. 61, 216 (1976).

19. Owyoung, A., Opt. Lett. 2, 91 (1978).

20. Hall, R.J., Verdieck, J.F., and Eckbreth, A.C., Opt. Commun. 35, 69 (1980).

21. Rahn, L.A., Farrow, R.L., Koszykowski, M.L., and Mattern, P.L., Phys. Rev. Lett. 45, 620 (1980).

22. Esherik, P., and Owyoung, A., this book, page 499.

COHERENT ANTI-STOKES RAMAN SCATTERING OF GASES

Joseph W. Nibler*

Department of Chemistry
Oregon State University
Corvallis, Oregon 97330

*Humboldt U.S. Senior Scientist Awardee-Tech. Univ. of Munich

1. INTRODUCTION

The first measurements involving Coherent Anti-Stokes Raman Scattering (CARS) in gases were by Rado (1967) [1], who determined the nonresonant susceptibilities of a number of gases by using a ruby laser and a H_2 Stokes beam stimulated in a high pressure hydrogen cell. By employing pressure tuning of the Stokes beam, DeMartini (1972) [2] used CARS to scan the H_2 Q_1 line and measure the linewidth and Dicke narrowing as a function of pressure. However, the major impetus for development of the technique came from the pioneering experiments of Taran and coworkers (1973) [3] in which the important potential application of CARS to concentration measurements in combustion systems was demonstrated.

With the subsequent development of tunable dye lasers of high power and narrow linewidth, the CARS experiment has become much more convenient and much progress has been made in its application. Its success in the study of flames and practical combustors such as automobile and jet engines has been quite impressive and will be discussed in other lectures. The use of CARS in studies of electric discharges and plasmas has been demonstrated [4] and it has also been utilized to measure rotational temperatures and relaxation rates in free jet expansions [5-8]. With picosecond sources, CARS has been applied to the study of gas phase reaction dynamics [9] and relaxation mechanisms [10].

The above applications are based on <u>intensity</u> measurements

W. Kiefer and D. A. Long (eds.), Non-Linear Raman Spectroscopy and Its Chemical Applications, 261–280.

of CARS signals for molecules with known vibrational resonances. One can also take advantage of the high resolution capabilities of CARS to make accurate determination of Raman frequencies and linewidths. With pulsed lasers, high resolution (~0.01 cm^{-1}) vibrational [11,12] and rotational [13] CARS spectra have been reported and even higher resolution (~0.001 cm^{-1}) has been achieved using CW sources [14-17]. A more detailed discussion of many of these results is presented in a later section of this paper and in review articles [18-21].

2. THEORETICAL ASPECTS IMPORTANT FOR GASES

As discussed in earlier talks, the intensity (I) generated at the anti-Stokes frequency (ω_3) by plane wave pump (ω_1) and Stokes (ω_2) beams is given by [20-22].

$$I_3 = \frac{16\pi^4\omega_3^2}{n_1^2 n_2 n_3 c^4} |3\chi^{(3)}|^2 \, I_1^2 \, I_2 \, L^2 \left[\frac{\sin(\Delta kL/2)}{\Delta kL/2}\right]^2 \qquad (1)$$

where n_i is the index of refraction of the medium at ω_i, $|3\chi^{(3)}|$ is the third order susceptibility, L is the length of interaction in the medium and $\Delta k = 2k_1-k_2-k_3$ is the momentum mismatch of the beams. The term in brackets is a maximum for $\Delta k = 0$, a condition achieved for liquids and solids by crossing the ω_1 and ω_2 beams at a small angle (<5°). For gases at low pressure, the dispersion is so small that collinear beams can be used and the term in brackets is unity for lengths of many centimeters.

Because of the $I_1^2 I_2$ power density dependence, large signal gains are obtained by focusing the beams. One treatment [22] of the effect of focusing assumes the focal region to be a cylinder of plane waves of diameter $W_o = 4\lambda f/\pi W$ and length equal to the confocal parameter $b = \pi W_o^2/2\lambda_1$. Here f is the lens focal length and W is the beam diameter at which the intensity has dropped to $1/e^2$ of the peak value. Taking L = b, $\Delta k = 0$, $n_i = 1$ and $P_i = I_i \pi(W_o/2)^2$, eq. (1) becomes

$$P_3 = \left(\frac{2}{\lambda_1}\right)^2 \left(\frac{4\pi^2\omega_3}{c^2}\right)^2 |3\chi^{(3)}|^2 \, P_1^2 P_2 \qquad (2)$$

We note the fortuitous cancellation of beam diameter and lens focal length in this expression. More exact numerical calculations [23] show about a three fold increase in signal as f goes from 50 cm to 3 cm, a variation in accord with experimental observations [20]. The same calculations show that about 75% of the signal is generated within a length of 6b, implying that good resolution is possible for spatial probing of flames, plasmas, expansion jets and other inhomogeneous media.

The bulk susceptibility $|3\chi^{(3)}|$ contains the frequency resonances and concentrations of the species in the sample and is written as

$$|3\chi^{(3)}(\omega_3)| = \sum_i \left(\chi_{NR_i} + \sum_v \chi_{iv}\right) \tag{3}$$

Here χ_{NR_i} is the nonresonant electronic susceptibility of species i and $\sum_v \chi_{iv}$ measures the vibrational-rotational contributions to the susceptibility at frequency ω_3. For a particular resonance ω_v, the latter term can be written [21]

$$\chi_{iv} = \frac{A_v}{2\delta\omega - i\Gamma_v} \tag{4}$$

where

$$A_v = \frac{2N\Delta_v c^4}{h\omega_2^4}\left(\frac{d\sigma}{d\Omega}\right)_v \tag{5}$$

In these equations $\delta\omega$ is a measure of the detuning $(\omega_v - \omega_1 + \omega_2)$, Γ_v is the full width at half maximum of the Raman resonance, $N\Delta_v$ is the population difference between lower and upper states of the transition and $\left(\frac{d\sigma}{d\Omega}\right)_v$ is the spontaneous Raman cross section. These relations, or slight modifications of them, have been derived by both quantum mechanical and semiclassical methods and all yield the Raman cross section as a factor in χ_{iv}. Thus CARS and Raman spectroscopy have the same selection rules and both give the same basic frequency and polarization information [24,26].

Eqs. (3-5) also reveal the basic quadratic dependence of the CARS signal on number densities which means a signal decrease of $\sim(10^3)^2$ for gases compared to liquids or solids. Fortunately much of this loss can be recovered due to the $1/\Gamma^2$ linewidth dependence of P_3 since Γ is typically 10 to 100 times smaller for gases compared to liquids. The narrower resonances in the gas phase also stand out more easily against the background produced by χ_{NR} so that lower relative concentrations can be measured for gases compared to liquids. (These advantages are, of course, obtained only with ω_1 and ω_2 beams which are narrower than Γ). Finally, much higher laser powers (1-10 MW) can be used for gases compared to liquids (0.01 to 0.1 MW) without sample breakdown or generation of a stimulated Raman Stokes beam.

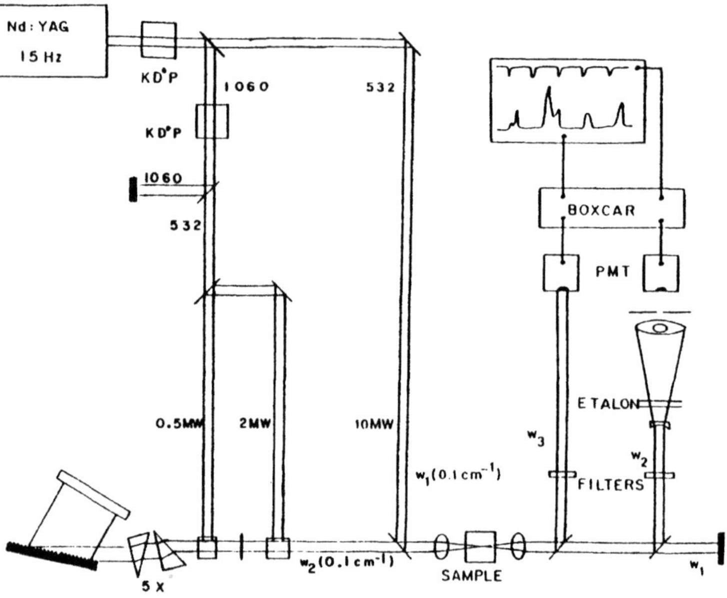

Fig. 1. Apparatus used at OSU to record CARS spectra of gases.

3. EXPERIMENTAL ASPECTS

Although most of the early CARS studies utilized a ruby laser, the higher repetition rate, reliability and more convenient spectral region of the doubled Nd-YAG source has made this the preferred CARS source for pulsed work. Figure 1 is a schematic of the system at Oregon State University which is based on a Nd-YAG oscillator-amplifier (Quanta-Ray) which has been modified by addition of a second etalon and an electronic line narrowing accessory to permit "nearly" single axial mode operation (~60% of the shots). The 1060 nm pulse (10 ns) is doubled in type II KD*P to provide ~50-100 mJ of 532 nm radiation for ω_1, and the residual 1060 nm light is doubled to pump the oscillator (10 mJ) and the amplifier (20 mJ) cells of a dye laser. With a 35% reflective output mirror, a 5x prism expander and a four inch, 2500 ℓ/mm holographic grating used in grazing incidence (Littman design), we obtain ω_2 powers of 0.1 to 2 mJ with a linewidth of 0.10 cm^{-1}. This laser output is expanded with a telescope (not shown in Fig. 1) to match the diameter of ω_1. The beams are made collinear with a dichroic mirror and focused with a 50 to 300 cm focal length lens. After recollimation, ω_3 is separated from ω_1 and ω_2 by dichroic mirrors and filters (and sometimes a small monochromator) and the signal is detected with an RCA 31034A photomultiplier. The resultant pulse is amplified and measured with an analog pulse integrator or else is digitized after capture by a sample and hold amplifier under computer control. Simulaneous measurement is made of the transmission of the ω_2 beam through a 1.00 cm^{-1} etalon to provide relative frequency calibration as the dye laser is scanned in 0.02 cm^{-1} steps.

For broadband or multiplex CARS studies [8,25,26] a simple mirror or low dispersion prism-mirror combination is substituted for the dye laser grating to give an ω_2 width of 50 to 300 cm^{-1}. The sample then selects those frequencies which satisfy the molecular resonance conditions and the ω_3 output consists of several bands. We use a double monochromator (Spex 1402) to disperse these prior to detection by a optical multichannel array (PAR SIT OMA-2) which gives a resolution of ~0.4 cm^{-1} for each of 500 detector elements. The multiplex CARS measurements have the important advantage that broad spectral regions can be examined in one 10 nsec laser flash so that transient events such as explosions or other pulsed chemical reactions can be studied. This method also is useful for measuring turbulence in combustion systems and in aerodynamic flows.

4. DETECTION LIMITS AND NOISE SOURCES

Because of its high efficiency, CARS spectroscopy is very useful for measurements of pure gases at low densities. Fig. 2

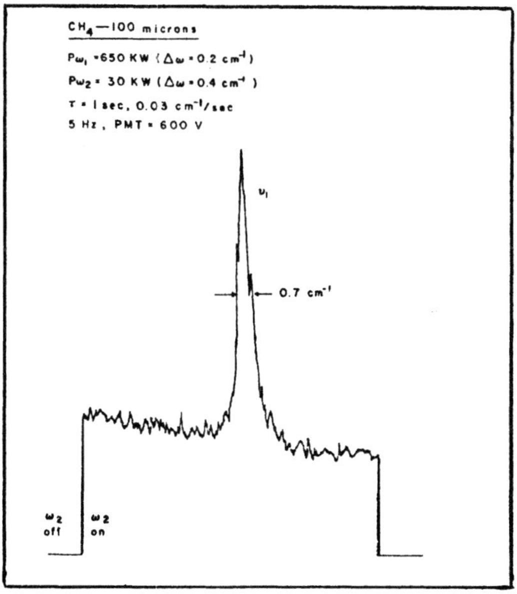

Fig. 2
CARS spectrum of methane at 0.1 torr.

shows the strong ν_1 mode of methane at a pressure of 0.1 torr for relatively low power conditions and modest resolution. The peak signal corresponds to 10^5 photons per 10 nsec laser pulse, much larger than the predicted Raman signal (370 photons), but about 60 times smaller than predicted by equations 2-5. Similar discrepancies of 10 to 100 have been noted by others [15,18] and this is attributed mainly to the poor approximation of real lasers to the monochromatic plane waves used in deriving the theoretical expressions. Closer agreement is expected for CW lasers with higher spectral purity and mode quality and Barrett found this to be the case in his work on H_2 and CH_4 [14].

The high background level of Fig. 2 is produced by χ_{NR} of the gas cell windows and this can be reduced by use of larger beams and longer cells. With these changes and higher laser powers, we estimate that 10^{-3} torr or less of CH_4 could be detected by CARS. In the case of H_2, spectra have been obtained in plasmas at 0.1 torr and a detection limit of 10^{12} cm^{-3} ($\sim 10^{-4}$ torr) has been estimated for a given ro-vibrational state [18]. Unfortunately, addition of other gases degrades the detection capability due to interferences between the resonant and non-resonant susceptibilities. The latter quantities are in the ratio of about 10^4 for CH_4 and air so that a concentration ratio of about 100 ppm is a realistic lower limit for most species. Some improvement may be possible by cancellation of χ_{NR} using polarization schemes [18,27] but these add to the experimental difficulty.

It is, of course, the noise associated with the CARS signal which causes the resonant signal to be lost in the nonresonant background. Zebrowski has recently studied this problem and has examined the contribution of different noise sources to the standard deviations of resonant and nonresonant CARS signals [28]. He assumes that P_3 takes the functional form:

$$P_3 = K \frac{\overline{P}_1^{\,2}\,\overline{P}_2}{4\delta\omega^2 + \Gamma^2} \cdot T(t) \cdot X(q) \cdot D$$

where $\overline{P}_1$, $\overline{P}_2$ represent the average pulse energies of the ω_1 and ω_2 beams, T(t) is some function describing the time overlap of ω_1 and ω_2 pulses, X(q) gives the spatial overlap of ω_1 and ω_2 beams at the focus and D measures the variation due to photon statistics, detector noise, and electronic noise in amplifying and digitizing the signal. In the simplest case, ignoring correlation effects, the percent variance (S_3^2) in P_3 can be expressed in terms of percent variances for each experimental variable as

$$S_3^{\ 2} = 4S_1^{\ 2} + S_2^{\ 2} + 4S_\omega^{\ 2} + S_t^{\ 2} + S_q^{\ 2} + S_d^{\ 2} \tag{6}$$

Measurements or estimates of each of the standard deviations (S_i) were made using sample and reference legs with two separate detectors. For one series of 2500 measurements S_1 and S_2 were measured directly and found to be 1.8 and 18.8% respectively (this value for S_2 was higher than normal due to the use of 355 nm radiation as a pump at energies below the saturation level of the dye laser amplifier). S_d was estimated to be 3.7% by taking one-half the standard deviation of the ratio of P_1 measured simultaneously in both sample and reference legs. S_ω was found to be 15.2% by comparing the difference between S_3 at the sample detector for a resonant sample (NO gas) and for a nonresonant sample (glass). S_q was deduced to be 18.7% by placing nonresonant samples in both legs and comparing the standard deviations in the ratio. S_t was not directly measurable but was calculated from eq. 6 using the measured value of S_3 and the

Table 1 % Noise Contributions in a CARS signal for NO

S_3^2CARS	=	$4\cdot S_1^{\ 2}$	+	$S_2^{\ 2}$	+	$S_d^{\ 2}$	+	$4\cdot S_\omega^{\ 2}$	+	$S_q^{\ 2}$	+	$S_t^{\ 2}$
$(53.8)^2$	=	$4(1.9)^2$	+	$(18.8)^2$	+	$(3.7)^2$	+	$4(15.2)^2$	+	$(18.7)^2$	+	$(36.4)^2$
1.000	=	.005	+	.122	+	.005	+	.319	+	.121	+	.458

other standard deviations. Table 1 summarizes the results and the last line shows the relative importance of the various noise terms. It is concluded that the dominant noise contributions come from time (~46%) and frequency jitter (~32%) associated with the multimode character of the Nd:YAG and dye lasers.

No special effort was made to optimize the experimental conditions for these measurements and the ω_1 laser linewidth (0.1 cm^{-1}) encompassed four cavity modes. Single mode operation of either ω_1 or ω_2 would greatly reduce the time jitter noise but this is difficult to achieve experimentally. A reduction in S_3 can also be achieved by sacrificing resolution in both ω_1 and ω_2 however this has the effect of reducing the peak amplitude of the resonant signal. We have found that a factor of two reduction in S_3 can be achieved by ratioing signals obtained in the resonant and nonresonant legs but this requires considerable care in achieving good beam overlap in both paths. A simpler way to achieve this factor of two improvement in S/N is to average four times as many shots for each resolution element.

It should be emphasized that the 54% standard deviation in the resonant P_3 signal reported here is somewhat high and more typical values with our CARS apparatus are ~30% without ratioing. Taran has reported deviations of 30% or less and suggests that 5% variations are achievable with a well engineered system [18,29]. The use of "robust statistics" to reject the extremes of a measurement set may also be advantageous for nonlinear experiments such as CARS [28].

5. EXAMPLES OF APPLICATIONS

A. Molecular Spectroscopy

Although CARS is considerably less convenient than Raman spectroscopy, it is attractive because of its high resolution capability and sensitivity at sample pressures low enough to reduce collision broadening of lines. Fig. 3 shows an early scan of the CH_4 spectrum obtained at modest resolution in about 30 minutes [21]. Much vibrational-rotational structure of the ν_1 and ν_3 CH stretching modes is seen and the large relative reduction of the A_1 ν_1 mode is apparent in the lower trace in which the ω_1 and ω_2 polarizations were perpendicular. Scans of the ν_1 Q branch of Figures 1 and 2 have been obtained with CARS at much higher resolution (~0.01 cm^{-1}) by Boquillon and Moret-Bailly [12] and with quasi-CW Raman gain spectroscopy (~0.004 cm^{-1}) by Owyoung [30]. These spectra reveal rich detail which is unavailable by conventional Raman spectroscopy and discussion of the assignments and interpretation of this structure may be presented in later lectures at this institute.

In principle CW lasers offer the ultimate in resolution and these have been used to make very accurate measurements of Q branch frequencies and linewidths of H_2 and D_2 by Byer and co-workers [15]. Fig. 4 shows CARS results obtained for the Q branch of C_2H_2 by Fabelinsky et al. [16] using CW lasers with linewidths of 0.001 cm^{-1}. Here the lines are collision broadened and the low signal levels of the CW experiment make it difficult to go to much lower pressures to take full advantage of the laser resolution. Substantial increases in sensitivity can be realized by amplification of the CW lasers with Nd:YAG

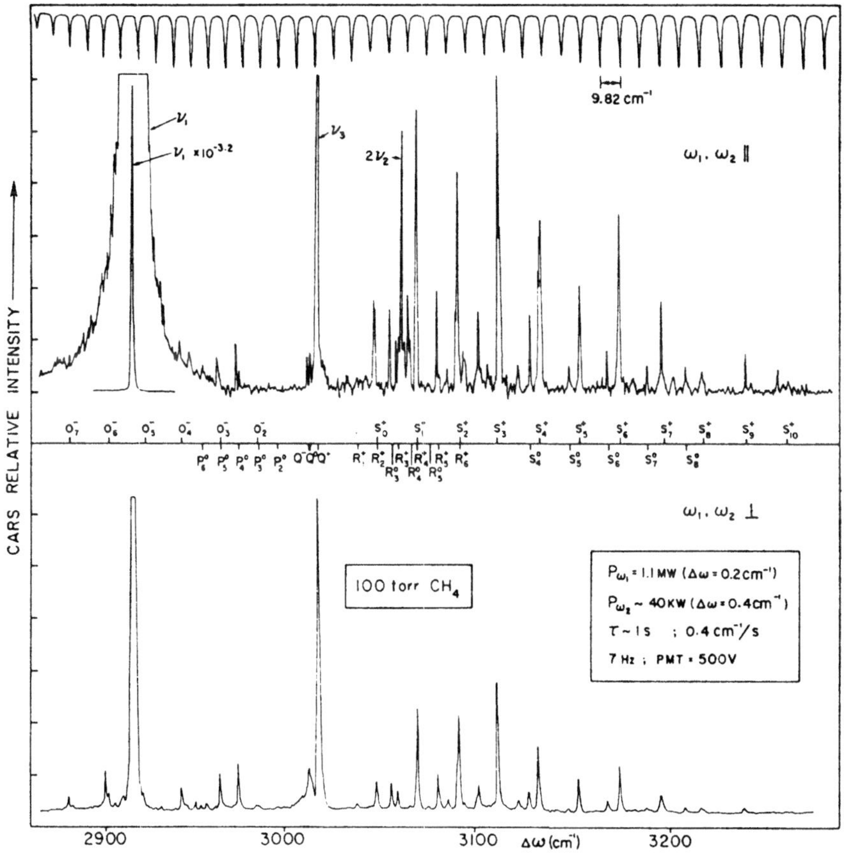

Fig. 3. CARS spectra of CH_4 ν_1 and ν_3 modes for two polarization arrangements.

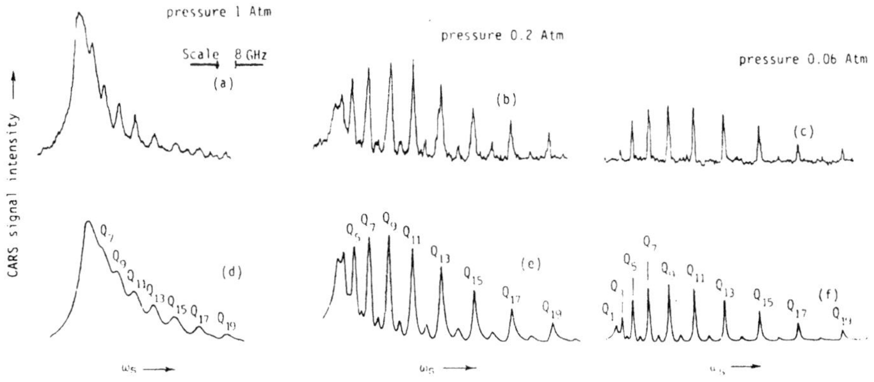

Fig. 4. High resolution CW CARS spectra of the ν_2 Q-branch of C_2H_2 at different pressures. a,b,c = obs., d,e,f = calc. spectra. From Ref 16.

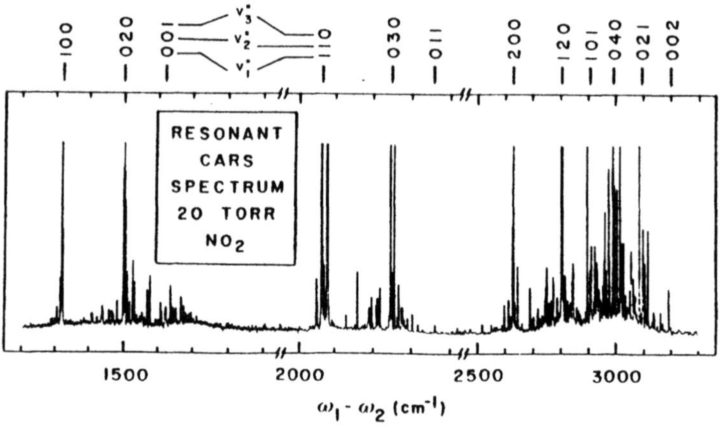

Fig. 5. CARS spectrum of nitrogen dioxide from 1200 to 3300 cm^{-1}.

pumped dye cells and this quasi-CW form of CARS or Raman gain spectroscopy looks the most promising for high resolution molecular spectroscopy. These experiments are expensive and slow since continuous scans of CW lasers are limited to about 1 cm^{-1} intervals. Use of computers can be of assistance in matching overlapping spectral regions so that extended scans such as that of NO_2 in Fig. 5 may be possible. This trace, recorded at modest resolution, is resonantly enhanced so that many molecular overtone and combinations bands are seen. Detailed examination [31] of the structure of these bands shows them to be very sensitive to the ω_1 frequency, a complication which makes the spectral interpretation more difficult but which also offers the promise of yielding details of the energy levels in the upper electronic states.

B. CARS Spectra of Molecules in Free Jet Expansions

Recently, it has been demonstrated that CARS and CW Raman gain spectra can be obtained under the low density-temperature conditions which occur in supersonic expansions [5-8,32]. In the free flow regions of such jets one can obtain collision-free spectra at very low temperatures and at gas densities far in excess of equilibrium values. As an example, Fig. 6 shows CARS spectra obtained for the vibrational Q branch of O_2 under static and

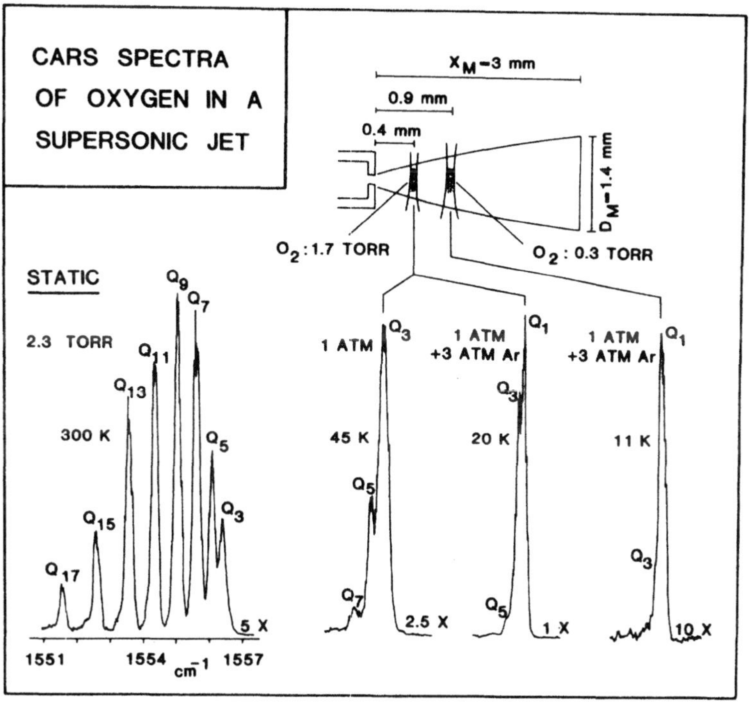

Fig. 6. CARS spectra of the Q branch of oxygen under static conditions and in a supersonic jet. Stagnation pressures and the deduced rotational temperatures are indicated. The nozzle diameter was 0.060 mm and the laser energies were P_1 = 33 mJ, P_2 = 0.2 mJ. The shaded region in the laser focus represents the volume over which ~90% of the CARS signal is generated. The sample densities at these points are the same as those for static room temperature samples at the indicated pressures (i.e., 1.7 and 0.3 Torr). Relative expansion factors of the vertical intensity scale of this figure are indicated.

expansion conditions. The dramatic cooling effect is quite obvious and by comparison with calculated curves it is possible to determine rotational temperatures with an accuracy of a few degrees. Such information is not available for O_2 and many other molecules by the more common laser induced fluorescence and ionization methods.

It is noteworthy that ~90% of the CARS signal is generated in the shaded focal volume of Fig. 6 so that detailed spatial probing of density and temperature within the jet is possible. This nonlinear dependence on laser intensity also means that only modest pumping systems are needed since the free flow zone (defined by the Mach disc parameters X_M and D_M of Fig. 6) need not be large. Because of the quadratic dependence on number density, the CARS intensity increases dramatically as the Boltzmann distribution is narrowed in cooling. This is illustrated in Fig. 6 where the addition of 3 atm. of argon as a driving gas to 1 atm. O_2 drops the rotational temperature from 45 K to 20 K and the peak signal increases by a factor of 2.5. This intensity gain will be much larger for more complex molecules where jet cooling will produce even greater population increases in the lowest few rotational states.

Similar CARS spectra have been obtained with N_2 jets, a special case in which an independent measurement of T_{ROT} is possible by electron beam fluorescence techniques. The agreement is excellent for comparable expansion conditions and, for both N_2 and O_2, the rotational temperature is significantly higher than the predicted translational temperature. For example, at the 0.4 mm position of Fig. 6, T_{ROT} is measured as 45 K, nearly twice the calculated translational temperature of 26 K. Using a simple relaxation model, one can calculate T_{ROT} for various assumed values of an average rotation-to-translational collision number (Z_{ROT}). In this manner, we have obtained values of 10 ± 4 (N_2) and 14 ± 4 (O_2) for the number of collisions for rotational energy loss for these gases [8].

By means of high resolution CW or quasi-CW CARS methods Byer and coworkers have obtained excellent spectra of CH_4 and C_2H_2 in continuous and pulsed jets [5,6]. Substantial reductions in the Doppler width occur in the cooled jets and linewidth measurements offer a means of deducing translational temperatures. By crossing laser and molecular beams at angles of about 45°, CARS measurements of Doppler shifts can also yield an accurate measurement of the velocities of supersonic jets [33]. Similar applications are possible using Raman gain techniques [30,32,34].

Expansions of hydrogen and deuterium are of particular interest since rotational to translational energy (R → T) changes in collisions yield information about the anisotropic terms in the $H_2 \ldots H_2$ interaction potential. R → T cross sections thus can provide a useful basis for judging the accuracy of theoretical potential surfaces of these prototypical collision problems. Since the rotational spacing of D_2 and H_2 is quite large, only a few rotational levels are populated and relaxation is much less efficient than for N_2 or O_2. Fig. 7 shows CARS spectra for

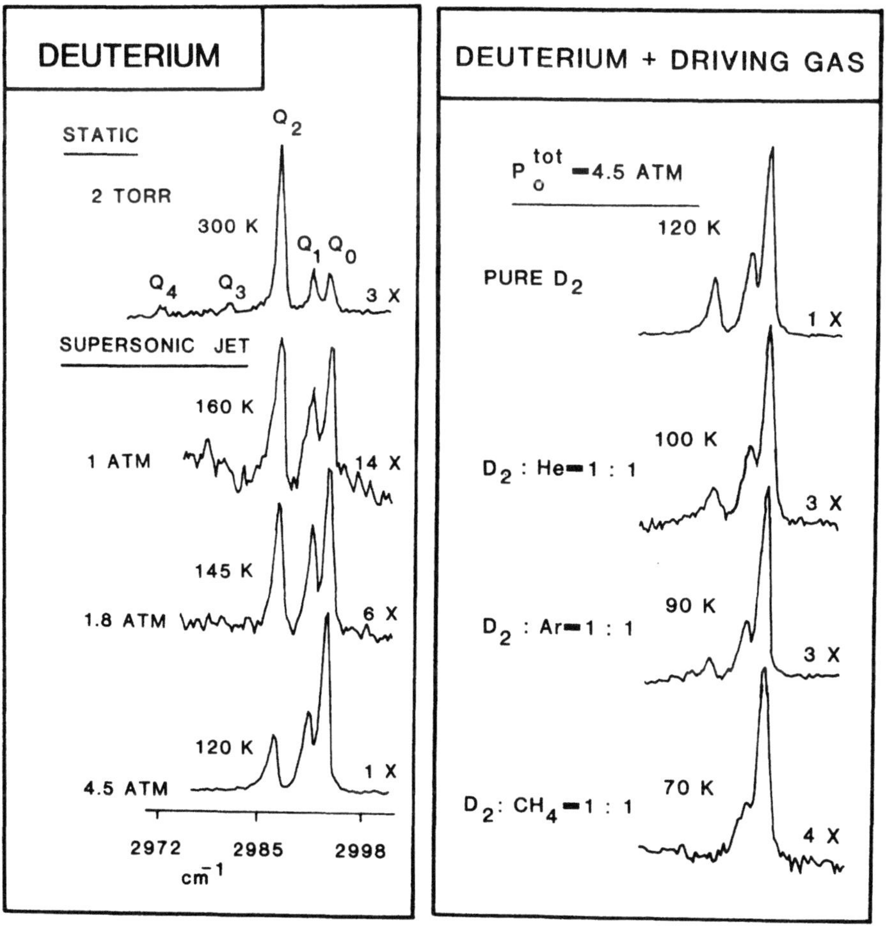

(a) (b)

Fig. 7. Broadband CARS spectra of the Q branch of deuterium. The nozzle diameter was 0.080 mm and X = 0.60 mm. Laser energies were P_1 = 33 mJ and P_2 (~120 cm^{-1} linewidth) = 0.7 mJ. The spectra are the result of averaging 4000 shots obtained with an OMA.

deuterium expansions with the resultant values of rotational temperatures. The simple relaxation analysis gave a value of about 130 for Z_{ROT} for $D_2...D_2$ collisions but the variation of this value with driving pressure was large. A more detailed analysis using different cross sections for the various level changes seems to be required. Nonetheless, the general trends in the relaxation efficiencies can be seen directly in the CARS spectra of $D_2...X$ mixtures. This is illustrated in Fig. 7b in which it is clear that the efficiency of transfer of rotational energy increases in the order $X = D_2 < He < Ar < CH_4$.

It should be mentioned that the D_2 spectra of Fig. 7 were obtained by the use of a broadband ω_2 source and an optical multichannel analyzer. Although the detector sensitivity is about a factor of 10 lower than with a photomultiplier, the signals are sufficient to permit rapid accumulation of spectra with excellent signal to noise ratios. More accurate relative intensity ratios can be obtained in the broadband CARS measurements and, because of the low ω_2 power density at each frequency, there is less probability of saturation broadening which can affect intensity measurements at high powers [35]. Fig. 8 shows broadband spectra recently recorded for hydrogen jets in 1.5 minutes with a S/N of about 100 for the weaker Q_3 lines. The spectra show expansions of n-H_2 and of a mixture equilibrated over an ortho-para conversion catalyst at 77 K prior to expansion. By such studies of even purer para H_2 and ortho D_2, it may be possible to deduce accurate state-to-state rotational relaxation cross sections for hydrogen.

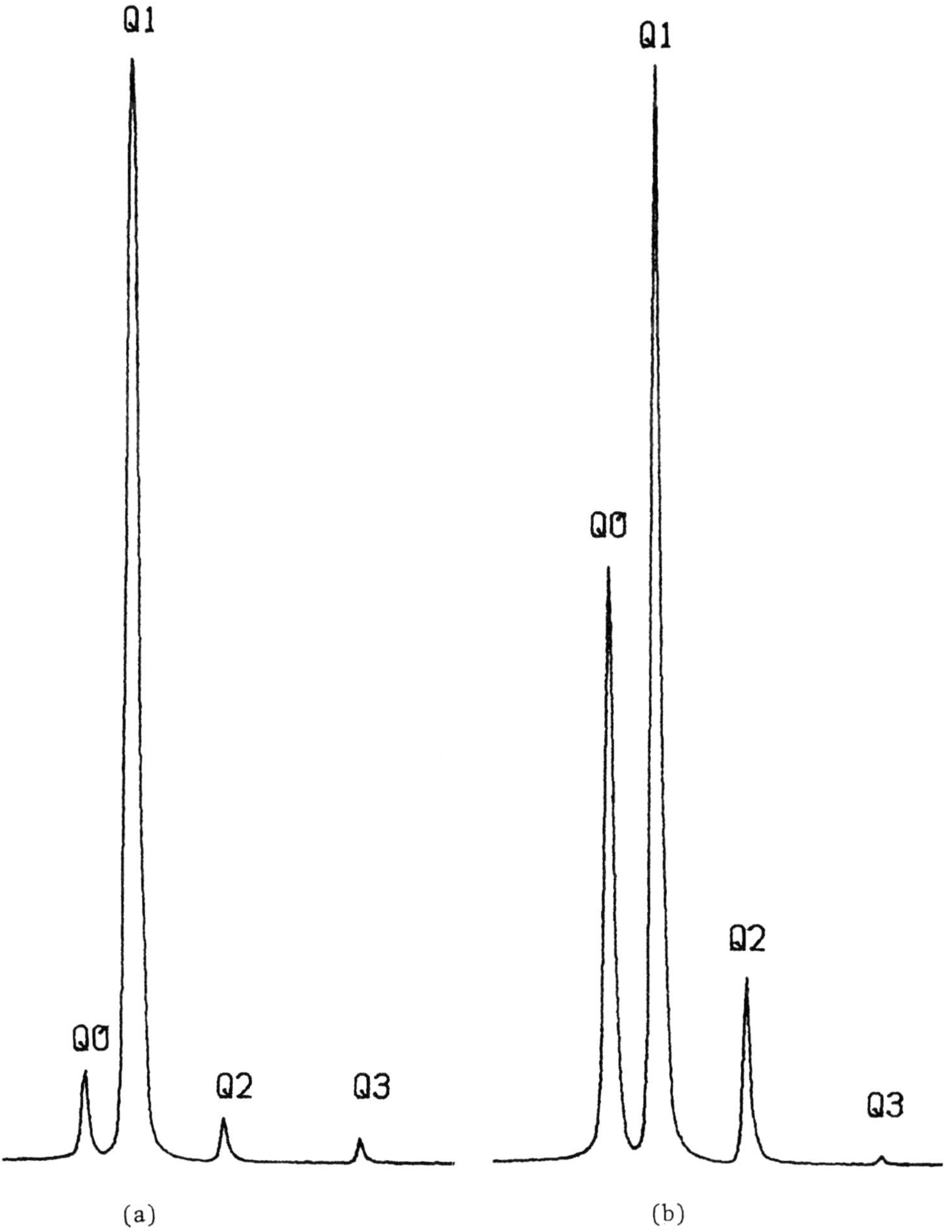

Fig. 8. Broadband CARS Spectra of n-H_2 in a free jet. P_o = 2000 Torr. Spectrum (a) is of n-H_2; spectrum (b) is of H_2 equilibrated over ortho-para conversion catalyst at 77 K prior to expansion at T_o = 295 K. Spectra are for 1000 scans recorded in 1.5 minutes.

C. CARS Spectra of Molecular Photofragments

Because of the efficiency and short duration of the CARS measurement with pulsed lasers, the technique holds promise for the study of transient species such as those generated by laser or flash discharges. The first such application was to the multiphoton fragmentation of benzene and other aromatic molecules [21,36]. Using the 266 nm fourth harmonic of the Nd-YAG laser as a photolysis source, one obtains a complex spectrum which is especially rich in the region near $\omega_1 - \omega_2 = 3000\ cm^{-1}$. The

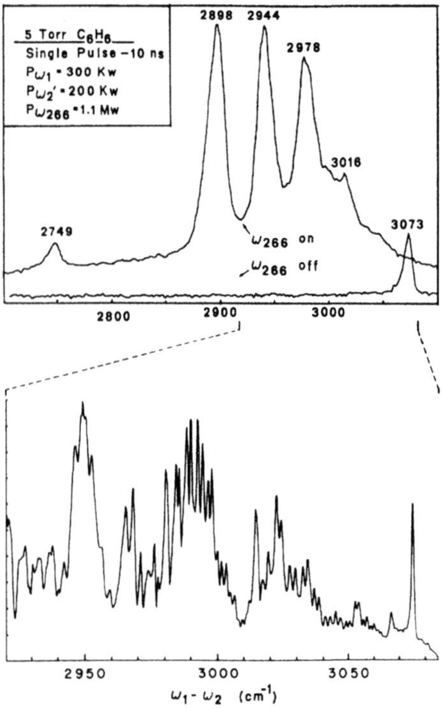

Fig. 9. Broadband CARS spectra of 266 nm photolysis products of benzene. The upper traces are for a single laser pulse. The bottom trace is for 500 scans at higher resolution.

upper traces of Fig. 9 show broadband spectra of this region obtained in a single laser flash with and without the 266 nm beam on. The spectrum is recorded before the products have undergone any collisions and hence represents the energy distribution of the photofragments at the moment of birth. With signal averaging over 500 scans, a higher resolution spectrum reveals much rovibronic structure. Detailed studies show that most of this structure remains for C_6D_6 and C_2H_2 samples and that these features correspond to overtone transitions of the C_2 molecule in the triplet a state [36, 37]. This species is formed as a minor product of multiphoton dissociation of benzene and other hydrocarbon parents and the enormous signal intensities observed are characteristic of enhancements due to electronic resonances of ω_1, ω_2 and ω_3 in C_2 [37].

Recently, Valentini et al. [38] have used CARS to determine the nascent rotational and vibrational state distribution of O_2 fragments produced by photolysis of O_3 in the 530-580 nm region. Even without resonance enhancement, they obtain excellent spectra which show non-Boltzmann rotational distributions peaked at high J values (~30) for v = 0-4 vibrational states of O_2. These results are particularly impressive in view of the fact that the cross section for visible photo dissociation of ozone is orders of magnitude smaller than for most dissociation processes. By using U.V. photolysis lasers and time delaying the CARS probe beams, collisional relaxation or chemical reactions of atomic and molecular fragments could be examined.

6. SUMMARY AND CONCLUSIONS

The objective of the previous sections has been to indicate some of the basic theoretical and experimental features of the CARS spectroscopy of gases. Although the experiment is still non-trivial and difficult, the CARS method has obvious advantages for many non-routine problems. Of necessity, the examples discussed here were mainly ones in which we have been involved but many additional applications of the CARS technique will be presented in other lectures. It seems clear that CARS and other coherent Raman techniques have advanced significantly in the past few years and it is expected that even more exciting developments will come in the next decade.

7. ACKNOWLEDGEMENTS

The financial support through equipment and research grants by the National Science Foundation is gratefully acknowledged. We are also thankful to the Alexander von Humboldt Stifting for the Humboldt Award. Finally, most of the examples presented here came from the excellent and dedicated work of P. Huber-Wälchli, D. Guthals, K. Gross, S. Hou, J. Zebrowski, T. Lundeen and G. Pubanz and to them I am especially indebted.

REFERENCES

1. Rado, W.G., Appl. Phys. Lett., 11, 123 (1967).

2. DeMartini, F., Giuliani, G.P., and Santamato, E., Opt. Comm., 5, 126 (1972).

3. Regnier, P. and Taran, J-P.E., Appl. Phys. Lett., 23, 240 (1973).

4. Nibler, J.W., McDonald, J.R., and Harvey, A.B., Opt. Commun., 18, 371 (1976); Shaub, W.M., Nibler, J.W., and Harvey, A.B., J. Chem. Phys., 67, 1883 (1977).

5. Duncan, M.D., Oesterlin, P., and Byer, R.L., Opt. Lett., 6, 90 (1981).

6. Duncan, M.D. and Byer, R.L., J. Chem. Phys., 74, 2174 (1981).

7. Huber-Wälchli, P., Guthals, D.M., and Nibler, J.W., Chem. Phys. Lett., 67, 233 (1979).

8. Huber-Wälchli, P. and Nibler, J.W., J. Chem. Phys., 76, 273 (1982).

9. Greene, B.I., Weisman, R.B., and Hochstrasser, R.M., Chem. Phys. Lett., 59, 5 (1978).

10. Laubereau, A., Wochner, G., and Kaiser, W., Phys. Rev., A13, 2212 (1976).

11. Nitsch, W. and Kiefer, W., Opt. Commun., 23, 240 (1977).

12. Boquillon, J.P., Moret-Bailly, J., and Chaux, R., C. R. Acad. Sc. Paris, B284, 205 (1977); Boquillon, J.P. and Bregier, R., Appl. Phys., 18, 195 (1979).

13. Beattie, I.R., Gilson, T.R., and Greenhalgh, D.A., Nature, 276, 378 (1978).

14. Barrett, J.J. and Begley, R.F., Appl. Phys. Lett., 27, 129 (1975).

15. Henessian, M.A., Kulevsky, L., and Byer, R.L., J. Chem. Phys., 65, 5530 (1977).

16. Fabelinski, V.I., Krynetsky, B.B., Kulevsky, L.A., Miskin, V.A., Prokorov, A.M., Savel'Ev, A.D., and Smirnov, V.V., Opt. Commun., 20, 389 (1977); 21, 225 (1977).

17. Hirth, A. and Volrath, K., Opt. Commun., 18, 213 (1976).

18. Druet, S.A.J. and Taran, J.P., "CARS Spectroscopy", Progress in Quantum Electronics 7, 1 (1981).

19. Eckbreth, A.C. and Schreiber, P., Chap. 2 of "Chemical Applications of Nonlinear Raman Spectroscopy", ed. by A.B. Harvey, Acad. Press (1981).

20. Nibler, J.W. and Knighten, G.V., Chap. 7 in "Topics in Current Physics", ed. by A. Weber, Springer Verlag (1977).

21. Tolles, W.M., Nibler, J.W., McDonald, J.R., and Harvey, A.B., Appl. Spectrosc., 31, 253 (1977).

22. Regnier, P., Moya, F., and Taran, J.P., AIAA J., 12, 826 (1974).

23. Regnier, P., ONERA Tech. Note No. 215, 92320 Chatillon, France (1973); Shaub, W.M., Harvey, A.B., and Bjorklund, G.C., J. Chem. Phys., 67, 2547 (1977).

24. Yuratich, M.A. and Hanna, D.C., Mol. Phys., 33, 671 (1977).

25. Roh, W.E., Schreiber, P., and Taran, J.P., Appl. Phys. Lett., 29, 174 (1976).

26. Nibler, J.W., Shaub, W.M., McDonald, J.R., and Harvey, A.B., Chap. 3 of "Vibrational Spectra and Structure", ed. by J.R. Durig, Vol. 6, Elsevier (1977).

27. Rahn, L.A., Zych, L.J., and Mattern, P.L., Opt. Commun., 30, 249 (1979).

28. Zebrowski, J.A., "Experimental Aspects of Optical Three-Wave Mixing", M.S. Thesis, Oregon State University, 1981.

29. Druet, S. and Taran, J.P., in "Chemical and Biochemical Applications of Lasers", ed. by C.B. Moore, Vol. 4, Academic Press (1979).

30. Owyoung, A., Chap. 7 of "Chemical Applications of Nonlinear Raman Spectroscopy", ed. by A.B. Harvey, Acad. Press (1981).

31. Guthals, D.M., Gross, K.P., and Nibler, J.W., J. Chem. Phys., 70, 2393 (1979).

32. Valentini, J.J., Esherick, P., and Owyoung, A., Chem. Phys. Lett. 75, 590 (1980).

33. Gustafson, E.K., McDaniel, J.C., and Byer, R.L., IEEE J. Quantum Electron., QE-17, 2258 (1981).

34. Herring, G.C., Fairbanks, W.M., Jr., and She, C.Y., IEEE J. Quantum Electron., QE-17, 1975 (1981).

35. Duncan, M.D., Osterlin, P., Konig, F., and Byer, R.L., to be published.

36. Gross, K.P., Guthals, D.M., and Nibler, J.W., J. Chem Phys., 70, 4673 (1979).

37. Guthals, D.M., "Coherent Anti-Stokes Raman Spectroscopy of Gases". PhD Thesis - Oregon State University, 1981.

38. Valentini, J.J., Moore, D.S. and Bomse, D.S., Chem. Phys. Lett., in press.

COHERENT ANTI-STOKES RAMAN SCATTERING

J.P. Taran

Office National d'Etudes et de Recherches Aérospatiales
29 avenue de la Division Leclerc
92320 Châtillon, France

1. INTRODUCTION

The possibility of carrying out temperature and concentration measurements in gases by a Raman spectroscopy analysis was suggested and demonstrated about a decade ago. Following some early publications on this subject [1,2], a massive effort was undertaken in order to evaluate the potential of spontaneous Raman scattering in the important areas of atmospheric sounding and combustion diagnostics. In these experiments, the concentrations of the molecular species are deduced from the intensities of their respective Raman bands and the temperature is obtained from the contour of any one of these bands. Detailed accounts of early experimental work can be found, among other publications, in several Project SQUID and AIAA workshop proceedings [3-5]. Important instrumental developments were accomplished, with improved collection efficiencies and signal to noise ratio enhancement. The fields of applications were rapidly delineated and it appeared that spontaneous Raman scattering could prove valuable for the investigation of such easily analyzed samples as cold or warm aerodynamic jets, but was of limited potential in low pressure gases, fluorescent samples or in luminous reactive media.

A non-linear optical technique capable of performing Raman spectroscopy with much improved signal strength was then proposed as a competitor to spontaneous Raman scattering in the specific area of combustion diagnostics [3-6]. This technique is based on a four wave mixing process called Coherent Anti-Stokes Raman Scattering, or CARS. CARS, which is one of many well-known third-order processes, was actually observed as early as 1963 [7,8], and has been since then applied to crystal spectroscopy [9-11] and to the

W. Kiefer and D. A. Long (eds.), Non-Linear Raman Spectroscopy and Its Chemical Applications, 281–323.

measurement of third-order susceptibilities in gases [12-14]. Raman spectroscopy by CARS received a considerable impetus in the early seventies when reliable tunable sources of good optical quality were developed. Progress was made in three important areas, which we shall review in turn.

The first area is that of the theoretical understanding of CARS. Particular attention has been paid to the creation of the non-linear source polarization, to the birth and growth of the signal electromagnetic wave and to energy exchange processes within the material.

In the second chapter, we discuss the application of CARS to practical temperature and concentration measurements in reactive media.

In the last chapter, we present a prospective study of electronic resonance enhancement in CARS, which shows great promise for a novel form of molecular spectroscopy and for sensitive detection of trace species.

2. CARS THEORY

A. General presentation

In gases, CARS is observed when two collinear light beams with frequencies ω_1 and ω_2 (hereafter called laser and Stokes respectively, with $\omega_1 > \omega_2$), traverse a sample with a Raman active vibrational mode of frequency $\omega_v = \omega_1 - \omega_2$. A new wave is then generated at the anti-Stokes frequency $\omega_3 = \omega_1 + \omega_v = 2\omega_1 - \omega_2$ in the forward direction, and collinear with the pump beams (fig. 1).

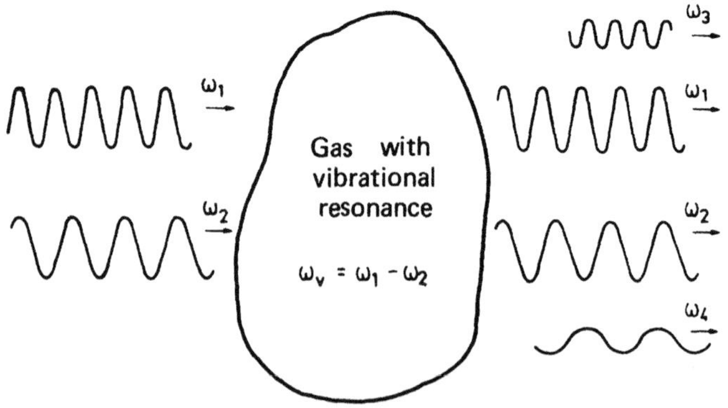

Fig. 1 : CARS and CSRS.

This new wave results from the inelastic scattering of the wave at ω_1 by the molecular vibrations, which are coherently driven by the waves at ω_1 and ω_2 (hence the name of the effect). We note that the same mechanism creates, for reasons of symmetry, a similar wave at $2\omega_2 - \omega_1$ (CSRS for Coherent Stokes Raman Scattering). This wave has been observed and is sometimes used for spectroscopic purposes, in spite of the difficulties connected with background light rejection and poorer detector efficiencies. In the following, we shall give a presentation of the CARS theory and a description of the physical mechanisms.

We recall that, because of non-linearities, it is customary to write the polarization created in a medium by intense optical beams in terms of a power series in the field amplitudes. We have at point $\underline{r}$:

$$\underline{\mathcal{P}}(\underline{r},t) = \underline{\mathcal{P}}^{(1)}(\underline{r},t) + \underline{\mathcal{P}}^{(2)}(\underline{r},t) + \ldots + \underline{\mathcal{P}}^{(n)}(\underline{r},t) + \ldots, \quad (1)$$

The main linear contribution $\underline{\mathcal{P}}^{(1)}(\underline{r},t)$ and those of higher order (which are smaller in general) all can be given an explicit expansion. To this end, we take the applied radiation in the form of monochromatic plane waves. Under steady state conditions, we can expand the total field vector $\underline{\xi}$ as a function of its m distinct frequency components ω_i of wave vector $\underline{k}_i$:

$$\underline{\xi} = \Sigma_i \, \underline{\xi}_i \, (\underline{r},t) \quad (2)$$

$$\underline{\xi}_i \, (\underline{r},t) = \frac{1}{2} \, \underline{E} \, (\underline{r}, \omega_i) \, e^{-i \, (\omega_i t - \underline{k}'_i \underline{r})} + cc. \quad (3)$$

The higher-order frequency components of the polarization expansion (1) now can be written. If ω_s is the frequency of one such component, we write the latter as :

$$\underline{\mathcal{P}}^{(n)}(\underline{r}, t, \omega_s) = \frac{1}{2} \, \underline{P}^{(n)} \, (\underline{r}, \omega_s) \, e^{-i \, (\omega_s t - \underline{k}'_s \underline{r})} + cc, \quad (4)$$

with the phenomenological expansion :

$$\underline{P}^{(n)} \, (\underline{r}, \omega_s) = (\tfrac{1}{2})^{(n-1)} \, \underline{\underline{X}}^{(n)} \, (-\omega_s, l_1\omega_{j_1}, l_2\omega_{j_2} \, l_n\omega_{j_n})$$

$$\times \, \underline{E}_{l_1} \, (\underline{r}, j_1) \, \underline{E}_{l_2} \, (\underline{r}, j_2) \ldots \underline{E}_{l_n} \, (\underline{r}, j_n) \quad (5)$$

and with $\omega_s = \sum_{i=1}^{n} l_i \, \omega_{j_i}$, $\underline{k}'_s = \sum_{i=1}^{n} l_i \, \underline{k}_{j_i}$; here, $\underline{\underline{X}}^{(n)}$ is the

susceptibility tensor of order n (the rank of this tensor is n+1) ; we also specify $l_i = \pm 1$ and $1 < j_i < m$, and

$$\underline{E}_{l_i}(\underline{r}, j_i) = \underline{E}(\underline{r}, \omega_i) \text{ if } l_i = +1$$
$$\underline{E}^*(\underline{r}, \omega_i) \text{ if } l_i = -1$$

$\underline{\mathcal{P}}^{(1)}$ is associated with linear effects (dispersion and absorption) ; $\underline{\mathcal{P}}^{(2)}$, which is responsible for such effects as frequency doubling or parametric conversion, vanishes in media possessing inversion symetry, e.g. centrosymmetric crystals, gases and liquids ; all other even order terms also vanish in these media ; $\underline{\mathcal{P}}^{(3)}$ stands for a large class of effects such as third harmonic generation, and three-wave mixing via two photon and Raman non-linearities.

The source polarization component of frequency ω_s gives birth to an electromagnetic wave at ω_s. This wave is a solution of the wave equation, which can be written :

$$\left(\nabla^2 - \frac{n^2}{c^2}\frac{\partial^2}{\partial t^2}\right) \underline{\xi}_s(\underline{r},t) = \frac{4\pi}{c^2}\frac{\partial^2}{\partial t^2}\underline{\mathcal{P}}^{(n)}(\underline{r},t,\omega_s) \quad (6)$$

for a non magnetic homogeneous medium. We assume here that $\underline{\mathcal{P}}^{(n)}(\underline{r}, \omega_s)$ as a given in Equations 4, 5 is the only source term at frequency ω_s. Its spectral properties or, in other words, the spectral dependence of the nonlinear optical susceptibility tensor $\underline{\underline{X}}^{(n)}(-\omega_s, l_1\,\omega_{j_1}, l_2\omega_{j_2}, \ldots\, l_n\,\omega_{j_n})$ as a function of the applied field frequencies ω_1, ω_2, etc..., are directly reflected in the rate of growth of the signal wave. By tuning the applied field frequencies and monitoring the resultant changes in the signal amplitude, one performs a non-linear optical spectroscopy of the medium. For instance, three-wave mixing spectroscopy when carried out in the proper frequency domain gives information about the Raman active vibrational modes of the medium ; this is the basis for a whole line of experiments comprising pure Raman spectroscopy, analytical chemistry and pressure induced resonances.

There are numerous instances where the source polarization component of Eq. 6 is not unique. In effect, one can often find other terms at the same frequency, with the same order in the non-linearity, but which proceed from a distinct physical mechanism and are associated with a different susceptibility tensor.

Terms with a higher order in the nonlinearity are also possible. All such terms giving contributions at ω_s have to be added to $\underline{\mathcal{P}}^{(n)}(\underline{r}, \omega_s)$ in the right hand side of Eq.(6).

For instance, if two fields at ω_1 and ω_2 are applied with $\omega_1 > \omega_2$, one has two third-order polarization terms at the frequency $\omega_3 = 2\omega_1 - \omega_2$

$$\underline{\mathcal{P}}^{(3)}(\underline{r}, t, \omega_3) = \underline{\mathcal{P}}^{(3)CARS}(\underline{r}, t, \omega_3) + \underline{\mathcal{P}}^{(3)SRS}(\underline{r}, t, \omega_3) \quad (7)$$

The first one is the CARS component :

$$\underline{\mathcal{P}}^{(3)CARS}(\underline{r}, t, \omega_3) = \frac{1}{2}\underline{P}^{(3)CARS}(\underline{r}, \omega_3)\, e^{i(\underline{k}'_3 \underline{r} - \omega_3 t)} + cc \quad (8)$$

with

$$\underline{P}^{(3)\,CARS}(\underline{r}, \omega_3) = \frac{1}{4}\,\underline{\underline{X}}^{(3)CARS}(-\omega_3, \omega_1, \omega_1, -\omega_2)\,\underline{E}^2(\underline{r}, \omega_1)\,\underline{E}^{*}(\underline{r}, \omega_2)$$

and $\underline{k}'_3 = 2\underline{k}_1 - \underline{k}_2$, while the second one reflects the stimulated Raman scattering (inverse Raman scattering) interaction between the waves at ω_3 and ω_1 :

$$\underline{\mathcal{P}}^{(3)\,SRS}(\underline{r}, \omega_3) = \frac{1}{4}\underline{\underline{X}}^{(3)\,SRS}(-\omega_3, \omega_1, -\omega_1, \omega_3)\,|E(\underline{r}, \omega_1)|^2 \quad (9)$$

$$\times\, \underline{E}(\underline{r}, \omega_3) \times e^{i(\underline{k}_3 \cdot \underline{r} - \omega_3 t)} + cc$$

The latter is negligible in a CARS experiment, since the susceptibility components are of comparable magnitude and since

$$|\underline{E}(\underline{r}, \omega_3)| \ll |\underline{E}(\underline{r}, \omega_1)|\,, |\underline{E}(\underline{r}, \omega_2)|\ ; \quad (10)$$

$\underline{k}_3$ is here the wave vector of the anti-Stokes wave. Similarly, the third-order polarization terms at ω_1 and ω_2 can also be broken down into equations similar to Eq. (6). For these, however, the stimulated Raman scattering term is the stronger.

In conclusion, we have two separate problems to solve in non-linear optical spectroscopy :

- derivation of all the relevant non-linear susceptibility terms,
- calculation of the electric field solution of the wave equation.

The derivation of the non-linear susceptibility terms, which is essential in predicting the spectral properties of a medium, can be done through several distinct approaches. In the case of third-order Raman-type non-linearities, the classical Placzek model of molecular polarizabilities leads to a rapid calculation of essential results. It gives good insight into the physical mechanisms, but is inadequate for the case where one or more of the light waves is in resonance with one-photon absorption frequencies of the Raman-resonant species. Quantum mechanical derivations are more accurate. Those based on a wavefunction representation are often sufficient and have been employed extensively using a perturbative treatment of the electric field interactions. We prefer the density operator formalism which, in association with a Feynman-like diagrammatic representation, leads to a rapid derivation of all relevant susceptibility terms and to an easy interpretation of the physical mechanisms involved. The tensor properties of the susceptibility components also follow easily.

The search for the wave equation solution is the second major problem. This solution reveals the important properties of the signal generation : phase-matching, energy exchange between the light waves and the matter, pulse shape characteristics and spatial resolution of CARS measurements using focused beams.

B. Derivation of the susceptibility

Our purpose in this section is not to give a complete derivation of the susceptibility, but only an outline of the principles. The quantum state of the scattering molecules is represented, at point r, as is conventional, by the density operator ρ with the well-known equation of motion :

$$\frac{\partial}{\partial t} \rho (\underline{r}, t) = -\frac{i}{h} \left[H_o + V (\underline{r}, t), \rho (\underline{r}, t)\right] + \left.\frac{\partial \rho}{\partial t}\right|_{damp} \quad (11)$$

H_o is the free molecule Hamiltonian with a discrete spectrum of eigenstates $|n>$ corresponding to eigenenergies $\hbar \omega_n$; the Hamiltonian describing the interaction of the molecules with the radiation field is $V (\underline{r}, t) = -\underline{p}. \underline{E} (\underline{r}, t)$ in the dipolar approximation ; $\underline{p}$ is the dipole moment operator ; $\left.\frac{\partial \rho}{\partial t}\right|_{damp}$ is the damping term, which is determined by stochastic processes such as spontaneous emission of light and collisions between molecules.

We assume the perturbation $V(\underline{r},t)$ to be weak enough to allow the solution of Eq.(11) to be expanded in successive powers of $V(\underline{r},t)$. The density operator is then obtained to any order l by the familiar series expansion

$$\rho(\underline{r}, t) = \rho^{(0)}(\underline{r}, t) + \rho^{(1)}(\underline{r}, t) + \ldots + \rho^{(l)}(\underline{r}, t) \quad (12)$$

The lth order term $\rho^{(l)}(r, t)$ is proportional to $V^{l}(\underline{r},t)$ and is obtained by l iterative applications of Eq. (11). The term responsible for the CARS polarization is or order 3, and the polarization is given by :

$$\mathcal{P}^{(3)CARS}(\underline{r}, t, \omega_3) = N \, \mathrm{Tr}\left[\rho^{(3)CARS}(\omega_3, \underline{r}, t)\, \underline{p}\right] \quad (13)$$

where $\rho^{(3)CARS}(\omega_3, \underline{r}, t)$ labels the CARS Fourier component of $\rho^{(3)}$ $(\underline{r}, t)$ at frequency ω_3. Identification between Eqs (8) and (13) eventually yields the expression for the CARS susceptibility tensor.

The entire derivation is straightforward but time consuming. Recently, diagrammatic representations of all possible density operator evolutions have been introduced for the treatment of nonlinear optical processes. These representations give useful insight into the microscopic physical mechanisms [15-24]. They are applied with a set of simple rules which allow one to rapidly calculate all the relevant susceptibility terms [17-24].

Similar representations are used in nuclear physics. In our representation, we use the fact that the density operator at any specified order can be shown to result from a number of contributions ; each of these is associated with a specific time sequence of perturbations to the density operator, or to the ket vector $|\psi\rangle$ and its complex conjugate $\langle\psi|$ (in the pure state case) ; the time-ordering of the perturbations to $|\psi\rangle$ with respect to those to $\langle\psi|$ is of crucial importance in the case of collisional relaxation. Each of these elementary time-ordered contributions can be visualized by means of a double-sided Feynman-like diagram. Ordinary Feynman diagrams [25] have been used in non-linear optics [26-29]. However, their application is limited to the case where simplifying assumptions on collisional rate are made [17,18,30] and they do not depict the physical processes as clearly. In a double-sided diagram, the time evolution of the density matrix is depicted along two parallel vertical bars (one for each subscript of the density matrix) with time increasing upwards. Each interaction with the electromagnetic field is represented by a segment pointing downwards from a vertex if it corresponds to a term oscillating as $e^{-i\omega_j t}$ in the interaction Hamiltonian V (t) and pointing upwards if the term oscillates as $e^{+i\omega_j t}$. The vertex is on the left or right hand side vertical bar depending on whether the left or right hand side subscript of the density matrix element is changed through the interaction. The eigenstates between which the interaction Hamiltonian is operating are indicated below and above each vertex.

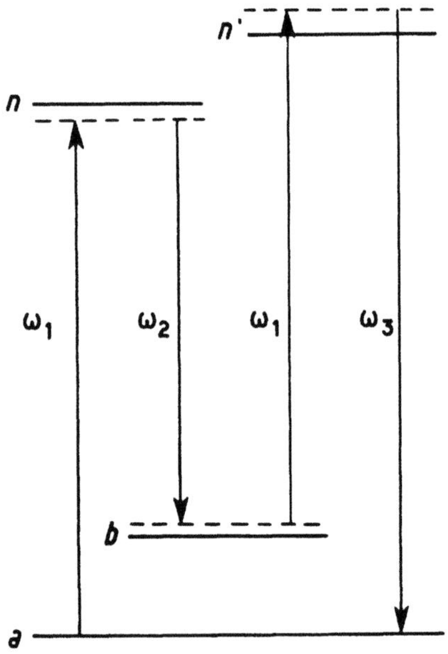

Fig. 2 : CARS energy level diagram showing the transitions in closest resonance with the fields.

In CARS, one must combine two vertices at ω_1 and one at ω_2 in order to get the polarization component $\underline{P}^{(3)}$ (as given in Eq.(8)); its $e^{-i\omega_3 t}$ dependence implies two segments pointing down from the vertex for the interactions with ω_1 and one pointing up for ω_2. If we use a set of four molecular levels a, b, n, n' as shown in the energy level diagram of Fig. 2 (n and n' being of parity opposite to that of a and b), and if the sequence of interactions is applied to be unperturbed density operator $\rho^{(0)}_{aa}$ then there are 24 time-ordered possibilities for this sequence of interactions. Each possible time sequence of interactions gives a distinct contribution to the susceptibility. Another set of 24 terms proportional to $\rho^{(0)}_{bb}$ is also found if state b is populated. It is beyond the scope of this overview to present a detailed account of the rules one uses to derive the susceptibility term from its associated diagram. These rules are found in Refs. [17-24] together with their justification.

The CARS susceptibility can be written

$$\underline{\underline{X}}^{(3)\,CARS}(-\omega_3, \omega_1, \omega_1, -\omega_2) = \underline{\underline{X}}_{NR} + \underline{\underline{X}}_R^{a,\,b,\,n,\,n'} \qquad (14)$$

where $\underline{\underline{X}}_R^{a,\,b,\,n,\,n'}$ is the Raman resonant part associated with the Raman active transition of frequency ω_{ba} between a and b, for the set of two levels n and n'. The tensor element pertaining to a set of particular field polarizations $\hat{e}_1, \hat{e}_2, \hat{e}_3$ is :

$$\hat{e}_3\, \underline{\underline{X}}_R^{a,\,b,\,n,\,n'}\, \hat{e}_1\hat{e}_1\hat{e}_2 = \frac{N}{\hbar^3}\, \frac{1}{(\omega_{ba} - \omega_1 + \omega_2 - i\Gamma_{ba})} \times (A + B)$$

$$\times \left[\rho^{(0)}_{aa}\, (\alpha + \beta) - \rho^{(0)}_{bb}\, (\gamma + \delta)\right] \qquad (15)$$

with
$$A = \mu_{an'}\, \mu_{n'b}\, (\omega_{n'a} - \omega_3 - i\Gamma_{n'a})^{-1},$$
$$B = \mu_{an'}\, \mu_{n'b}\, (\omega_{n'b} + \omega_3 + i\Gamma_{n'b})^{-1},$$
$$\alpha = \mu_{bn}\, \mu_{na}\, (\omega_{na} + \omega_2 - i\Gamma_{na})^{-1},$$
$$\beta = \mu_{bn}\, \omega_{na}\, (\omega_{na} - \omega_1 - i\Gamma_{na})^{-1},$$
$$\gamma = \mu_{bn}\, \mu_{na}\, (\omega_{nb} - \omega_2 + i\Gamma_{nb})^{-1}.$$
$$\delta = \mu_{bn}\, \mu_{na}\, (\omega_{nb} + \omega_1 + i\Gamma_{nb})^{-1}$$

Here, N is the number density of active molecules. The absorption frequencies from states |a> and |b> to state |n> are ω_{na} and ω_{nb} respectively, and the Γ's are the corresponding damping factors ; μ_{an} is the matrix component of the dipole moment operator μ_{an} =

$\langle a|\ \underline{p}\ .\ \hat{e}_1|n\rangle$ where $\hat{e}_1$ is the unit vector in the direction of polarization of the ω_1 field ; μ_{bn}, $\mu_{n'b}$, $\mu_{an'}$ involve interactions with ω_2, ω_1, ω_3 fields respectively. If more than four levels are present, a summation must be taken and the vibrationally resonant part becomes

$$\underline{\underline{X}}_R = \sum_{a,\, b,\, n,\, n'} \underline{\underline{X}}_R^{a,\, b,\, n,\, n'}$$

Molecular spectroscopy by CARS consists in carrying out an analysis of the spectral properties of $\underline{\underline{X}}^{(3)}(-\omega_3^{CARS}, \omega_1, \omega_1, -\omega_2)$. As shown in Eq. (14), the latter contains two parts. In mixtures, part $\underline{\underline{X}}_{NR}$, which is called non-resonant, is contributed both by probed molecules and by the non-Raman-resonant molecular species (diluent molecules). It is composed of terms analogous to those of $\underline{\underline{X}}_R^{a,\, b,\, n,\, n'}$ but with non-resonant two-photon sum or difference denominators in place of the Raman resonance denominator. In the usual case where the number density N of the probed molecules is small compared to that of the dilent molecules, $\underline{\underline{X}}_{NR}$ is mainly contributed by the latter and is therefore a frequency-independent real tensor (provided that there are no one- or two-photon electronic resonances in the diluent molecules). The presence of this non-resonant part is one of the most severe problems in the application of CARS spectroscopy.

We are particularly interested in the spectral properties of the Raman-resonant part (Eq. (15)). Off electronic resonance, i.e. $\omega_1, \omega_2, \omega_3 \ll \omega_{na}, \omega_{nb}, \omega_{n'a}, \omega_{n'b}$, all the coefficients A, B, α, β, γ, δ are of similar magnitude and depend only weakly on the electric field frequencies. If we assume for simplicity, that all fields have the same polarization $\hat{e}_1$, the relevant tensor element then reduces to

$$\hat{e}_1 \underline{\underline{X}}_R^{(3)}\ \hat{e}_1\hat{e}_1\hat{e}_1 = \sum_{ab} \frac{Nc^4}{\hbar^3\omega_1\omega_2{}^3} \left(\rho_{aa}^{(0)} - \rho_{bb}^{(0)}\right) \frac{d\sigma}{d\Omega} \frac{1}{\omega_{ba}-\omega_1+\omega_2-i\Gamma_{ba}} \tag{16}$$

where $\frac{d\sigma}{d\Omega}$ is the spontaneous Raman scattering cross section. The spectral analysis thus reveals the Raman resonances ω_{ba} contained in the denominator of Eq. (16). Identification of these resonances and monitoring of their amplitude are active research areas for analytical chemistry. Furthermore, the other tensor elements of

$\underline{\underline{\chi}}_R$ can be measured by an adequate choice of field polarizations. All these properties will be discussed in chapter III.

When the electronic resonances are approached, only two terms in Eq.(15) become large. Thus we have

$$\underline{\underline{\chi}}_R \simeq \frac{N}{\hbar^3} \frac{1}{\omega_{ba} - \omega_1 + \omega_2 - i\Gamma_{na}} (\rho_{aa}^{(0)} A\ \beta - \rho_{bb}^{(0)} A\ \gamma). \quad (17)$$

We give the time-ordered diagrams from which these two terms have been obtained in Fig. 3. Since A, β and γ all undergo large enhancements and large variations as the field frequencies are varied, the spectral analysis is complicated somewhat. This particular problem of resonance-enhanced CARS is treated in chapter 4.

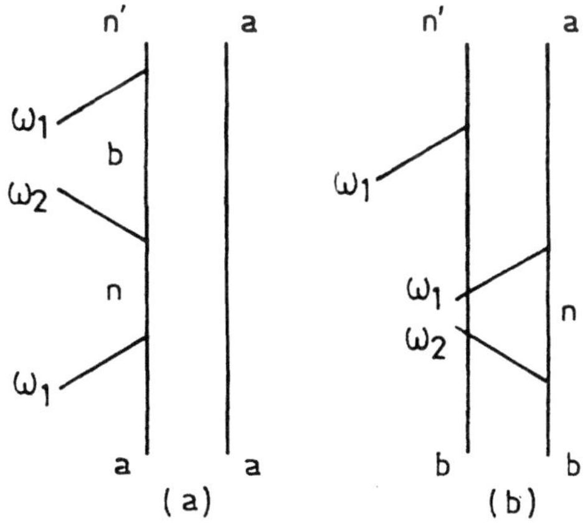

Fig. 3 : Time ordered diagrams representing the density operator evolutions which lead to the two most important resonant CARS susceptibility terms shown in equation 16

a) term proportional to $\rho_{aa}^{(0)}$;

b) term proportional to $\rho_{bb}^{(0)}$.

Indices a, b, n, n' refer to molecular eigenstates as depicted in Fig. 2.

In addition to these main vibrationally resonant terms given in Eq.(15), one has to consider terms which contain vibrational resonances in the excited electronic state and which are generally left out with $\underline{\underline{\chi}}_{NR}$. These terms have been given attention recently. They have been identified and treated as corrections to the main terms in Eq.(15) by Druet et al. and Yee et al. [17,18,24]. Their spectroscopic and physical importance has been recognized by Bloembergen and coworkers [30-33] who drew attention to the fact that they represent vibrational contributions from states that have no initial population, and who experimentally demonstrated their existence in Na vapor ; yet this existence has been disputed by Eesley [29] and Carreira et al. [34] who used a simpler quantum mechanical derivation.

An example of these corrective terms has been treated by Druet and Taran [24, Appendix I]. This example is different from that of Bloembergen et al. [33] but also lends itself to experimental checks in molecular spectroscopy. It is based on the two terms shown in Fig. 4 (with their corresponding resonance denominators) which can be combined with the one in $\rho_{bb}^{(o)}$ of Fig. 3, yielding a susceptibility contribution of the form :

$$N\rho_{bb}^{(o)} \frac{\mu_{an'}\,\mu_{n'b}\,\mu_{bn}\,\mu_{na}}{\hbar^3} \frac{1}{(\omega_{bn}+\omega_2-i\Gamma_{bn})(\omega_{ba}-\omega_1+\omega_2-i\Gamma_{ba})}$$

$$\times \frac{1}{\omega_{n'b}-\omega_1-i\Gamma_{n'b}} \tag{18}$$

$$\times\left[1+\frac{i(\Gamma_{n'a}-\Gamma_{n'b}-\Gamma_{ba})+i(\Gamma_{nn'}-\Gamma_{n'b}-\Gamma_{bn})\dfrac{\omega_{ba}-\omega_1+\omega_2-i\Gamma_{ba}}{\omega_{n'n}-\omega_1+\omega_2-i\Gamma_{n'n}}}{\omega_{n'a}-\omega_3-i\Gamma_{n'a}}\right]$$

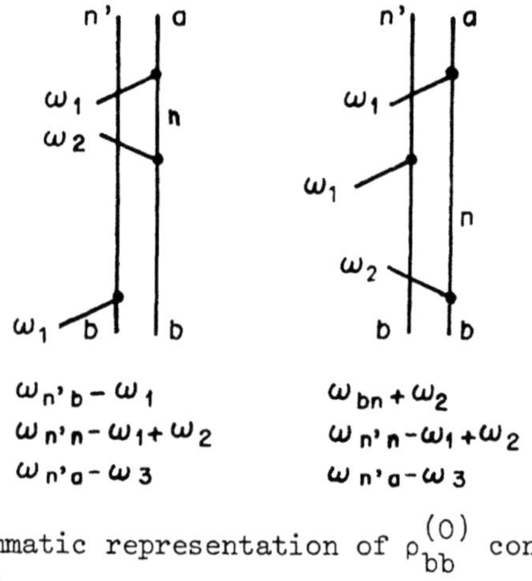

Fig. 4 : Diagrammatic representation of $\rho_{bb}^{(0)}$ contributions to $\chi^{(3)CARS}$ having the resonant denominator $\omega_{n'n} - \omega_1 + \omega_2$.

where the terms have been grouped as in Ref. 30.

We can see that the 3 terms in (18) sum to give one term when there is no damping ($\Gamma = 0$) or if :

$$\Gamma_{n'a} - \Gamma_{n'b} - \Gamma_{ba} = \Gamma_{nn'} - \Gamma_{n'b} - \Gamma_{bn} = 0 \tag{19}$$

This particular equation is what we call the damping approximation. It is satisfied when there is no collisional elastic broadening and if the lifetimes of states $|a\rangle$ and $|b\rangle$ are much longer than those of $|n\rangle$ and $|n'\rangle$. Combining Eq. (18) and Eq. (19). We then obtain :

$$N\rho_{bb}^{(0)} \frac{\mu_{an'}\mu_{n'b}\mu_{bn}\mu_{na}}{\hbar^3} \tag{20}$$

$$\times \frac{1}{(\omega_{bn} + \omega_2 - i\Gamma_{bn})(\omega_{ba} - \omega_1 + \omega_2 - i\Gamma_{ba})(\omega_{n'b} - \omega_1 - i\Gamma_{n'b})}$$

We notice that the resonances at $\omega_{nn'} - \omega_1 + \omega_2$ vanish and that the terms have combined to merely result in a shift of the electronic resonance in the $A\gamma\rho_{bb}^{(0)}$ term of eq. (15) from $\omega_{n'a} - \omega_3$ to $\omega_{n'b} - \omega_1$. The latter expression (20) is usually obtained from perturbation theory in the absence of damping, which had led Carreira and Eesley to their conclusions [29,34].

The interest for these terms is not purely "zoological". They allow interesting Raman spectroscopic information to be collected about unpopulated states, i.e. without having to populate those states beforehand ; they differ in the nature of their broadening by the Doppler effect [23]; they finally pose delicate problems about their true nature and their relation to collisional broadening, as claimed by Grynberg [35]. For all these reasons, they will constitute an active research area in four-wave mixing spectroscopy for the coming years.

C. Wave propagation

We here treat the problem of gas phase CARS. The amplitude of the signal wave in CARS is obtained from the wave equation (6). We take the pump fields in the form of collinear, travelling plane waves as given by equation (3). Looking for an anti-Stokes wave also travelling in the forward direction along the z axis, we can reduce the degree of equation (6), to obtain a steady state equation :

$$\frac{\partial}{\partial z} E_3 = \frac{i\pi\omega_3}{2c} \left| \underline{\underline{X}}^{(3)}{}^{CARS}(-\omega_3, \omega_1, \omega_1, -\omega_2)\, \underline{E}_1^2\, \underline{E}_2^{*}\, e^{i\delta kz} + \underline{\underline{X}}^{(3)}{}^{SRS}(-\omega_3, \omega_1, -\omega_1, \omega_3)\, \underline{E}_1\, \underline{E}_1^{*}\, \underline{E}_3 \right| \tag{21}$$

with $\delta k = k'_3 - k_3 = 2k_1 - k_2 - k_3$; we have taken the refractive index $n \simeq 1$ and we assume the gas to be homogeneous ($\underline{\underline{X}}^{(3)}$ is independent of $\underline{r}$). We have written for simplicity $\underline{E}_1 = \underline{E}(\underline{r}, \omega_1)$, $\underline{E}_2 = \underline{E}(\underline{r}, \omega_2)$, $\underline{E}_3 = \underline{E}(\underline{r}, \omega_3)$ and assumed that $\underline{E}_3$ is a slowly varying function of z :

$$\left|\frac{\partial}{\partial z} \underline{E}_3\right| << k_3 \left|\underline{E}_3\right|$$

Similar equations also hold at ω_1 and ω_2 :

$$\frac{\partial}{\partial z} \underline{E}_1 = \frac{i\pi\omega_1}{2c} \Big[\underline{\underline{X}}^{(3)}{}^{CARS}(-\omega_1, \omega_2, \omega_3, -\omega_3)\, \underline{E}_1^{*}\, \underline{E}_2\, \underline{E}_3\, e^{-i\delta kz} + \underline{\underline{X}}^{(3)}{}^{SRS}(-\omega_1, \omega_3, -\omega_3, \omega_1)\, \underline{E}_1\, \underline{E}_3\, \underline{E}_3^{*} + \underline{\underline{X}}^{(3)}{}^{SRS}(-\omega_1, \omega_2, -\omega_2, \omega_1)\, \underline{E}_1\, \underline{E}_2\, \underline{E}_2^{*} \Big] \tag{22}$$

$$\frac{\partial}{\partial z} \underline{E}_2 = \frac{i\pi\omega_2}{2c} \Big[\underline{\underline{X}}^{(3)}{}^{CARS}(\omega_2, \omega_1, \omega_1, -\omega_3)\, \underline{E}_1^2\, \underline{E}_3^{*}\, e^{i\delta kz} + \underline{\underline{X}}^{(3)}{}^{SRS}(\omega_2, \omega_1, -\omega_1, \omega_2) \left|\underline{E}_1\right|^2 \underline{E}_2 \Big] \tag{23}$$

The set of Equations (21-23) can only be solved numerically. However, if we assume the coupling to be weak, $\underline{E}_1$ and $\underline{E}_2$ can be taken as constants and the SRS term in Eq. (21) can be neglected. Then the latter equation is integrated readily. With boundary condition $\underline{E}_3\big|_{z=0} = 0$, we have

$$\underline{E}_3 \simeq \frac{i\pi\omega_3}{2c} \underline{\underline{X}}^{(3)}{}^{CARS}(\omega_3, \omega_1, \omega_1, -\omega_2)\, \underline{E}_1^2\, \underline{E}_2\, e^{i\delta kz} \frac{\sin(\delta kz/2)}{\delta k/2} \tag{24}$$

$$I_3 = \frac{16\pi^4 \omega_3^2}{c^4} \left| \underline{\underline{X}}^{(3)\,CARS} (-\omega_3, \omega_1, \omega_1, -\omega_2)\hat{e}_1\ \hat{e}_1\ \hat{e}_2 \right|^2 I_1^2\ I_2 \times \left(\frac{\sin(\delta kz/2)}{\delta k/2} \right)^2 \quad (25)$$

where $\hat{e}_1$ and $\hat{e}_2$ are the unit polarization vectors of the pump waves. Equations 24 and 25 constitute the basis for the interpretation of the anti-Stokes wave properties. The most important ones are the following :

1 - the anti-Stokes field polarization vector is oriented along the vector

$$\hat{f}_3 = \underline{\underline{X}}^{(3)\,CARS} (-\omega_3, \omega_1, \omega_1, -\omega_2)\ \hat{e}_1\ \hat{e}_1\ \hat{e}_2,$$

which depends on the applied field polarizations as well as on the tensor properties of the susceptibility. This vector has two independent components associated with the non-resonant and the Raman-resonant parts of the susceptibility (as we have mentioned at the end of part B). This property can be used for non-resonant background cancellation in the spectra.

2 - The CARS signal intensity, which is the parameter directly measured using photodetectors, is proportional to $|\hat{f}_3|^2$ (Eq. (25)), and therefore to the squared number density of the medium. It also has a sinusoidal dependence on z ; in gases, however, we have $\delta k \simeq 0$ because the dispersion is weak, so that the behavior is parabolic over long distances. Pump depletion would eventually limit this parabolic growth.

Energy exchange between the light waves can be analyzed by recasting Eqs.(21-23) into equations for the rates of change of photon number per unit volume and also considering the rate equation for the molecular population change $N \frac{\partial}{\partial t} (\rho_{aa}^{(4)} - \rho_{bb}^{(4)})$. This discussion has been conducted in detail elsewhere [20-24] and we only summarize the conclusions here for the off-electronic resonance case.

The stronger process is the SRS coupling between the ω_1 and ω_2 waves : one photon at ω_1 is converted into a Stokes photon at ω_2 and a quantum of molecular vibration. The rate of this process is proportional to the imaginary part of $\underline{\underline{X}}^{(3)SRS}$. Although it is not specifically a CARS interaction, this process is important because large vibrational population changes can result ; this can in turn bring higher order corrections to $X^{(3)CARS}$ and bias the results.

The CARS generation mechanism per se is made up of two distinct processes :

(i) a "parametric" process whereby two laser photons at ω_1 are converted into a Stokes photon at ω_2 and an anti-Stokes photon at ω_3 ; the molecules are, on the average, returned to their ground state after the interaction ; this process can be reversed if the phases of the waves are changed and its rate is proportional to the real part of the susceptibility ;

(ii) a "Raman like" process whereby a Stokes photon is converted into an anti-Stokes photon and two vibrational quanta, on the average, are taken away from the molecules ; this process has a rate proportional to the imaginary part of the susceptibility and can be reversed by changing the phases of the waves.

It is noteworthy that the second process is the only one responsible for the anti-Stokes generation exactly on vibrational resonance, since the real part of the susceptibility then vanishes. Yet, the so-called "parametric" process has often been erroneously cited as being the only CARS mechanism. This belief has originated from the fact that the energy level diagram of Fig. 2 gives the misleading impression that the CARS interaction returns the molecules to their initial state. It should be emphasized that such energy level diagrams should be used in nonlinear optics to only depict the establishment of polarizations. It is only in the case of processes like Raman scattering or multiphoton absorption that they can also be used to depict energy exchange without ambiguity. Finally the above-mentioned considerations on net energy exchange and molecular population changes cannot be dissociated from the quantum processes themselves. In effect, on the microscopic scale, molecules can undergo sequences of interactions which either return them to their initial state after the final interaction with the anti-Stokes field (e.g. Fig. 3a) or place them in a different vibrational state (e.g. Fig. 3b).

3. PRACTICAL APPLICATION OF CARS

A. General considerations

The laws governing the signal growth and the spectral properties of CARS have been established in the preceding chapter. We here show how CARS can be used for practical measurements and what level of performance can be obtained.

Spatial resolution

Unfocussed parallel beams with large diameters are seldom used because no spatial resolution is possible in this geometry.

Since the growth of the power density I_3 is proportional to $I_1^2 I_2$, it seems advantageous to focus the beams to a small diameter and to use high peak power sources. If the condition $\delta k = 0$ is assumed, then it can be shown that :

(1) the anti-Stokes flux is contained within the same cone angle as the pump beams emerging from the focal region ;

(2) this flux is generated for the most part within the focal region (where $I_1^2 I_2$ is large) ;

(3) the total power in the anti-Stokes beam some distance beyond the focus is independent of beam diameter and focal length and is approximately given by

$$P_3 = \left(\frac{2}{\lambda}\right)^2 \left(\frac{4\pi^2 \omega_3}{c^2}\right)^2 \left| \hat{f}_3 \right|^2 P_1^2 \; P_2 \, , \qquad (26)$$

where refractive indices were taken as unity, where $\lambda = 2\pi\ c/\omega$ with $\omega \simeq \omega_1 \simeq \omega_2 \simeq \omega_3$ and where P_1 and P_2 are the powers at ω_1 and ω_2 respectively. This expression was obtained by assuming that all the signal is generated from a small cylindrical volume about the focus having a length equal to the confocal parameter l of the beams. If Gaussian beams are used, the beam waist at the focus is $\phi = 4\lambda f/\pi d$ where f is the focal length of the lens and d the beam diameter in the plane of the lens : we also have $l = \pi\phi^2/2\lambda$. In reality, 75% of P_3 are generated from a volume of length $6l$ as shown by numerical calculations.

In practical experiments, the spatial resolution is on the order of 1 to 20 mm with laser beams of good optical quality. This may still be insufficient in some experimental situations where higher spatial resolutions are needed. A particular beam arrangement called BOXCARS has been proposed for better resolution [36]. In this arrangement, the beams are crossed at a small angle so that the polarization wave vector k'_3 remains equal to the anti-Stokes signal wave vector k_3 (Fig. 5). The beam configuration is shown in Fig. 6. The spatial resolution is well under 2 mm.

Spectral information

CARS spectroscopy can be accomplished in various manners depending on the application envisioned (e.g. high-resolution spectroscopy or chemical analysis). The spectra are usually retrieved by holding ω_1 fixed, varying ω_2 so that $\omega_1 - \omega_2$ is swept across the resonances of interest while monitoring the anti-Stokes flux. In gas mixtures, the following information is obtained from the spectra using Eq. (16).

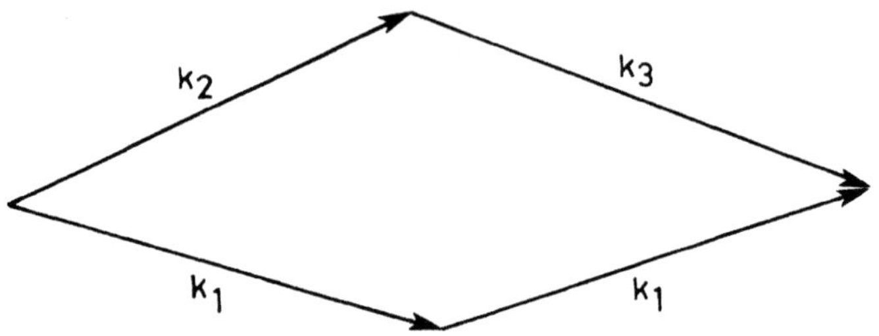

Fig. 5 : Wave vector diagram for crossed-beams phase-matched CARS or BOXCARS.

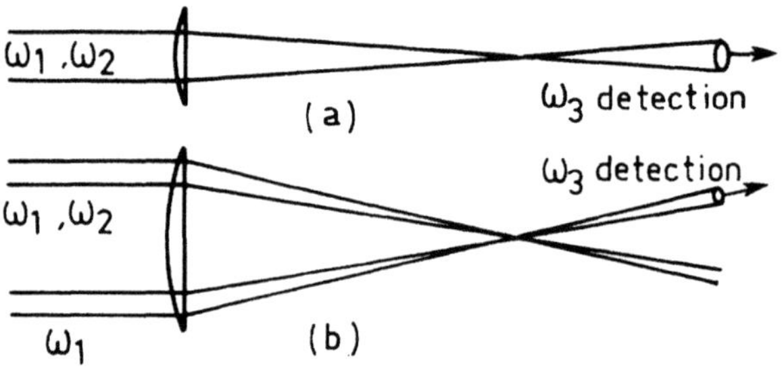

Fig. 6 : Experimental beam arrangement : a) conventional CARS ; b) BOXCARS. In BOXCARS, a conventional CARS beam is also emitted in the direction of ω_2 ; this beam is actually 10 to 50 times stronger than the BOXCARS beam.

- composition since each molecular species has a particular set of vibrational resonances which can seldom be confused with that of other species ;

- temperature since the frequency of any particular vibrational mode depends slightly on the rotational quantum number (in Q-branch transitions) and on the vibrational quantum number ; the resulting splitting can be used to monitor populations in distinct rovibrational states and deduce the rotational and vibrational temperatures using the corresponding Boltzmann coefficients.

We note, however, that the existence of the nonresonant susceptibility poses a problem with the detection of trace species in mixtures, since the nonresonant contribution from the diluent gases may swamp the Raman-resonant part of the trace species of interest. As a matter of fact, detection sensitivities are in the range of 10^2 to 10^4 ppm for most cases of interest. These figures can be improved by a factor of about 30 if advantage is taken of the different tensor properties of $\underline{\underline{X}}_{NR}$ and $\underline{\underline{X}}_R$ (polarization CARS [37]).

Advantages of CARS

CARS offers many advantages over other optical methods for nonintrusive spatially resolved diagnostics of gases and reactive media :

- spatial resolution is excellent,
- the signal is emitted in a well collimated beam, which makes discrimination against stray light easier,
- the spectral position of the signal, to the anti-Stokes side of the pump, makes it easier to reject fluorescence interference (which usually lies to the Stokes side of the pump),
- the signal strength is considerable ; using conventional pulsed solid state and dye lasers, the number of photons collected in a typical experiment is about 10^5 to 10^{15}, i.e. 5 to 10 orders of magnitude larger than that collected in a spontaneous Raman scattering experiment.

All these advantages justify the introduction of CARS as a diagnostics tool in reactive media. This application is to date the most important one. Other applications, such as high resolution molecular spectroscopy or chemical analysis of biological samples, also are attractive but shall not be discussed here.

B. CARS Instrumentation

We here describe the CARSspectrometer in use at ONERA. This spectrometer was developed jointly with Quantel. The optical components for the laser sources and the beam combining optics are bolted directly onto a light-weight, portable, rigid 50cm x 150 cm cast aluminium table (Fig. 7). The passively Q-switched Yag oscillator with two amplifiers and one frequency doubler delivers over 150 mJ of radiation at 532 nm in 10 ns pulses at 1 to 10 Hz (ω_1 beam). The output is single frequency over 95% of the shots and presents a spectral jitter under ± 0.01 cm^{-1}. These characteristics are possible only through the use of a stable cavity design for the Yag oscillator. A second doubler is used to convert the remaining infrared energing from the first doubler, thus producing an additional 40 mJ of green to pump the dye chain. This one is composed of a dye laser and one amplifier stage and produces the "Stokes" beam at ω_2. The dye laser can be tuned with a fixed, high incidence grating and a rotating mirror. The linewidth is 0.7 cm^{-1} ; it can be reduced to 0.07 cm^{-1} through insertion of a prism beam expander. This operation maintains the cavity alignment precisely and causes a slight shift of the line center.

With a 40 mJ pump energy, the dye chain delivers from 1 to 5 mJ of tunable radiation in a diffraction-limited beam over the useful CARS range of 560-700 nm, regardless of the linewidth.

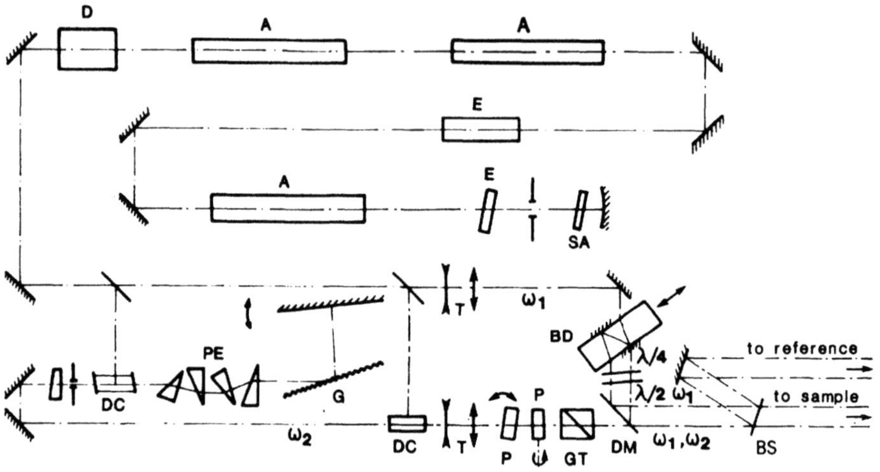

Fig. 7 : Laser source assembly. A : Nd yag amplifier. BD : parallel plate for production of parallel beams for BOXCARS (BOXCARS arrangement shown, translation of the plate allows passage to collinear arrangement without loss of alignment). BS : beam splitter for reference channel. D : KDP doubler . DC : dye cell. DM : dichroic mirror. E : Fabry-Perot etalon. G : grating. GT : Glan-Thompson prism. P : AR coated parallel plate for beam translation. PE : prism expander. SA : saturable absorber. T : telescope . $\lambda/4$ et $\lambda/2$ are quarter-wave and half-wave plates respectively. ω_1 : "laser" beam. ω_2 : "Stokes" beam.

The tuning is driven by a stepping motor which allows both a continuous sweep from 500 to 800 nm in coarse steps of 0.07 cm^{-1}, and limited sweeps of 6 nm about the coarse drive setting in fine steps of 0.007 cm^{-1}. A broadband mode of operation is also provided for the dye laser, giving about 100 cm^{-1} linewidth ; this mode is used for multiplex CARS experiments in conjunction with a spectrograph and an optical multichannel analyser [38]. In this case an interference filter is used for the tuning.

A space of 60 cm x 30 cm is left on the table for the mounting of various beam handling optics. This space is occupied by components for beam-matching and superposition, and for simultaneous non-resonant background cancellation and improved spatial resolution using BOXCARS [39]. We thus use telescopes to adapt the divergences of the beams, parallel plates to translate the beams and to split them for BOXCARS, one dichroic mirror for beam superposition and alignment, as well as polarization rotators and polarizers. The dichroic mirror is held on a sturdy mount having better than 10 μrad alignment sensitivity. A small fraction (≃5%) of the laser beams is subsequently split off to pump the reference channel. In the following, we shall concentrate on the detection system and on the problem of background cancellation.

The detection assembly, including the reference leg, is installed on a separate table of 50cm x 150 cm (Fig. 8). All focusing lenses are AR coated air-spaced achromats. The anti-Stokes signals are filtered by means of compact double monochromators preceded by dichroic filters to prevent breakdown on the entrance and intermediate diaphragms ; detection is done with PM tubes which are mounted in the same rack as the signal processing electronics (see below) to avoid RF interference. Light is piped to them by means of 1 mm diameter optical fibers. The anti-Stokes signal levels in the sample and reference channels are adjusted at approximately 10^4 photoelectrons per shot, which corresponds to a Poisson uncertainty of about 1 %. Higher fluxes may cause saturation, lower fluxes result in unacceptable uncertainty levels. The sample and reference channels are matched carefully, especially in BOXCARS experiments. The reference cell contains 50 b of argon. The reference signal is usually much stronger than needed and has to be attenuated ; it remains adequate even when a combination of BOXCARS and background cancellation is used (see below).

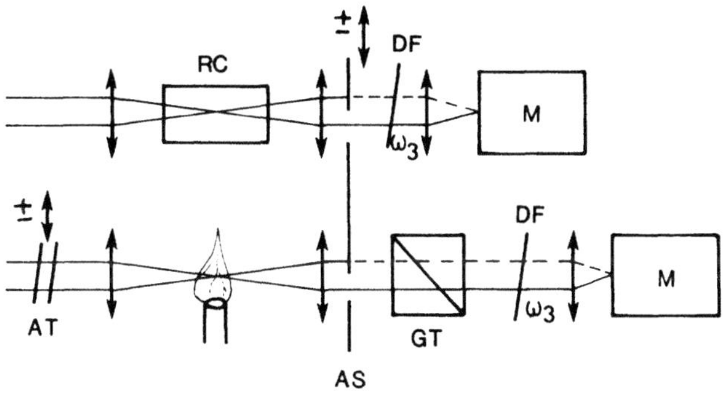

Fig. 8 : Schematic of sample and reference channels :
AS : movable aperture stop for operation with parallel beams or crossed beams (cross-beam position shown here) ; AT : movable attenuators ; DP : dichroic filters ; M : monochromator and detector ; RC : reference cell.

In both channels, the signal level is maintained at its prescribed level of 10^4 photoelectrons by adjusting the pump powers with attenuators, which is a safeguard against Raman saturation at line center. The photocurrent pulses are treated by an electronic device that gates them, calculates their ratios, square roots and average for a fixed number n of shots (n = 1 to 10 in practice). The electronics unit also rejects shots which do not fall with ± 35% of the mean in the reference leg, and tunes the dye laser after the n shots have been collected.

For multiplex CARS, a spectrograph and an optical multichannel analyzer (OMA2 from PAR) are used. The dispersive element in the spectrograph is a 2100 lines/mmn, aberration-corrected concave holographic grating with f = 750 mm. The net spectral resolution is 0.7 cm^{-1}. Both signal and reference spectra are recorded simultaneously on the vidicon and ratioed channel by channel ; square roots, and averages if necessary, are subsequently calculated. Recording the reference spectrum is a vital requirement since the dye laser spectrum is not reproducible and exhibits appreciable modulation.

The technique of background suppression using the tensor properties of the nonlinear susceptibility has been studied in detail for collinear beams, and several possible polarization arrangements have been described [37,40]. In BOXCARS, some flexibility is afforded by the availability of two spatially distinct "laser" beams at ω_1, which can have different polarizations. A discussion of that problem will be found in reference [39].

C. Results

The feasibility of concentration measurements in flames by CARS was shown in 1973 [41]. Since then, many experiments have been carried out and many species have been detected. Here we shall concentrate on results obtained at ONERA in recent times.

1. <u>Measurement of rovibrational excitation of H_2 in a discharge</u>. The study of rovibrational populations in tenuous discharges is an impossible task using mechanical probes. This study is also exceedingly difficult using absorption/emission methods. CARS, however, offers an interesting measurement potential over an appreciable temperature and density range. This was demonstrated recently [42] in an H_2 discharge (Fig. 9) designed for H^- production [43]. The spectra are shown in Figures 10 and 11 for a total pressure of 0.13 mbar, without the discharge (Fig. 10a) and with a discharge voltage of 90 V and current of 3 A (Fig. 10b). Collinear CARS without background cancellation was used for maximal signal strength and sensitivity. Note that, since the spatial resolution of collinear CARS is not exelllent, the amplitude of the v = 0 lines may come out slightly stronger because of some signal contribution from the cold H_2 surrounding the generator. Only the central portions of the first four Q lines were plotted. Horizontal bars on the plots give the theoretical heights of these lines assuming Boltzmann equilibria at 290 K (Fig. 10a) and 475 K (Fig. 10b) for the rotation. Uncertainties in temperature measurements are 5 K and 15 K respectively. From

the line intensities with the discharge turned on, we deduce that molecular H_2 constitutes approximately 90% of the gaseous mixture, the rest being composed of ions, radicals and electronically excited H_2. The comparison of the relative amplitudes of Q (1) lines in v = 0, v = 1 and v = 2 (Fig. 11) also gives a measure of the vibrational excitation.

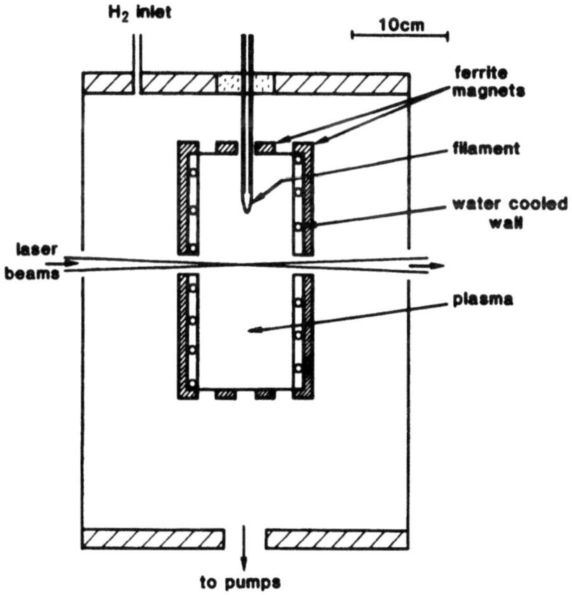

Fig. 9 : Schematic of the plasma generator, drawn to scale.

The line associated with v = 3 could not be detected. In this preliminary work, the detection sensitivity on v = 2 is about 10^{12} cm^{-3}. A sensitivity of 10^{11} cm^{-3} is technically feasible and should be demonstrated in the near future.

The present results give :

$$N\ (v=0) = 1.9\ 10^{15}\ cm^{-3}$$

$$N\ (v=1) = 4.5\ 10^{13}\ cm^{-3}$$

and

$$N\ (v=2) = 4.4\ 10^{12}\ cm^{-3}$$

assuming

$$N\ (v=3) = 10^{12}\ cm^{-3}$$

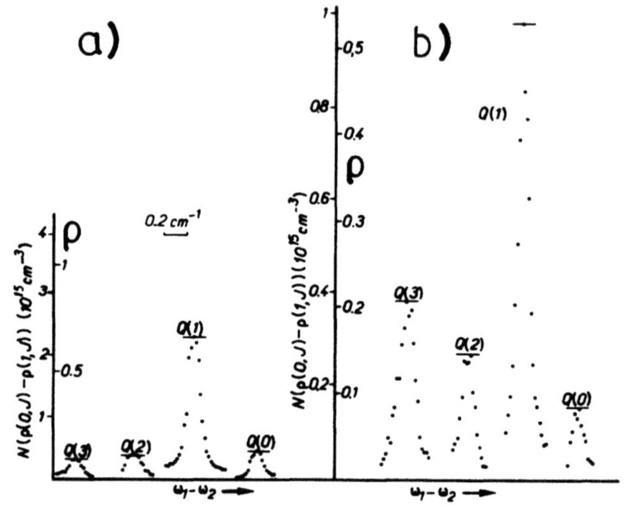

Fig. 10 : Profiles of lines Q(0) to Q (4) for the $\nu = 0 \rightarrow 1$ fundamental transition of neutral H_2 :

a) without the discharge ;
b) with a discharge of 90V, 3A.
The strongest portions of the lines only are shown, in steps of $0.02cm^{-1}$, and with 10 consecutive measurements averaged at each point. The spectral resolution of the dye laser was $0.07\ cm^{-1}$.

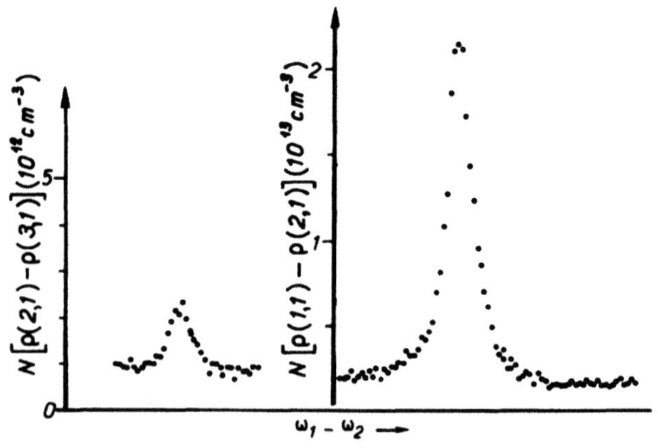

Fig. 11 : Profiles of excited states lines, with 30 measurements averaged at each point.

The assumption on N (v=3) is necessary since it is the difference between the number densities of the Raman levels which is measured in CARS experiments (Eq. (15)). These results clearly show that the vibrational equilibrium established is non-Boltzmann (Fig. 12). Similar effects were also seen in D_2 and N_2 using CARS [44].

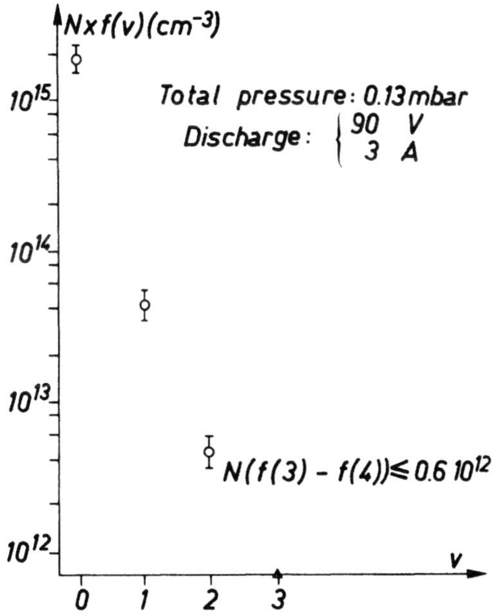

Fig. 12 : Distribution of vibrational populations in the discharge.

2. Temperature and H_2 concentration measurements in an ethylene-air bunsen flame. H_2 gas plays an important role in the chemistry and energy balance of flames. This was demonstrated recently in a premixed Bunsen flame burning ethylene in air [45]. By recording CARS spectra of the Q-branch of the H_2 formed by the pyrolysis of ethylene, we were able to obtain the static temperature profile and the H_2 concentration distribution. A BOXCARS optical arrangement was used to give a spatial resolution of about 1 mm. The spectral analysis was similar to that done in the plasma, although we only monitored peak line intensities. In the data processing, special precautions had to be taken with regard to line broadening mechanisms and concentration calibration. The burner itself was designed so as to give good flame stability. It has a diameter of 10 mm. It was operated at a C/O ratio of 0.57 and mass flow rates of 8.9 mg/s for C_2H_4 and 76.2 mg/s for air ; this gave a flame cone height of 26 mm at 296 K and 0.98 bar, corresponding to a burning velocity close to 17 cm/s. The results are presented in Figs. 13 and 14.

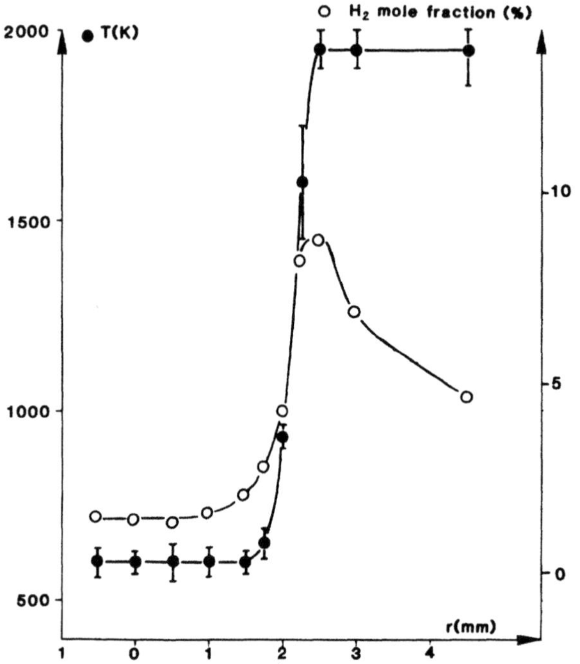

Fig. 13 : Rotational temperatures and hydrogen molar percentages O as a function of radial distance from burner axis. Height above burner exit plane : 11.5 mm.

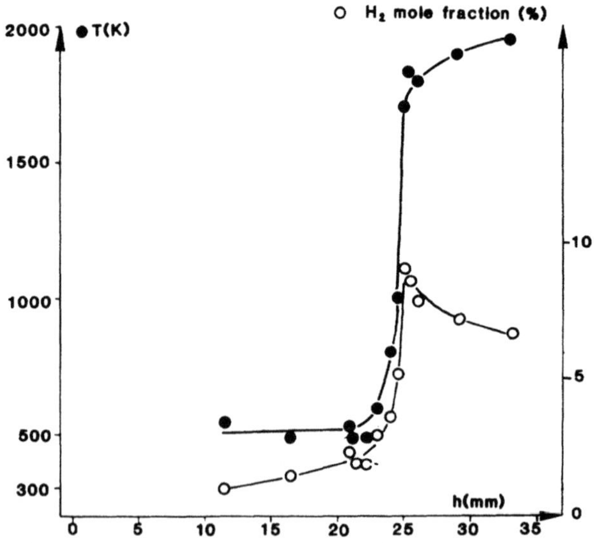

Fig. 14 : Rotational temperatures and hydrogen molar percentages O as a function of distance from burner exit plane along burner axis.

Most interesting is a comparison of these results with thermodynamic equilibrium calculations of adiabatic flame temperature and stable product composition. Using the water-gas equilibrium and neglecting dissociation, we find for our flame an adiabatic flame temperature of 1940 K and a composition of 14.9 mol% CO, 4.0 mol% CO_2, 10.2 mol% H_2O and 8.6 mol% H_2. There is strikingly good agreement between the final measured temperature in the burnt gas and the calculated temperature, and,in particular, between the hydrogen molar fraction measured just beyond the reaction zone and its thermodynamic equilibrium value of 8.6%. This agreement leaves no doubt that the CARS technique is indeed suitable for spatially resolved concentration measurements in reactive media. However, more checks have to be carried out, e.g. CARS measurements of CO, CO_2 and H_2O to find out if water-gas equilibrium is established or not. Furthermore, the hydrogen rotational temperatures should be compared to N_2 rotational temperatures. Hence, at the present moment, we regard our measured hydrogen concentration as semi-quantitative ; it will be improved upon once better knowledge of lineshapes and widths becomes available.

As to now, our results demonstrate remarkably well, besides oxidation of hydrogen in the burnt gas, the effect of hydrogen diffusion from the reaction zone into the fresh gas. It is quite surprising that the hydrogen molar fraction is about 1 % at h = 11.5 mm (i.e. 14.5 mm downwards from the flame top and at a radial distance of 2.5 mm from the flame cone) and as high as 3.7 % at a distance of 2 mm below the flame top. Hydrogen is thus enriched in the central flow line. A similar effect, the enrichment of heavier hydrocarbons in the central flow line of a Bunsen flame, was described by Flossdorf et al. [46]. However, the enrichment of hydrogen in the fresh gas has some implications on the heat balance in the fresh gas. It is quite conceivable that the pre-reaction zone hydrogen originates from the recombination of hydrogen atoms. This would, as a quick calculation shows, for a hydrogen molar fraction of 1% give a temperature rise of the fresh gas of about 80 K and hence partially account for our observed temperature rise of about 200 K, compared to ambient, at h = 11.5 mm. Even if the molecular hydrogen itself was the main diffusing species, it would still carry some enthalpy and thus still act as a "heat recirculating chemical carrier", though, of course, in a much less efficient way than the hydrogen atom. The rise in H_2 concentration near the reaction zone with increasing height above the burner exit plane could also be an explanation for the well established fact that the burning velocity is not a constant over the total flame area [46].

3. Instantaneous measurement of temperature in a well-stirred reactor. The well-stirred reactor presents interesting challenges to the CARS diagnostician : can one measure the magnitude of the temperature fluctuations and the gas residence time ? None of these measurements can be undertaken properly using conventional probing techniques, yet they are of extreme interest to the theorist.

A well-stirred reactor was built for these studies [47]. The reactor is cylindrical in shape with 5.5 cm id. and 3cm height. A mixture of methane and air is injected through an array of small holes placed on a central injection column. Exhaust is through five 1 cm-diameter holes drilled in the top flange.

The CARS beams are passed horizontally through holes of 6 mm diameter which are diametrically opposed in the cylinder walls. The temperature study was done in the multiplex mode, using a BOXCARS configuration which gave under 1 mm spatial resolution. Numerous instantaneous spectra were recorded. That presented in Fig. 15 is typical. This spectrum is shown with a theoretical spectrum calculated for a temperature of 1700 K. The single shot temperature measured is close to the adiabatic temperature (2100 K), the drop being easily accounted for by losses to the walls. Thermocouples and IR pyrometry both give temperatures close to 1800 K. The standard deviation in the measured peak intensity (16%) receives contributions from the local temperature fluctuations as well as from our measurement uncertainty. The measurement uncertainty comes largely from beam wandering caused by the gradients in the flame. Originally, a large source of imprecision was the mixing of hot flame gases with cold air in the vicinity of the optical ports. This caused amplitude fluctuations in the range of 1 to 4, and distortions of the spectral contour which seriously affected the temperature measurement accuracy. The cure was to apply fused silica windows to the ports. Heating of the walls limits burner operation to 5 minutes, which enables one to collect about 500 spectra.

The residence time was monitored using a tracer injected in the fresh gases immediately before the burner for periods of 30 ms separated by 30 ms. The tracer which we intended to use originally was CF_4, but this gas was rapidly abandoned because of its chemical reactivity in the burner above 1500 K, and also because of the rapid drop in CARS line intensity as a function of temperature due to broadening (Fig. 16). Instead, we used CO_2 which offers two sharp and intense Q branches which do not broaden as a function of temperature (Fig. 17). The dye laser was operated narrowband and was tuned approximately 0.2 cm^{-1} away from the ν_1 mode so that a small frequency jitter would not cause excessive signal change. For a sample of fixed concentration, the standard deviation on the CO_2 concentration measurement is thus 15%. Fig. 18 shows the exponential rise of the excess CO_2 concentration in the burner following beginning of its injection. In the well-stirred reactor, the rise and decay times are in principle equal. The residence time deduced from this curve (1.8 ms) is in good agreement with that (1.7 ms) obtained by dividing the burner volume (71 cm^3) by the volumetric flow rate used in that particular run (41 $10^3 cm^3/s$ at 1700 K using 7.82g/s of air and 0.404 g/s of CH_4) ; 0.627 g/s of CO_2 was introduced during the injection periods.

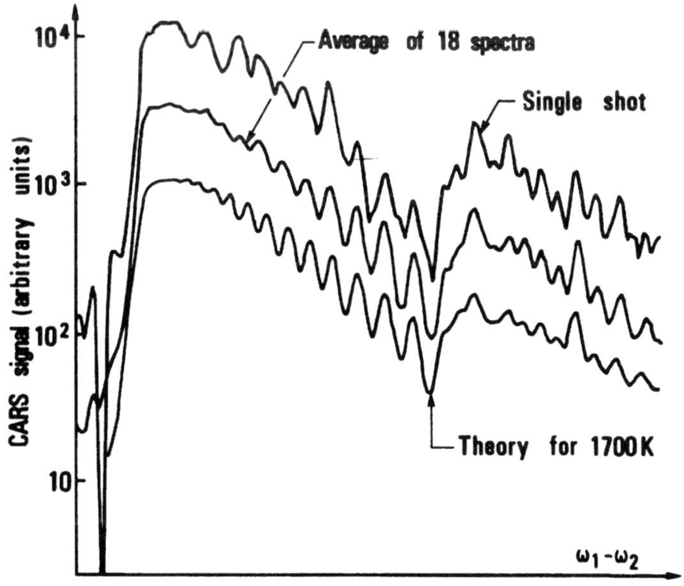

Fig. 15 : Single shot BOXCARS spectrum of N_2 in well-stirred reactor and calculated spectrum assuming T = 1700 K. The peak amplitude corresponds to about 5.10^3 photoelectrons per OMA channel. Spectral resolution is about 1 cm^{-1}. Gated intensifier stage.

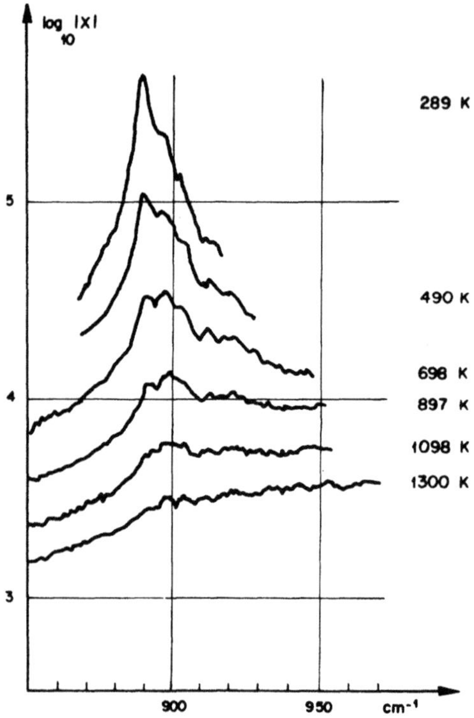

Fig. 16 : Spectrum of pure CF_4 in furnace at atmospheric pressure and for several temperatures, recorded at 0.1 cm^{-1} resolution. Scanning CARS was used, with 4 measurements averaged at each point.

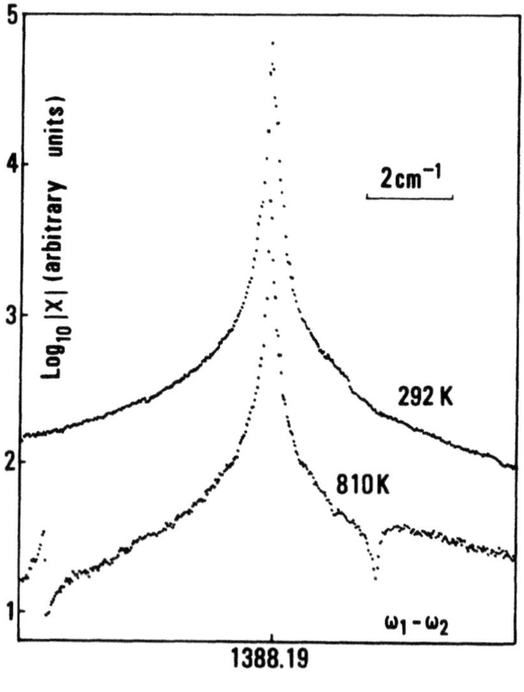

Fig. 17 : Spectrum of pure CO_2 at atmospheric pressure for two values of temperature. Same conditions as for Fig. 16.

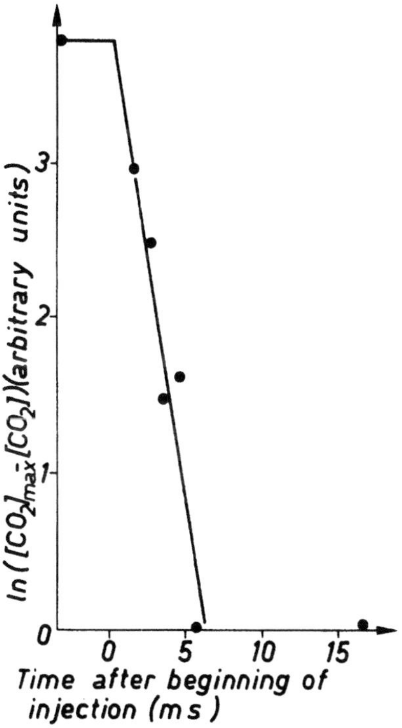

Fig. 18 : Rise in CO_2 concentration in reactor following beginning of injection ; 100 measurements are averaged for each data point ; repetition rate : 2 Hz ; the data are presented after subxtration of maximal CO_2 concentration during injection ; opening and closing times are about 0.3 ms.

4. <u>Temperature measurements in a simulated turbomachine combustor</u>. Large scale combustors are difficult to tackle by CARS, since environment is harsh and optical access is limited. Therefore, many precautions have to be taken to protect the optical set-up from noise, dust and oil vapors. The feasibility of making temperature and concentration measurements by CARS in a kerozene-fueled burner was tested at SNECMA [48]. The burner produces a flow which is approximately rectangular, with a size of 15 x 50 cm, and a mass flow rate approaching 2 kg/s for atmospheric pressure operation. The CARS optical tables were installed in a test room near to that housing the combustor (Fig. 19), along with the power supplies and electronics. Single shot BOXCARS spectra were taken at varous locations in the burner, with a quality approaching, but not as good as, that of Fig. 15. A small percentage of spectra was lost in the primary zone (near the injectors) because of fuel droplets and turbulence, which cause spectral envelope distortion and attenuation.

Figure 20 presents a temperature profile recorded in the primary zone. This temperature is a time average over about 20 consecutive single shot measurements recorded in a 20-minute period (the data acquisition rate then was close to 2 per minute). The profile clearly shows the drop in temperature, which tends to a limiting value of 500 K close to that of the preheated air used in that particular run.

We have also studied O_2 and CO, trying to obtain their spatial distributions. The detectivity using BOXCARS without background cancellation was about 3% in mole fraction for these species in the flame, at the time of these experiments. With background cancellation, the detectivity was further degrated by a factor of two. These results will be published elsewhere [48].

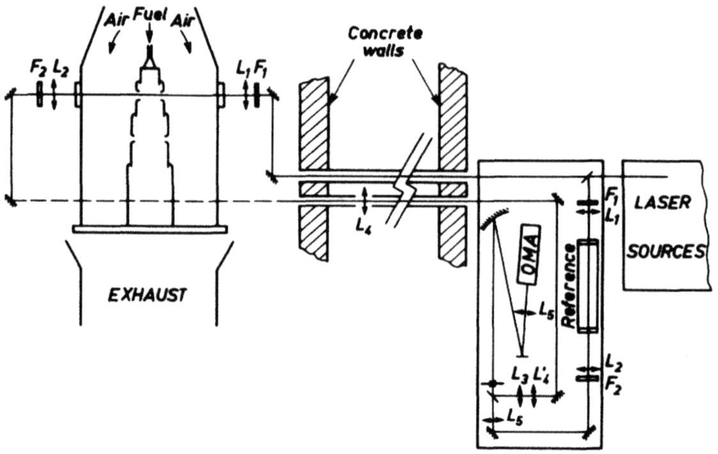

Fig. 19 : Lay-out of optical set up near kerozene fueled combustor. The details of the laser source assembly (Fig. 7) are not given here. The optical path to the burner is about 10 m. The beams travel several meters through open air and had to be enclosed in metal pipes for safety. F_1 : Schott OG 515 filter ; F_2 : short-pass dichroic filter for rejection of pump beams ; L_1, L_2 : focusing and recollimating achromats : L_3 : achromat used to focus anti-Stokes beams into entrance slit of spectrograph ; L_4,L'_4 : a focal lens combination used to reduce the effect of beam steering in the burner ; L_5 gives a magnification of 2 on the OMA face. The anti-Stokes beam is passed under the combustor rig (dashed line). Access to the primary reaction zones is through air dilution holes in the can.

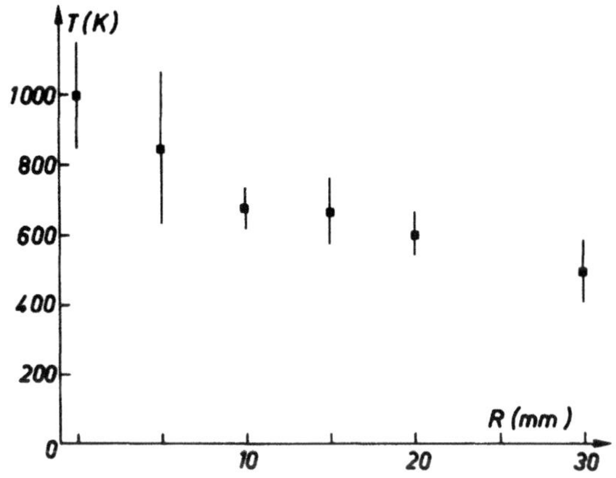

Fig. 20 : Radial temperature profile in primary zone. R is measured from burner midplane ; note that the pump beams cross the plane of the air injection port at R = 50 mm. The error bars represent the standard deviation in the measurements and reflect contributions from real temperature fluctuations and from measurement uncertainty.

5. Conclusion. Only a few of the potential applications of CARS have been presented here. Many more have been explored and demonstrated in other laboratories. Piston engines, jet engines, chemical vapor deposition (CVD), laser media all have been or are currently being probed by conventional CARS. See the contribution by Stenhouse in this series for a review.

The situation is far different in resonance-enhanced CARS which, in spite of its extremely promising characteristics, remains an extremely difficult technique to apply. The next chapter is devoted to some of its aspects.

4. RESONANCE-ENHANCED CARS

Conventional CARS offers a detection sensitivity of about 1000 ppm in usual gas mixtures near STP. Improvements up to 30 are possible using cancellation of the nonresonant background (polarization CARS) as we have seen in the previous chapter. Much larger improvements on the order of 10^2 to 10^4 are actually obtained using resonant CARS, as we showed with I_2 vapor [20]. This has been the major motivation for several other research programs in resonant CARS throughout the world. Observations in

C_2 [49-50] and NO_2 [51-52] have been reported recently. A related area of research, resonance CARS in absorption continua, is treated as a separate topic by Kiefer in this course. This form of spectroscopy has been performed recently in Iodine [53] ; such studies are delicate since the susceptibility is much weaker than when enhancement is obtained with discrete lines. We here restrict our discussion to the latter case.

1. Spectral properties

Resonance CARS spectroscopy is usually carried out by holding ω_1 fixed near a one-photon absorption and by varying ω_2. There are three types of lines associated with the lower state $|a\rangle$ (term proportional to $\rho_{aa}^{(0)}$), as depicted in Fig. 21. We call these laser-enhanced Raman resonances (Fig. 21a), double-electronic resonances (Fig. 21b) and anti-Stokes-enhanced Raman resonances (Fig. 21c). Both fundamental and overtone vibrational transitions are possible. In addition, one may find similar sets of resonant transitions associated with $|b\rangle$ if that state is significantly populated ; their spectral properties are easily derived from the corresponding resonance denominators.

A notable character of the double electronic resonances is that their positions in the spectrum, which are given by the condition $\omega_1 - \omega_2 = \omega_{n'a} - \omega_1$, are a function of ω_1 (contrary to Raman resonances of the 21a or 21c type). This feature is observed in I_2 gas [20] when ω_1 is shifted slightly about ω_{na}. It facilitates the assignment of the lines. Further, the observation of double electronic resonances is of considerable importance to the spectroscopist since one can easily interpret the corresponding $\omega_{n'a}$ absorption lines if the ω_{na} transition in resonance with ω_1 is known. This can facilitate the analysis of unknown portions of absorption spectra. A set of handy notations has been introduced in order to label the double or triple resonances. The three molecular transitions appearing in the Raman, laser and anti-Stokes denominators of the $\rho_{aa}^{(0)}$ term of Eq. 15 are listed between brackets, in that order ; the J quantum number of $|a\rangle$ is indicated to the right of the last bracket. Capital letters are used for the transitions in resonance with the fields, lower case letters otherwise. Recently, resonance enhanced CARS spectra of C_2 in a discharge have been obtained under controlled conditions by precisely tuning the "laser" and "Stokes" frequencies into resonance with Swan band absorption lines and have been interpreted unambigously [54].

For that interpretation, we had to obtain the spectral

positions of the Swan emission lines to a precision of better than 0.01 cm^{-1} using Fourier transform spectroscopy ; this work was carried out in collaboration with P. Luc and C. Amiot at CNRS [55].

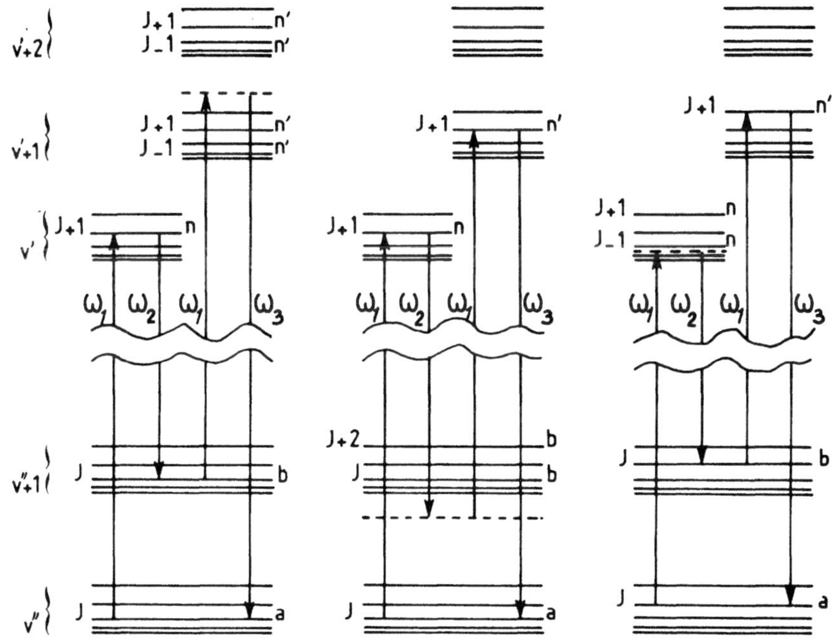

Fig. 21 : Energy level diagrams representing the states contributing to resonance-enhanced CARS in a diatomic molecule

a) ω_{na} enhanced fundamental (Δv = 1) vibrational transition ;

b) double electronic resonance with $\omega_1 = \omega_{na}$ and $\omega_3 = \omega_{n'a}$;

c) $\omega_{n'a}$ enhanced vibrational transition.

We first undertook the recording of the emission spectrum of C_2. A stable microwave discharge source for the Swan band was developed for that purpose. With a spectral resolution of 0.07cm^{-1}, the characteristic triplet splitting and the staggering of the triplet components caused by the Λ doubling are clearly visible in the spectra. The latter study has provided a thorough and accurate set of emission frequencies together with the appropriate set of spectroscopic constants which were not available from earlier work [56]. If v, J and Ω are respectively the vibrational, rotational and spin quantum numbers, an a priori calculation shows that a nearly triple resonance is achieved for one type of resonance CARS line which is (in the notation or ref. 24) :

$[Q_{\Omega}\ (1\text{-}0),\ R_{\Omega}\ (0\text{-}0),\ P_{\Omega}\ (1\text{-}0)]\ J$ with J = 18, 19 and 20, for

Ω = 3, 2 and 1 respectively.

These C_2 CARS lines have been searched for and observed by means of a microwave discharge in a flowing mixture of 3% acetylene in He at 40mb total pressure. The "laser" and "Stokes" beams are focused into the discharge by a 30 cm focal length lens. The anti-Stokes signal is filtered by a monochromator and detected by a photomultiplier. Fluorescence in Iodine was used to calibrate the laser frequency ω_L ; final accuracy is about 0.05 cm^{-1}. The spectrum of figure 22 has been recorded with ω_L = 19493.85± 0.05cm^{-1}.

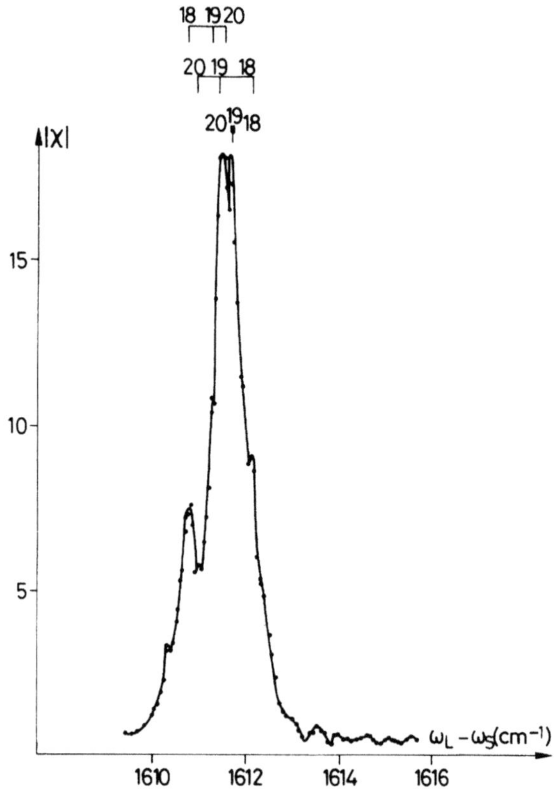

Fig. 22 : Resonance CARS spectrum of C_2 in a discharge.

Because of the high vibrational temperature, additional contributions are involved in the resonant CARS susceptibility. These are associated with the terms proportional to $\rho_{bb}^{(0)}$ of Eq. (15). Thus our CARS spectra present previously unobserved characteristics, particularly with regard to spectral content. Resonance CARS lines of the type [(Q_1 (1-0), R_1(0-1), P_1 (1-0))] 20, [(Q_2 (1-0), R_2 (0-0), P_2 (1-0))] 19 and [(Q_3 (1-0), R_3 (0-0), P_3 (1-0))] 18 are seen. Different features of the type [(Q_1 (1-0), R_1 (0-1), P_1 (1-0)] 20, [(Q_2 (1-0), R_2 (0-1), P_2 (1-0))] 19 and [(Q_3 (1-0), R_3(0-1), P_3 (1-0))] 18 are probably mixed to the above ones. All these lines are lumped together in the profile of Fig. 22. Their widths are about 0.12 cm^{-1} and contain Doppler and collisional contributions.

2. Resonance CARS in the Doppler regime

Resonance CARS in dilute samples, where the Doppler effect is the main cause of line broadening, is a problem of considerable interest. An intuitive, but fallacious argument, leads to the erroneous conclusion that resonance CARS lines should be Doppler-free. We take, for example, a laser-enhanced Raman resonance (Fig. 21a). One particular velocity group in the Doppler-broadened ω_{na} absorption contour is precisely resonant ; we call ν its velocity component along the wave vector axis $\nu = c(\omega_{na}-\omega_1)/\omega_1$. One expects this group to give a dominant contribution in the CARS spectrum as $\omega_1 - \omega_2$ is being swept past ω_{ba}, thus giving a Doppler-free resonance line, which is shifted from the Raman resonance line center by the amount $\omega_{ba}\nu/c$. This simple-minded argument actually turns out to be wrong, because we have neglected the influence of adjacent velocity groups which, in this case, interfere destructively with the main group [23].

The correct approach is to recast Eq. 15 into its Doppler form. The molecules in the above-mentioned velocity group see the laser and anti-Stokes frequencies with a Doppler shift $-k_i\,\nu$, where k_i is the wave vector. Their elementary contribution to the susceptibility is shown to be

$$X_r(\nu) = \frac{\rho_{aa}^{(0)}}{\hbar^3}\sum_b \frac{1}{\omega_{ba} - \omega_1 + \omega_2 + (k_1 - k_2)\,\nu - i\Gamma_{ba}} \times \sum_{n'} \frac{\mu_{an'}\,\mu_{n'b}}{\omega_{n'a} - \omega_3 + k_3\,\nu - i\Gamma_{n'a}} \times \sum_n \frac{\mu_{bn}\,\mu_{na}}{\omega_{na} - \omega_1 + k_1\nu - i\Gamma_{na}} \tag{27}$$

where we have neglected the $\rho_{bb}^{(0)}$ contribution. The net susceptibility is finally obtained by integrating over the velocity distribution function F (ν) :

$$\chi_r = \int_{\infty}^{\infty} N\, F(\upsilon)\, \chi_r\, (\upsilon)\, d\upsilon \qquad (28)$$

Equation 27 can be integrated numerically. A closed form solution in terms of complex error functions is also easily derived if F (ν) is taken as a Gaussian. In this manner, it is possible to show that the $\rho_{aa}^{(0)}$ term in Eq.(15) indeed has the full Doppler width, whereas the $\rho_{bb}^{(0)}$ term is Doppler free. Therefore, the latter may eventually give rise to the strongest lines in the spectrum under certain thermodynamic conditions, although $\rho_{bb}^{(0)} < \rho_{aa}^{(0)}$. A few of the other susceptibility terms in CARS are also Doppler free ; some contain at least one anti-resonance and should be small ; others contain a resonance of the type $\omega_{n'n} - \omega_1 + \omega_2$, which under certain conditions on the collision rates can be quite large [17,24] (see discussion at the end of 2B).

In resonant CSRS, the Raman-resonant term which dominates in the collision regime is Doppler-free. We have also investigated the Doppler nature of the susceptibility terms in Stimulated Raman Gain Spectroscopy (SRGS) under a variety of experimental situations (co or counter-propagating laser waves, and Stokes or inverse Raman gain measurements). Depending on these conditions, a large number of terms can become Doppler-free (ref. 23, addendum). Note that a partial discussion of this same problem had already been given by Hänsch and Toschek [57]. In short, we conclude from this analysis that CSRS and SRGS in principle are better suited than CARS for Doppler-free spectroscopy ; however, we must bear in mind that they are prone to fluorescence interference.

5. CONCLUSION

We have examined in detail the theory of CARS, we have presented the key instrumental requirements and have reviewed some experimental results obtained to date. In brief, CARS :

1) gives the same spectroscopic information as spontaneous Raman scattering,

2) is insensitive to fluorescence interference in resonance Raman work,

3) gives excellent spatial resolution (1mm),

4) can give good time resolution (10 ns) by sacrificing some detectivity,

5) is capable of excellent spectral resolution (0.03 cm^{-1} on ordinary set-ups, $10^{-3} cm^{-1}$ in special applications using *cw* sources,

6) is extremely luminous.

CARS have a few disadvantages :

1) the major disadvantages is the presence of the non-resonant background, which limits the detectivity at a value of 10 ppm to 1% depending on thermodynamic conditions and species studied,

2) it is sensitive to laser instabilities,

3) it is subject to saturation at the higher power levels (1 MW or more).

For these reasons, CARS will often be preferred to normal Raman Scattering for the following measurements :

1) study of reactive media, plasmas, gas laser amplifiers, etc...,

2) analysis of media containing particulate matter (e.g. sooting flames) or investigation of flows near solid obstacles (as close as 50μm to the surface),

3) high resolution spectroscopy,

4) resonance-enhanced Raman spectroscopy of liquids and gases.

Due to the cost of a state of the art CARS set-up (200 000 to 300 000$ for flame diagnostics), the decision to prefer CARS to normal Raman should be carefully weighed. However, fundamental research in all these areas will remain one of the most active branches of molecular spectroscopy.

REFERENCES

1. Widhopf, G.F. and Lederman, S., AIAA J., 9, 309 (1971).

2. Lapp, M., Goldman, L.M. and Penney, C.M., Science, 175, 1112 (1972).

3. "Laser Raman Gas Diagnostics", Proceedings of the Project SQUID Laser Raman Workshop on the Measurement of Gas Properties, May 10-11, 1973, Schenectady, edited by M. Lapp and C.M. Penney Plenum Press, New York, London (1974).

4. Proceedings of Project SQUID Workshop on Combustion Measurements in Jet Propulsion Systems, Edited by R. Goulard, Purdue University, Lafayette, Indiana (1975).

5. "Experimental Diagnostics in Gas Phase Combustion Systems", Progress in Astronautics and Aeronautics, Vol. 53, Edited by B.T. Zinn, Martin Summerfield Series Editor (1977).

6. Régnier, P.R. and Taran, J.P.E., Appl. Phys. Letters, 23, 240, (1973).

7. Terhune, R.W., Bull. Amer. Phys. Soc., 8, 359 (1963).

8. Maker, P.D. and Terhune R.W., Phys. Rev., 137, A801 (1965).

9. Yablonovitch, E., Bloembergen, N., and Wynne, J.J., Phys. Rev., B3, 2060 (1971).

10. Akhamanov, S.A., Dmitriev, V.G., Kovrigin, A.I., Koroteev, N.I., Tunkin, V.G. and Kholodnykh, A.I., JETP Letters, 15, 425 (1972).

11. Levenson, M.D., Flytzanis, C. and Bloembergen, N., Phys. Rev., 6, B3962 (1972).

12. Rado, W.G., Appl. Phys. Letters, 11, 123 (1967).

13. Hauchecorne, G., Kerhervé, F., and Mayer G., J. de Physique, 32, 47 (1971).

14. De Martini, F., Giuliani, G.P. and Santamato, E., Optics Comm., 5, 126 (1972).

15. Fiutak, J., and Van Kranendonk, J., Can. J. Phys. 40, 1085 (1962).

16. Omont, A., Smith, E.W. and Cooper, J., The Astroph. J., 175, 185 (1972).

17. Druet, S., Attal, B., Gustafson, T.K. and Taran, J.P.E., Phys. Rev. A18, 1529 (1978).

18. Yee, S.I., Gustafson, T.K., Druet, S.A.J. and Taran, J.P.E., Opt. Commun. 23, 1 (1977).

19. Yee, S.Y. and Gustafson, T.K., Phys. Rev. A18, 1597 (1978).

20. Druet, S.A.J. and Taran, J.P.E., "Coherent anti-Stokes Raman Spectroscopy", in Chemical and Biochemical Applications of Lasers, vol. 4, Edited by C.B. Moore (Academic Press, New York) (1979).

21. Bordé, C.J., Hall, J.L., Kunasz, C.V. and Hummer, D.G., Phys. Rev. A14, 236 (1976).

22. Bordé, J. and Bordé, C.J., J. Mol. Spectrosc. 78, 3530 (1979).

23. Druet, S.A.J., Taran, J.P.E. and Bordé, C.J., J. de Physique, 40, 819, (1979), addendum, ibidem, 41, 183 (1980).

24. Druet, S.A.J. and Taran, J.P.E., Progress in Quant. Elec. 7, 1 (1981).

25. Feynman, R.P., Quantum Electrodynamics, Benjamin, New York (1962).

26. Ward, J.F., Rev. Mod. Phys. 37, 1 (1965).

27. Yariv, A., IEEE J. Quant. Elect. QE13, 943 (1977).

28. Hanna, D.C., Cotter, D., and Yuratich, M., "Non linear Optics of Free Atoms and Molecules", Edited by D.L. McAdam, Springer Series in Optical Sciences, Vol. 17, Springer Verlag, Berlin, Heidelberg, New York (1979).

29. Eesley, G.L., J.Q.S.R.T., 22, 507 (1979).

30. Bloembergen, N., Lotem, H., and Lynch, R.T., Indian J. Pure Appl. Phys. 16, 151 (1978).

31. Prior, Y., Bogdan, A.R., Dagenais, M. and Bloembergen, N. Phys. Rev. Letters, 46, 111 (1981).

32. Bogdan A.R., Prior, Y. and Bloembergen, N., Opt. Letters, 6, 82 (1981).

33. Bloembergen, N., Bogdan, A.R. and Downer, M.W. in Laser Spectroscopy V (McKellar, Oka and Stoicheff, ed.) Springer Verlag, Berlin, Heidelberg, New York (1981).

34. Carreira, L.A., Gross, L.P. and Malloy, T.B., J. Chem. Phys. 69, 855 (1978).

35. Grynberg, M., J. Phys. B. Atom Mol. Phys. 14, 2089 (1981)

36. Eckbreth, A.C. Appl., Phys. Letters, 32, 421 (1978).

37. Rahn, L.A., Zych, L.J., and Mattern, P.L., Opt. Comm. 30, 249, (1979).

38. Roh, W.B., Schreiber, P.W. and Taran, J.P.E., Appl. Phys. Letters, 29, 174 (1976).

39. Attal, B., Péalat, M. and Taran J.P.E., J. Energy 4, 135 (1980).

40. Song, J.J., Eesley, G.L. and Levenson, M.D., Appl. Phys. Letters, 29, 567 (1976).

41. Régnier, P.R. and Taran, J.P.E., Appl. Phys. Letters, 23, 240, (1973).

42. Péalat, M., Taran, J.P.E., Taillet, J., Bacal M., and Bruneteau, A.M., J. Appl. Phys. 52, 2687 (1981).

43. Bacal, M. and Hamilton, G.W., Phys. Rev. Letters 42, 1538 (1979).

44. Nibler, J.W., Shaub, W.M., McDonald, J.R. and Harvey, A.B., "Coherent Anti-Stokes Raman Spectroscopy" in Vibrational Spectra and Structure (Durig J.R., Editor), Vol. 6, Elsevier, Amsterdam, New York, (1977).

45. Müller-Dethlefs, K., Péalat, M. and Taran, J.P.E., Ber. Bunsenges, Phys. Chem. 85, 803 (1981).

46. Flossdorf, J., Jost, W. and Wagner, H. Cg., Ber. Bunsenges, Phys. Chem. 78, 378 (1974).

47. Bouchardy, P., Gicquel, P., Péalat, M. and Taran, J.P.E., To be published.

48. Bédué, J., Gastebois, P., Bailly, R., Péalat, M, and Taran, J.P.E. To be published.

49. Gross, K.P., Guthals, D.M. and Nibler, J.W., J. Chem. Phys. 70, 4673 (1979).

50. Hetherington III, W.M., Korenowski, G.M., and Eisenthal, K.B., Chem. Phys. Letters 77, 275 (1981)

51. Guthals, D.M., Gross, K.P., and Nibler, J.W., J. Chem. Phys. 70, 2393 (1979).

52. McIlwain, M.E. and Hindman, J.C., J. Chem. Phys. 73, 68 (1980).

53. Beckman, A., Fietz, H., Baierl, P., and Kiefer, W., Chem. Phys. Letters 86, 140 (1982).

54. Attal, B., Müller-Dethlefs, K., Débarre, D., and Taran, J.P.E., to be published.

55. Luc, P., Amiot, C., Debarre, D., Attal, B., Müller-Dethlefs, K., "New Analysis of the C_2 Swan system", to be published.

56. Phillips, J.G., and Davis, S.P., The Berkeley analysis of molecular spectra , Vol. 2 (a), University of California Press, (1968).

57. Hänsch, Th. and Toschek, P., Z. Phys. 236, 213 (1970).

HIGH RESOLUTION CARS

J. MORET-BAILLY

Laboratoire de Spectronomie moléculaire
de l'Université de DIJON
DIJON - FRANCE

1. INTRODUCTION

Standard spectroscopy of stable molecules uses transitions induced by the electric dipole moment. In the search for a high resolution in gases, a limitation was, for a long time, the Dopplerwidth of the lines, of the order of 100 MHz ; this is the limit of grating or Fourier transform spectrometers which remain the most usual and productive ones.

The use of lasers induced large improvements in Raman spectrometry, allowing to reach a Doppler limited resolution, mostly by stimulated Raman scattering (SRS) and coherent anti-stokes Raman spectroscopy (CARS).

The resolution of electric dipole spectroscopy has been increased by several orders of magnitude using a non-linear process : saturated absorption. It is tempting to try to reach similar results in Raman spectroscopy ; this has been shown to be possible by SRS in the easiest case, that is for the deuterium molecule (1). The main problems are :

- To get a sufficient sensitivity in SRS or CARS, one needs to use pulsed lasers ; but it is difficult to obtain sharp and stable lines from these lasers.

- In many sub-Doppler experiments, the class of molecules, the radial velocity of which is negligible, is selected by a saturation ; so, the proportion of useful molecules is low ;

W. Kiefer and D. A. Long (eds.), Non-Linear Raman Spectroscopy and Its Chemical Applications, 325–333.

simultanously, the pressure must be lowered to get a narrow collisional linewidth. Shall we reach a sufficient sensitivity ?

- Using CARS or SRS, the mixture of two light beams in the studied gas is critical ; how is it possible to get rid of the instability of the lasers to measure the light influence of the saturation ?

I will study these three problems.

2. NARROW-LINEWIDTH PULSED LASERS

We need, at least, a tunable frequency laser. It must be powerful, the energy of the pulses must be of the order of a fraction of Joule.

Non tunable lasers

The laser must be :

- Single mode transverse ; this condition is necessary, both for a good frequency control and for a reliable use of the emitted light beam which must be well mixed with the beam of another laser.

- Single mode longitudinally : indeed, we want a coherence length of a few meters, that is larger than the distance between the modes of a regular-sized laser.

- Stable ; the optical length of the cavity must be precisely tuned.

Let us look at these problems in the reversed order :

a) The geometrical length of a cavity may be stable if it is built from materials such as invar. If the laser is not under vacuum, which is very inconvenient, the optical length of the laser depends on the temperature and the pressure of the atmosphere. The amplificating medium of the laser is never very stable so that a static stabilization of the optical length of the cavity is impossible. We may only hope that the change of the refractive index of this medium be low enough during a laser pulse to obtain a sharp line. After a pulse, the amplificating medium must recover its initial state before the next shot ; if the recovery is fast enough, the evolution of the remainder of the laser may be weak between two shots, so that a feed-back stabilization of the frequency is possible. Thus, the slow cooling of the solid laser crystals appears as a fundamental limitation of the use of these lasers in very high resolution spectroscopy.

b) We suppose now that the optical length of the cavity is well tuned ; it remains to select the right longitudinal mode ; this is not easy if the gain bandwidth of the laser is large because the setting of mode selectors is difficult in pulsed lasers, and it induces a great loss of energy ; this is the second inconvenient of solid state lasers. So a gas laser appears more convenient, but the following conditions must be fulfilled :

- A low pressure for a sharp collisional bandwidth. But, consequently, the density of energy is low.

- The efficiency of the laser must be high to get little perturbations in the laser medium during the pumping.

- The linewidth must not be stretched by Zeeman, Stark...effects.

There is unhappily no laser equivalent to the CO_2 laser in the visible or near infrared. It seems to me that the laser which uses the atomic iodine made by photolyse of an organic compound should be the most convenient (2).

c) To get a transverse monomode laser, unwished modes must be damped. This is difficult for a low pressure gas laser, such as the iodine laser : as the volume of gas is large, the diameter of the pipe which contains it is large too, so that the geometry of the system is almost the same for many modes. It is difficult to introduce a telescope in the cavity of the laser ; so, it seems easier to use a ring laser geometry.

Tunable lasers

Dye lasers are the most used tunable lasers in the visible. As it is very difficult to build a pulsed laser stable spacially, in frequency and intensity, it is needed to amplify a continuous wave laser. The amplification of a beam in a small cell of dye pumped by a laser is easily done ; for good results, the beam of the pumping laser must be optically good ; so, a Nd-YAG laser, with frequency doubling, is very convenient. But the pulses of these lasers are usually short ($\simeq$ 20 ns) so that the sharpness of the lines is limited. These lasers are very expensive too if some power is needed.

In the injection dye pulsed laser, the selection of a mode results partly from the initial presence of a sufficient energy density created by a cw laser. The cw laser and the cavity length must be tuned so that there is no frequency shift during the pulse. To get the tune, the length of the cavity is generally modulated and the shot is started at the convenient time. For CARS spectroscopy, we need simultaneous pulses of the two lasers and a more sophisticated tuning system is necessary.

The dye is excited by flashes. The duration of a pulse is of the order of 1 μs so that the lines may be sharp.

3. INTENSITIES OF CARS LINES

To increase the resolution of a gas spectrum, the pressure must be lowered. It is important that the observed CARS intensity does not decrease too fast versus the pressure. An elementary theory leads to disturbing results :

In absorption or classical Raman spectroscopy, the lineshapes (of simple transitions) are simple and easy to compute in two fields of pressure : pressure broadening and Doppler broadening. In the first case, the linewidth is roughly proportional to the pressure, so that, using monochromatic lasers, the number of excited molecules does not depend on the pressure ; the polarization which gives a CARS signal (CARS polarization) and the CARS intensity are, so, independant of the pressure.
In the second case, the linewidth does not depend on the pressure we may think that a constant proportion of the molecules is tuned and creates a CARS polarization and,so, a CARS amplitude proportional to the pressure, i.e. an intensity proportional to the square of the pressure. Happily, the experiment shows that this answer is wrong because it implicitly assumes that one may convolute Doppler and collisional linewidths in a non-linear process !

The experiment shows that if the lasers are almost monochromatic the CARS scattered intensity is, in the Doppler region almost proportional to the pressure. The dipole induced in a tuned molecule depends on the pressure : indeed the amplitude of an oscillator increases with time until a collision stops it or until a change of phase (due to an approximate tuning) decreases it ; if the collisions are not frequent the mean value of the dipole is larger. An approximate computation leads to the linear rule (3).

4. SATURATION IN CARS PROCESS

When two lasers of pulsations ω_p and ω_s excite a gas, many phenomena appear ; in particular, beyond CARS, molecules are pumped through stimulated Raman. This pumping is proportional to the intensity of each laser and to the difference of populations ; it leads to make the populations of the upper and lower levels equal and consequently to decrease the CARS signal. So, we obtain a saturation of the CARS signal which is easily observed using powerful sources.

Let us show that this saturation may be used to do a Doppler-free CARS spectroscopy.

Two monochromatic plane waves propagating along the x axis are normally reflected by a mirror. The fields $\mathcal{E}_p$ and $\mathcal{E}_s$ are given versus the amplitude E_p and E_s by

$$\mathcal{E}_p = 2\, E_p \sin \omega_p t \cos \frac{\omega_p x}{c}$$

$$\mathcal{E}_s = 2\, E_s \sin \omega_s t \cos \frac{\omega_s x}{c}$$

Let us assume first that the molecules are steady and that the pulsations of an isolated Raman transition is $|\omega_s - \omega_p|$. The polarization of the molecules is :

$$P = a(1 - 2b\, |\mathcal{E}_p\, \mathcal{E}_s|^2)\, \mathcal{E}_p^2\, \mathcal{E}_s$$

where b is a coefficient connected to the saturation ; writing this formula, we suppose that the Raman pumping is low ; the "two" before b takes into account the pumping by the forward and backward waves. From the polarization P, we extract the CARS component at the pulsation $\omega_d = 2\,\omega_p - \omega_s$:

$$P_d = 2a\, E_p^2\, E_s \left(1 - 32b\, E_p^2\, E_s^2 \cos^2 \omega_p \frac{x}{c} \cos^2 \omega_s \frac{x}{c}\right)$$

$$\sin\,(2\omega_p - \omega_s)\,t \cos^2 \frac{\omega_p x}{c} \quad \cos \frac{\omega_s x}{c}$$

If the polarized medium between h_1 and h_2 is homogeneous, the CARS field, for x large, is given by

$$\mathcal{E}_d = \int_{-h_2}^{h_1} + \int_{h_1}^{h_2} 2\, a\, f\, E_p^2\, E_s \left(1 - 32b\, E_p^2\, E_s^2 \cos^2 \frac{\omega_p x}{c}\right.$$

$$\left.\cos^2 \frac{\omega_s x}{c}\right) \cos^2 \frac{\omega_p x}{c} \cos \frac{\omega_s x}{c} \left(\sin \omega_d t \cos \frac{\omega_d x}{c}\right.$$

$$\left. - \cos \omega_d t \sin \frac{\omega_d x}{c}\right) dx$$

where f is a coefficient connected with the geometry of the system. The first integral is related to the amplitude reflected on the mirror. The term containing cos $\omega_d t$ may be cancelled because we have an even integral of an odd function. As h_1-h_2 is large compared to the wavelengths the only terms giving a significant contribution are those which have a non-zero mean value (in x). So, the field may be written

$$E_d = a\ f\ g\ E_p^2\ E_s\ (1 - 24b\ E_p^2\ E_s^2)\ \cos\ \omega_d t$$

where g is a constant.

Let us consider now the thermal velocity of the molecules if $|\omega_p - \omega_s|$ is equal to the Raman transition pulsation, the preceding computation may be applied to the molecules the speed component of which along Ox (radial speed) may be neglected (that is when the component of the displacement between two collisions is less than the wavelength $\frac{2\pi c}{\omega_d}$). But if the Doppler effect is needed for the resonance, the active molecules are only excited by one of the progressive waves : the factor 2 which we introduced before b must be suppressed.
So, the Doppler shape of the CARS line has a dip, the width of which is a collisional width (or the natural width if the pressure is very low).

It is easy to observe a CARS signal because its wavelength is different from the wavelengths of the two lasers. But it is difficult to mix two laser beams, and the mixture is easily disturbed by very small changes in a pulsed laser or in an amplifier or by thermal changes in the sample ; so, the CARS signal is noisy and it seems difficult to extract a small Doppler-free signal from the remainder of the CARS signal. Happily, it is easy to find methods which allow to get the pure Doppler-free signal. We propose the following method :
we suppose, to simplify, that the three beams with pulsations ω_s, ω_p, and ω_d have the same polarization ; we introduce a second system of standing waves equivalent to the preceding, but which may be distinguished by the polarization. The two systems are shifted by a half fringe by a phase change of π when the reflexion of one of the polarizations takes place.

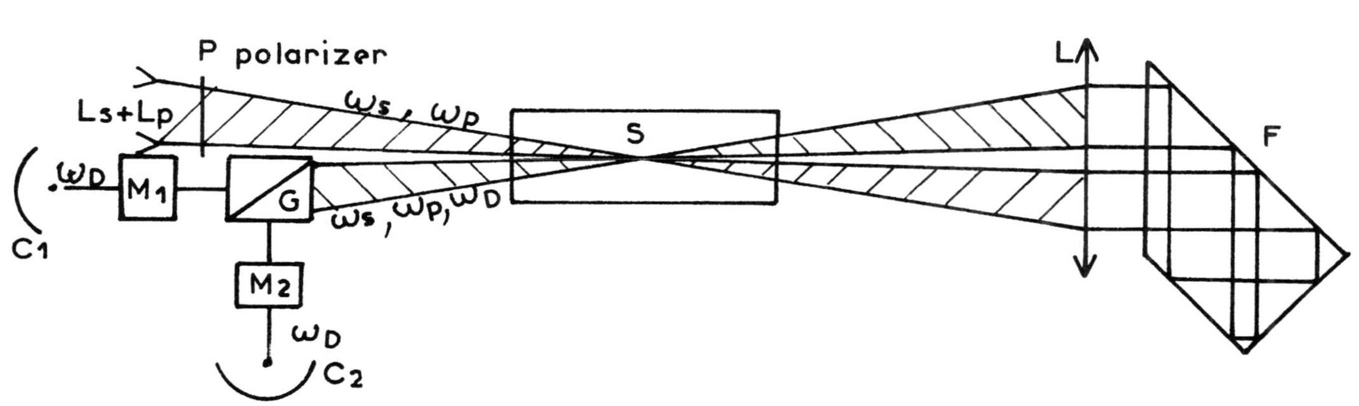

Figure 1 : Experiment for saturated (Doppler-free) CARS Spectroscopy

$L_s + L_p$: common incident beam of the two source-lasers (pulsations ω_s, ω_p)

M_1 M_2 : Monochromators filtering the CARS frequency ω_d

P : Polarizer (if necessary)
S : Sample (gas cell)
L : Lens

F : Glass prism
G : Glan prism
C_1C_2: Photomultipliers

The four reflexions in the prism F introduce a phase shift of π between the two polarizations if its refractive index is 1,554 and if its angles are $\pi/2$, $\pi/4$, $3\pi/4$. One of the cells C_1, C_2 gives a Doppler-free signal, the other the regular signal, if P and G are conveniently rotated.

If the intensity ratio for the two polarizations is large, the saturation may be neglected for one of the CARS signals : so, a polarizer may produce a cancellation of the Doppler-shifted signals. A straightforward computation gives the best polarization angles, and the precise tuning may be set using low power laser pulses. Figure 1 is the drawing of a simple experimental apparatus.

5. CONCLUSION : COMPARISON OF CARS AND SRS SPECTROSCOPY.

Both techniques give spectra using standard cw lasers (power of the order of 1 W) (4) (5). The detection of the signal is sophisticated in both cases, but, however the signal is so low that the resolution is limited by pressure broadening, not by Doppler broadening or by the instability of the lasers. So, it appears that the methods are somewhat equivalent in the case of strong Raman transitions.

The use of pulsed lasers leads to a much more sensitive spectroscopy. So better results will be obtained, but probably in different areas because it seems to me that the two methods have different advantages and inconvenients :

a) leaving constant the average power, the use of pulsed lasers increases CARS signal ; on the contrary, it allows only a better detection of SRS signal.

b) the intensities of SRS signals are proportional to the Raman intensities (those in regular Raman scattering). CARS signals are proportional to the square of these intensities : SRS appears better for the study of weak lines.

For relatively strong lines, CARS is very manageable and the only problem to obtain an useful Doppler-free spectroscopy is to have convenient lasers. Indeed, the saturation is very commonly observed in CARS and the observation of Doppler-free spectra does not need three sources as the observation of SRS saturation. I have tried to explain why J.P. BOQUILLON (6) and H. BERGER develop respectively CARS and SRS experiments in my lab. I am sure that within a few years, many labs will obtain sub-Doppler spectra !

REFERENCES

1. Owyoung, A., Esherick, P., in Lasers and Applications , page 67, Springer, Heidelberg 1981.

2. Hohla, K., Brederlow, G., Volk, R., Witte, K.J., Opt. Commun. 18, 168 (1976).

3. Moret-Bailly, J., Boquillon, J.P., J. Physique 40 (1979).

4. Fabelinsky, V.I., Krynetsky, B.B., Kulevsky, L., Mishin, V.A., Prokhorov, A.M., Savelev, A.D., Smirnow, V.V., Opt. Commun. 20, 389 (1977).

5. Owyoung, A., in Laser Spectroscopy IV, page 175, Springer, Heidelberg 1981.

6. Boquillon, J.P., Bregier, R., J. Appl. Phys. 18, 195 (1979).

CHEMICAL APPLICATION OF GAS PHASE CARS

H. Fietz[a], H. Wieser[b] and W. Kiefer[a]

[a]Physikalisches Institut der Universität Bayreuth D-8580 Bayreuth, F.R. Germany
[b]Department of Chemistry, The University of Calgary, Calgary, Alberta, Canada

1. INTRODUCTION

Even though CARS is a relatively new technique, its advantages over conventional Raman spectroscopy have been described on many occasions [1-4]. These virtues, it is claimed, make it a particularly powerful method for a variety of chemical applications. Inspite of the promising capabilities, however, relatively few reports have appeared in the literature utilizing CARS for studying chemical phenomena in the gas phase. Of those that have been published most concern various diagnostics of processes occuring in plasmas or during combustion, and the resolution of pure rotational or rotation - vibration transitions in small molecules. As far as we are aware, there have been virtually no applications of CARS to the classical chemical problem of conformational analysis. Yet two features of CARS in particular could be exploited advantageously for that purpose, namely the intensity enhancement and the higher resolution capability. The latter aspect gains special importance when viewed with the recent developments in the domain of infrared spectroscopy, where equal or better resolution and detection of lower signal levels are becoming available routinely through the technique of FT interferometry. One reason, and possibly the major one we suspect, for the delay in realizing CARS' promise is the fact that it still is very much the specialist's tool.

One area of conformation analysis which in the past has generated considerable interest from the infrared and Raman spectroscopists is the study of the skeletal deformations of four-, five-, and six-membered ring compounds. Examples of this intensive

W. Kiefer and D. A. Long (eds.), Non-Linear Raman Spectroscopy and Its Chemical Applications, 335–365.

activity are given in another section in this book [5]. The vibrations generally are pictured as slow ponderous motions that require relatively little excitation energy. Because of the type of intramolecular forces that come into play, namely a balance of torsions about single bonds and the strain of restricting the skeletal bond angles, the vibrations usually also are anharmonic in nature. As a result one observes sequences of transitions either between about 50 to 300 cm^{-1} or as combinations with other modes of vibration from thermally populated energy levels. The spacing of the transitions may vary from <0.1 to 100 or more cm^{-1}, depending largely upon the anharmonicity of the vibration. The chemical interest in the structural and dynamic aspects of the phenomenon has a long and exciting history. As far as these ring compounds are concerned, it is important to know whether one or more stable conformations can exist in normal circumstances, what their molecular structure would be, and what energy may be required to go from one to another conformation. In favourable cases one could hope to learn about the relative forces that determine the structures and the dynamic path via which the structures are converted from one to another. All these goals, of course, can be cited for conformational analysis in general, and a number of physical-chemical methods have been and still are being used for that purpose. Infrared and Raman spectroscopy, however, occupy a special place as an analytical tool in the respect that the sequences resulting from the skeletal deformations in most cases show up rather prominently. Indeed, vapour phase infrared and Raman spectroscopy remain the most direct way of gaining the desired information for these ring compounds in the gaseous phase.

Finally, one needs to ask how CARS in particular could contribute. As will be seen, resolution frequently limits one's ability to establish the more subtle aspects of the skeletal deformations. It stands to reason, therefore, that at least the hope would linger that CARS might provide the desired and necessary information, complementary to infrared.

To the best of our knowledge, our recent work on the four-membered ring compound trimethylene oxide (also known as oxetan) at the moment constitutes the only application of CARS to the conformational analysis of cyclic systems. In this article, therefore, after a very brief review of previous work we will bring into focus the necessary intensity considerations and then describe examples of the results we obtained, concluding with our thoughts about the future of CARS for this kind of applications.

2. BRIEF REVIEW OF RING PUCKERING

The name ring puckering has been coined to describe the out-of-plane deformation of a four-membered ring compound.

Depending on the relative magnitudes of the intramolecular forces, the ring skeleton either will be planar, or more or less puckered. In the latter event, and for molecules which are symmetric about their molecular plane, there exist two equivalent conformations separated by a barrier to inversion. Although it was suggested already in 1945 that this inversion vibration should have a fourth power dependence on a suitable coordinate, later spectroscopic investigations established that a quadratic-quartic potential function best describes the molecular motion. The simplest and generally most successful model, therefore, has as its basis the function

$$V(X) = aX^4 + bX^2, \qquad (1)$$

X being a suitable variable that describes the motion. In the case of a puckered ring, "b" is a negative quantity and its magnitude relative to "a" determines the height of the inversion barrier (equal to $b^2/4a$). The function then is said to describe a double minimum situation.

For trimethylene oxide, $\overline{CH_2CH_2CH_2O}$ (abbreviated as TMO from here on), the spectroscopic results established a near planar ring with an inversion barrier below the zero point vibrational level. The situation is graphically depicted in Figure 1. The puckering potential function is drawn for the condition when all other modes of vibration are in their ground state (i.e. all $v'' = 0$), and with one particular mode in its first excited state ($v' = 1$). The puckering energy levels in each case are designated by v''_p and v'_p, respectively. Also shown are the pure puckering transitions that are observed in the conventional Raman spectrum (i.e. $\Delta v = 0$ and $\Delta v_p = 2$), and possible combination transitions with the second mode as one might observe by either infrared or Raman methods, depending on selection rules (in this case, $\Delta v = 1$, $\Delta v_p = 0$).

The discussion thus far is necessarily simple and brief. In order to be able to appreciate complications that will appear later in the article, some more subtle points should be noted. One is that rotational transitions always accompany vibrational transitions. Although the selection rules are quite specific, the effective rotational constants in the different vibrational states may in fact differ, so much so that the combined transitions may distort the band contours that one would normally expect. The second point is that the ring puckering coordinate which formulates the potential in Equation (1) may be too simple. While for reasons of simplicity it is usually pictured as a displacement of the skeletal atoms out of the molecular plane, the CH_2 groups thereby maintaining local C_{2v} symmetry, the real coordinate of course could be a combination of motions of the heavy atoms and the hydrogen atoms in an undetermined way. A number of

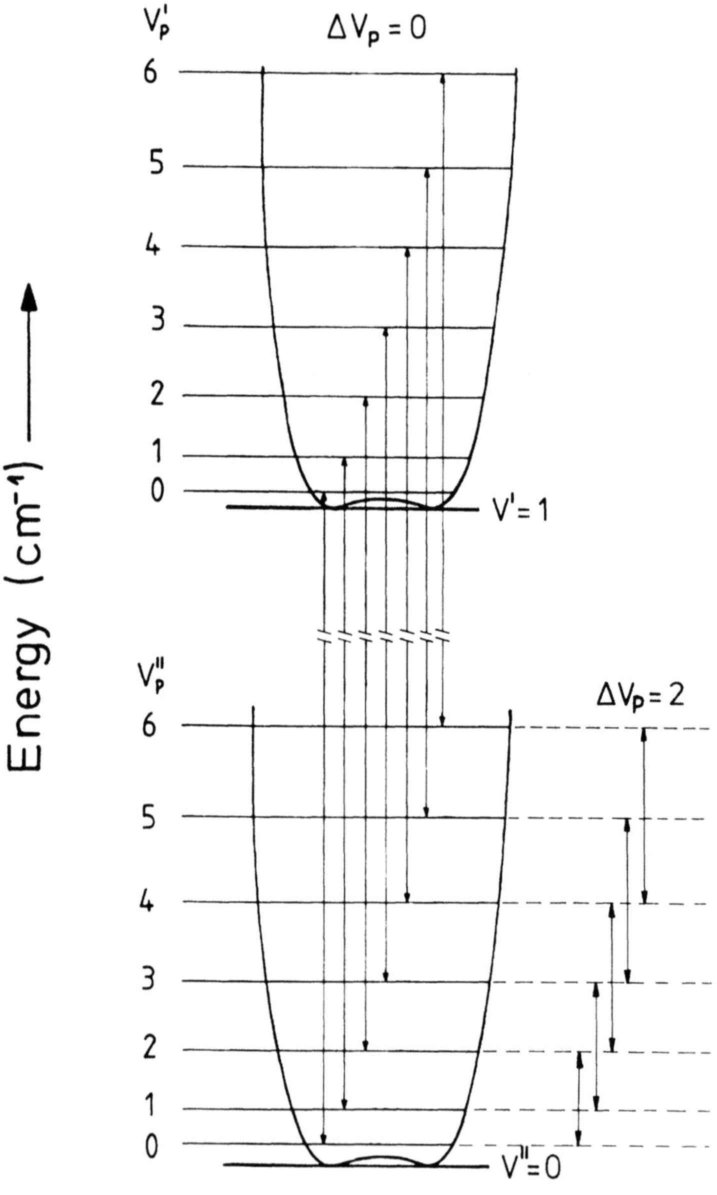

Fig. 1 Schematic of puckering potential functions and energy levels of TMO for the ground state of all other vibrations (v"=0) and the first excited state of a second mode (v'=1); v_p'' and v_p' designate the puckering quantum numbers in the two states, respectively; the pure puckering Raman transitions (Δv_p=2) and Δv_p=0 combination transitions are also shown.

refinements have been attempted with varying degrees of success. For the excited state in an other mode it is customary to refer to the potential as an effective function, indicating that one is not certain to what extent the puckering vibration couples with any one or more of the other vibrations. In general one can judge the veracity of the approximations made by the model in individual cases by how close one can calculate the puckering transitions with a one-dimensional Hamiltonian that incorporates the potential function (1) in the usual quantum mechanical way. Finally, the reduced mass for the puckering motion which enters into the computation through the kinetic energy part of the Hamiltonian is assumed to remain constant during the vibration. This need not be the case and improvements have been demonstrated when some functional dependence of the reduced mass is included.

Over the past few years a number of excellent reviews have been written, to which the reader is referred for a less cursory introduction and more detailed information [6-11]. The foregoing brief discussion is intended to set the stage for the interpretation of the results that follow.

3. SELECTION RULES AND INTENSITIES

In this section we shall first review the general Raman selection rules particularly as they are affected by the puckering vibration. We will then consider the factors that dictate the intensities so that we may be able to predict qualitatively the relative CARS intensities that would be expected. All the information needed to formulate the first part, namely the selection rule, is contained in Herzberg [12], although with respect to the puckering motion it was first thought out clearly in Ref. 13 to which the reader is referred if the following discussion is not sufficiently clear.

According to Placzek's classical Raman theory [14] vibrational Raman scattering becomes observable when a change of the polarizability, α_{ij}, accompanies the vibration (α_{ij} being an element in the molecular polarizability tensor). Since it is found that the α_{ij}'s need not be linear functions of the vibrational coordinates, one customarily expands the tensor elements in a Taylor series about a suitable reference configuration of the molecule which in the puckering case is usually taken to be the planar form. In the following expression we separate the puckering coordinate, X, from the other normal modes:

$$\alpha_{ij} = \alpha_{ij}^{\circ} + \left(\frac{\partial\alpha_{ij}}{\partial X}\right)_{\circ} X + \sum_{s}\left(\frac{\partial\alpha_{ij}}{\partial Q_s}\right)_{\circ} Q_s + \left(\frac{\partial^2\alpha_{ij}}{\partial X^2}\right)_{\circ} X^2 +$$

$$\sum_s \left(\frac{\partial^2\alpha_{ij}}{\partial X \partial Q_s}\right)_\circ XQ_s + \sum_{r,s}\left(\frac{\partial^2\alpha_{ij}}{\partial Q_r \partial Q_s}\right)_\circ Q_r Q_s + \left(\frac{\partial^3\alpha_{ij}}{\partial X^3}\right)_\circ X^3 +$$

$$\sum_s \left(\frac{\partial^3\alpha_{ij}}{\partial X^2 \partial Q_s}\right)_\circ X^2 Q_s + \sum_{r,s}\left(\frac{\partial^3\alpha_{ij}}{\partial X \partial Q_r \partial Q_s}\right)_\circ XQ_r Q_s + \text{higher terms} \qquad (2)$$

The summations extend over all normal coordinates r,s = 1 to 3N - 7, except the puckering coordinate, X.

In its planar form TMO belongs to the C_{2v} (mm2) symmetry point group. Within that group the polarizability elements span the following symmetry species:

A_1: α_{xx}, α_{yy}, α_{zz}

A_2: α_{xy}

B_1: α_{xz}

B_2: α_{yz}

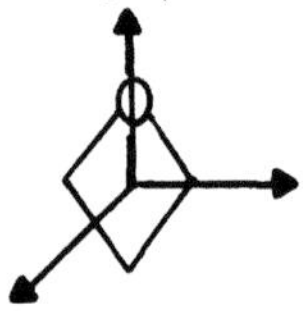

Since the expansions cannot alter the overall symmetry of each polarizability element, only certain terms are permitted. For example, the symmetry allowed terms for α_{ii} (i = x,y, or z), remembering that the puckering coordinate belongs to B_1, are:

$$\alpha_{ii} = \sum_{s(A_1)} \left(\frac{\partial\alpha_{ii}}{\partial Q_s}\right)_\circ Q_s + \left(\frac{\partial^2\alpha_{ii}}{\partial X^2}\right)_\circ X^2 + \sum_{s(B_1)}\left(\frac{\partial^2\alpha_{ii}}{\partial X \partial Q_s}\right)_\circ XQ_s +$$

$$\sum_{\substack{r,s \\ \text{(same symmetries)}}} \left(\frac{\partial^2\alpha_{ii}}{\partial Q_r \partial Q_s}\right)_\circ Q_r Q_s + \sum_{s(A_1)}\left(\frac{\partial^3\alpha_{ii}}{\partial X^2 \partial Q_s}\right)_\circ X^2 Q_s +$$

$$\sum_{\substack{r(A_1) \\ s(B_1)}} \left(\frac{\partial^3\alpha_{ii}}{\partial X \partial Q_r \partial Q_s}\right)_\circ XQ_r Q_s + \sum_{\substack{r(A_2) \\ s(B_2)}} \left(\frac{\partial^3\alpha_{ii}}{\partial X \partial Q_r \partial Q_s}\right)_\circ XQ_r Q_s +$$

$$\text{higher terms} \qquad (3)$$

The leading term of Equation (2) is not included in this expression since, being responsible for Rayleigh scattering only, it does not contribute to the Raman intensities. The expansions of the remaining polarizability tensor elements can be written correspondingly.

In order to determine the quantum mechanical selection rules we first write the corresponding transition moments. For example for α_{ii}:

$$\langle\psi'\phi'|\alpha_{ii}|\psi''\phi''\rangle = \sum_{s(A_1)}\left[\left(\frac{\partial\alpha_{ii}}{\partial Q_s}\right)_\circ \langle\psi_s'|Q_s|\psi_s''\rangle\langle\phi'|\phi''\rangle\right]$$

$$+\left(\frac{\partial^2\alpha_{ii}}{\partial X^2}\right)_\circ \langle\phi'|X^2|\phi''\rangle\langle\psi'|\,\psi''\rangle$$

$$+\sum_{s(B_1)}\left[\left(\frac{\partial^2\alpha_{ii}}{\partial X\partial Q_s}\right)_\circ \langle\psi_s'|Q_s|\psi_s''\rangle\langle\phi'|X|\phi''\rangle\right]$$

$$+\sum_{(r,s\ \text{same symmetries})}\left[\left(\frac{\partial^2\alpha_{ii}}{\partial Q_r\partial Q_s}\right)_\circ \langle\psi_r'|Q_r|\psi_r''\rangle\langle\psi_s'|Q_s|\psi_s''\rangle\langle\phi'|\phi''\rangle\right]$$

$$+\sum_{s(A_1)}\left[\left(\frac{\partial^3\alpha_{ii}}{\partial X^2\partial Q_s}\right)_\circ \langle\psi_s'|Q_s|\psi_s''\rangle\langle\phi'|X^2|\phi''\rangle\right]$$

$$+\sum_{\substack{r(A_1)\\ s(B_1)}}\left[\left(\frac{\partial^3\alpha_{ii}}{\partial X\partial Q_r\partial Q_s}\right)_\circ \langle\psi_r'|Q_r|\psi_r''\rangle\langle\psi_s'|Q_s|\psi_s''\rangle\langle\phi'|X|\phi''\rangle\right]$$

$$+\ \text{same term for } r(A_2) \text{ and } s(B_2)$$

$$+\ \text{higher terms} \qquad (4)$$

The ψ's and ϕ's in this expression represent the wavefunctions for the normal modes r or s, and for the puckering vibration, respectively. Moreover, the factorization of the integrals assumes that the vibrations are independent of one another. While this may not be strictly correct, the error is likely to be small particularly in our present example, and we may still proceed to demonstrate the principle involved.

The next factor to consider is the anharmonicity of the puckering motion. The wavefunctions ϕ' and ϕ'' then must be also anharmonic. They are formed by summing an appropriate number of harmonic oscillator functions. The end result is that the integral $\langle\phi'|X|\phi''\rangle$ has finite values for $\Delta v_p = \pm 1$ and $\langle\phi'|X^2|\phi''\rangle$ for $\Delta v_p = 0, \pm 2$, whereas only $\Delta v_p = \pm 1$ would be possible if X were purely harmonic [13].

The transition moments for the off-diagonal elements of the polarizability are written in the same fashion as above for α_{ii}, although of course they will contain different terms based on symmetry. We will not consider their expansion explicitly here for the following reason. In addition to the vibrational selection rules one needs to consider also those for the rotational transitions. Without further discussion we shall simply state (see [12] for details) that only rotational levels of the same symmetry and the same rotational quantum number J can combine for the totally symmetric polarizability elements (in a given symmetry point group). This means that the Raman lines for our α_{ii} components will be prominent rotation-vibration band contours of many closely spaced Q-branches clustered about the band origin, with the proviso that the rotational constants for the two states not be too different and other perturbing resonances do not occur. The selection rules for the off-diagonal tensor elements in general will lead to broad and hence less prominent features in the Raman spectrum. In practice, therefore, we can hope to find the rotation-vibration transitions for α_{ii} but not for α_{ij}.

We can now summarize the information that can be gleaned from the Equation (4). Namely, we can expect to see the following Raman Q-branches, alone from the point of view of being allowed:

1. term: The A_1 fundamentals, modulated by a Franck-Condon type integral $\langle\phi'|\phi''\rangle$; the latter allows $\Delta v_p = 0$ transitions so that sequences will appear if the puckering separations in the ground and excited states of the A_1 fundamental differ.
2. term: Pure puckering transitions for which $\Delta v_p = 2$, in the ground or excited state of any other mode.
3. term: Puckering sequences for which $\Delta v_p = \pm 1$, for B_1 fundamentals.
4. term: Combinations of any two vibrations (except puckering) of the same symmetry with $\Delta v = \pm 1$ for each, or the first overtone of any vibration (except puckering), all modulated by $\langle\phi'|\phi''\rangle$ as in the first term.
5. term: Puckering sequences for which $\Delta v_p = \pm 2$, for A_1 fundamentals; note that this term in principle can also contribute to the $\Delta v_p = 0$ combinations of the first term, depending on the relative magnitudes of the coefficient and, of course, the relative value of the integral $\langle\phi'|X^2|\phi''\rangle$ for $\Delta v_p = 0$ and ± 2.
6. term: Puckering sequences for which $\Delta v_p = \pm 1$, for combinations of A_1 and B_1 fundamentals.
7. term: Same as in the sixth term, except for A_2 and B_2 fundamentals.

For all the terms except the second, one must recall that the wavefunctions ϕ' and ϕ'' describe the puckering motion in two different states of another mode. A substantially different effective puckering potential in the excited state for whatever reason, therefore, can have a significant effect on the intensities.

Table 1 C_{2v} Symmetry allowed expansion coefficients of polarizability tensor elements

Element \ Coefficient	$\frac{\partial\alpha_{ij}}{\partial X}$	$\frac{\partial\alpha_{ij}}{\partial Q_s}$	$\frac{\partial^2\alpha_{ij}}{\partial X^2}$	$\frac{\partial^2\alpha_{ij}}{\partial X\partial Q_s}$	$\frac{\partial^2\alpha_{ij}}{\partial Q_r\partial Q_s}$	$\frac{\partial^3\alpha_{ij}}{\partial X^3}$	$\frac{\partial^3\alpha_{ij}}{\partial X^2\partial Q_s}$	$\frac{\partial^3\alpha_{ij}}{\partial X\partial Q_r\partial Q_s}$
α_{xx}, α_{yy}, α_{zz} (A_1)	No	A_1 fundtls $\Delta v_p=0$	$\Delta v_p=2$	B_1 fundtls $\Delta v_p=\pm1$	Q_r and Q_s same symmetries $\Delta v_p=0$	No	A_1 fundtls $\Delta v_p=0,\pm2$	A_1+B_1 or A_2+B_2 fundtls $\Delta v_p=\pm1$
α_{xz} (B_1)	$\Delta v_p=1$	B_1 fundtls $\Delta v_p=0$	No	A_1 fundtls $\Delta v_p=\pm1$	A_1+B_1 or A_2+B_2 fundtls $\Delta v_p=0$	$\Delta v_p=1$	B_1 fundtl $\Delta v_p=0,\pm2$	Q_r and Q_s same symmetries $\Delta v_p=\pm1$
α_{xy} (A_2)	No	A_2 fundtls $\Delta v_p=0$	No	B_2 fundtls $\Delta v_p=\pm1$	A_1+A_2 or B_1+B_2 fundtls $\Delta v_p=0$	No	A_2 fundtl $\Delta v_p=0,\pm2$	A_1+B_2 or A_2+B_1 fundtls $\Delta v_p=\pm1$
α_{yz} (B_2)	No	B_2 fundtls $\Delta v_p=0$	No	A_2 fundtls $\Delta v_p=\pm1$	A_1+B_2 or A_2+B_1 fundtls $\Delta v_p=0$	No	B_2 fundtl $\Delta v_p=0,\pm2$	A_1+A_2 or B_1+B_2 fundtls $\Delta v_p=\pm1$

The coefficients of the symmetry allowed components of the transition moments for all the polarizability elements are summarized in Table 1. As is well known, these components enter into the normal Raman intensity as the squared magnitude. For CARS the situation is somewhat different. The derivation of the expression for CARS is outlined in another section of this book [15]. For the present purposes it will suffice to quote only the pertinent factors. Namely, the power of the generated CARS signal is proportional to

$$P_{CARS} \propto (N\Delta)^2 \, [\alpha_{ij}]^4, \tag{5}$$

where $[\alpha_{ij}]$ is the appropriate transition moment and N is the molecular number density. The product NΔ gives the average difference in population between the two states involved in a transition. The comparable relation for the normal Raman effect is the Raman intensity

$$I_{RAMAN} \propto N \, [\alpha_{ij}]^2. \tag{6}$$

The fourth power dependence in the transition moment in CARS leads to the well documented observation that strong lines appear to be stronger than their weaker neighbours, as compared to the conventional Raman spectrum. By the same token, already prominent lines in the latter will be even more enhanced. The factor NΔ in CARS likely would seriously diminish the intensity of pure puckering transitions since excited puckering states are appreciably thermally populated, but will virtually have no effect on the combination sequences with higher energy vibrations.

The only remaining question in this brief survey is the one concerning the selection rules. Fortunately, as was shown in an explicit fashion [16], they are exactly the same as for the conventional Raman effect. We can therefore use the foregoing outline as a helpful guide.

4. LINEAR RAMAN AND CARS SPECTRA AND THEIR INTERPRETATION

Of the 24 normal vibrations of TMO, 8 span the A_1 symmetry species in the C_{2v} point group. The modes of interest in particular are the in-phase symmetric C-H stretching vibrations of the two α-CH_2 groups (designated as ν_1), the same mode for the single β-CH_2 group (ν_2), the ring breathing mode where all skeletal bonds are pictured to stretch in phase (ν_6), and the β-CH_2 scissoring mode (ν_4) (see Figure 2). Except for the last mentioned, they all exhibit strong Raman lines; in fact they are the most prominent lines in the conventional Raman spectrum. The β-CH_2 scissoring mode ν_4 will be mentioned only incidentally in connection with ν_6. As representative examples we select the spectra of

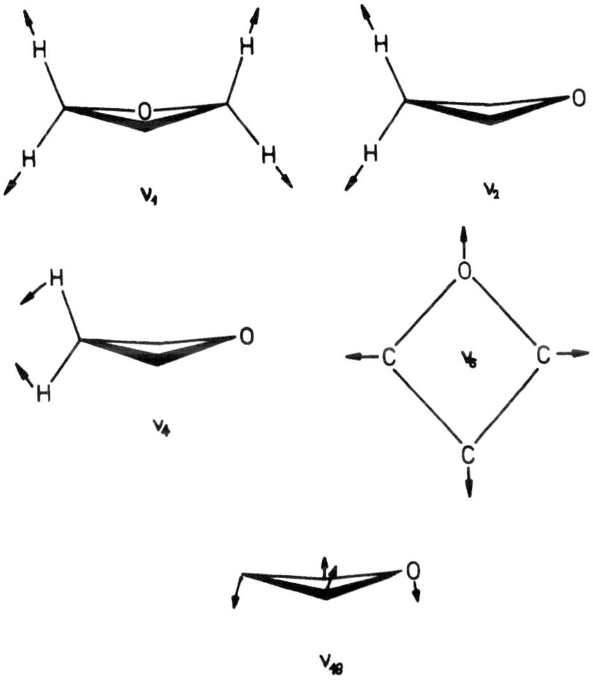

Fig. 2 Pictorial representation of the approximate coordinates for ν_1 (in-phase α-CH_2 symmetric stretch), ν_2 (β-CH_2 symmetric stretch), ν_4 (β-CH_2 scissor), and ν_{18} (ring puckering).

the hydrogenated parent compound, n TMO, and the compounds deuterated in the β position, β-d_2 TMO, and the α positions, α,α'-d_4 TMO. Survey spectra for n and β-d_2 TMO are reproduced in Figure 3, where the strong lines for ν_1, ν_2 and ν_6 (near 2900, 2980, and 1000 cm^{-1}, respectively) are clearly seen [17]. The spectra also show very distinctly the pure puckering transitions at low wavenumbers [18].

While the linear Raman lines in the CH stretching region have been published previously [19], those in the ring breathing region have not [20]. The CARS spectra were recorded recently [21]. The experimental set-up is described in another article in this book [22]. The infrared positions reported here constitute recent more accurate measurements with a Nicolet 8000 interferometer[23].

In Figure 4 the conventional Raman spectra of ν_1 for n and β-d_2 TMO are displayed. The corresponding CARS spectra are shown in Figure 5. Although qualitatively the linear Raman and the CARS spectra are similar in the sense that all the features are present

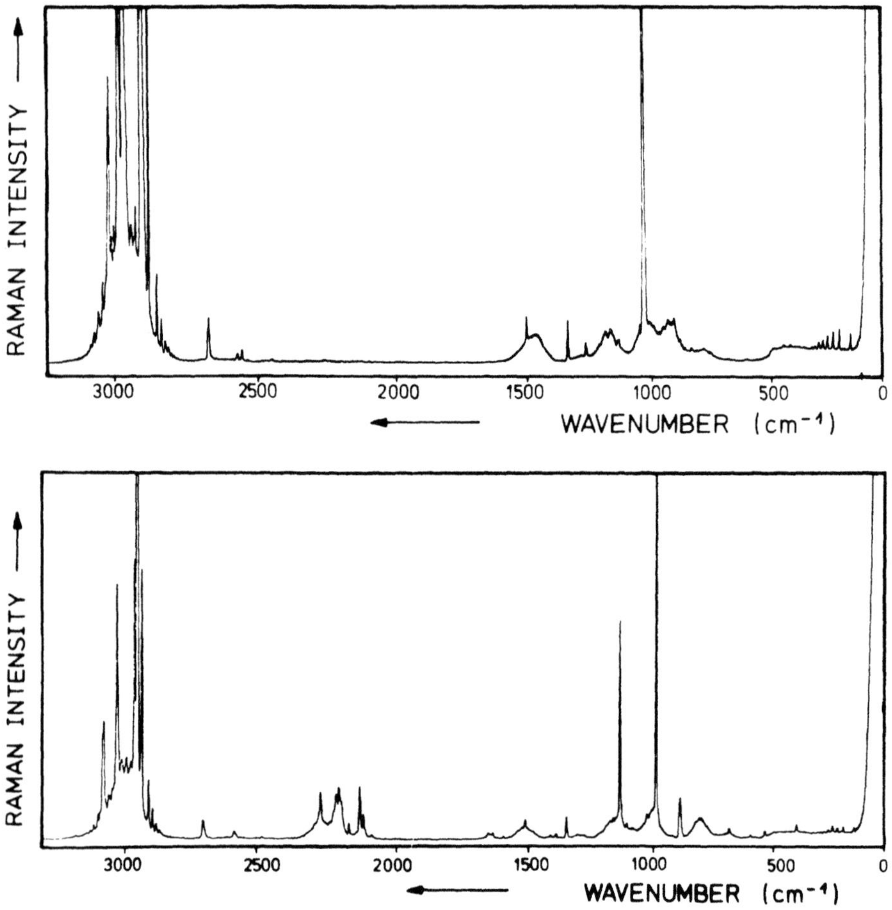

Fig. 3 Linear Raman survey spectra of normal (upper field) and β-d_2 TMO (lower field). Vapor pressure 33 kPa; slit width 3 cm^{-1}.

in both, the resolution obviously is much improved in the latter. Clearly evident is also the apparent intensity decrease of the weaker lines relative to the strongest one (e.g. 2896 vs. 2894 cm^{-1} in n TMO). In the earlier work the assignment of the main features was based on the extensive and well separated infrared difference absorptions, i.e. those for which $\Delta v_p = -1$ [19]. Although the pure ν_1 infrared fundamental ($v''_p = v'_p = 0$) was difficult to locate precisely, due to the extensive overlapping of the broad rotation-vibration contours, its expected peak position could be determined quite accurately from the sum of the first pure puckering and the first difference transitions (53.5+2840.5cm^{-1},

Table 2 Puckering combinations of the α-CH_2 symmetric stretching vibration

Mid IR			CARS				Mid IR			CARS			
Transition		Peak position	Transition		Peak position $[cm^{-1}]$		Transition		Peak Position	Transition		Peak position $[cm^{-1}]$	
v'_p ←	v''_p	$[cm^{-1}]$	v'_p ←	v''_p	Predicted[a]	Measured	v'_p ←	v''_p	$[cm^{-1}]$	v'_p ←	v''_p	Predicted[a]	Measured
		normal TMO							β-d_2 TMO				
0	1	2841.05	0	0	2894.09	2894.49	0	1	2845.58	0	0	2894.87	2894.98
1	2	2814.12 2786.10	1	1	2903.80 2875.78	not measured by CARS	1	2	2823.40 2796.35	1	1	2906.56 2879.51	not measured by CARS
2	3	2793.35	2	2	2897.84	2898.98 2898.55	2	3	2801.99	2	2	2898.88	2898.97
3	4	2778.18	3	3	2896.11	2896.04	3	4	2787.70	3	3	2897.00	2897.05
4	5	2767.28	4	4	2896.19	2896.26	4	5	2778.20	4	4	2897.64	2897.86
5	6	2756.77	5	5	2895.24	not measured or obscured by main peak	5	6	2768.06	5	5	2896.29	2896.43
6	7	2747.68	6	6	2894.66		6	7	2759.80 2758.80	6	6	2895.76 2894.76	not measured or obscured by main peak
7	8	2739.60	7	7	2894.28		7	8	2752.76	7	7	2895.98	
8	9	2732.17	8	8	2893.93		8	9	2745.44	8	8	2894.97	

[a] predicted by summing the appropriate pure puckering and difference transitions in infrared spectrum.

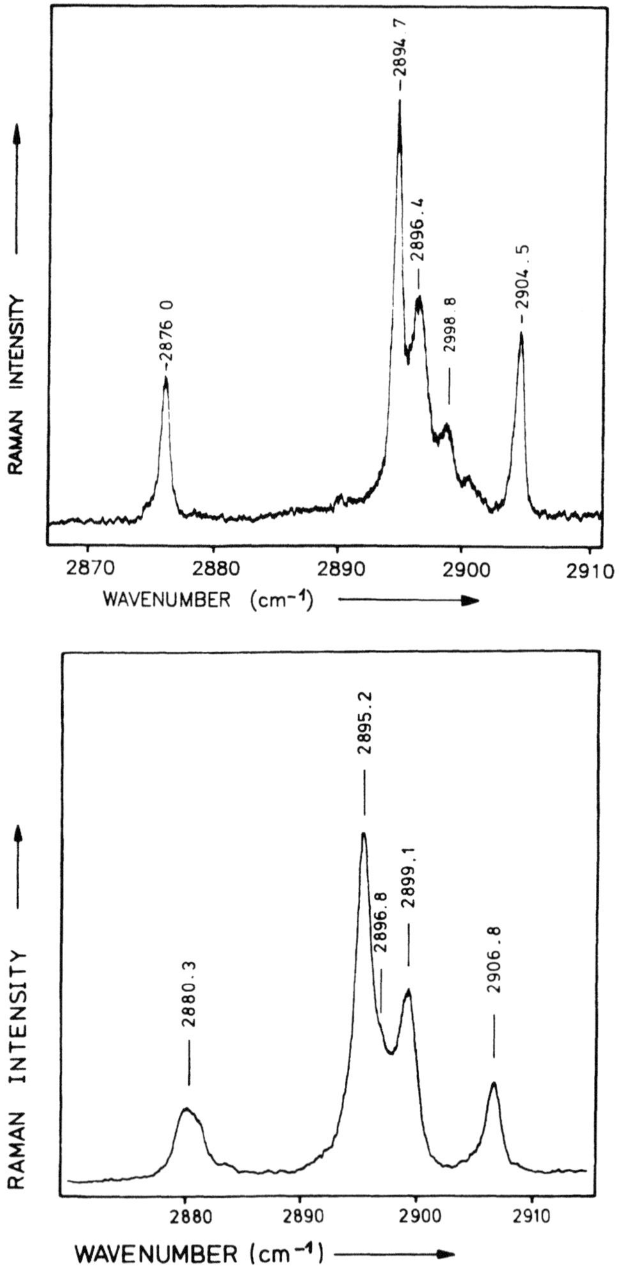

Fig. 4 Linear Raman spectra of ν_1 in n TMO (upper field) and β-d_2 TMO (lower field). Vapor pressure: 33 kPa; slit widths: 0.5 cm^{-1} for n TMO and 1.2 cm^{-1} for β-d_2 TMO.

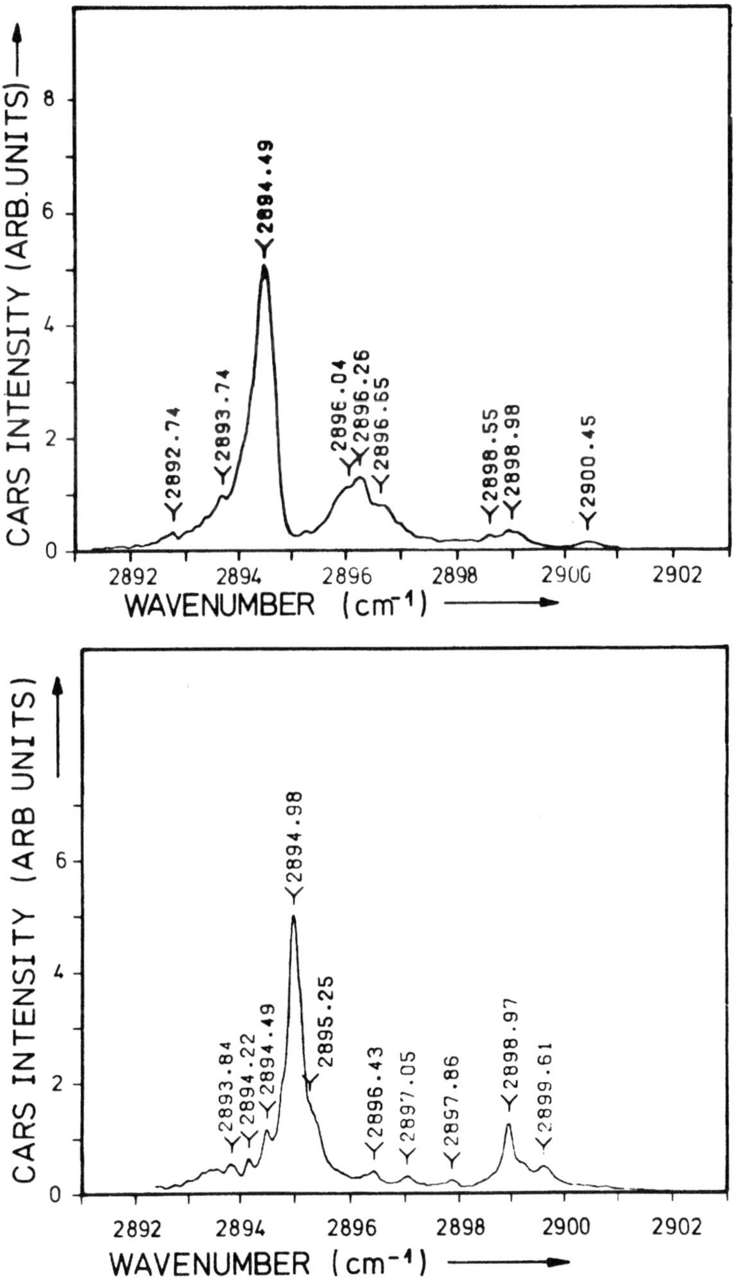

Fig. 5 CARS spectra of ν_1 in n TMO (upper field) and β-d_2 TMO (lower field). Vapor pressure 27 kPa; resolution 0.05 cm^{-1}.

respectively, in TMO). The assignment of the fundamental to the main Raman line (2894.7 cm^{-1} in n TMO) was therefore assured. The ($v_p' = 1$) ← ($v_p'' = 1$) transition [24] is undoubtedly split by Fermi resonance [19], so that the 1 ← 1 lines (the doublet 2876.0/2904.5 cm^{-1} in n TMO) would not show in the relatively narrow range we investigated by CARS. The 2 ← 2 transition was also located unambiguously (at 2898.8 cm^{-1} in n TMO) by comparison with the infrared absorptions. Higher $\Delta v_p = 0$ transitions, however, were not resolved but were lumped into one band (2896.4 cm^{-1} in n TMO, the shoulder at 2896.8 cm^{-1} in β-d_2 TMO). Their assignment is now confirmed by CARS, especially for β-d_2 TMO where they are seen as distinct if weak lines. The CARS positions are compared to positions predicted from infrared absorptions in Table 2.

Gratifying as it may seem at a first glance, there are some problems. First, the infrared and CARS peak positions do not uniformly correspond to each other, even if the assignment is clear. This observation is not entirely unexpected since the apparent maxima of the rotation-vibration contours for the two methods may not coincide, governed as they are by different selection rules and the effective rotational constants. Second, some bands exhibit an apparent doublet structure, as for instance the 2898 cm^{-1} line of n TMO which clearly must correspond to the 2 ← 2 transition. It is interesting to note in this case that the corresponding infrared band, 2 ← 3 at 2793.35 cm^{-1} [23], also has an irregular contour under high resolution. One explanation could be the occurence of an unresolved resonance with another vibrational energy level which again would affect the rotational contours. Finally, there are a number of lines that are not accounted for, notably to lower wavenumbers of the main peak. From the foregoing discussion it is certain that they cannot be puckering combinations on the ν_1 fundamental. One tempting possibility is that these lines are $\Delta v = +1$ transitions from the first and second levels of ν_1, displaced to lower wavenumbers as a result of the C-H stretching anharmonicity. From Equation (5), however, we would expect the second transition to be weaker by a factor of about 3×10^{-13} ($\Delta \approx 5 \times 10^{-7}$, the Boltzmann population of an energy level at 3000 cm^{-1}). In order to become visible, the coefficient $(\partial^3\alpha_{ii}/\partial X\partial Q^2)_0$ would have to compensate by about as much. A second explanation could be isotopic shifts due to ^{13}C with its natural abundance of 1.1 %. A further suggestion may be combinations of other fundamentals, but nothing more definitive can be ascertained.

Figures 6 and 7 allow a comparison of the linear Raman spectra of ν_2 of n and α,α'-d_4 TMO with the CARS spectra in the same region. There evidently is qualitative agreement between them. All the features observed with conventional Raman are reproduced by CARS, although with different apparent intensities and, of course, much improved resolution. The earlier interpretation [19] saw the unresolved lines as puckering sequences on the ν_2 fundamental, whereby all levels were to be perturbed by Fermi resonance with the corresponding puckering levels on the second excited state of ν_4 (the CH_2 scissoring mode). The same assignment could be made still on the basis of the CARS spectra, but unfortunately there are not infrared absorptions of any kind that could be invoked to either confirm or alter the interpretation. Here is an instance, therefore, where CARS undoubtedly has provided better resolved spectra and hence more accurate peak positions, but for the moment no new information.

The third and last example to be discussed is ν_6, the ring breathing mode in n and β-d_2 TMO. The conventional and CARS spectra are compared in Figures 8 and 9, respectively. Again the improved resolution and the qualitative agreement are clearly seen. The sharp line at 1033 cm^{-1} in n TMO, apparently consisting of at least two components, is beautifully resolved by CARS. Unfortunately, the assignment is not immediately obvious. Alone on the basis of the CARS spectrum, several possibilities are possible. Assuming for example that there are no resonances which would perturb the intensities, we may place the $0 \leftarrow 0$ transition at 1034.28 cm^{-1}, $1 \leftarrow 1$ at 1033.18 cm^{-1}, $2 \leftarrow 2$ at 1034.04 cm^{-1}, $3 \leftarrow 3$ at 1032.92 cm^{-1}, etc., up to $6 \leftarrow 6$ at 1031.74 cm^{-1}. At least such a sequence would make sense if the Boltzmann factor were the main contributor. We could then determine the puckering separations in the excited state of the ring breathing mode, remembering that the pure puckering levels are known. One way of testing an assignment is to fit a set of calculated separations to the observed ones by means of the one-dimensional Hamiltonian with a quadratic-quartic potential function mentioned earlier in this article. After all, the ground state puckering levels can be reproduced very well with this model, and as long as the puckering mode does not couple with the second vibration, the model should work well also for the excited state, all be it with "effective" potential constants since one has no knowledge of the dynamic form of the puckering coordinate in that state. It is not necessary here to outline the details of such calculations as the method is well documented [6-11]. Nor, as will become clear later, is it meaningful to quote the potential constants and the inversion barriers for these tests. We will attempt to judge the success of the assignment by the quality of the fit of the calculated to observed puckering separations, expressed as the standard deviation, σ, of an observed "transition".

The ground state separations fitted in this manner up to the sixth energy level yield a $\sigma = 0.06$ cm^{-1}, i.e. an excellent fit. The same separations determined for the excited state as noted above are not reproduced as well, σ now being worse by factor of 10, though still reasonable. We then try another assignment.

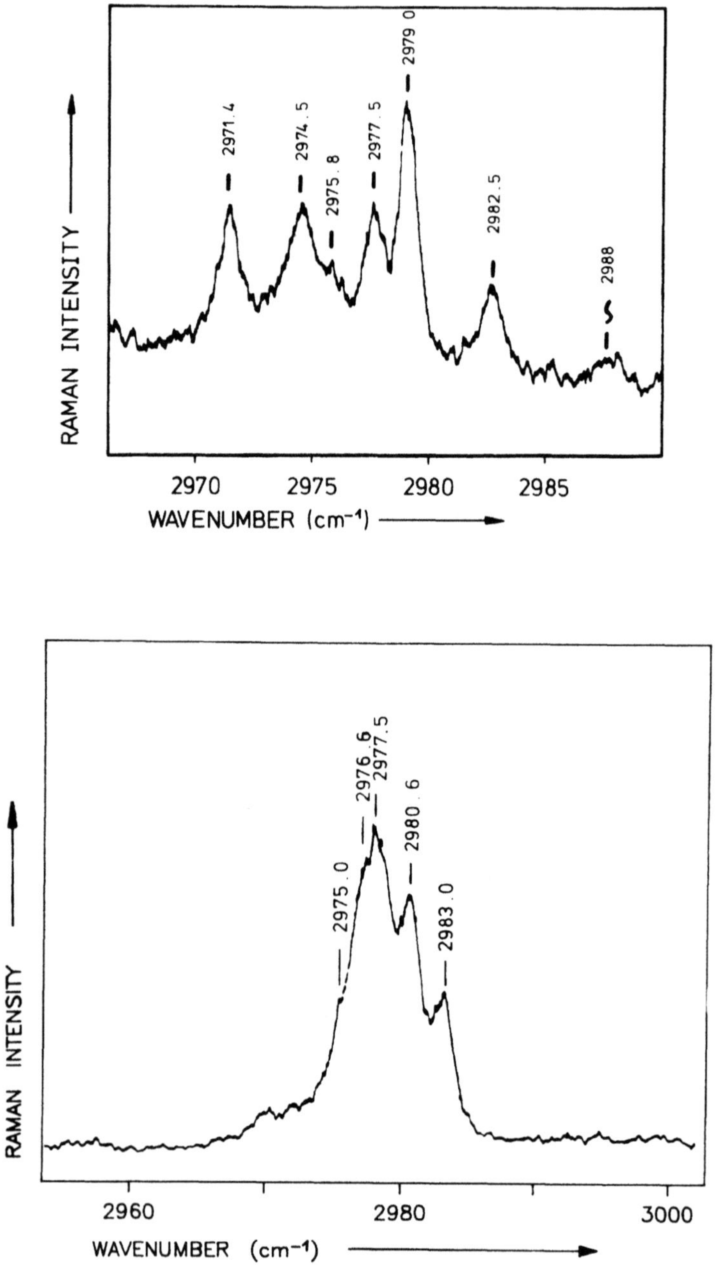

Fig. 6 Linear Raman spectra of ν_2 in n TMO (upper field) and α,α'-d_4 TMO (lower field). Vapor pressure: 33 kPa; slit widths: 0.5 cm^{-1} for n TMO and 1.2 cm^{-1} for α,α'-d_4 TMO.

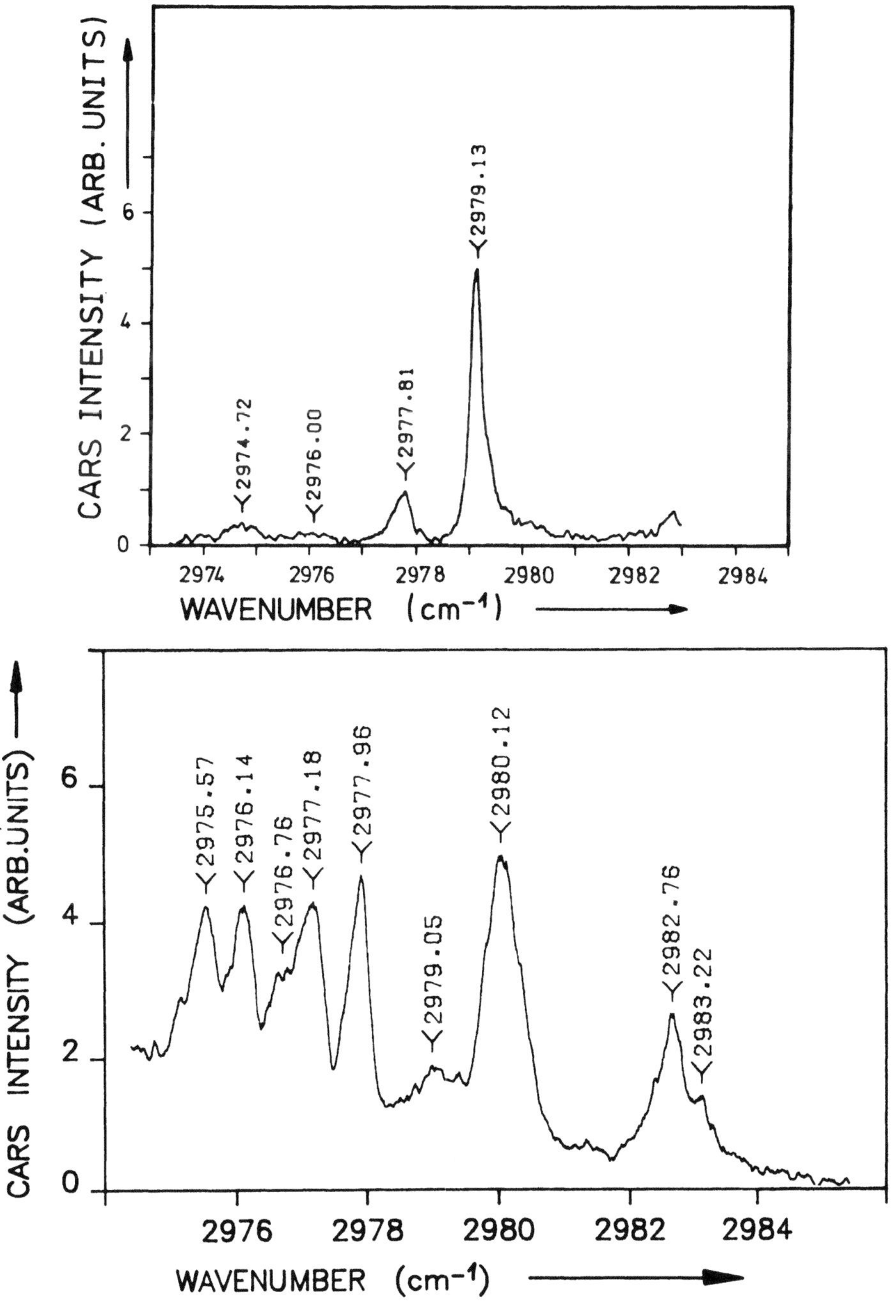

Fig. 7 CARS spectra of ν_2 in n TMO (upper field) and α,α'-d_4 TMO (lower field). Vapor pressure 27 kPa; resolution 0.05 cm^{-1}.

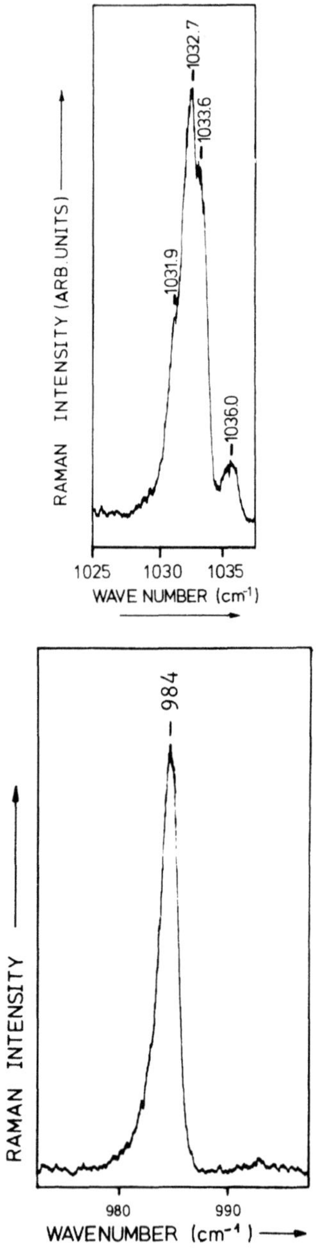

Fig. 8 Linear Raman spectra of ν_6 in n TMO (upper field) and β-d_2 TMO (lower field). Vapor pressure 33 kPa; slit widths: 0.5 cm^{-1} for n TMO and 1.0 cm^{-1} for β-d_2 TMO.

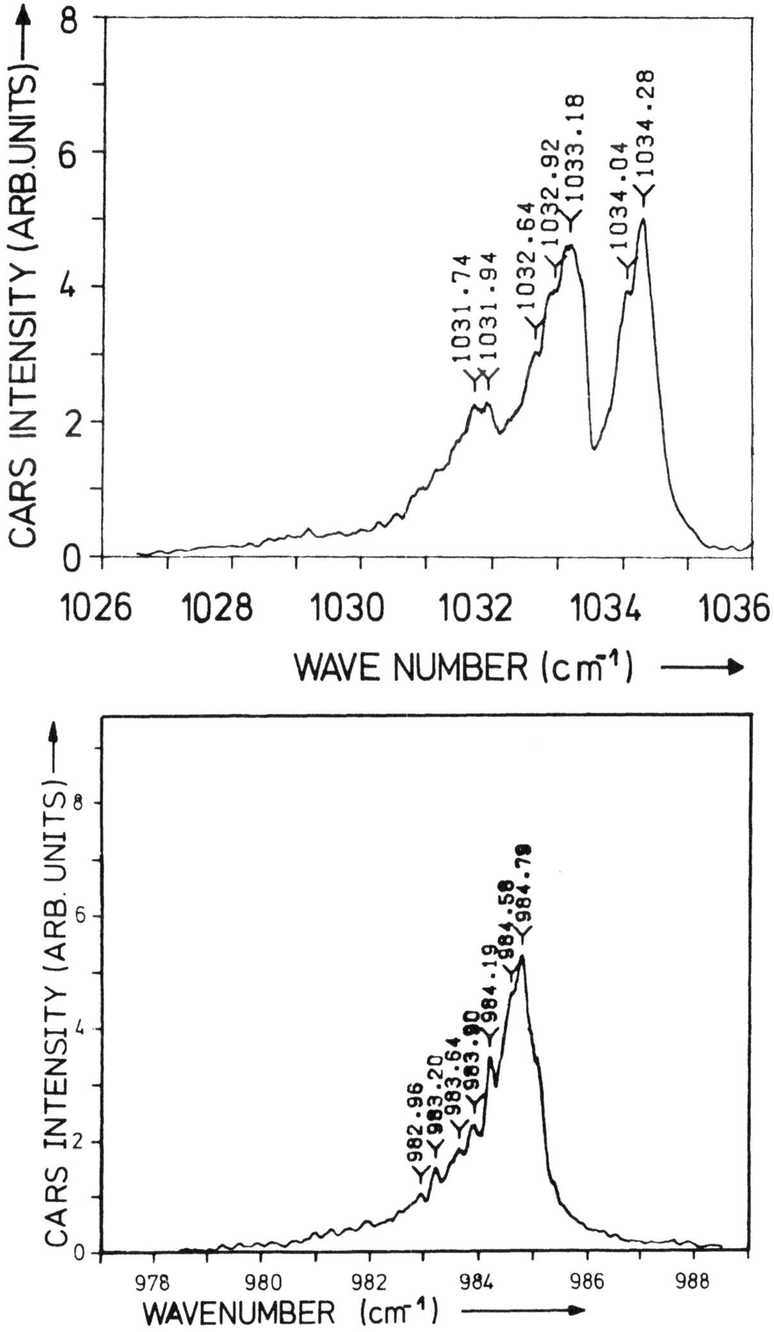

Fig. 9 CARS spectra of ν_6 in n TMO (upper field) and β-d_2 TMO (lower field). Vapor pressure 27 kPa; resolution 0.05 cm^{-1}

Namely, we use the positions from higher to lower wavenumbers in sequence, without regard for the relative intensities and beginning with 1034.28 cm^{-1} as the 0 ← 0 transition. The fit in this case is slightly improved, namely σ = 0.3 cm^{-1}. It is clear that an assignment of the CARS lines on this basis is not meaningful. The model simply is not sufficiently sensitive to distinguish between the small differences in line positions resulting from the high resolution.

Alternatively and preferably we must take recourse to the infrared spectra. With the aid of the CARS spectrum we can unambiguously assign a very weak infrared absorption at 944.60 cm^{-1} as the second difference band, i.e. 1 ← 2, which then places the 1 ← 1 Raman line at 1034.28 cm^{-1}. The infrared spectrum further shows weak but clear sum bands based on the ν_6 fundamental. They are displayed as an example in Figure 10. Using these absorptions we can finally arrive at plausible assignment for the CARS lines. It is summarized in Table 3. The agreement is quite remarkable, although it is not without some sacrifice. We again find some features for which we can offer no explanation, and we again find the apparent doublet structure for some lines. In addition, however, the infrared spectrum raises questions for which there are no immediate answers. The first three sum bands, whose assignment seems fairly certain, exhibit pronounced C-type contours as symmetry arguments in fact would predict for n TMO. The fourth sum band is then completely absent, while further sum bands are again present though with much reduced intensity. It appears as though the first four puckering levels in the excited state are resonance perturbed in some manner. This kind of interaction is certainly not out of the question. Within the same region there are a number of fundamentals, such as a CH_2 rocking mode (B_1) near 1140 cm^{-1}, a CH_2 wagging mode (A_1) near 1340 cm^{-1}, and a CH_2 scissoring mode (A_1) near 1450 cm^{-1} [17] all of which together with their puckering sequences are potential candidates for exactly that kind of resonance. While there are indications in the infrared spectra that these are not well behaved, no clear correlation has emerged thus far [23].

Nevertheless, we could again test the assignment, to which the infrared sum bands have led us with reasonable certainty, with the quadratic-quartic puckering model. The quality of the fit in terms of observed-calculated values is given in Table 4 for n TMO for both the ν_1 and ν_6 excited state puckering separations. It is evident for both cases that the model is incapable of reproducing the observed numbers. A discussion of this failure is postponed until later. For the moment we can say that a resonance perturbation of the puckering levels is certainly plausible, which in turn could explain the curious band contours in the Raman spectrum, and the apparent intensity enhancement of some absorptions and the absence of the 4 ← 3 sum band in the infrared spectrum.

Table 3 Puckering combinations of the totally symmetric ring stretching vibration

Mid IR			CARS				Mid IR			CARS			
Transition		Peak position	Transition		Peak position $[cm^{-1}]$		Transition		Peak position	Transition		Peak position $[cm^{-1}]$	
v'_p ←	v''_p	$[cm^{-1}]$	v'_p ←	v''_p	Predicted[a]	Measured	v'_p ←	v''_p	$[cm^{-1}]$	v'_p ←	v''_p	Predicted[a]	Measured
normal TMO							β-d_2 TMO						
1	2	944.60											
			0	0	-	1033.18	0	1	935.11	0	0	984.40	984.58
1	0	1087.43	1	1	1034.39	1034.28	1	0	1033.97	1	1	984.68	984.79
2	1	1123.80	2	2	1034.12	1034.04	2	1	1067.68	2	2	984.52	obscured by main peak
3	2	1136.37	3	3	1031.88	1031.94 1031.74	3	2	1082.45	3	3	985.56	obscured by main peak
							4	3	1092.82	4	4	983.52	983.90
5	4	1165.09	5	5	1036.18	1036.64 1036.27	5	4	1104.95	5	5	985.51	not measured
6	5	1173.98	6	6	1035.51	not measured	6	5	1112.05	6	6	983.82	984.19
7	6	1182.53	7	7	1035.55	not measured							

[a]Predicted by summing the appropriate pure puckering and difference transitions in the infrared spectrum.

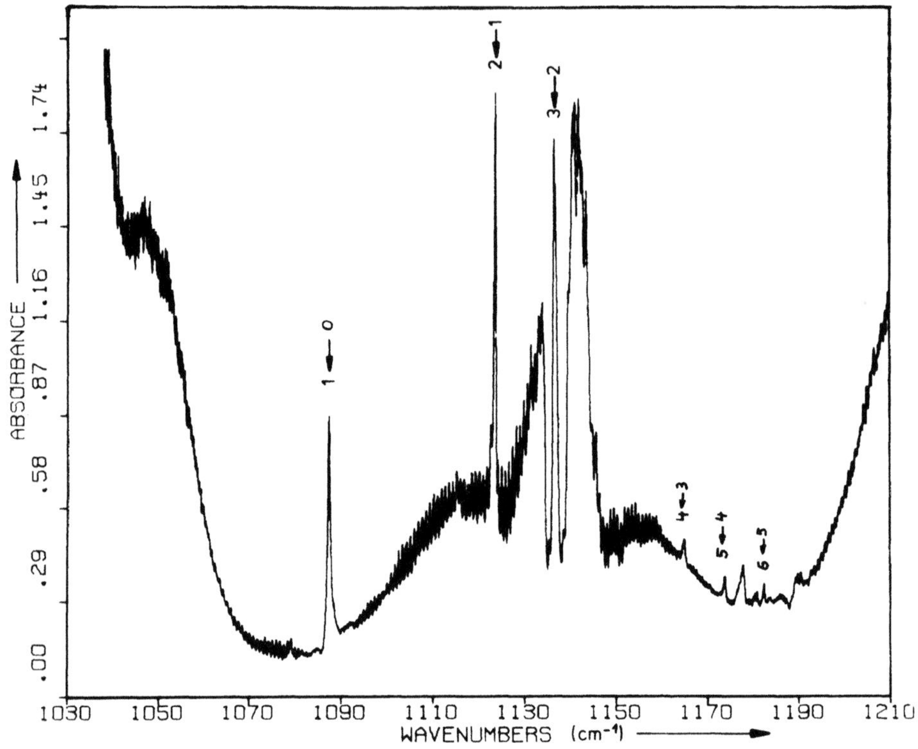

Fig. 10 Infrared spectrum of ν_6 sum bands in n TMO; pressure: ~300 Pa; path length: 20 m; resolution: 0.06 cm^{-1}

The spectra of the same region in β-d_2 TMO pose much the same problems. The sharp and almost featureless line at 984 cm^{-1} in the linear Raman spectrum (Figure 8) is hardly resolved by CARS (Figure 9). Although an assignment of the features from higher to lower wavenumbers in sequence certainly is feasible, its correctness is questionable. Fortunately, infrared spectra prove themselves helpful again in this case. The weak 0 ← 1 absorption at 935.11 cm^{-1} suggest the 0 ← 0 Raman line to be near 984.40 cm^{-1}, slightly off the peak maximum of the complex CARS band. Clear and unambiguous infrared sum absorptions then place the higher transitions within that band. Although not all the CARS features can be assigned thereby without some degree of uncertainty, one can at least justify the complexity of the Raman line and the failure of resolving it by CARS. The assignments are also summarized in Table 3.

As before we attempt to reproduce the excited state puckering separations. The results of the calculations are listed in Table 5

Table 4 Puckering separations (in cm^{-1}) in the first excited states of other vibrations in n TMO, derived from infrared transitions.

Transition		Ground state		ν_1 Excited state		ν_6 Excited state	
v'_p	$\leftarrow v''_p$	Observed	Obs.-Calc.	Observed	Obs.-Calc.	Observed	Obs.-Calc.
1	0	53.04	0.05	146.47	0.24	54.25	0.43
2	1	89.68	-0.03			89.41	-0.91
3	2	104.49	-0.09	102.76	-1.67	102.25	-3.06
4	3	117.93	-0.08	118.01	0.85	251.14	2.68
5	4	128.91	-0.05	127.96	0.32		
6	5	138.47	0.03			137.80	-1.41
7	6	146.98	0.15			147.02	-0.60

Table 5 Puckering separations (in cm^{-1}) in the first excited states of other vibrations in β-d_2 TMO, derived from infrared transitions

Transition		Ground state		ν_1 Excited state		ν_4 Excited state		ν_6 Excited state	
v'_p ←	v''_p	Observed	Obs.-Calc.	Observed	Obs.-Calc.	Observed	Obs.-Calc.	Observed	Obs.-Calc.
1	0	49.29	0.00	136.46	0.22	53.75	0.50	49.57	0.20
2	1	83.16	0.03			80.85	-2.52	83.00	0.03
3	2	96.89	-0.02	95.01	-1.82	97.32	0.20	97.93	1.19
4	3	109.30	-0.02	109.94	1.42	109.41	0.55	107.26	-1.81
5	4	119.44	-0.08	118.09	-0.07	119.42	0.88	121.43	2.23
6	5	128.23	0.02					126.54	-1.36

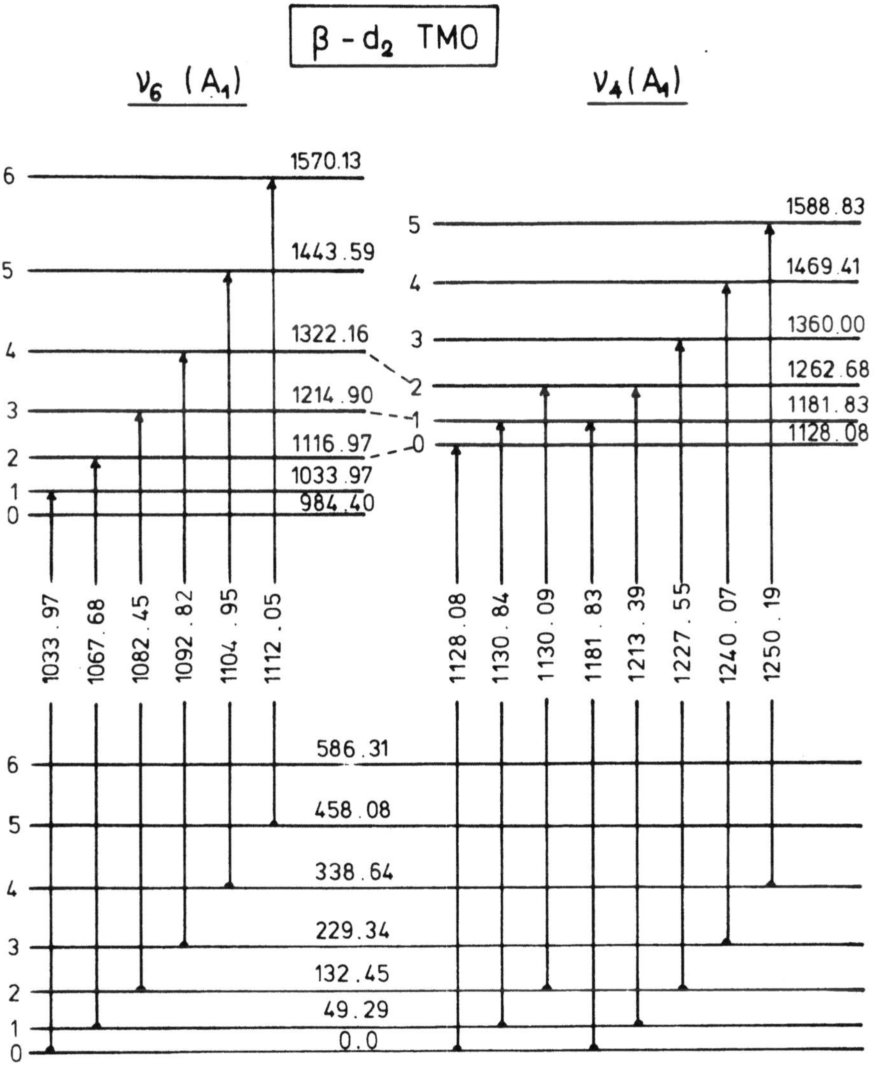

Fig. 11 Energy level diagram of ν_4 and ν_6 in β-d_2 TMO showing observed infrared transitions

for the ν_1, ν_4 and ν_6 fundamentals and compared with those for the pure puckering transitions. It is apparent in all cases but the last, that some kind of resonance interaction could be taking place. The ν_6 fundamental in particular shows large alternating deviations from a well behaved puckering situation. There is in addition another A_1 fundamental, ν_4, the β-CD_2 scissoring mode

(A_1), which not only absorbs very near to ν_6 but also may couple with it. Fortunately, this ν_4 mode gives rise to distinct infrared bands with $\Delta v_p = 0$ and +1 so that an energy level scheme can be readily determined. Such a scheme is sketched in Figure 11, where it is compared with a similar scheme for ν_6 and where the observed infrared transitions are also shown. It again becomes clear that resonance interaction between the puckering levels of the two vibrations is not only plausible but also probable.

Before closing this section it seems appropriate to examine the implications of the above observations on the puckering potential functions and the purpose for which the calculations were used. Some of the complications that arise from these calculations were mentioned briefly in an introductory section. Although instances have been cited where effective puckering (or puckering-like) potential constants were determined reasonably successfully for the excited state of another mode, they usually were cases where the second mode is another low energy vibration. On some occasions when it seemed that the one-dimensional puckering Hamiltonian was not adequate even for the ground state, two-dimensional Hamiltonians which describe the two vibrations simultaneously were employed. Still another approach is to provide for a reduced mass for the puckering vibration in the kinetic energy that varies with the coordinate [10]. All these refinements manifest the uncertainty about the true nature of the out-of-plane deformation. When we use the one-dimensional quadratic-quartic Hamiltonian for testing our assignments therefore, we have to be aware of the pitfalls. Namely, the observation that excited state puckering separations do not comply with the one-dimensional model may well mean that the puckering coordinate is coupled with the second mode either through the kinetic or the potential energy part of the Hamiltonian or both. By the same token, however, the observation may also mean near coincidences of one or several selected energy levels. To sort out the true situation for the moment seems an intractable problem.

5. CONCLUSION

In this article we have demonstrated an application of CARS to the problem of the inversion of a cyclic compound through its planar configuration, in particular of the near-planar, four-membered ring molecule trimethylene oxide. Extensive earlier spectroscopic information (both linear Raman and infrared), and the apparent relative simplicity of the spectra compared to related systems make this a good starting point. The earlier conventional Raman spectra proved to be indispensible for interpreting the infrared spectra, thereby contributing to our understanding of the anharmonic ring inversion (puckering). The resolution available at that time, however, only hinted at the complexity of some lines, leaving the hope that better spectra may be obtainable some day.

Indeed, CARS has fulfilled some of its promise. We were able to separate some of the strong Raman lines in question into their puckering components. With this information we could locate and assign a number of weak infrared combination bands. The combined data then allowed us to deduce puckering separations in the first excited state of several other vibrations. At the same time the CARS spectra brought out the likelihood that extensive resonances occur among these excited state puckering levels. The rotation-vibration contours of both the CARS and infrared bands consequently are distorted making it difficult to recognize and assign them. Moreover, by displacing the puckering levels in an unpredictable way these resonances make it impossible for the moment to draw any conclusions about the puckering mode itself and its interaction with other modes of vibration. In the end one wishes for better resolution still. From our point of view, therefore, further progress is to be sought in other non-linear Raman spectroscopic methods, such as inverse Raman or Raman gain spectroscopy which promise very high resolution [25].

Already with our CARS measurements it became apparent that the resolution has outgrown the sensitivity of the simple puckering model. The subtle improvements of the model that are already documented in the literature will be put to the test more critically. It is also predictable that further refinements of the theory will become necessary.

The prospects for applying CARS to the out-of-plane deformation of other molecules we feel are generally not very promising. The greatest benefit would accrue if the deformation vibrations themselves could be measured at low wavenumbers. Such measurements, however, may be seriously hampered by the low intensity these modes have inherently and as a result of the nature of CARS. The combination sequences for either more puckered or larger molecules are likely to be even more complicated than for TMO. Finally, it is dubious whether CARS will become a routine method in the near future.

ACKNOWLEDGEMENTS

The authors wish to thank Mr. A. Beckmann who has been involved in setting up the CARS system at the University Bayreuth and who also participated in recording the CARS spectra of TMO. Financial support from the Deutsche Forschungsgemeinschaft is highly acknowledged. HW gratefully acknowledges the extension of a Dozentenstipendium from the Alexander von Humboldt Foundation. Technical assistance by Mr. W. Griessl, Mr. R. Haagen and Mr. P. Hurych is highly appreciated.

REFERENCES

1. Tolles, W.M., Nibler, J.W., McDonald, J.R., and Harvey, A.B., Appl. Sepctrosc. 31, 253-271 (1977).

2. Nibler, J.W.,and Knighten, G.V., in Raman Spectroscopy of Gases and Liquids, in Topics in Current Physics, Vol. 11, Weber, A., ed., Springer-Verlag, Berlin, 1979, Chptr. 7.

3. Harvey, A.B., Chemical Applications of Nonlinear Raman Spectroscopy, Academic Press, New York, 1981.

4. Druet, S.A.J, and Taran, J.-P.E., Progr. Quant. Electr. 7 1-71 (1981).

5. Durig, J.R, and Sullivan, J.F., this book, page 51.

6. Blackwell, C.S. and Lord, R.C., in Vibrational Spectra and Structure, Vol. 1, Durig, J.R., ed., Marcel Dekker, New York, 1972, Chptr. 1.

7. Wurrey, C.J., Durig, J.R., and Carreira, L.A., in Vibrational Spectra and Structure, Vol. 5, Durig, J.R., ed. Elsevier Scientific Publishing Co., Amsterdam, 1976, Chptr. 4.

8. Robiette, A.G., in Molecular Vibrations and Structure, Specialist Periodical Reports of the Chemical Society, Vol. 1, The Chemical Society, London, 1972, Chptr. 4.

9. Gwinn, W.D., and Gaylord, A.S., in Spectroscopy, Vol. 3 in MTP International Review of Science, Physical Chemistry Series 2, Ramsay, D.A., ed., Butterworths and University Park Press, London, 1976.

10. Carreira, L.A., Lord, R.C., and Malloy, Jr., T.B., in Large Amplitude Motion in Molecules II, in Topics in Current Chemistry, Vol. 82, Springer-Verlag, Berlin, 1979.

11. Legon, A.C., Chem. Rev. 80, 231 (1980).

12. Herzberg, G., Molecular Spectra and Molecular Structure; II. Infrared and Raman Spectra of Polyatomic Molecules, Van Nostrand Reinhold Co., New York, 1945.

13. Stone, J.M.R., and Mills, I.M., Mol. Phys. 18, 631 (1970).

14. Placzek, G., Handb. d. Radiologie, Vol. 6, Marx, E., ed., Akad. Verlagsgesellschaft, Leipzig, 1934.

15. Kiefer, W., this book, page 241.

16. Yuratich, M.A., and Hanna, D.C., Mol. Phys. 33, 671 (1977).

17. Kydd, R.A., Wieser, H., and Kiefer, W., Spectrochim. Acta, in press.

18. Kiefer, W., Bernstein, H.J., Wieser, H., and Danyluk, M., Chem. Phys. Letters, 12, 605 (1972),and J. Molec. Spectrosc. 43, 393 (1972).

19. Wieser, H., Danyluk, M., Kydd, R.A., Kiefer, W., and Bernstein, H.J., J. Chem. Phys., 61, 4380 (1974).

20. Kiefer, W., and Wieser, H. unpublished results.

21. For preliminary results see: Fietz, H., Beckmann, A., Kiefer, W., and Wieser, H., Chem. Phys. Letters, 87, 491 (1982).

22. Beckmann, A., Baierl, P., and Kiefer, W., this book, page 393. see also: Beckmann, A., Fietz, H., Kiefer, W., and Laane, J., Phys. Rev. A, 24, 2518 (1981); and Fietz, H., Beckmann, A., Hohenberger, H.P.,and Kiefer, W., Appl. Spec., 35, 609 (1981).

23. Wieser, H.,unpublished results.

24. Throughout this article the designation $v_p' \leftarrow v_p''$ implies a transition from the initial puckering level in the ground state of all other vibrations to the final puckering level in the first excited state of the second mode.

25. Owyoung, A., in Chemical Applications of Nonlinear Raman Spectroscopy, Harvey, A.B., ed., Academic Press, New York, 1981, p. 281.

COHERENT ANTI-STOKES RAMAN SPECTROSCOPY IN CONDENSED MEDIA

L.A. Carreira and M.L. Horovitz

Department of Chemistry
University of Georgia
Athens, Georgia 30602

Coherent anti-Stokes Raman spectroscopy (CARS), a relatively new but extremely powerful technique, has been used to obtain Raman spectra with high efficiency. The first observation of the CARS process was reported by Maker and Terhune in 1965 [1]. Since these initial experiments, many investigations demonstrating the application of CARS in solids, liquids and gases have been reported. Publications describing CARS investigations are appearing at an ever increasing rate and the technique has been successfully applied to such diversified studies as combustion diagnostics, biological systems, low frequency Raman modes and excited state vibrational spectroscopy.

The CARS technique utilizes two laser beams, with angular frequencies ω_1 and ω_2, focused together in a sample result of the beams mixing, a coherent signal resembling a low intensity laser beam at a frequency $\omega_3 = 2\omega_1-\omega_2$ is generated in the medium. Conversion of the two laser beams into the anti-Stokes shifted component at ω_3 is a consequence of the third order nonlinear properties of the system. By use of tunable dye lasers, the frequency interval $\omega_1-\omega_2 = \Delta$ can be continuously adjusted. When Δ equals ω_r, the frequency of a Raman active mode, the signal is greatly enhanced. The CARS spectrum is obtained by holding the frequency at ω_1 fixed and scanning ω_2 such that Δ covers the Raman frequencies of interest.

The CARS experiment offers several distinct advantages over the spontaneous Raman experiment. First, the spontaneous Raman process yields only one inelastically scattered photon in 10^6 - 10^8 incident photons. This low efficiency makes efficient optics and sensitive detection system necessary. On the ohter hand, the CARS

W. Kiefer and D. A. Long (eds.), Non-Linear Raman Spectroscopy and Its Chemical Applications, 367–389.

process can yield one anti-Stokes photon in 10^2-10^3 incident photons. This represents a gain of 10^4, or a conversion efficiency approaching 1 %. Second, the radiation produced by the spontaneous Raman experiment is scattered over 4πsr., so only a small portion of the total scattered radiation can be collected. Whereas in CARS the signal appears as a laser-like beam, with divergence characteristics like those of the incident beams. The signal can therefore be separated from the incident beams and collected with about 90 % efficiency. Third, the radiation produced in the spontaneous process consists of photons with frequencies spanning all the internal degrees of freedom, so that a large monochromator is required. The CRAS signal consists of photons of only one frequency ω_3 so no monochromator is required for detection. Finally, fluorescence may make spontaneous Raman spectra troublesome, if not impossible, to obtain. Fluorescence signals are often several orders of magnitude greater than Raman signals and propagate isotropically, making collection unavoidable. Fluorescence interference is minimized in the CARS experiment because the signal is anti-Stokes shifted and thereby frequency isolated from fluorescence emission. The spatial characteristics of the CARS signal also make it easy to minimize fluorescence collection.

Another advantage of CARS is a direct result of the type of lasers used in the experiment. For condensed-phase studies pulsed-dye lasers with high peak powers ($\sim$ 100 kW), short pulse widths ($\sim$ 10 nsec), and low average powers ($\sim$ 10 mW) are used instead of the CW gas lasers usually employed in the spontaneous Raman experiment. Pulsed lasers can be used in the CARS experiment because the process depends on the instantaneous power rather than the average power. The low average powers associated with pulsed lasers in CARS help to prevent sample degradation and allow small sample volumes (1μl) to be used. This is particularly important in resonance work where the exciting laser is close to the electronic absorption of the molecule of interest. In the spontaneous Raman experiment, resonance work usually entails spinning a rather large sample to prevent degradation.

I. EXPERIMENTAL SETUP FOR CONDENSED PHASES

CARS is considered a technically difficult experiment in the condensed phase. The difficulty arises from the pase-matching conditions of the process. Since the CARS process begins and ends in the same energy level, energy and momentum are conserved. The momentum vectors depend, not only on the wavelength of the two beams, but also on the index of refraction of the media at the particular wavelength. The equation for the momentum matching condition is given in [10].

Figure 1 shows the momentum vector matching for the process. Since the index of refraction varies with frequency (dispersive medium) the vectors K_1, K_2 and K_3 will not lie in a line but must be oriented at certain angles with respect to each other. The angle θ between the K_1 and K_2 vectors is called the crossing angle whereas the angle θ' between the K_3 and K_1 vectors is usually called the collection angle. In most cases these two angles are nearly equivalent. As a result of this momentum matching criterion the crossing angle as well as the collection angle must be varied as the difference interval $\omega_1-\omega_2 = \Delta$ is scanned to obtain a spectrum. It has been found that the total wavenumber range that can be scanned without adjustment is determined by the dispersion of the sample, which changes with frequency. In a highly dispersive medium like benzene, the scan range is limited to approximately 50 cm^{-1} before the crossing angle must be adjusted. In water, the scan range is 150-250 cm^{-1} [2].

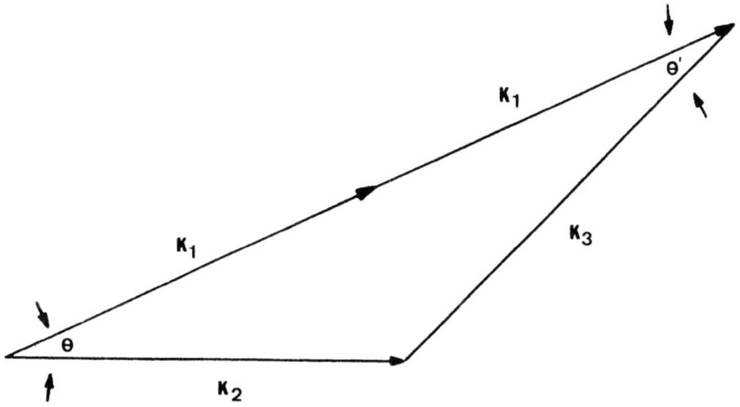

Figure 1. Momentum matching vectors for the CARS process.

Spectra were thus obtained in small overlapping segments and pieced together before computer control was implemented. Under computer control the spectra are obtained in a manner of minutes instead of hours and digitally stored.

A vast number of experimental arrangements have been used to generate CARS signals. The earliest experiments [3-6] utilized stimulated Raman scattering to create an ω_2 beam which was then mixed with the pump laser beam to generate the CARS signal. However, this could be used only for measuring resonant CARS signal from benzene or nonresonant cross sections of other materials at a frequency shift of 922 cm^{-1}. Needless to say, this was not a very convenient method for obtaining spectra. Another experimental arrangement employed a single nitrogen laser pumping two dye lasers and generated two individually tunable wavelengths at power levels of 10 - 100 kW each. This arrangement has been used in several experiments [7-8]. The two dye laser beams can be split, forming a double, nonlinear mixing arrangement to generate a reference signal for the purpose of reducing amplitude and mode fluctuations.

The kind of dye laser utilized for experiments depends on the spectral properties of the sample to be studied and the type of laser available for pumping. In general, spectra in the condensed phase have line widths on the order of a few cm^{-1} or greater. A Hänsch-type dye laser [9] typically employed with a nitrogen laser pump, a low-power beam expander and grating, readily give line widths of 0.3 - 1.0 cm^{-1}. An amplifier stage may be added if higher powers are required in the output beam.

It should also be mentioned that CARS need not be performed with narrow band lasers. If one of the lasers is fixed in frequency and spectrally narrow, the other laser may be broad band. CARS signals will be formed only with the spectral portions of the broad dye laser output that correspond to Raman shifts. The resulting CARS lines, however, will need to be separated and analyzed by means of a monochromator. Such an arrangement could be useful for kinetic, photochemical and concentration studies.

Figure 2 shows the CARS setup used at the University of Georgia. It consists of a nitrogen laser split into two beams by a 2:1 beam splitter. These beams are then used to pump two DL200 dye lasers which create the pump (ω_1) and probe (ω_2) beam for the CARS experiment. Scanning of the dye lasers is achieved by a sine bar driven grating. The two beams are crossed and focused by lens 2. The resulting CARS signal (ω_3) is generated at an angle θ' with respect to the ω_1 beam.

In order to detect this signal the detector and monochromator must be able to follow the θ' angle as the scan is taken. This was accomplished by mounting them on a movable stage which pivots about the crossing point of the pump and probe beams. This stage allows the angle of collection to be varied synchronously with the crossing angle. The signal beam ω_3 passes through an iris 30 cm from the sample. The beam is then focused onto the

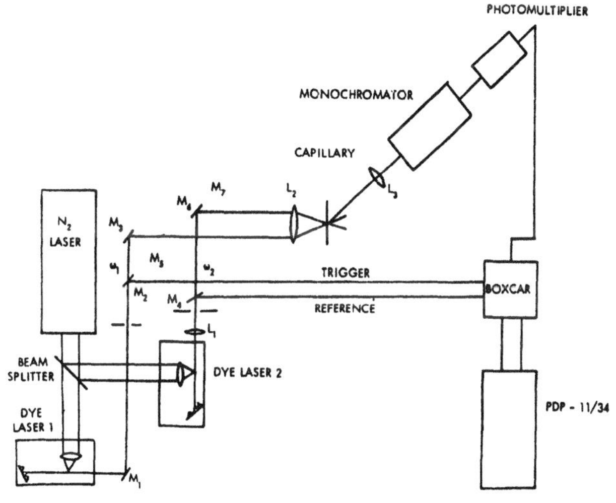

Figure 2. Experimental CARS setup used in the studies.

entrance slit of a JY-DH-205 1200 UV monochromator. The monochromator is chiefly used for background fluorescence rejection since the resolution of the experiment is determined solely by the linewidth of the dye lasers. In order to be compatible with the wide wavelength range available when using nitrogen-pumped dye lasers (260-740 nm), an RCA C31034 gallium arsenide photomultiplier tube, which has a 10 % or greater quantum efficiency throughout this range, was used. The signal from the photomultiplier was averaged using a boxcar integrator (EG and G Princeton Applied Research Model 162). The integrator was triggered by the output from a pin diode (Molectron LP-141) monitoring the reflection of the ω_1 beam from a microscope slide. A reference was provided by a reflection of ω_2 using a similar arrangement as that in ω_1. A PDP-11/34 computer controlled positioning of the beams, scanning of the probe laser and collection of the data.

Computer control of the CARS experiment required the incorporation of a minimum of five stepping motors. Three were required for adjustment of the crossing angle. They controlled mirror translation and vertical and horizontal rotation. The horizontal and vertical rotation stepping motors were 400 step/revolution whereas all other motors employed were 200 step/revolution. The collection stage position as well as the monochromator scanning are controlled by stepping motors.

In order to extend the CARS technique to wavelength shorter than 360 nm, we have used angle-tuned KDP (potassium dihydrogen phosphate) frequency-doubling crystals [10]. Figure 3 shows that a UV-CARS spectrum could be obtained for toluene using a pump frequency of 295 nm. Under computer control the tuning angle can be kept in phase with ω_2 and the power variance kept at a minimum.

In order to obtain small path lengths and minimize absorption of the various beams, standard melting point capillaries were used to contain the sample. This proved more convenient and considerably less expensive than the standard cuvettes used in past CARS experiments. This setup allows samples to be flowed in case of photochemical decomposition. The capillary was placed in the plane of ω_1 and ω_2 (Fig. 2) and perpendicular to their bisector. The capillary was then raised or lowered until the ω_1 beam passed undeviated. This placement was critical since the rounded walls greatly displaced the beams unless they travelled along the diameter. This displacement actually made the alignment easier. Once the ω_1 beam passed undeviated the steering mirror for the ω_2 beam (see Fig. 2) was adjusted to make it pass undeviated. This visual alignment is much more sensitive than the knife edge test [11] for vertical placement. Once aligned vertically, the steering mirror for the ω_2 beam was adjusted horizontally to obtain the CARS signal. The loss in signal as compared to a cuvette is only on the order of 10 %. Quartz capillary tubing had to be adopted for the UV experiments.

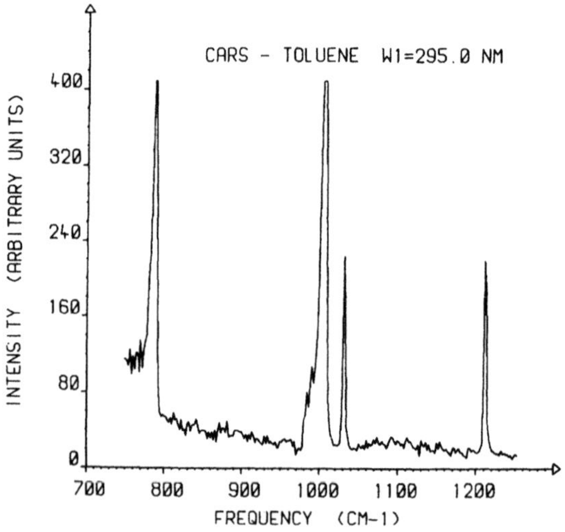

Figure 3. Coherent anti-Stokes Raman spectrum of toluene with ω_1 = 295.0 nm.

II. INTERFACE OF THE CARS EXPERIMENT

Hardware

The CARS experiment is a multiparameter process which requires precise control of mirror and micrometer movements. To obtain large frequency scans in short periods of time it is necessary to control the various parameters with a computer. The minicomputer chosen for this purpose was a PDP 11/34 with such options as a 20 MB hard disk drive, two general pupose I/O ports (DR11K and DR11C), analog-to-digital converter (AD11K) with multiplexer (AM11K), and digital-to-analog converter (AA11K). Included with this package was Digital's RT-11 or RSX realtime Fortran software package for real-time control.

Four types of interfaces had to be built in order to make the CARS unit compatible with the minicomputer. They were (a) busing system for the I/O ports (peribus), (b) analog-to-digital converter interface (c) DL-040 A Molectron scan unit interface, and (d) stepping motor interfaces. The peribus (peripheral bus) was constructed to increase the number of possible I/O ports. This is achieved by busing all of the bits from the DR11 port to many accessible plug-in ports. This allows more devices to have access to the same I/O interface. In order for the computer to discriminate between devices each device or operation has its own address. The peribus consists of open collector drivers which boost and distribute the signal from the DR11 I/O port to the various interfaces.

The scan unit has three types of operations that had to be interfaced to the computer. They included data input (latching), data output, and pulse generation. A simplified scheme for handling these operations is shown in Fig. 4. The operations that required latching were those in which a level (high or low) had to be held for long periods of time. Examples of such operations are the scan and direction inputs in which the data levels had to be held constant to scan in a particular direction. Data output was accomplished by two input open collector nand gates in which one input was the address decode and the other actual data being sent to the computer. The data being sent to the computer from the scan unit was the BCD output of the wavelength of the dye laser. Certain of the functions performed by the scan unit required a pulse for activation. The pulses were produced by a monostable multivibrator.

The stepping motor interface is also controlled by the DR11 I/O port. This interface consists of two operations, stepping motor control, and motor status monitoring. The general scheme for this interface is shown in Fig. 5. The information required to make a particular stepping motor run is the address of that motor, the number of steps to be made, and the direction in which to step. Thus, a single output word can run a stepping motor leaving the computer free for other operations. The run status of each motor and whether a limit has been reached by a motor are monitored by the motor status card and sent to the computer through the input word.

The stepper control card can operate in one of two possible modes, local or computer. Part A of Fig. 6 depicts a simplified scheme of what is involved in the local mode. The operations controlled are run (free run), direction (clockwise or anticlockwise), and step (single step). The run switch turns on an oscillator which puts out a pulse train to the motor in either the clockwise or counterclockwise direction. The step switch puts out a single pulse to the motor. If a limit has been reached by the stepping motor a microswitch in that direction is shorted to ground, blocking the pulse train at the three input nand gate. The computer mode of operation is depicted in Part B of Fig. 6. An output word containing the address, direction, and number of steps is put on the output bus. The address decodes, allowing a binary counter to be loaded. The counter is used only in the down

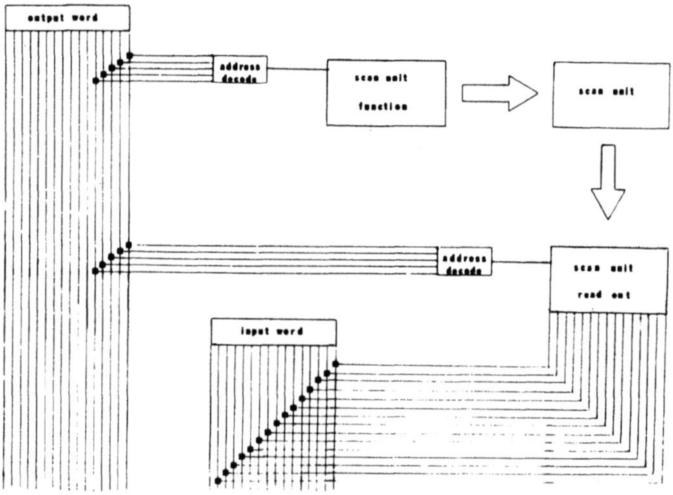

Figure 4. Simplified schematic of the basic operation of the scan unit interface. Both the input ports of the DR11K were required for communication with the computer.

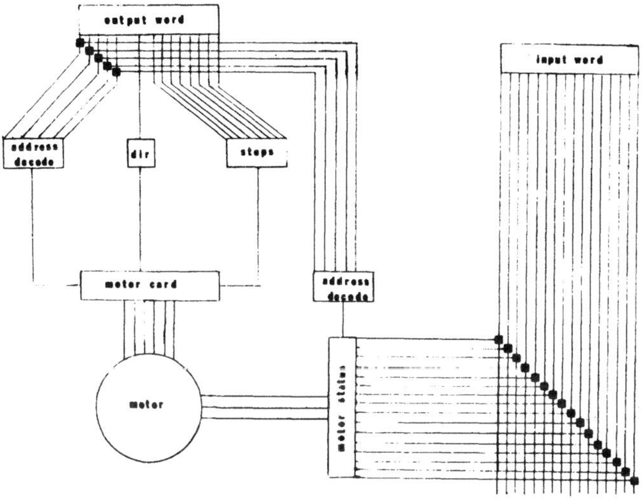

Figure 5. Simplified schematic of the basic operation of the stepping motor interface. The DR11C I/O port was used for controlling this interface.

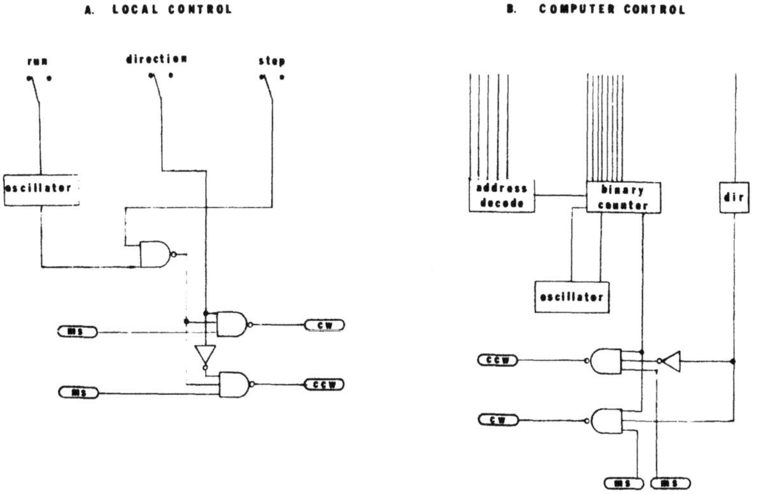

Figure 6. More detailed schematic of the stepping motor interface under local (a) and computer (b) control.

count mode for this interface. Once the counter has been loaded an external oscillator is switchted on and a pulse train sent to the count down of the counter. The binary counter counts the number loaded down, with its output first being coupled with the direction and then sent to the motor. The actual circuit that was used is given in ref. [12]. The run signal of each motor is monitored by this card and when the correct address is given the information is sent to the input port and subsequently to the computer. The microswitches are monitored collectively. This is accomplished by two-eight input nand gates. If a limit is reached by a motor, the program stops all operations and rings an alarm warning the operator that a limit has been reached.

The backbone of the real-time software package are the Fortran callable support routines for laboratory peripherals (LDPLIB) ["FORTRAN RT-11: Extensions Manual", Digital Equipment Corp. 3-1 (1976)]. These routines allow the user to initiate control on the following types of hardware devices:(1) programmable real-time clock, (2)analog-to-digital conversion, (3)16 bit digital I/O port, (4) CRT display controller, and (5) direct memory access option.

The three major operations that the computer had to perform were (1) control of the scan unit, (2) stepping motor control, and (3) data collection from the boxcar. In order to accomplish this task Fortran subroutine handlers were written and stored in an accessible library file. These subroutines were based on the Fortran-callable subroutines of the LDPLIB package. The main scanning routines were then developed around these subroutine handlers. This made it rather easy to write and modify the main routine as was needed.

The three parameters that must be varied to change the crossing angle and collect the signal at θ' are shown in Fig. 7. They consist of (a) lateral translation of the ω_2 mirror, (b) horizontal rotation of the ω_2 mirror-to keep the two beams crossing at the same point in space as the crossing angle is varied, and (c) rotation of the collection state allowing the detector to track ω_3. Any scanning algorithm for CARS must be able to vary these three parameters. The correlation between these parameters was the first step in obtaining a scanning CARS unit.

The correlation between the mirror translation and horizontal rotation was obtained in the following manner. A computer program was written in which the mirror translation and horizontal rotation stepping motors would be accessed. Then the following experiment was carried out. First, the focus of ω_1 was found by a knife edge test. In the CARS experiment ω_1 is held constant and ω_2 is scanned and translated to vary the crossing angle. Thus, it is important to align the system to the ω_1 beam. At this point the ω_1 beam was extinguished and ω_2 was monitored by a pin diode. The ω_2 beam was then purposely adjusted to pass the knife edge.

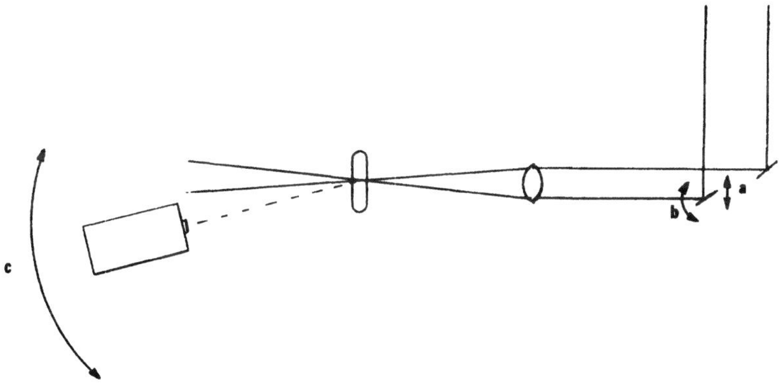

Figure 7. Schematic of the crossing and collection angle scheme depicting the three parameters that must be controlled by the computer. They include (a) crossing angle (micrometer translation), (b) horizontal mirror rotation, and (c) movement of the collection stage.

The horizontal rotation stepper was allowed to step the beam toward the knife edge until the signal was extinguished by a factor of two. The translation stage containing the ω_2 mirror was then allowed to move to a larger crossing angle and the ω_2 beam was again rotated onto the knife edge. This procedure was continued until the maximum crossing angle possible with our single lens system was obtained. The results of this calibration are shown in Fig. 8. The correlation was found to be linear with a slope of 0.352 (horizontal steps per thousandths inch mirror translation).

From previous work with the CARS setup it was known that θ varied as a function of the frequency difference between ω_1 and ω_2 as well as the wavelength of the pump. This variation in the crossing angle is the result of the dependence of the index of refraction upon the wavelength. Since most of our work is being done in resonance, which requires only small amounts of solute, the index of refraction is largely determined by the solvent. Thus, if one knew the variation in θ for each solvent it would be a simple matter to devise a scanning algorithm. To determine the variation of θ for different solvents a computer program was developed. The program worked in the following manner. First, an initial alignment was made to insure that the correct crossing angle was found for the initial point. The ω_2 dye laser was then scanned by a predetermined amount, the collection stage swept to determine the position of ω_3, and the horizontal mirror rotation varied to find the maximum signal. The crossing angle was then increased and the stage mirror varied to find the maximum signal again. At this point the two maxima were compared. If the previous maximum was greater, the crossing angle, stage position, and mirror were returned to this position. If the last maximum was larger, the crossing angle was increased again and the stage and mirror varied for maximum intensity. This position was recorded and ω_2 was scanned to a new frequency and the above repeated. What was obtained from such a procedure is the stage position and micrometer setting (crossing angle) as a function of the frequency difference between ω_1 and ω_2 at a particular pump wavelength. Figure 9 shows the results obtained for benzene at a pump wave-

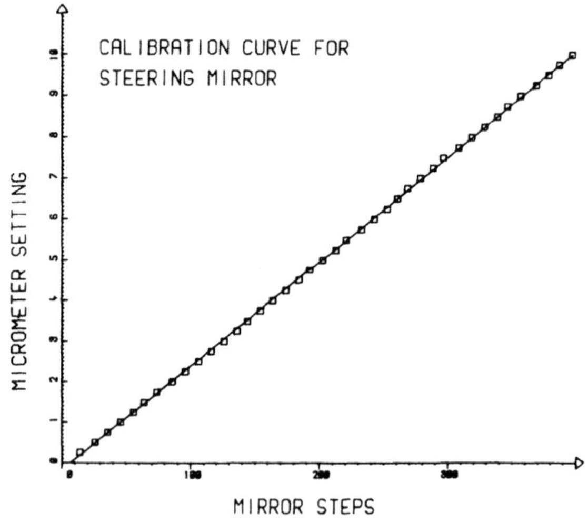

Figure 8. Horizontal mirror calibration for various micrometer positions (crossing angles). The correlation was linear with a slope of 0.352 steps/ths.

length of 430 nm. The correlation was found to be linear for all solvents tried. A list of some of the more common solvents and their micrometer and stage slopes is given in Table 1. The next step was to determine how these slopes varied as a function of the pump wavelength. The above procedure was used in this case at various pump wavelengths. For all solvents tested except benzene the slope was found to increase as the exciting wavelength was decreased toward the near ultraviolet.

To explain these results a theoretical procedure [12] was developed for the calculation of the micrometer slope as a function of the pump wavelength. The index of refraction can be ex-

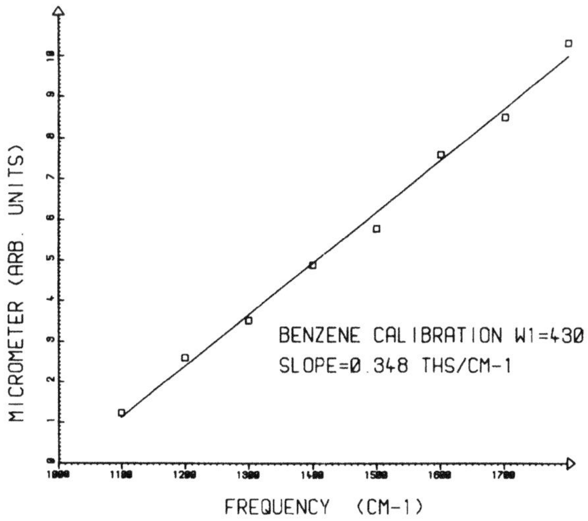

Figure 9. Calibration diagram of the solvent benzene. The correlation was linear with a slope of 0.348 ths/cm.

pressed by

$$n(\omega_1) - 1 = \sum_i \frac{A_i}{B_i - \omega_1^2} . \tag{1}$$

TABLE I

Experimental Slopes for Various Solvents at Two Pumps[a]

Solvent	THS (cm^{-1})	INS (cm^{-1})	Pump (nm)
H_2O	0.152	$6.25x10^{-4}$	505
H_2O	0.153	$5.68x10^{-4}$	365
C_2H_5OH	0.174	$5.70x10^{-4}$	505
C_2H_5OH	0.158	$6.34x10^{-4}$	365
CH_3OH	0.136	$4.70x10^{-4}$	505
CH_3OH	0.142	$5.9x10^{-4}$	365
$CHCl_3$	0.250	$7.168x10^{-4}$	505
$CHCl_3$	0.222	$7.68x10^{-4}$	365
p-Dioxane	0.181	$6.203x10^{-4}$	505
p-Dioxane	0.181	$7.08x10^{-4}$	365
C_6H_6	0.309	$9.84x10^{-4}$	505
C_6H_6	0.378	$1.30x10^{-3}$	365

[a]The micrometer slope which is directly related to the crossing angle which is in units of thousands per wavenumber. The collection stage slope is in units of inches per wavenumber. This slope is directly related to the collection angle θ.

Equation (1) was expanded in a power series in ω_1 and the first three terms were kept yielding

$$n(\omega_i) = C_o+C_2\omega_i^2+C_4\omega_i^4 \quad (2)$$

By fitting the three parameters in the above equation the slope of the micrometer movement to the ω_1 wavelength can be determined. Figure 10 shows the fit for benzene. The other solvent slopes were essentially constant in the region that was probed because this region corresponded to a linear part in their curves. If one were to probe these solvents closer to their absorption, one would find their slopes increasing much the same as benzene. With all the necessary correlations obtained, a scanning algorithm could be developed for the CARS unit. All the computer needed to know was which solvent was being used for the experiment and the spectrum could then be taken. The program that was developed for this task consisted of three sections 1) initialization of parameters, 2) initial alignment, and 3) actual scanning of the spectrum.

The initialization of the experimental parameters was required for the computer to be able to control the scan unit. This

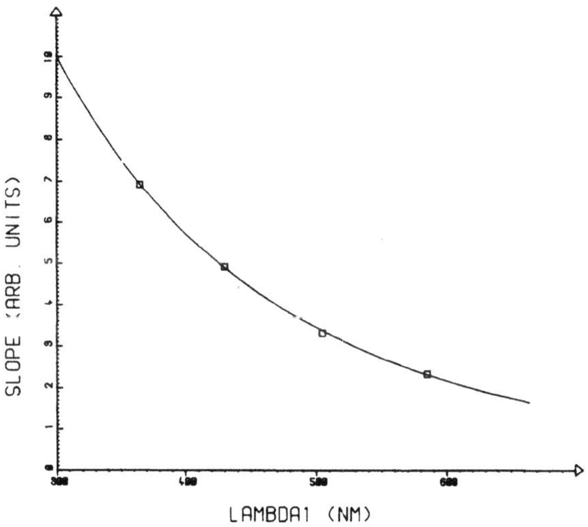

Figure 10. Calculated fit to the slope of benzene as a function of the pump wavelength. The squares are the experimental observed slopes.

included all parameters for the DL040 and stepping motors. Previously, the CARS unit had to be aligned by hand at the first point. The incorporation of the initial alignment procedure allowed the computer to find the correct alignment, freeing the operator from this task.

The spectrum is obtained in the following manner. First, the computer collects a predetermined number of data points at the initial frequency difference. When this is accomplished, the scan unit is scanned and the stepping motors controlling the crossing and collection angle are then stepped according to the solvent parameters initially fed to the computer. The next point is then taken and the above procedure continued until the last point is taken. This section also has the capability of correcting itself if the initial slopes are not correct. This is accomplished by having the computer correct the alignment by varying the stage, mirror and micrometer positions until the maximum signal is found. The frequency interval for having this done is set in the initialization part of the program. By making this interval small one would not need to know the correct slopes. The only problem with this procedure is that it is time consuming and if done often, slows the speed at which the scan can be taken.

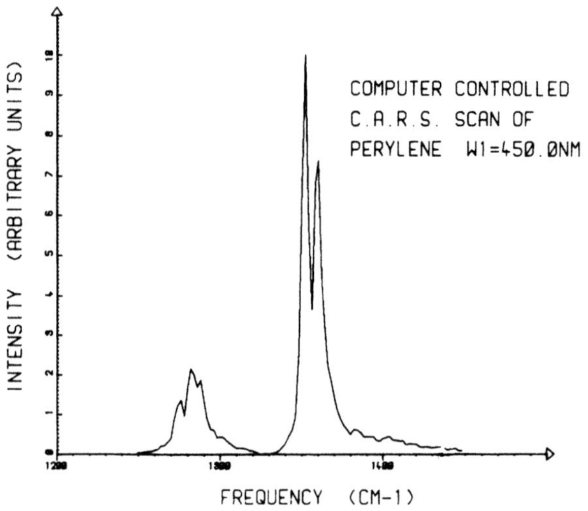

Figure 11. First spectrum taken with the scanning algorithm developed for the CARS experiment. The spectrum is of the compound perylene.

On completion of the scan the data is stored on a permanent file for latter analysis. The first spectrum to be taken with this setup is shown in Fig. 11.

III. LOW FREQUENCY CARS

Although CARS is an highly efficient process which has found successful applications in diverse fields, the momentum matching condition limits its effectiveness in standard two-laser CARS, the phase-matching requirement leads to an angular separation of the signal from the laser beams, which is determined by both the size of the Raman shift and the dispersion of the medium. This angular resolution approaches zero for low dispersion media and/or small Raman shifts and is completely lost in nondispersive media such as gases. Since gases are virtually dispersionless, the photon energy conservation condition $\Delta K = 0$ indicates that phase matching occurs when the input and signal beams are collinear. A method that permits large angular separation of the input frequencies, while still satisfying the phase matching requirement was developed by Eckbreth [13]. Based upon the shape of the phase-matching diagram (Fig. 12) this technique has been termed BOXCARS. However,

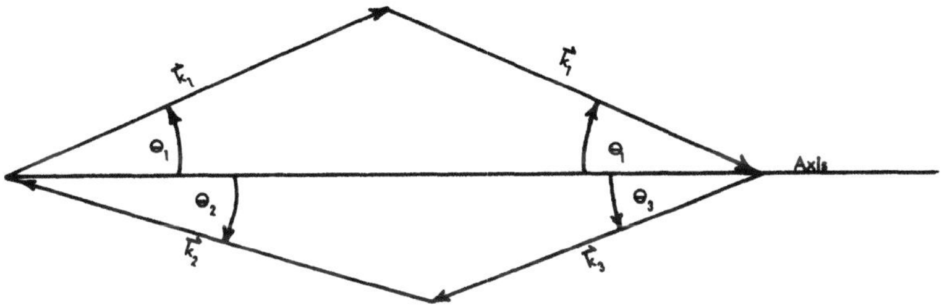

Figure 12. Phase-matching diagram for BOXCARS and Folded BOXCARS with the ω_2-ω_3 plane rotated 90°.

for small Raman shifts this geometry reduces to a parallelogram, and the signal is propagated collinearly with one of the ω_1 beams. Thus spatial and frequency resolution vanishes as Δ approaches zero.

Recently a new phase-matching technique, Folded BOXCARS (FBCARS) has been successfully applied to obtaining low frequency luminescence-free gas phase Raman spectra [14,15]. One of the major problems of applying this technique to condensed phase spectra has been the proper phase-matching due to the greater dispersion of the sample. We have incorporated a FBCARS scanning algorithm along with optical modifications to the computer controlled CARS spectrometer described in section (1). Low frequency CARS spectra of several organic liquids have been obtained.

The computerized experimental setup is similar to that described in section (2). The major modification is the incorporation of a 25 mm pellicle beam splitter in ω_1. Parallel ω_1 beams are achieved by reflecting the split portion of ω_1 with a 25 mm mirror.

The beam geometry of our FBCARS setup is shown in Fig. 13.

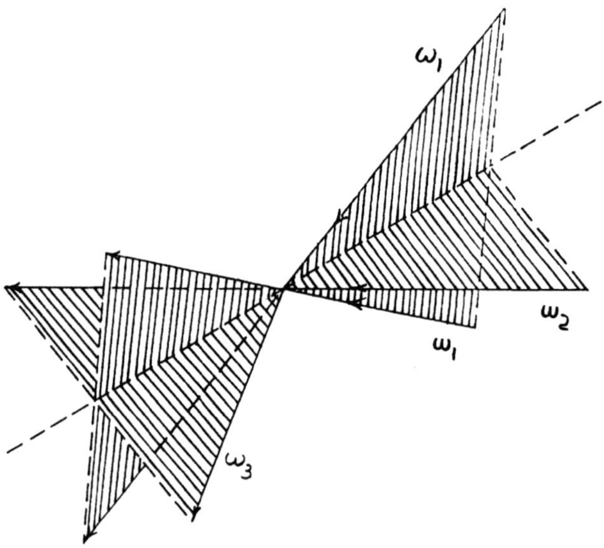

Figure 13. Geometry of the optical beams for Folded BOXCARS.

The pump beam, ω_1, is split and focused onto the sample, in a plane perpendicular to the optical table, with an intersection angle of $2\theta_1$. The probe beam, ω_2, in a plane perpendicular to the ω_1 plane is focused onto the sample at an angle θ_2 to the ω_1 plane. The sample beam ω_3 propagates from the point of intersection of ω_1 and ω_2 and exits in the ω_2 plane with an angle θ_3 with respect to the ω_1 plane. A drawback to this system is that the sample volume illumination is reduced in Folded BOXCARS with respect to CARS, and in practice the signal is attenuated by approximately a factor of ten. The resulting signal, however, is still very intense.

In order to obtain a CARS or FBCARS spectrum, θ_2 and θ_3 must vary with ω_r $(=\theta_1-\theta_2)$, in order to assure phase matching. A vector diagram with the ω_2-ω_3 plane rotated 90° is shown in Fig. 12. A scanning algorithm has been developed for the phase matching geometry of FBCARS [16].

During a CARS or FBCARS scan, the computer utilizes this algorithm to evaluate θ_2 and θ_3, thus maintaining optimum beam crossing angles and monochromator orientation.

A FBCARS spectrum in the region of the 459 cm^{-1} peak of carbon tetrachloride is shown in Fig. 14. The isotopic splitting of

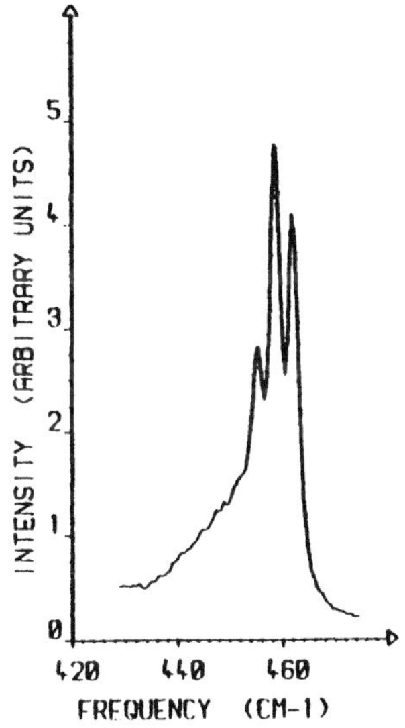

Figure 14. Folded BOXCARS spectrum of carbon tetrachloride in the region of the 459 cm^{-1} peak. ω_1 is 495 nm.

the chlorines is evident. A lower frequency FBCARS spectrum of carbon tetrachloride over the range of 140 cm^{-1} to 400 cm^{-1} is shown in Fig. 15. The greater spatial separation of the ω_3 beam in FBCARS has allowed Raman spectra to be recorded in a frequency range lower than that obtainable in regular CARS; this is readily observable in Fig. 15. The spatial isolation of ω_3 is the result of two factors: ω_3 is not coplanar with ω_1 and the variability of the θ_1 angle allows much larger θ_2 and θ_3 angles than are possible in CARS. The observed increase in the background towards 140 cm^{-1} is due to leakage of ω_1 into the low dispersion monochromator. A second spatial filter was added to the system and spectra to and below 100 cm^{-1} could be easily obtained with little or no stray light. A good test of any CARS scanning algorithm is a relatively long scan of a highly dispersive (large thetaslope) solvent. The FBCARS spectrum of bromoform in the region 70-570 cm^{-1} (Fig. 16)demonstrates the ability of the scanning algorithm to phase match the beams in a highly dispersive medium (n_d = 1.60) and the ability of FBCARS to scan in the low frequency region. The increase in signal at $\sim$ 100 cm^{-1} is not due entirely to stray light but has contributions from librational components and an extremely strong zero frequency component.

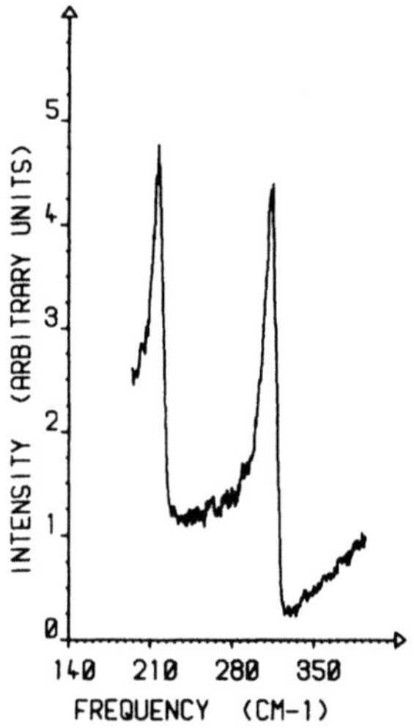

Figure 15. Low frequency Folded BOXCARS spectrum of carbon tetrachloride. ω_1 is 495 nm.

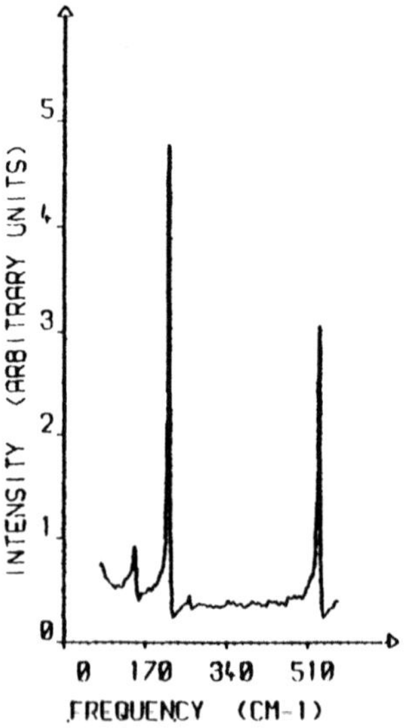

Figure 16. Extended Folded BOXCARS spectrum of bromoform. ω_1 is 495 nm.

IV. ZERO FREQUENCY

In order to observe very low frequency Raman modes we have utilized the optical setup and Folded BOXCARS beam geometry with improved spatial filtering of the ω_3 beam and have obtained spectra containing zero Raman shifts. At zero frequency $\omega_1 = \omega_2 = \omega_3$ and $\theta_3 = \theta_2 = 2\theta_1$, therefore all the beams are symmetrically arranged on a circle. The diameter of the circle can be varied as the separation of the ω_1 beams. Therefore ω_3 remains spatially isolated even at zero frequency shifts. The major experimental difficulty arises from the fact that ω_3 has the same frequency components as ω_1 and ω_2 and spontaneous Rayleigh scattering as well as reflections from ω_1 and ω_2 can very easily mask the ω_3 beam.

To minimize interference from the ω_1 and ω_2 beams particulate matter in the solvent was removed from the samples by filtering with 0.22 μm Millipore filter. The capillary tubes containing samples were thoroughly cleaned with ethanol. These procedures along with the isolation of the ω_3 beam with an 2.0 mm iris helped reduce the amount of stray light at zero frequency to approximately 2 % of the total signal.

We have recorded low noise near zero frequency spectra of benzene and carbon disulfide and have fit the data to try our theory and fitting scheme. Figure 17 shows the experimental data for carbon disulfide along with our fit.

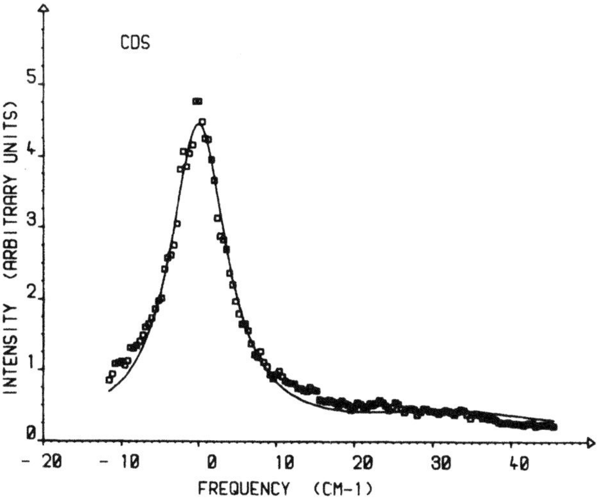

Figure 17. Low frequency carbon disulfide spectrum and theoretical fit.

The fit is composed of two Lorentzian terms at 0 cm^{-1}. This is in agreement with Carreira [17] who used Raman Gain/Loss spectroscopy to study carbon disulfide in this region. The spectrum of benzene is shown in Fig. 18. An additional lineshape had to be added in order to fit the more detailed spectrum of benzene. This is a low frequency (.3 cm^{-1})Lorentzian term which has been attributed to Brillouin scattering [18]. More detailed lineshape analysis of low frequency CARS spectra are being carried out in our lab and the results will be published shortly.

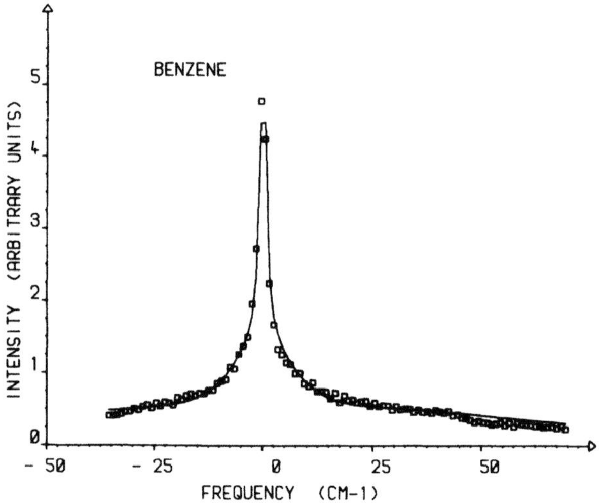

Figure 18. Low frequency benzene spectrum and theoretical fit.

REFERENCES

1. Maker, P.D., Terhune, R.W., Phys. Rev. A 137, 801 (1965).

2. Carreira, L.A., Goss, L.P., Malloy, T.B., Chem. Phys. 68, 280 (1978).

3. Maker, P.D. and Terhune, R.W., Phys. Rev. A137, 801 (1965).

4. Wynne, J.J., Phys. Rev. Lett. 29, 630 (1972).

5. Yablonovitch, E., Flytzanis, C. and Bloembergen, N., Phys. Rev. 29, 285 (1972).

6. Levenson, M.D., Flytzanis, C. and Bloembergen, N., Phys. Rev. 86, 3962 (1977).

7. Begley, R.F., Harvey, A.B., Byer, R.L. and Hudson, B.S., Appl. Phys. Lett. 25, 387 (1974).

8. Begley, R.F., Harvey, A.B., Byer, R.L. and Hudson, B.S., Am. Lab. 6, 11 (1974).

9. Hänsch, T.W., Appl. Optics, 11, 895 (1972).

10. Carreira, L.A., Goss, L.P. and Malloy, T.B., J. Chem. Phys. 66, 2762 (1977).

11. Chabey, I., Klauminzer, G.K. and Hudson, B.S., Appl. Phys. Lett. 28, 27 (1976).

12. Carreira, L.A., Rogers, L.B., Goss, L.P., Martin, G.W., Irwin, R.M., Von Wandruszka, R. and Berkowitz, D.A., Chem., Biomed. and Environ. Instrumentation, 10, 249 (1980).

13. Eckbreth, A.C. and Schreiber, P.W., in Chemical Applications of Nonlinear Raman Spectroscopy, p. 27, (H.B. Harvey, ed.) Academic Press (1981).

14. Shirley, J.A., Hall, R.J. and Eckbreth, A.C., Opt. Lett. 5 380 (1980).

15. Prior, Y., Appl. Opt. 19, 1741 (1980).

16. Carreira, L.A., Nelson, R., Horovitz, M. and Azarraga, to be published in Appl. Opt. in June 1982.

17. Carreira, L.A., Antcliff, R. and Horovitz, M., to be published.

18. Jacobson, A. and Shen, Y.R., Appl. Phys. Lett. 34, 464 (1979).

RESONANCE CARS

CONTINUUM RESONANCE CARS

A. Beckmann, P. Baierl and W. Kiefer

Physikalisches Institut der Universität Bayreuth
D-8580 Bayreuth, F.R. Germany

1. INTRODUCTION

CARS has already been shown to be a superior tool for Raman spectroscopy when performed under off resonance conditions. Resonance enhancement can be obtained when one or both of the intermediate levels of the CARS process are close to or even coincide with real levels of excited electronic states. If these levels are discrete , e.g. the rotational-vibrational levels of small molecules, then the resonant CARS signal can have enhancements of several orders of magnitude. Such discrete state resonances in CARS have been reported recently for I_2 [1-4], C_2 [5-7] and NO_2 [8-9].

We have reported very recently on preliminary results of CARS in iodine where the excitation occurs with frequencies in resonance with absorption continua [10]. As Taran [7] has pointed out, such studies are more delicate to perform than CARS studies in resonance with discrete levels, since the susceptibility is much weaker for continuum resonance excitation.

While the motivation for carrying out discrete state resonance CARS is to improve the detection sensitivity, our interests have been to study a simple scattering system, which allows to perform an experimental as well as a detailed theoretical treatment. Since the rotational-vibrational eigenstates of iodine are well known or can be calculated quantum-mechanically even for continuous states, this system is ideal for such studies and can therefore serve as a basis for more complicated systems. Furthermore, resonance enhanced CARS spectra of liquid systems are usually of such a type.

W. Kiefer and D. A. Long (eds.), Non-Linear Raman Spectroscopy and Its Chemical Applications, 393–427.

In this article we shall first briefly review experimental and theoretical results of continuum resonance (linear) Raman scattering in halogen gases. Then we present the appropriate theory for coherent anti-Stokes continuum resonance Raman scattering and after a short description of a CARS instrumentation we finally shall report on experimental results and on comparisons with first-principle calculated continuum resonance CARS spectra.

2. BRIEF REVIEW OF CONTINUUM RESONANCE (LINEAR) RAMAN SPECTROSCOPY OF DIATOMIC MOLECULES

A. Experimental observations

Continuum resonance (linear) Raman scattering has been observed already in 1970 by Holzer et al [11] in halogen gases. Kiefer and Bernstein [12] showed that the overtones observed in resonance Raman scattering above the dissociation limit of the $B(^3\Pi_{0+u})$ state of the iodine molecule exhibit a complex structure. The details of this structure were attributed to S, Q, and O ground-electronic-state $X(^1\Sigma_{0+g})$ vibrational-rotational branches of several thermally populated hot bands. In Figure 1 we show spectral regions for the fundamental ($\Delta v = 1$) and several overtones ($\Delta v = 2$ to $\Delta v = 6$) of the I_2 vibration when excited with $\lambda_o = 488$ nm. One realizes how the complexity increases with higher vibrational quantum number changes (Δv).

Because of experimental difficulties in continuum resonance CARS (e.g. dye fluorescence, background of glass, etc.), we have only been able to study experimentally overtones whose frequencies are high enough, that this handicaps do play only a minor role. Most CARS studies have been performed for the $\Delta v = 6$ transition. We therefore concentrate first on the linear continuum resonance Raman spectrum of this particular transition. Figure 2 shows in the upper and middle field, the $I_{\parallel}$ and $I_{\perp}$ spectra, respectively, whereas in the lower field of Figure 2 the $I_{\parallel} - 4/3\ I_{\perp}$ spectrum is displayed [13]. All three spectra have been recorded simultaneaously with a special technique described elsewhere [14]. It is quite clear, that by means of this technique the isotropic part of Raman scattered light can be directly recorded and separated from the anisotropic part, allowing to experimentally distinguish between S- and O-branches (anisotropic scattering) and Q-branches (isotropic scattering). For 488 nm excitation, which corresponds to an energy about 450 cm^{-1} above the dissociation limit of the iodine molecule, one observes besides transitions originating from the vibrational ground level ($v' = 6 \leftarrow v'' = 0$), a series of hot band transitions ($v' = 7 \leftarrow v'' = 1$, $v' = 8 \leftarrow v'' = 2$, $v' = 9 \leftarrow v'' = 3$, etc.). These hot band transitions occur for both, S- and Q-branches, as indicated in Figure 2. The asymmetry

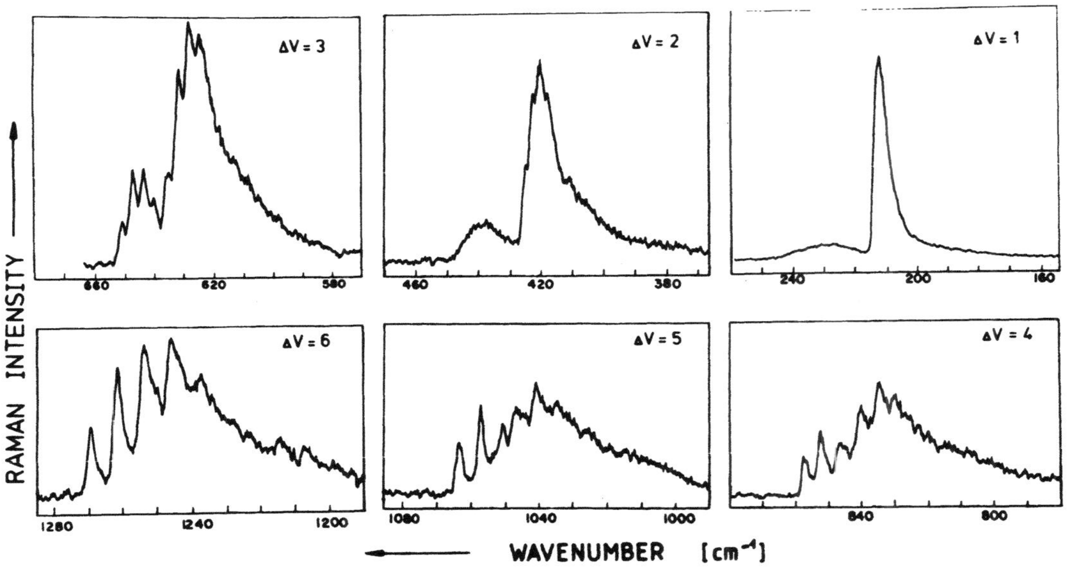

Fig. 1 Continuum resonance (linear) Raman spectra of iodine vapor. Displayed are the fundamental vibrations ($\Delta v = 1$) and several overtones ($\Delta v = 2$ to $\Delta v = 6$), as indicated. Excitation wavelength is $\lambda_o = 488.0$ nm. Slit widths: $s = 1.3$ cm^{-1} for $\Delta v = 1$ and 2, $s = 1.5$ cm^{-1} for $\Delta v = 3$–5 and $s = 1.7$ cm^{-1} for $\Delta v = 6$. The intensities between different overtones and fundamental cannot be compared to each other, since the various spectral regions have been recorded with different gain settings. (Adapted from Reference 12).

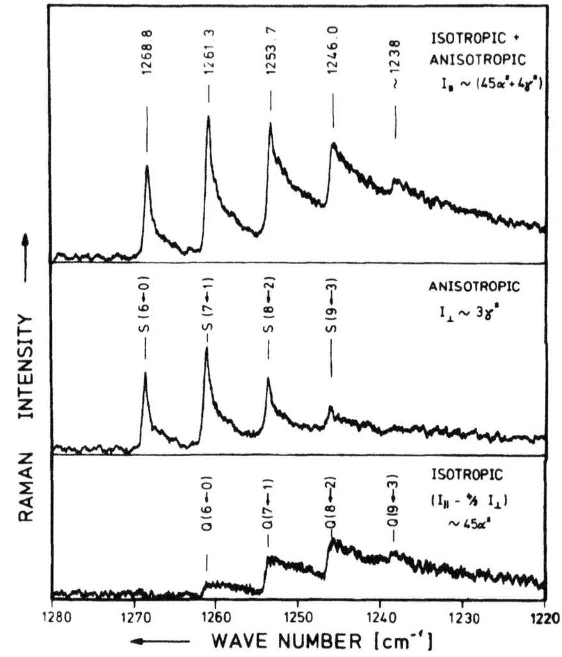

Fig. 2 Continuum resonance (linear) Raman spectra of the $\Delta v = 6$ transitions in iodine vapor; excitation wavelength $\lambda_o = 488.0$ nm $\hat{=}$ 20 492 cm^{-1} power P = 1 Watt; slit width s = 0.5 cm^{-1}. Upper field: the $I_{\parallel}$ spectrum (isotropic plus anisotropic part); middle field: the $I_{\perp}$ spectrum (anisotropic part); lower field: the $I_{\parallel}-4/3\ I_{\perp}$ spectrum (isotropic part only) [13].

of the bands is due to unresolved rotational structure and can simply be explained by calculation of Fortrat diagrams (rotational quantum number versus Raman shift) for the various $\Delta J = 0,\pm 2$ transitions [12]. Such diagrams are shown in Figure 3 together with the observed spectra. In this figure we can see, that the $\Delta J = +2$ transitions (S-branch) form a band-head which is shaded towards lower wavenumbers (Raman shifts), whereas the Q-branch traces are quite flat. The Q-branch traces are very steep at lower J values but then again turn to a flat curve. Measureable intensities in an unresolved rotational-vibrational Raman spectrum can only be observed when the various rotational lines are very close together as is the case for the S-branch heads or the Q-branch transitions at low J values. Then an intense, unresolved Raman line is observed. The shading of the various vibrational Raman lines can nicely be seen for the S-, as well as for the Q-branches (see Figures 2 and 3). The Fortrat diagrams also immediately explain why no peaks assigned to O-branches are observed.

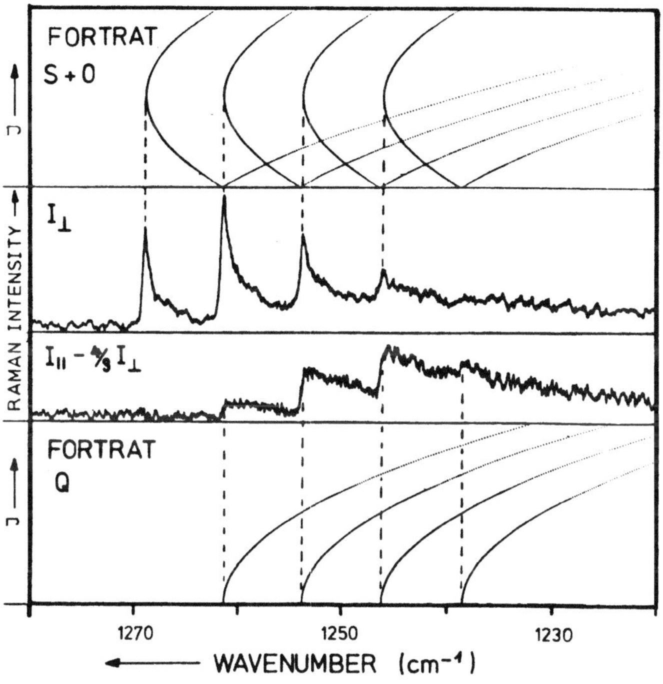

Fig. 3 The $\Delta v = 6$ continuum resonance Raman transitions in iodine vapor: explanation of observed bands by means of Fortrat diagrams [13].

It should be mentioned, that the coincidence of spectral positions of $v' + 1 \leftarrow v'' + 1$ S-branch and $v' \leftarrow v''$ Q-branch transitions for $\Delta v = 6$ is accidental (e.g. the S(7 ← 1) band head has approximately same wavenumber as the Q(6 ← 0) transition; see vertical lines in Fig. 3). For other Δv transitions the S- and Q-branch peaks can be separated [12].

Depending on the energy of the exciting laser frequency the intensities of the various vibrational transitions can vary dramatically [15-18]. Similar observations were also made in bromine vapor [19-20]. The exciting frequency dependence of the observed spectral band shape variations were explained for iodine [15,16,21,22] and bromine [20-22] by a theory in which the scattering intensity depends on Franck-Condon overlap integrals. Since these theories combined with the theory on CARS succeptibilities serve also to explain the observations made on continuum resonance CARS in iodine, we shall briefly summarize them in the next chapter.

B. Theory for continuum resonance (linear) Raman scattering in halogen molecules.

The total Raman intensity scattered by a macroscopic number of molecules which undergo a transition from an initial (i) to a final (f) state may be given by the standard expression [21]

$$I^{if}_{\rho\sigma} = I_o N_i L \Omega_m \frac{\omega_f^4}{c^4} |(\alpha_{\rho\sigma})_{if}|^2 \quad (1)$$

The meaning of the various symbols in Equation 1 is as follows:

$I^{if}_{\rho\sigma}$ = scattered intensity of one particular Raman transition from initial state i to final state f for the polarization arrangement ρ(polarization of scattered Raman light) and σ (polarization of laser beam). Further explanation see Figure 1 of Reference 21.

I_o = intensity of incident light.

N_i = density of scattering molecules in state i; for numerical calculation of N_i see Equation 7 of Reference 21.

L = length from which scattered photons are collected

Ω_m = solid angle of collection

ω_f = frequency of scattered light.

c = velocity of light

$(\alpha_{\rho\sigma})_{if}$ = polarizability tensor component for the transition $f \leftarrow i$ with incident and scattered polarizations indicated by σ and ρ, respectively.

In general, the components of the polarizability tensor can be calculated quantum-mechanically by second-order perturbation theory [23] to be

$$(\alpha_{\rho\sigma})_{if} = \frac{1}{\hbar} \sum\!\!\!\!\!\!\int_r \left[\frac{\langle f|\mu_\rho|r\rangle\langle r|\mu_\sigma|i\rangle}{\omega_{ri} - \omega_o} + \frac{\langle f|\mu_\sigma|r\rangle\langle r|\mu_\rho|i\rangle}{\omega_{rf} + \omega_o} \right] \quad (2)$$

Here, the sign $\sum\!\!\!\!\!\!\int$ indicates that a summation and/or integration has to be taken over all important eigenstates r of the unperturbed molecule depending on the presence of discrete and/or continuous states, respectively. Since in the following we are only considering continuum resonance Raman scattering and weak contributions from distant discrete states we omit a damping term $i\Gamma_r$ in Equation 2 for simplicity. (Damping is not relevant to continuum states and Γ_r can therefore be taken to be infinitesimal). The meaning of the additional symbols in Equation 2 are:

μ_ρ, μ_σ = components of the operator of the total electric dipole moment of the electrons

ω_{ri}, ω_{rf} = transition frequencies from initial and final eigenstates to state r of the molecule, respectively

ω_o = exciting laser frequency

In resonance Raman scattering, where the exciting frequency ω_o is very close to or actually coincides with transition frequencies ω_{ri}, the first term in Equation 2 is much larger in magnitude compared to the second term because of the frequency denominator. For the following discussion we therefore can neglect the second term. It is very difficult to directly calculate the polarizability tensor components by applying Equation 2 because the exact molecular eigenstates in principle depend on both, the nuclear and electronic coordinates in a complicated way and are very difficult to utilize. To simplify the matrix elements one can apply the well known Born-Oppenheimer approxomation [16], where the vibrational wavefunctions are separated from the electronic wavefunctions. Furthermore, one can expand the electronic matrix elements in a rapidly converging Taylor series (Herzberg-Teller expansion) and neglect higher order terms. This results for the polarizability tensor components in a product of the oscillator strength for the electronic transition e ← 0 and a factor for which we use the following abbreviation

$$(\chi_{oe})^{v'}_{v''} = \frac{1}{\hbar} \sum\!\!\!\!\!\!\int_{v_e} \frac{\langle v'|v_e\rangle\langle v_e|v''\rangle}{\omega_{ri} - \omega_o} \tag{3}$$

This factor which for a given electronic state e has to be calculated for a given vibrational transition v' ← v" by an integration (continuum) and/or summation (discrete levels) over all vibrational levels v_e of the excited state, simply contains the products of Franck-Condon overlap integrals weighted by frequency denominators.

For intermediate continuum states the pole in the denominator at exact resonance is accommodated by rewriting Equation 3 as

$$(\chi_{oe})^{v'}_{v''} = P \int_0^{\infty} \frac{<v'|v_e><v_e|v''>}{\hbar\omega_{ri} - \hbar\omega_o} \rho(h\omega_{ri}) d(h\omega_{ri})$$

$$- i \pi <v'|v_e><v_e|v''> \rho(\hbar\omega_{ri}) \Big|_{\hbar\omega_{ri} = \hbar\omega_o} \quad (4)$$

for which we simply abbreviate

$$(\chi_{oe})^{v'}_{v''} = Re(\chi_{oe})^{v'}_{v''} + iI_m(\chi_{oe})^{v'}_{v''} \quad (5)$$

Here, $\rho(\hbar\omega_{ri})$ is the density of continuum states, $P\int$ is the principle part of the integral and Re and Im denote real and imaginary part of the vibrational transition amplitude, respectively.

For rotational-vibrational transitions in resonance with excited electronic states one has also to calculate the appropriate rotational intensity factors. While for S- and Q-branch transitions in diatomics these factors are the same as derived by Placzek and Teller [24], they are different for Q-branches depending on the specific excited electronic state. Two of us [21], however, have derived these factors for the electronic states $B(^3\Pi_{o+u})$ and $^1\Pi_{1u}$, since the latter are the one's which are resonantly enhancing the Raman scattering intensity in halogen molecules when excitation frequency in the visible region are employed.

Finally, we give the basic formula, which allows a fist principle calculation of the continuum resonance Raman intensity of a single rotational-vibrational transition $f \leftarrow i(v', J' \leftarrow v'', J'')$ in a halogen molecule with two excited electronic states ($e = B(^3\Pi_{o+u})$ or $^1\Pi_{1u}$). The final expression, which has been derived in detail in Reference 21 and which can be used for numerical calculations [22], is given in the following:

$$I^{if}_{\rho z} = I_o N_i L\Omega_m \frac{\omega_f^4}{c^4} \frac{g_i}{2J''+1} \times$$

$$\times \Big[b^{B,\Delta J}_{\rho z} |M^B|^4 |(\chi_{OB})^{v'}_{v''}|^2$$

$$+ b^{\Pi,\Delta J}_{\rho z} |M^{\Pi}|^4 |(\chi_{O\Pi})^{v'}_{v''}|^2$$

$$+ b^{B+\Pi,\Delta J}_{\rho z} |M^B|^2 |M^{\Pi}|^2 \{Re(\chi_{OB})^{v'}_{v''} Re(\chi_{O\Pi})^{v'}_{v''}$$

$$+ Im(\chi_{OB})^{v'}_{v''} Im(\chi_{O\Pi})^{v'}_{v''}\}\Big] \quad (6)$$

with $\rho = x,z$; $\Delta J = 0,\pm 2$.
Note, that we have specified the polarization of the laser beam to be $\sigma = z$, but allow for two polarization states of scattered Raman light ($\rho = x,z$). Observation direction of Raman light is y (see Figure 1 of Reference 21). The meaning of the additional symbols in Equation 6 is as follows:

g_i = nuclear spin statistics factor

B,Π = abbreviation for states ${}^3\Pi_{o+u}$ and ${}^1\Pi_{1u}$, respectively

$b_{\rho z}^{B,\Delta J}$ = rotational intensity factor for resonance Raman scattering via state B (listed for $\Delta J = 0,\pm 2$ and $\rho = x$ and z in Table 1 of Reference 21).

$b_{\rho z}^{\Pi,\Delta J}$ = rotational intensity factor for resonance Raman scattering via state Π(listed for $\Delta J = 0,\pm 2$ and $\rho = x$ and z in Table 2 of Reference 21).

$b_{\rho z}^{B+\Pi,\Delta J}$ = rotational intensity factor for the interference terms of the scattering intensities for resonance Raman scattering via two simultaneously present states B and Π (listed in Table 3 of Reference 21).

M^B, M^Π = electronic transition moment from ground state to the excited states B and Π, respectively (first term in Herzberg-Teller expansion).

Scattering experiments in halogen gases with cw lasers [12,19] as excitation sources revealed that high spectroscopic temperatures are involved. Therefore, in order to do the correct calculation of the total band shape of the same vibrational transitions ($\Delta v = n$, n = integer) one has to sum over all populated vibrational and rotational levels in the electronic ground state.

In order to make a first-principle calculation of resonance Raman spectra excited with a particular frequency one has first to start in determining all relevant vibrational wavefunctions of ground and excited electronic states. Since the appropriate potential functions needed for this purpose are well known for most of the halogen molecules, this is a simple task. The next step would be the evaluation of the Franck-Condon overlap integrals associated with the various vibrational transitions. Then one has to carry out the integration over the continuum as shown in Equation 4. This results in determining real and imaginary part of the vibrational amplitude (Equation 5). The latter has to be calculated for both the dissociative part of the B-state, as well as of the repulsive state Π. Using the electronic transition moments for the two states, the appropriate rotational intensity factors [21] and the thermal population distribution of the initial states one can apply Equation 6 for the intensity evaluation of one particular rotational-vibrational Raman transition (I^{if}).

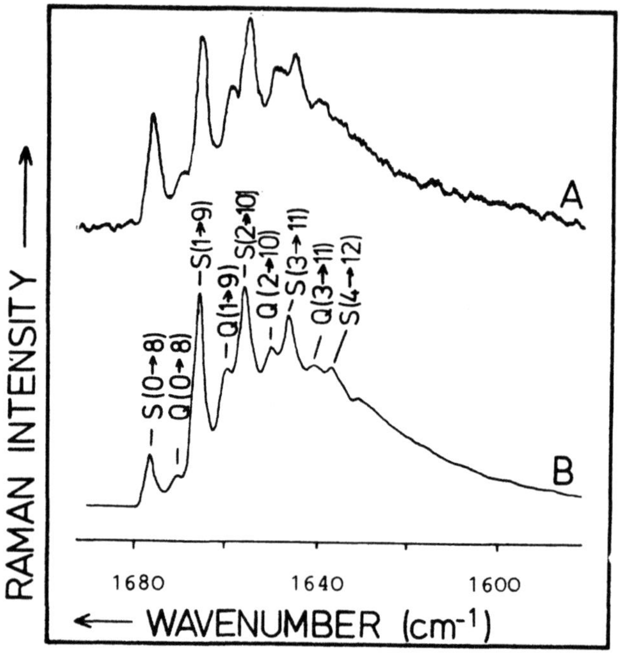

Fig. 4. Experimentally observed (A) and calculated (B) continuum resonance (linear) Raman spectrum of the seventh overtone ($\Delta v = 8$) of iodine vapor. Excitation wavelength was $\lambda_o = 488.0$ nm [22].

Summing over all populated initial states and convoluting with a slit function one finally arrives at a calculated resonance Raman spectrum.

In Figures 4 and 5 we present two examples of such first-principle calculations and compare the resulting theoretical resonance Raman spectra with observed data. Figure 4 shows the continuum resonance Raman spectrum of the seventh overtone ($\Delta v = 8$) of iodine vapor excited with $\lambda_o = 488$ nm. Figure 5 displays the spectrum of the first overtone ($\Delta v = 2$) of $^{79}Br_2$ for $\lambda_o = 457.9$ nm excitation. For both cases there is fairly good agreement between theory and experiment. Other examples with same good agreement can be found in Reference 22.

The success in reproducing theoretically continuum resonance (linear) Raman spectra in halogen molecules encouraged us to expand the theory also to a four-photon process (CARS) which is resonantly enhanced via the continuous excited electronic states. A brief summary of this theory is given next.

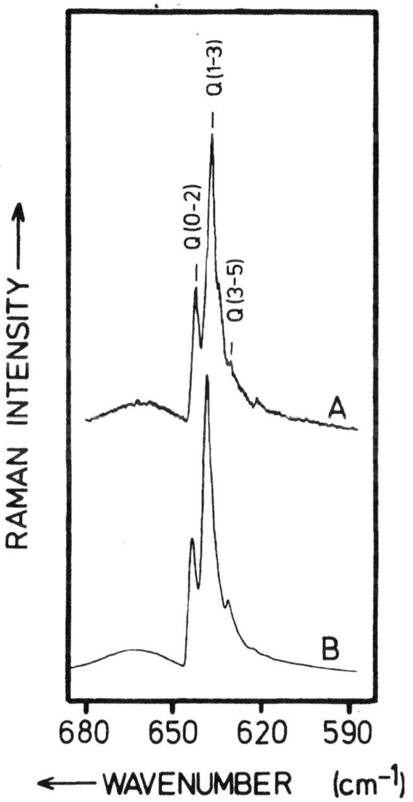

Fig. 5 Observed (A) and calculated (B) continuum resonance (linear) Raman spectrum of the first overtone of $^{79}Br_2$. Excitation wavelength λ_o = 457.9 nm [22].

3. Theory for continuum resonance CARS

In Figure 6 we display a schematic energy diagram for coherent anti-Stokes Raman scattering for resonance with the continua of states $B(^3\Pi_{o+u})$ and $^1\Pi_{1u}$. The four-photon CARS process is indicated for a Δv = 6 vibrational transition (initial state i = 0, final state f = 6) by the vertical lines. The laser frequency ω_L as well as the Stokes laser frequency ω_S are such that the intermediate states, labelled by r and r', respectively, are located above the dissociation limit of the B state, which is at 20043 cm^{-1} (note, that this value does not include the zero point energy). If one is interested in the spectral dispersion of the CARS intensity and not in the evaluation of absolute CARS intensities it is sufficient to calculate the third order non-linear susceptibility for the CARS process, $\chi^{(3)}_{CARS}$; it is well known,

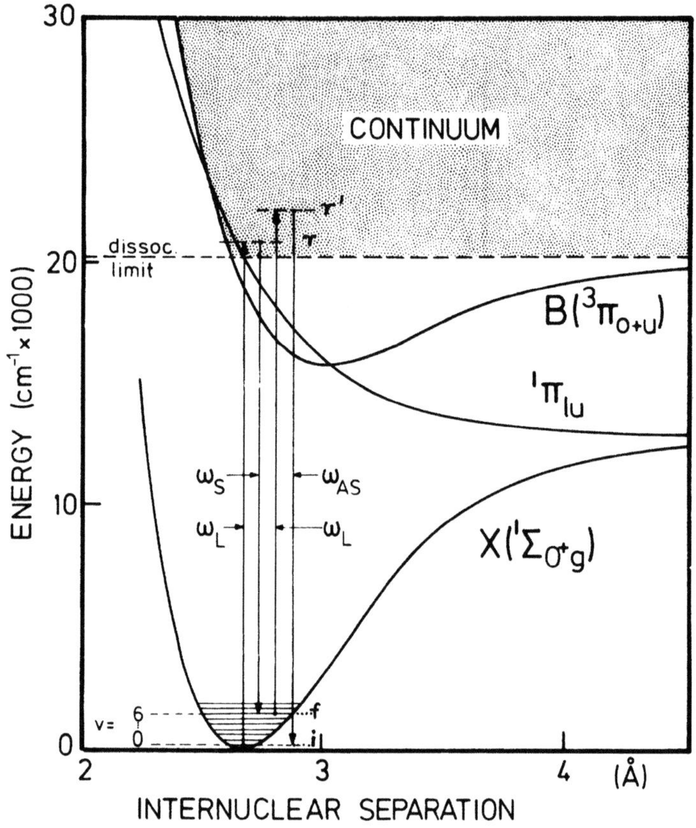

Fig. 6 Schematic diagram for coherent anti-Stokes continuum resonance Raman scattering in iodine. The vertical lines indicate the four-photon CARS process for a $\Delta v = 6$ vibrational transition which is in resonance with the dissociative continuum of the B- and the continuum of the repulsive $^1\Pi$-state. ω_L, ω_S and ω_{AS} denote frequency of the two lasers and of the generated CARS beam, respectively.

that

$$I_{CARS} \sim |\chi^{(3)}_{CARS}|^2 \tag{7}$$

Recently Taran's group [2-4] has derived comprehensively third-order susceptibilities using the density operator formalism. Applying their results we write the general expression for the CARS susceptibility (see for example Equation 18 in Reference 4) in the notation used by us in the preceding discussion (Section 2 and Figure 6) for one particular transition f ← i:

$$\left[\chi^{(3)}_{CARS}(-\omega_{AS},\omega_L,\omega_L,-\omega_S)\right]_{if} = \frac{N}{\hbar^3}\,\frac{1}{\omega_{if}-\omega_L+\omega_S-i\Gamma_{if}}$$

$$\times \sum_{r'}\left(\frac{\mu^{\tau}_{ir'}\,\mu^{\sigma}_{r'f}}{\omega_{r'i}-\omega_{AS}-i\Gamma_{r'i}} + \frac{\mu^{\sigma}_{ir'}\,\mu^{\tau}_{r'f}}{\omega_{r'f}+\omega_{AS}+i\Gamma_{r'f}}\right)$$

$$\times \sum_{r}\left[\left(\frac{\mu^{\rho}_{fr}\,\mu^{\sigma}_{ri}}{\omega_{ri}-\omega_L-i\Gamma_{ri}} + \frac{\mu^{\sigma}_{fr}\,\mu^{\rho}_{ri}}{\omega_{ri}+\omega_S-i\Gamma_{ri}}\right)\rho_i - \left(\frac{\mu^{\rho}_{fr}\,\mu^{\sigma}_{ri}}{\omega_{rf}-\omega_S+i\Gamma_{rf}} + \frac{\mu^{\sigma}_{fr}\,\mu^{\rho}_{ri}}{\omega_{rf}+\omega_L+i\Gamma_{rf}}\right)\rho_f\right]$$

$$+ \chi_{NR} \tag{8}$$

where $N\rho_i$ and $N\rho_f$ are initial number densities of the Raman active molecule in states $|i\rangle$ and $|f\rangle$, respectively. The absorption frequencies from states $|i\rangle$ and $|f\rangle$ to states $|r\rangle$ (or $|r'\rangle$) are ω_{ri} and ω_{rf} (or $\omega_{r'i}$ and $\omega_{r'f}$) respectively, and the Γ's are the corresponding damping factors; μ^{σ}_{ri} (as example) is the matrix component of the dipole moment operator $\mu^{\sigma}_{ri} = \langle r|\underline{p}\,\hat{e}_{\sigma}|i\rangle$, where $\hat{e}_{\sigma}$ is the unit vector in the direction of polarization of the ω_L field; μ^{ρ}_{fr} and μ^{τ}_{ir} involve interactions with ω_S and ω_{AS} fields, respectively ($\hat{e}_{\rho}$ and $\hat{e}_{\tau}$ are the unit vectors in the directions of polarisation of the ω_S and ω_{AS} fields, respectively). χ_{NR}, the non-resonant susceptibility, combines contributions from non-resonant two photon sum or difference denominators in place of the Raman resonance denominator and also from non-Raman-resonant molecular species (such as diluent molecules in liquids or glass from the cell in gas phase CARS).

We now apply Equation 8 to the special case of continuum resonance CARS in a diatomic molecule. First, due to the resonant frequency denominators we can neglect all terms which have a plus sign between the wave frequencies ω_{AS}, ω_S, ω_L and the absorption frequencies $\omega_{r'f}$, ω_{ri}, ω_{rf}, respectively (second terms within the round brackets). Second, as we have shown in chapter 2, there are many rotational-vibrational transitions resonant with the absorption continuum when laser frequencies are used whose energies are higher than the dissociation limit. Therefore, we have to sum over all contributions arising from transitions $f \leftarrow i$ with frequencies ω_{if}. Then we have to average over all degenerate initial states and carry out the summation over all degenerate final states. Finally, we have to account for the fact, that the predominant resonant state is a continuum, which leads us to perform an integration over all continuous states. We will also include weak resonances from nearly discrete levels of the B-state. The sign $\sum\!\!\!\!\!\!\int$ serves again for this purpose.

For continuum resonance CARS excitation in diatomic molecules in the gas phase Equation 8 can therefore be modified (according to the discussion above) to be

$$\chi^{(3)} = \frac{N}{\hbar^3} \sum_{\omega_{if}} \Bigg(\frac{1}{(\omega_{if}+\omega_S-\omega_L-i\Gamma_{if})} \times$$
$$\times \Bigg\{ \left[\frac{\rho_i\, g_i}{(2J''+1)} \sum_{m',m''} \left(\sum\!\!\!\!\!\!\int_{r'} \frac{<i|\mu_\tau|r'><r'|\mu_\sigma|f>}{\omega_{r'i}-\omega_{AS}-i\Gamma_{r'i}} \right) \times \left(\sum\!\!\!\!\!\!\int_{r} \frac{<f|\mu_\rho|r><r|\mu_\sigma|i>}{\omega_{ri}-\omega_L-i\Gamma_{ri}} \right) \right]$$
$$- \left[\frac{\rho_f\, g_f}{(2J'+1)} \sum_{m',m''} \left(\sum\!\!\!\!\!\!\int_{r'} \frac{<i|\mu_\tau|r'><r'|\mu_\sigma|f>}{\omega_{r'i}-\omega_{AS}-i\Gamma_{r'i}} \right) \times \left(\sum\!\!\!\!\!\!\int_{r} \frac{<f|\mu_\rho|r><r|\mu_\sigma|i>}{\omega_{rf}-\omega_S+i\Gamma_{rf}} \right) \right] \Bigg\} \Bigg)$$
$$+ \chi_{NR} \qquad (9)$$

Here, we have used the abbreviation $\mu_\tau = \underline{p}\hat{e}_\tau$ and similar signs for the other components of the electric dipole moment.

As mentioned in the section on continuum resonance (linear) Raman spectroscopy damping is not relevant to continuum states and the damping factors $\Gamma_{r'i}$, Γ_{ri} and Γ_{rf} can therefore all be taken to be infinitesimal. However, one has to be aware of the fact that when the integration over the continuum is performed, the imaginary parts have different signs, depending whether there is a plus or a minus sign in front of the correspondent damping factor Γ. The damping factor Γ_{if} cannot be set to zero, since it reflects the linewidth of the Raman line for a transition $f \leftarrow i$.

Inspection of Equation 9 shows, that the CARS susceptibility contains two products of two terms (the terms inside the round brackets) which are very similar to the first term in Equation 2 when damping is neglected. This first term in Equation 2 is, as we know, the dominant part for the amplitude of continuum resonance (linear) Raman scattering. Since we have demonstrated in chapter 2 how to calculate these Raman polarizabilities, Equation 9 can directly be employed for a first principle calculation of the dispersion of the third-order CARS susceptibility and hence by applying Equation 7 of a complete CARS spectrum. Such numerical calculations have been carried out by us [25]. Some of these results will be presented in chapter 5 of the present article together with experimentally obtained CARS spectra.

4. CARS INSTRUMENTATION

We briefly describe here the CARS spectrometer in use at the University Bayreuth. With this set-up several other investigations reported in this book have been carried out [26].

The experimental CARS set up is shown in Figure 7 and consists of a 1-MW nitrogen laser (Molectron UV-1000), which simultaneously pumps two dye-laser oscillators (Molectron DL-300), each equipped with an etalon. Only about 30 % of the nitrogen laser output power is used to pump the oscillators whereas 70% is taken for dye amplifiers (see below). The laser pulses generated in the two oscillators are amplified by dye-laser amplifiers which are also pumped by the nitrogen laser. In order to achieve maximum amplification and simultaneously minimum background radiation an optical-delay line (ODL) is installed between the nitrogen laser and dye amplifiers. The amplifiers produce 5-ns pulses with peak powers of about 20 and 10 kW for the dye laser with frequency ω_L and ω_S, respectively. Both have a bandwidth of about 0.03 cm^{-1}.

In order to separate the laser beams from the dye fluorescence background we use a small filter spectrometer (Anaspec model 300 S) for each laser beam. The absolute frequency of the dye-laser beams is determined using a simple method described elsewhere [27] where an iodine cell (I_2) is employed to produce a fluorescence spectrum which is then identified by means of high-resolution absorption spectra published by Gerstenkorn and Luc [28]. The absolute accuracy achieved by this method is on the order of 0.01 to 0.02 cm^{-1} which is sufficient for the resolution achieved with our CARS set up. The latter is about 0.05 cm^{-1} (convolution of the two dye lasers with bandwidths of 0.03 cm^{-1} each. However, in our continuum resonance CARS experiments in iodine, only the ω_S-laser can be determined in this way, because the frequency of the ω_L-laser is higher than the discrete levels of the excited electronic state and hence cannot produce fluorescence.

In order to control the peak power of the dye lasers a beam splitter is put in each laser beam and the reflected beams are monitored by means of photodiodes (D III, and D IV). The time coincident laser pulses are combined via a prism and focused into a cell with Brewster angle windows, using a 2.5-cm focal length lens. The power of the laser beams at the sample is about 10 kW for ω_L and about 5 kW for ω_S, respectively. The collinearly generated CARS signal is directed either with a prism or a mirror to the entrance slit of a Spex 14018 double monochromator and detected with a fast photomultiplier (PM) type RCA C31024. The signals are processed by a computer system consisting of two Zilog Z 80 microprocessors, 96-kilobyte memory, a two floppy-

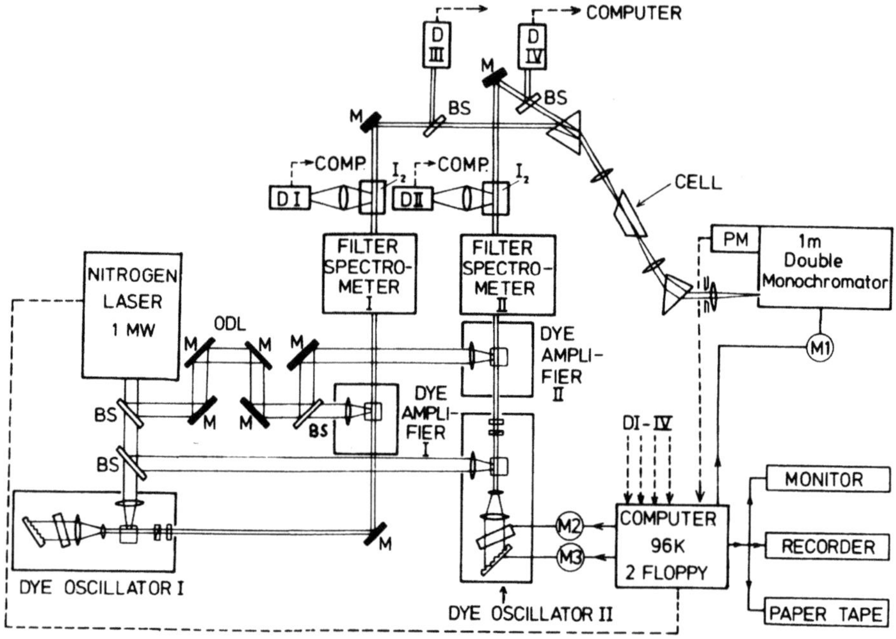

Fig. 7 CARS set up at the University Bayreuth: PM, photomultiplier; M1-M3, stepping motors; M, mirror; BS beam splitter; ODL optical delay line; D, photodiodes; I_2 iodine fluorescence cell. For further explanation see text; [10].

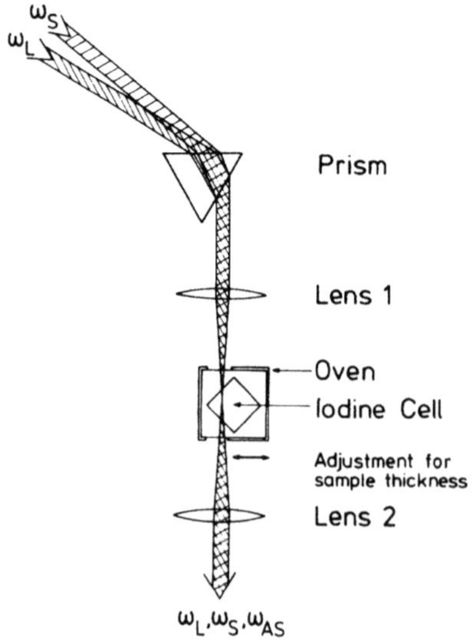

Fig. 8 (left) Sample arrangement for resonance CARS excitation in iodine vapor.

disk controller, a four-channel recorder, a lineprinter, and a paper-tape puncher. The computer also serves to scan simultaneously grating and etalon of the Stokes dye laser (ω_S) and is additionally interfaced to the spectrometer drive, thus allowing to set the spectrometer step by step precisely at the calculated frequency of the CARS signal. To manage this a multitask operating system has been written. It should be noted that the spectrometer whose slits were set to about 1-cm^{-1} spectral bandwidth serves only as a high-efficiency band-pass filter but does not influence the resolution of the CARS spectrum. By this method high discrimination between CARS and laser signals is achieved while at the same time the complete sensitivity of the detection system can be used, enabling us to also record spectra of weak CARS signals.

While the output frequency of the Stokes laser (ω_S) is tuned continuously within a range of approximately 24 cm^{-1} the CARS and reference signals are stored in a floppy disk. After a complete CARS spectrum has been measured, the latter is smoothed and plotted with the recorder.

Excitation of continuum resonance CARS spectra require a special sample arrangement. For the case of iodine we use the arrangement displayed in Figure 8. After the two laser beams are made collinear by means of a prism, they are focused with an f = 2.5 cm lens to a common focus inside a rectangular sample cell, whose plane surfaces are turned by 45° in respect to the direction of the laser beams (see Figure 8). The evacuated and sealed sample cell contains iodine as a solid. By heating the cell to definite and constant temperatures particular pressures of iodine vapor can be maintained in the cell. The effective thickness of the sample can be adjusted by moving the cell in direction perpendicular to the one of the laser beams.

The main difficulty to observe a CARS signal when the sample is highly absorbing, as is the case for excitation into the continuum of gaseous iodine, was to find the appropriate gas pressure, effective length of sample and power of the laser beams. It should be mentioned that only by careful optimization of these experimental parameters have we been able experimentally to obtain sufficient CARS signals for this type of experiment. The experimental results reported in the next chapter have been achieved with the following parameters: vapor pressure of iodine appr. 100 - 200 mbar; effective length of sample: 9.7 mm.
For the power of the laser beams as given above and with a repetition rate of 10 Hz typical CARS signals corresponding to 500 cps were obtained for these parameters when studies for the $\Delta v=6$ transitions in iodine were performed with excitation frequencies slightly higher than the dissociation limit of iodine.

5. CONTINUUM RESONANCE CARS SPECTRA OF IODINE

A. Variation of laser frequency (ω_L)

CARS experiments in iodine with resonance with the continua of the $B(^3\Pi_{o+u})$- and $^1\Pi_{1u}$-state have been carried out with excitation frequencies for ω_L in the range between the dissociation limit of the B-state and about 1000 cm^{-1} above this limit [10,25]. However, above about 20 700 cm^{-1} the CARS signal was found to be very weak.

Typical experimentally obtained high resolution continuum resonance CARS spectra of iodine for $\Delta v = 6$ vibrational transitions are displayed in the left field of Figure 9 for four excitation frequencies (ω_L = 20 667, 20 465, 20 301, and 20 168 cm^{-1}). Strong and sharp peaks at about 1269, 1261 and 1254 cm^{-1} are observed which have been assigned to S-branch transitions originating from the $v'' = 0,1,2$ vibrational levels of the ground electronic state, respectively. Comparing Figure 9 with Figures 2 and 3 we immediately recognize that these lines correspond to the rotational ($\Delta J = 2$) transitions at the turning point of the Fortrat parabola (S-band head). (Note that the wavenumber scales in Figures 2 and 3 are opposite to the one in Figure 9). However, due to the fact that CARS favors only strong lines and because we have used much higher resolution in CARS (0.05 cm^{-1}) compared with linear Raman excitation (0.5 cm^{-1}), one observes in the CARS spectra (Figure 9) only strong rotational transitions which have their rotational-vibrational transition frequencies very close together. This happens exactly at the band head. The right field of Figure 9 displays numerically calculated continuum resonance CARS spectra of the same spectral region. These spectra have been created applying the first-principle calculation described in chapter 3. The final expression for $\chi^{(3)}$ as given in Equation 9 was applied for this purpose. For the calculation of the wavefunction and the matrix elements we used data reported by Luc et al. [29] for the ground state, by Barrow et al. [30] for the excited bound state B, and by Williams et al. [17] for the repulsive Π state. The following parameters have been used too: spectroscopic temperature T = 1200 K and pressure-broadened Raman bandwidth Γ_{if} = 0.04 cm^{-1} (in the following we use Γ instead of Γ_{if}). For further details we refer to Reference 25. The influence of variations of these parameters to the intensity and band shape of the calculated CARS spectra will be discussed below. However, it should be mentioned, that all spectra shown in Figure 9 have been calculated with only one set of data as given above.

All spectra (observed and calculated) displayed in Figure 9 have been scaled (normalized) in such a way that the strongest line in each spectrum corresponds to full scale (4 arbitr. units).

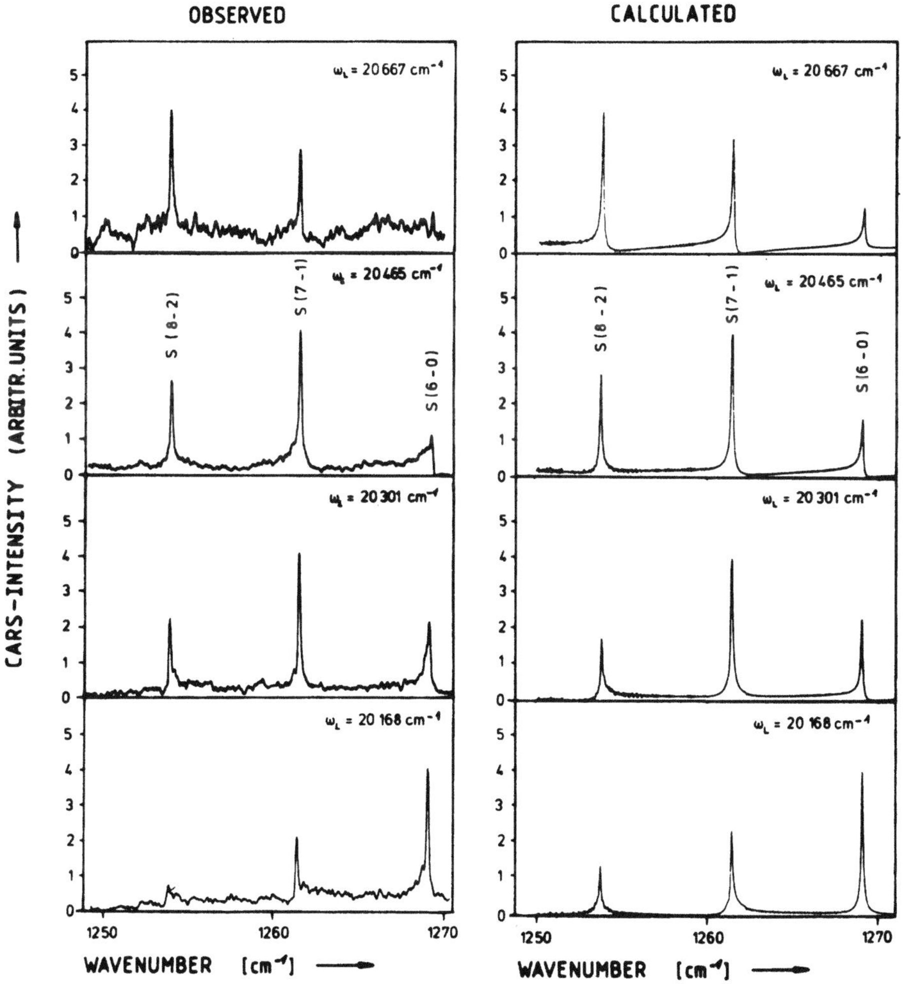

Fig. 9 Experimentally observed (left field) and theoretically calculated (right field) continuum resonance CARS spectra of the fifth overtone ($\Delta v = 6$) in iodine vapor for the pump laser frequencies ω_L = 20 667 cm^{-1}, 20 465 cm^{-1}, 20 301 cm^{-1}, and 20 168 cm^{-1} as indicated. S numbers refer to the initial and final vibrational-state assignments of the S-branches. The spectra for different ω_L are not scaled to each other.

This facilitates an easy comparison of relative intensity changes between S-band heads at same excitation frequency. We notice that the intensity of the S(6 ← 0) transition is most prominent for excitation (ω_L = 20 168 cm^{-1}) close to the dissociation limit of the B-state, while the one for the S(8 ←2) transition is nearly zero for this excitation frequency. The situation is quite opposite for higher excitation frequencies. For ω_L = 20667 cm^{-1} (upper spectrum of Figure 9) the second hot band (8 ← 2) is much more intense than the transition which starts at the vibrational ground level (6 ← 0).

While Figure 9 represents spectra which are not scaled to each other for different excitation frequencies, we show in Figure 10 observed and calculated spectra in such a way that their absolute intensities can be compared to each other. Note, that in this Figure only the spectral region between 1260 and 1270 cm^{-1} is displayed. Both, observed and calculated spectra show that the intensity of the S-band heads decreases rapidly for increasing excitation frequencies ω_L. Although the frequency of the ω_L-laser is changed by only about 600 cm^{-1} (this corresponds to a relative change of ω_L of approximately 3 % only!), the CARS signal changes dramatically. This observation, as well as the fact that the S-branches have different relative intensities compared to each other for particular excitation frequencies, are clearly due to the influence of the Franck-Condon overlap integrals. The latter are different for the different vibrational transitions and, of course, change in magnitude when ω_L is varied.

Quantitative calculations of the CARS intensity ($|\chi^{(3)}|^2$) for the three observed S branch transitions (6 ← 0, 7 ← 1, 8 ← 2) are plotted in Figure 11 together with experimentally obtained data for the frequency range 20 000 to 20 800 cm^{-1}. Again one can see, how the intensity of the S(6 ← 0) band (solid line) decreases rapidly when ω_L is changed away from the dissociation limit to higher frequencies. The intensity changes of the other bands (dashed line and pointed lines for the S(7 ← 1) and S(8 ← 2) transitions, respectively) are not so dramatic. The experimental points are plotted also in this figure.

In general, the intensities of the S band heads as demonstrated in Figures 9 - 11 are reproduced fairly well in the calcualted spectra, although there are some slight discrepancies. The observed overall trend in decreasing of the total intensity for increasing excitation frequency as well as the relative changes between different S-band heads for one particular excitation frequency is fully described by the theory.

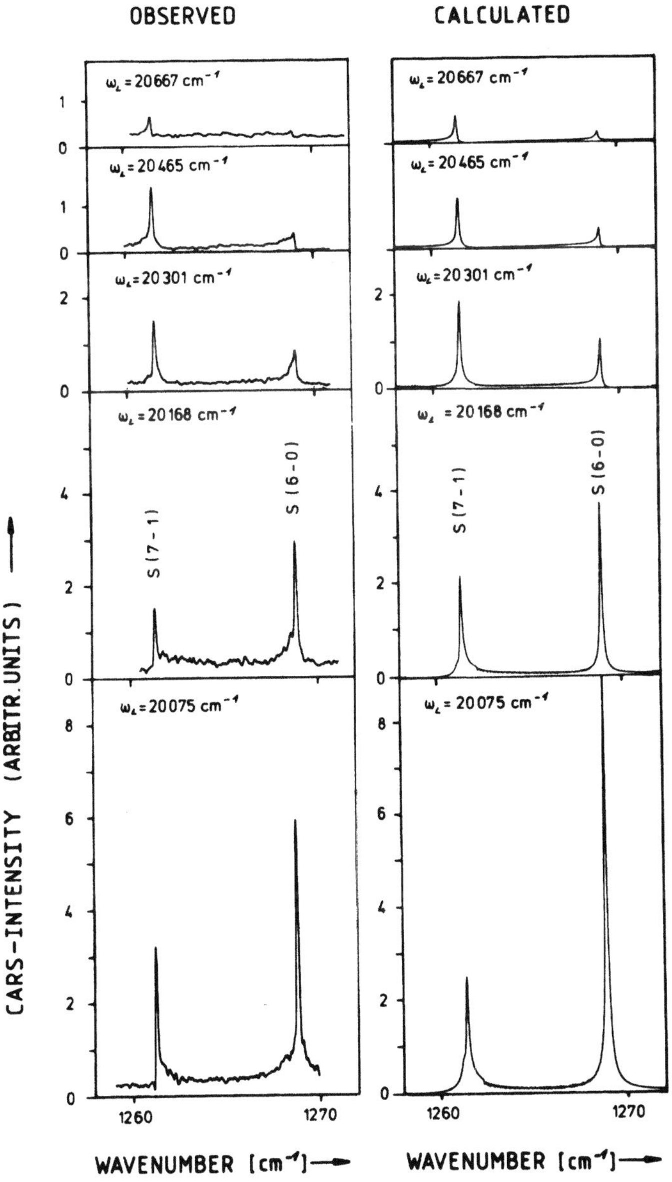

Fig. 10 Experimentally observed (left field) and numerically calculated (right field) continuum resonance CARS spectra of the Δv = 6 transitions in iodine vapor for five different values of ω_L as indicated. The spectra are scaled to each other. Shown are only the 6 ← 0 and 7 ← 1 vibrational transitions.

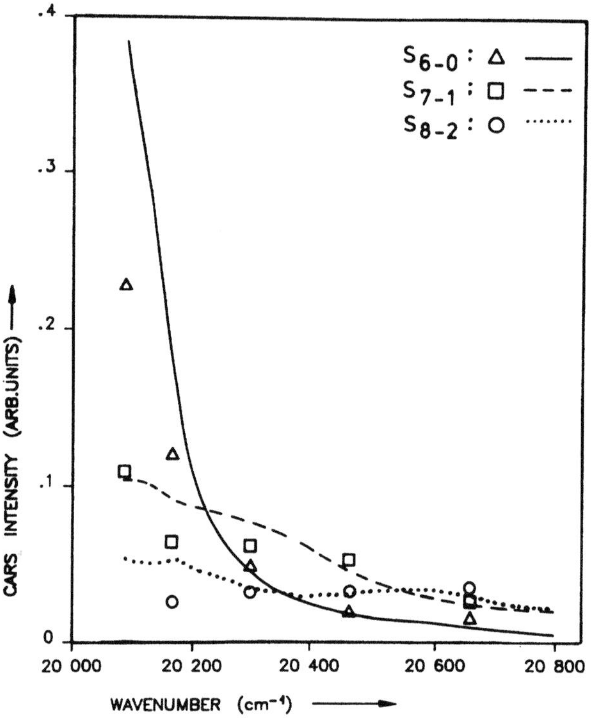

Fig. 11 Excitation profile (CARS intensity versus frequency of ω_L-laser) of continuum resonance CARS lines ($\Delta v = 6$) in iodine vapor. Displayed are calculated (curves) and measured (points) data.

B. Analysis of rotational transitions (S,Q,O)

In continuum resonance (linear) Raman excitation of iodine weak but still noticable bands from Q-branch transitions were observed besides the S-bands (see Figures 2 and 3). As also demonstrated there, only very weak and structureless contributions from O-branches were found experimentally. In order to verify theoretically, that in CARS mainly only the S-branches are observable, we have separately calculated the CARS spectra of rotational-vibrational transitions ($\Delta v = 6$) for the S-, Q- and O-branches. The latter was done for three excitation frequencies (ω_L) which are close to the excitation wavelength used for the continuum resonance (linear) Raman spectra shown in Figures 2 and 3. The calculated profiles for these $\Delta J = 0, \pm 2$ transitions in continuum resonance CARS of iodine are displayed in Figure 12. Nearly no O-branch contribution are to be seen and also the

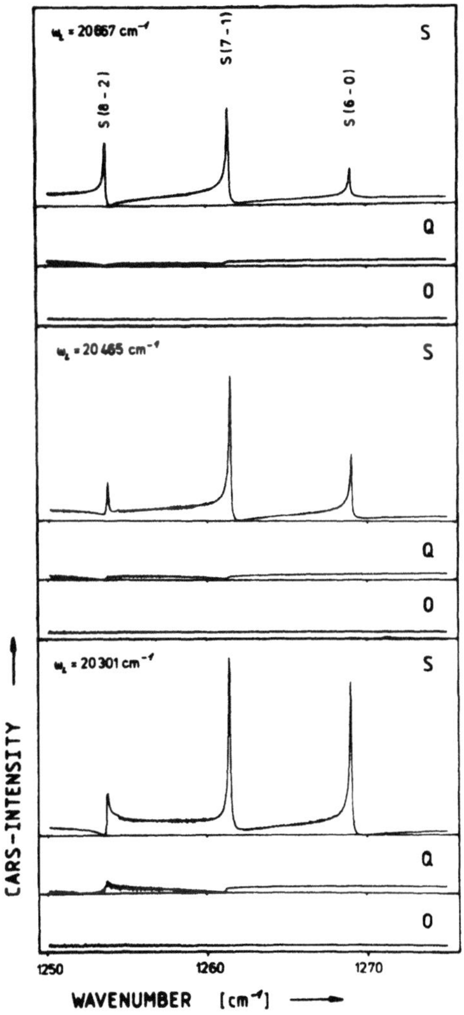

Fig. 12. Calculated profiles for S-, Q- and O-branch transitions in continuum resonance CARS of iodine for three different excitation frequencies (ω_L) as indicated.

contributions of the Q-branches are very weak compared to the one of the S-branches. This again reflects the fact that in CARS only the strong Raman transitions show up in the spectrum with high intensity. Note however, that the band profiles of the pure S-branches as displayed in Figure 12 differ slightly from those shown in Figure 9. This is due to weak interference effects with the remaining contributions from the Q-branches.

C. Contributions from different electronic states (B, $^1\Pi$)

In continuum resonance (linear) Raman excitation in iodine it was found [15-17] that the Raman intensity is determined predominantly through resonance excitation via the continuum of the dissociative part of the B-state. This is due to the quite small transition strength ratio $R = (M^{\Pi})^2/(M^B)^2$ between the two states involved (B, Π). Values for R between 0.17 and 0.20 have been reported in the literature [17,31,32]. For the two-photon process (linear Raman) this ratio comes into the expression for the scattering intensity as a squared quantity (see Equation 6), whereas in the four-photon-CARS process the contributions from the two different states is described by R^4; this dependence

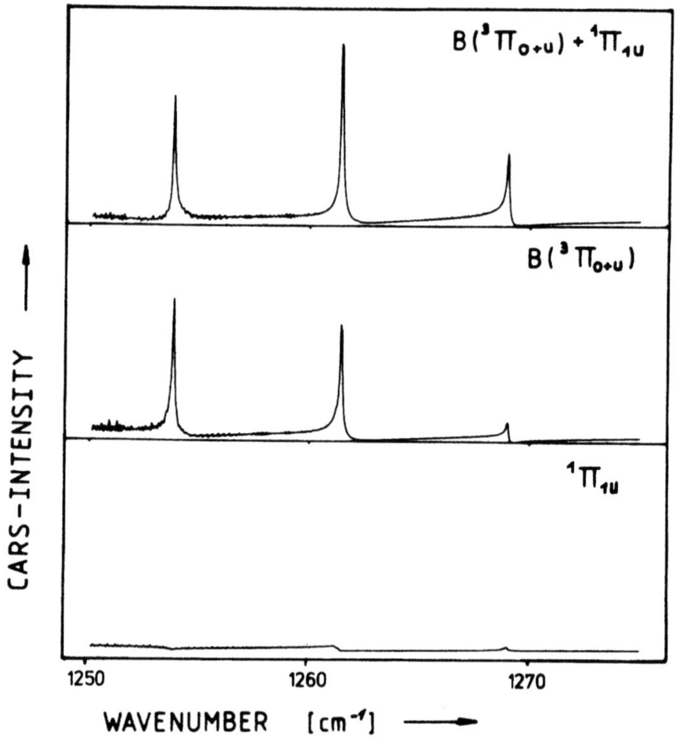

Fig. 13 Calculated continuum resonance CARS spectrum for the $\Delta v = 6$ transitions in iodine vapor ($\omega_L = 20465$ cm^{-1}). Shown is the spectrum arising from both electronic states B and $^1\Pi$ (upper field), the contribution from the B-state alone (middle field) and from the $^1\Pi$-state alone (lower field).

originates from the product of four matrix elements (see Equation 10). So, only very weak contributions from the Π-state can be expected theoretically. In Figure 13 we show in the upper field the calculated CARS spectrum (R = 0.17) when both states are involved. Of course, this situation is present when experiments are performed. Then, the two states are fundamentally and physically indistinguishable. In the middle field a synthetical CARS spectrum is shown, where we have performed our calculations such that the scattering occurs only via the B-state. In the lower field of Figure 13 the theoretical CARS spectrum is shown when only the Π-state would be present. The CARS spectra arising from contributions of the pure $B \leftarrow X$ or $\Pi \leftarrow X$ transitions alone are scaled relative to the combined spectrum (upper field). However, note that the CARS spectrum in the upper field of Figure 13 is not just the algebraic sum of the spectra displayed in the middle and the lower fields. Similar to the two-photon process (linear Raman) [20] also interference terms have to be included for the four-photon CARS process [25]. Nevertheless, Figure 13 clearly demonstrates that continuum resonance CARS in iodine occurs chiefly via the B-state.

D. Contributions from upper Raman states (ρ_f)

In non-resonant CARS studies of vibrational transitions with high-energy (> 1000 cm^{-1}) the contribution of the ρ_f-term (see Equation 10) can be neglected, since for this high energy the population of the upper Raman level is nearly zero at room temperature. In our preliminary report on CARS results of iodine we have also neglected this term for resonance excitation. Although the first-principle calculations of the spectral regions which have also been studied experimentally, in principle showed reasonable agreement, we meanwhile have found, that the inclusion of the ρ_f -term results in much better agreement between experiment and theory. We demonstrate this feature in Figure 14, where we reproduce the same experimental (upper field) and theoretical (lower field, B) CARS spectrum of the $\Delta v = 6$ transition in iodine as shown in Figure 2 of Reference 10) together with a calculated CARS spectrum (middle field, A) which includes the upper Raman state contribution (ρ_f). The spectra are all scaled to the intensity of the $S(7 \leftarrow 1)$ transition at about 1261 cm^{-1}. Comparing the relative intensities of the three S-bands among each other, the spectrum which includes the upper Raman state contributions (A) describes better the experimentally observed spectrum that does the spectrum given in the lower field (B) of Figure 14.
A spectroscopic temperature of T = 1200 K was taken for these calculations (see below).

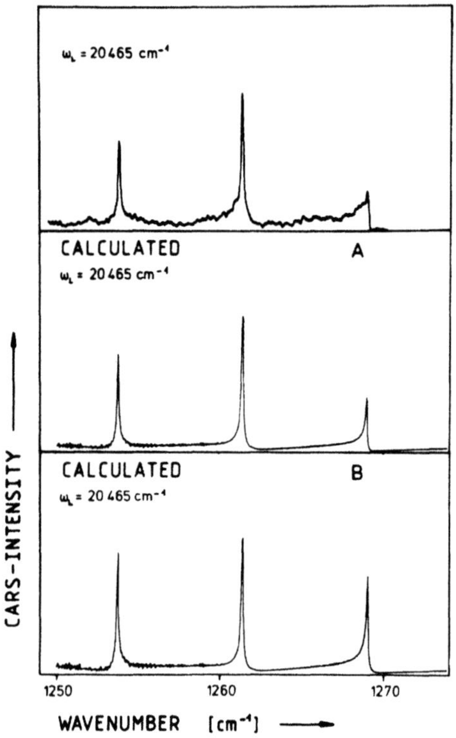

Fig. 14 Observed (upper field) and calculated (middle and lower fields) continuum CARS spectra of the $\Delta v = 6$ transition in iodine. Spectrum B differs from A by the neglection of the upper Raman state contributions (ρ_f) in Equation 10.

E. Variation of spectroscopic temperature (T)

Excitation of continuum resonance (linear) Raman spectra in strongly absorbing gases like I_2 with cw lasers of 1 Watt power revealed quite high spectroscopic temperatures (T). In iodine values for T were found in the order of 1100 K [12] while for the weaker absorbing Br_2 molecule T is of the order of only 700 K [19]. In pulsed CARS experiments with total power of approximately 15 kW the spectroscopic temperature is expected to be high as well, even so the pulses are quite short ($\sim$ 5 μs). Unfortunately , in CARS it is quite difficult to determine experimentally T for an absorbing gas. In principle, it could be derived from the envelope of the pure rotational spectrum of non-absorbing molecules which have been added to the absorbing molecules. This procedure is easy for linear Raman excitation [21], since the detectivity is high enough that only a small portion of foreign

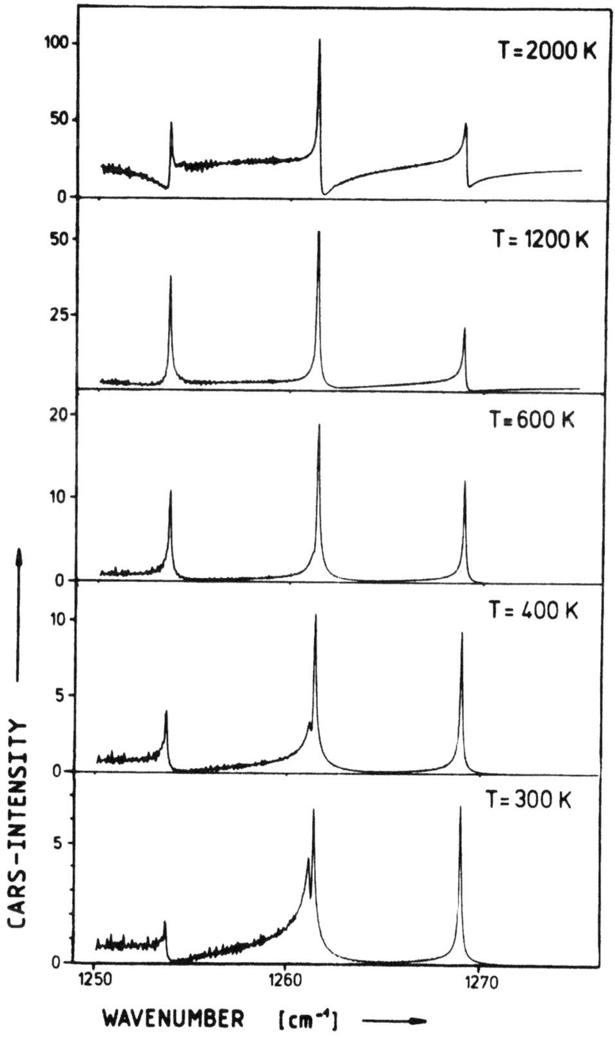

Fig. 15 Calcualted band profiles for continuum resonance CARS excitation of the $\Delta v = 6$ transitions in iodine vapor (ω_L = 20 465 cm^{-1}) as a function of spectroscopic temperature T. The CARS intensities are scaled such, that the strongest line of each spectrum has same value.

molecules has to be added. Presently we have not been able to determine exact values for T directly by a CARS experiment. However, we have performed calculations for different values of T. In Figure 15 we show, how the band profiles vary within the temperature range between 300 and 2000 K for some particular values of T.

Comparing the CARS spectra as shown in Figure 15 with the experimentally ovserved spectrum (Figure 9 second spectrum from above) generated by the same frequency (ω_L = 20 465 cm^{-1}) we can conclude, that the spectrum calculated for T = 1200 K best fits to the observed one. This spectrum fairly well reproduces the observed band shapes. Note for instance the asymmetry of the S(6 ← 0) transition with the characteristic tail on the low-frequency side in both, the observed and calculated spectra for ω_L = 20 465 cm^{-1} excitation (Figure 9).

F. Variation of non-resonant susceptibility (χ_{NR})

In Equation 9 we have lumped together in χ_{NR} all terms of $\chi^{(3)}$ which are due to contributions from other molecules than the one under investigation. The largest contribution to χ_{NR} certainly arises from the coherent signals generated in the glass material of the cell (mainly entrance windows). As mentioned in chapter 4, because of the high absorption of iodine, the effective cell length has to be quite short. Even so lenses with short focal lengthes have been used, it was unavoidable that relatively small contributions to $\chi^{(3)}$ from the window material were present. However, for the narrow spectral region (appr. 20 cm^{-1}) which have been explored experimentally, χ_{NR} is expected to be a constant over this range, although its absolute magnitude may vary depending on the special experimental arrangement. Again it would be difficult to derive reliable information on χ_{NR} from a CARS experiment. Instead a detailed theoretical treatment reveals approximately the magnitude of χ_{NR}. In Figure 16 we have plotted calculated spectra according to Equation 9 but with different values for χ_{NR} as indicated. Inspection of the bandshapes and a comparison with the total set of experimentally observed CARS spectra lead us to an estimation of χ_{NR} to be approximately 0.05. This value is in agreement with the one determined experimentally by measuring the background susceptibility far off any possible rotation-vibration resonance with frequency ω_{if} [10].

Although the value χ_{NR} = 0.05 is small, it still influences slightly the CARS band shapes (compare the spectra in Figure 16 labelled by χ_{NR} = 0.05 and χ_{NR} = 0). For further discussion of χ_{NR} we refer to Reference 25.

G. Variation of Raman bandwidth (Γ)

It would go beyond the scape of this article to give a full description of the dependence of shape and intensity of resonance CARS spectra from the Raman bandwidth Γ (2Γ = FWHM). Here, we only present first-principle calculations of the spectral region

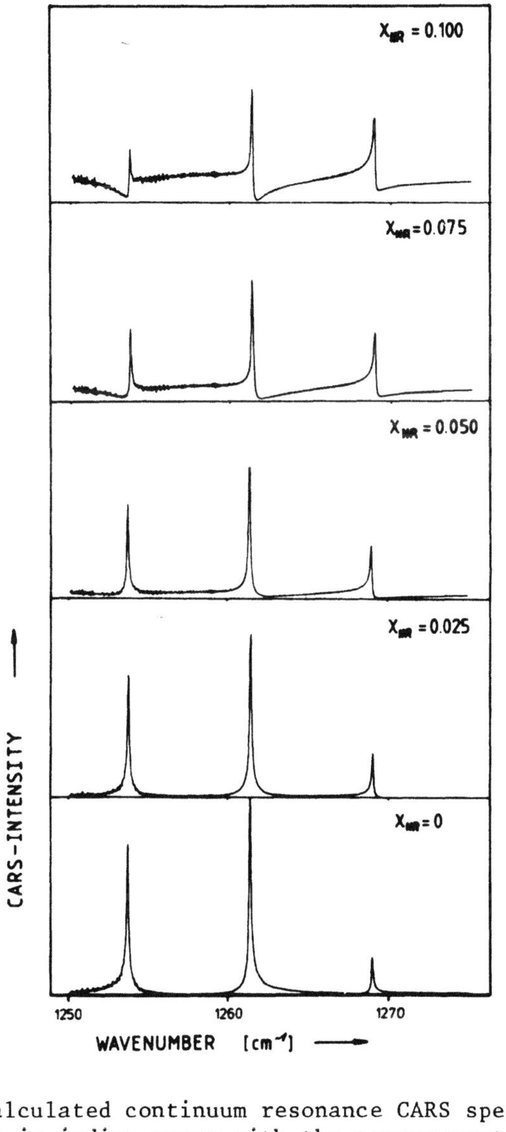

Fig. 16 Calculated continuum resonance CARS spectra of $\Delta v = 6$ transitions in iodine vapor with the nonresonant background susceptibility as a parameter. Spectra are calculated for $\omega_L = 20\ 465\ cm^{-1}$.

in iodine as discussed above as a function of Γ. Figure 17 demonstrates the increase in continuum resonance CARS intensity for decreasing bandwidth Γ. Note further how the bands get very narrow for small Γ's even so these lines represent band-heads which should have a shading towards lower Raman shifts. These

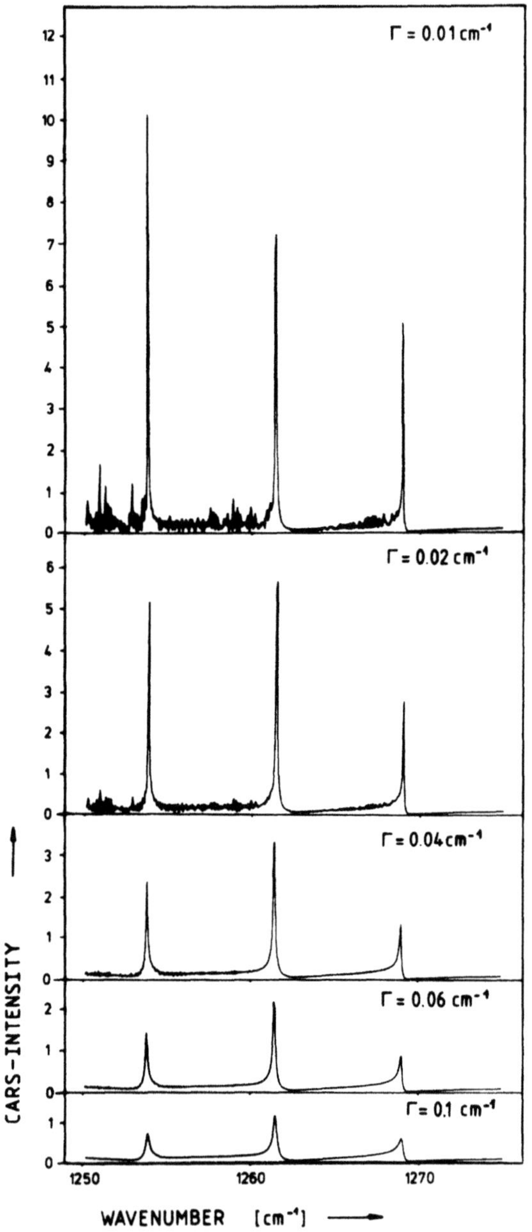

Fig. 17 Calculated continuum resonance CARS spectra of the $\Delta v = 6$ transitions in iodine vapor with the Raman bandwidth Γ as parameter. Spectra are calculated for $\omega_L = 20\ 465\ cm^{-1}$.

narrow lines apparently give the exact positions of the S-band heads for small values of Γ.

For the vapor pressure used in our experiments (see above) wer arrive at a Γ of 0.04 cm^{-1}. This value has been used in the calculation of all other synthetical spectra shown in this article. For further details we again refer to Reference 25.

H. Dispersion of real and imaginary part of $\chi^{(3)}$

In non-resonant CARS, it is customary to separate the CARS susceptibility into a real and an imaginary part [33]:

$$\chi^{(3)}_{CARS} = Re(\chi^{(3)}) + i\ Im(\chi^{(3)}) \tag{10}$$

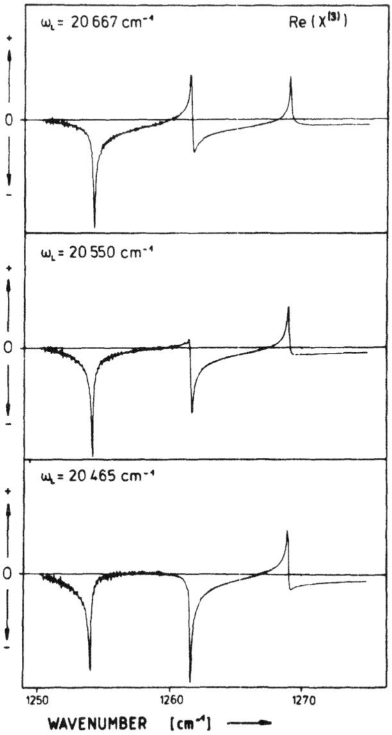

Fig. 18 Dispersion of the real part of $\chi^{(3)}$ in the $\Delta v = 6$ region of iodine vapor calculated for three different values of ω_L, as indicated.

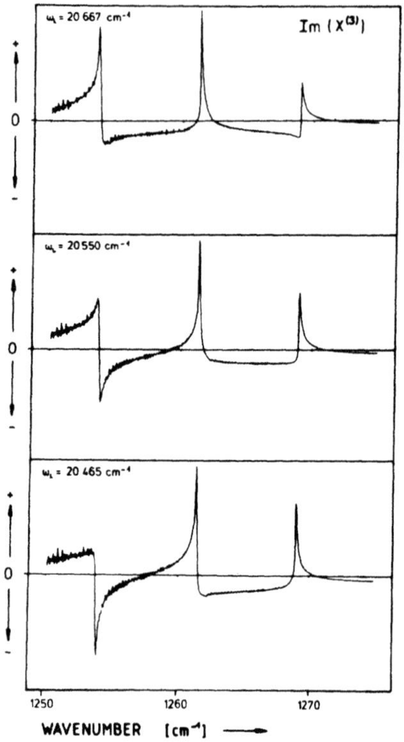

Fig. 19 Dispersion of the imaginary part of $\chi^{(3)}$ in the $\Delta v = 6$ region of iodine vapor calculated for three different values of ω_L, as indicated.

We can apply Equation 10 also for the expression given in Equation 9 of the present article. The final expressions for the functions $Re(\chi^{(3)})$ and $Im(\chi^{(3)})$ will be given elsewhere [25]. Here, we show in Figures 18 and 19 how the numerically calculated dispersion of these functions appear for the $\Delta v = 6$ transitions in iodine. Figure 18 represents the correspondent spectral region for the $Re(\chi^{(3)})$ function determined applying Equations 9 and 10 for three different laser frequencies ω_L, namely 20 667 cm^{-1}, 20 550 cm^{-1}, and 20 465 cm^{-1}. For the same ω_L's and the same spectral region we have plotted the imaginary part of $\chi^{(3)}$, $Im(\chi^{(3)})$, in Figure 19.

Although there is a total change of the value of ω_L of approximately 1 % only between the frequencies labelled in the upper and lower fields of Figures 18 and 19, one can observe in both figures quite a change in the displayed curves for the functions $Re(\chi^{(3)})$ and $Im(\chi^{(3)})$, respectively. As example, we draw attention to the S(7 ← 1) band at about 1261 cm^{-1} in Figure 19. Note how the amplitude of $Re(\chi^{(3)})$ changes from a derivative shaped profile for ω_L = 20 667 cm^{-1} to a complete negative peak when ω_L = 20 465 cm^{-1} is applied. Also the function $Im(\chi^{(3)})$ shows changes if different ω_L's are employed (compare as example the S(8 ← 2) band at about 1254 cm^{-1} in the upper and lower field of Figure 20. For further details see again Reference 25.

6. CONCLUSION

In this article we have outlined a theory for the numerical calculation of $\chi^{(3)}$ and the CARS intensity when excitation is such that both upper intermediate CARS levels are in resonance with continuum states. Also experiments have been performed in iodine vapor in order to check the theory. Fairly good agreement could be achieved for a variety of different excitation frequencies. The developed theory allowed us furthermore (i) to explain why only S-band heads were observed, (ii) to explain that scattering occurs mainly via the continuum of the dissociative B-state only, (iii) to study the influence of contributions from upper Raman states, (iv) to study the intensity and band shape changes with spectroscopic temperature, (v) to study the influence on band shapes due to the magnitude of the non-resonant susceptibility, (vi) to examine how the CARS spectra changes for different values of Raman bandwidth, and (vii) finally to derive the dispersion of real and imaginary part of the third-order susceptibility.

For the study described, iodine has been chosen because this is one of the few molecules which have continuous absorption in a spectral region easy accessable to pulsed dye-laser excitation and from which simultaneously the spectroscopic constants and potential functions are very well known. It is hoped that this model study in such a simple molecule can serve as a basis for more complicated systems.

ACKNOWLEDGEMENTS

The authors wish to thank Mr. H. Fietz who has been involved in setting up the CARS system at the University of Bayreuth and who also participated in recording the CARS spectra of iodine. Financial support from the Deutsche Forschungsgemeinschaft is highly acknowledged. We also thank the Leibnitz Rechenzentrum, Munich, for grants of computer time and W. Griessl, R. Haagen and P. Hurych for technical assistance.

REFERENCES

1. Attal, B., Schnepp, O.O., and Taran, J.P.E., Opt. Commun. 24, 77 (1978).

2. Druet, S.A.J., Attal, B., Gustafson, T.K., and Taran, J.P.E., Phys. Rev. A18, 1529 (1978).

3. Druet, S.A.J., and Taran, J.P.E., in Chemical and Biochemical Applications of Lasers, Vol. 4, p. 187, Moore, C.B., ed., Academic Press, New York, 1979.

4. Druet, S.A.J. and Taran, J.P.E., Progr. Quant. Electron. 7, 1 (1981).

5. Gross, K.P., Guthals, D.M., and Nibler, J.W., J. Chem. Phys. 70, 4673 (1979).

6. Hetherington III, W.M., Korenowski, G.M., and Eisenthal, K.B., Chem. Phys. Letters 77, 275 (1981).

7. Taran, J.P.E., this book, page 281.

8. Guthals, D.M., Gross, K.P., and Nibler, J.W., J. Chem. Phys. 70, 2393 (1979).

9. McIlwain, M.E., and Hindeman, J.C., J. Chem. Phys. 73, 68 (1980).

10. Beckmann, A., Fietz, H., Baierl, P., and Kiefer, W., Chem. Phys. Letters 86, 140 (1982).

11. Holzer, W., Murphy, W.F., and Bernstein, H.J., J. Chem. Phys. 52, 399 (1970).

12. Kiefer, W., and Bernstein, H.J., J. Mol. Spectrosc. 43, 366 (1972).

13. Kiefer, W., unpublished.

14. Kiefer, W., in Advances in Infrared and Raman Spectroscopy, Vol. 3, p. 1, Clark, R.J.H. and Hester, R.E., eds., Heyden & Son, London 1977.

15. Williams, P.F., and Rousseau, D.L., Phys. Rev. Lett. 30, 951 (1973).

16. Rousseau, D.L., and Williams, P.F., J. Chem. Phys. 64, 3519 (1976).

17. Williams, P.F., Fernandez, A., and Rousseau, D.L., Chem. Phys. Lett. 47, 150 (1977).

18. Rousseau, D.L., Friedmann, J.M., and Williams, P.F., in Raman Spectroscopy of Gases and Liquids, Weber, A., ed., Topics in Current Physics, Vol. 11, p. 203, Springer, Berlin 1979.

19. Baierl, P. and Kiefer, W., J. Raman Spectrosc. 3, 353 (1975).

20. Baierl, P., Kiefer, W., Williams, P.F., and Rousseau, D.L., Chem. Phys. Lett. 50, 57 (1977).

21. Baierl, P., and Kiefer, W., J. Raman Spectrosc. 10, 197 (1981).

22. Baierl, P., and Kiefer, W., J. Raman Spectrosc. 11, 393 (1981).

23. Placzek, G., in Handbuch der Radiologie, Marx, E., ed., Vol. VI, part II, page 205, Akad. Verlagsges., Leipzig 1934.

24. Placzek, G., and Teller, E., Z. Physik 81, 209 (1933).

25. Beckmann, A., Baierl, P., Fietz, H. and Kiefer, W., to be published.

26. Fietz, H., Wieser, H., and Kiefer, W., this book, page 335.

27. Fietz, H., Beckmann, A., Hohenberger, H.P., and Kiefer, W., Appl. Spectrosc. 35, 609 (1981).

28. Gerstenkorn, S., and Luc, P., Atlas du Spectre d'Absorption de la Molécule d'Iode, Editions du C.N.R.S., Paris, 1978.

29. Luc, P., J. Mol. Spectrosc. 80, 41 (1980).

30. Barrow, R.F., and Yee, K.K., J. Chem. Soc. Faraday II, 69, 684 (1973).

31. Tellinghuisen, J., J. Chem. Phys. 58, 2821 (1973).

32. Brith, M., Schnepp, O., and Stephens, P., Chem. Phys. Lett. 26, 649 (1974).

33. See references in the article by Kiefer, W. in this book, page 241.

RESONANCE COHERENT ANTI-STOKES RAMAN SPECTROSCOPY IN CONDENSED PHASES

L. A. Carreira and M. L. Horovitz

Department of Chemistry
University of Georgia
Athens, Georgia 30602

Coherent anti-Stokes Raman spectroscopy (CARS) has emerged as a highly efficient technique for obtaining Raman spectra of molecular species. A major drawback to the CARS technique has been a lower limit on solute concentrations of about 0.05 M due to the background signal generated by the nonresonant third order susceptibility of the solvent. This background signal is a very slowly varying function of Raman frequency [1] (see chapter on CARS in condensed phases) and remains nearly constant in intensity for typical experiments. The solute signal at ω_3 can be so small for solute concentrations of less than 0.05 M as to be 'buried' in the background emission. Resonance enhanced CARS has greatly increased the sensitivity of the technique in analytical and structural investigations.

In CARS, when the frequency ω_1 approaches the frequency of an allowed electric dipole transition, ω_e, the tensor elements with denominators containing the frequency differences $\omega_e - \omega_1$ and $\omega_e - \omega_3$ become large. This gives an enhanced scattering amplitude and, therefore, increased intensity at ω_3. Chabay, Klauminzer and Hudson [2] have observed that as ω_1 approaches ω_e the CARS signal was enhanced, whereas the background emission remained virtually unchanged. This allowed the authors to record a low noise, high intensity spectrum of diphenyloctatetraene at a concentration of 1.4×10^{-3} M. More recently [1], the spectrum and excitation profiles of β-carotene in the rigorous resonance region have been recorded at concentrations as low as 5×10^{-7}, which was considerably lower in concentration than previously observed by the CARS technique. Carreira and Goss [3] have demonstrated resonance enhancement of the CARS signals for fluorescent compounds. In addition Horovitz and Carreira [4] have observed resonance enhance-

W. Kiefer and D. A. Long (eds.), Non-Linear Raman Spectroscopy and Its Chemical Applications, 429–443.

ment of the CARS signal for β-carotene at zero frequency Raman shifts. The studies reported here serve to exemplify the power of resonance CARS in obtaining molecular information.

The intensity of a CARS signal is proportional to the square of the third-order susceptibility. The susceptibility is the sum of two terms

$$\chi^{(3)} = \chi_E^{(3)} + \chi_R^{(3)} \tag{1}$$

where χ_E is the electronic susceptibility and χ_R is the Raman susceptibility. $\chi_E^{(3)}$ a slowly varying function of the frequency difference $\omega_1-\omega_2 = \Delta$, and the sum is dominated by the Raman susceptibility when Δ is equal to an allowed Raman frequency ($\Delta = \omega_r$) of the molecule. As shown by Hudson et al. [14] when $\Delta = \omega_r$, $\chi_R^{(3)}$ is given by

$$\chi_R^{(3)} = \frac{N\pi C^4}{ih\omega_1\omega_2^3\Gamma_R}\,\alpha(\omega_1)\,\alpha(\omega_3) \tag{2}$$

where N is the number densitiy and Γ_R the Raman linewidth. The scattering tensor α is given in terms of the Tang and Albrecht theory, modified for CARS [5] by

$$\alpha(\omega_1)_{ij} = A + B \tag{3}$$

where A and B are defined in [5].

For totally symmetric Raman peaks only the A component of the scattering tensor is found to be important. In the case of a nontotally symmetric Raman peak, the B component must be taken into account. One can qualitatively understand the resonance effect by assuming only a single peak with a single upper electronic state. In this case, the summation is reduced to single term multiplied by a weighing factor, the Franck-Condon overlap factor. This results in

$$A \propto \frac{1}{\omega_e-\omega_1-i\Gamma_e} + \frac{1}{\omega_e+\omega_2-i\Gamma_e}\ . \tag{4}$$

It is apparent that if $\omega_1 \approx \omega_e$, the A component would drastically increase resulting in an increase in the intensity of the Raman peak.

The intensity of a Raman peak as a function of the excitation frequency is called an excitation profile. Such profiles have been extensively studied in resonance Raman work. In order to test the theoretical resonance enhancement results for CARS, β-carotene was chosen as a model system. This choice was made because of the extensive work carried out by Inagaki [6] on the excitation profiles for the spontaneous Raman effect. The parameters derived by these workers were found to be directly transferable to the CARS experiment [5]. They also determined that the contribution of the B term was negligible for the symmetric modes ν_1 and ν_2 in the rigorous resonance region. Using a harmonic oscillator approximation, they calculated the separations between the equilibrium positions of the ground and electronic excited states for the ν_1 and ν_2 modes by treating these as adjustable parameters to fit the experimental excitation profiles. These values then reproduced the observed low temperature absorption spectrum when a Lorentzian halfwidth of 250 cm^{-1} was used for electronic absorption lines.

In order to calculate the theoretical CARS excitation profiles, we have used the separation parameters of Inagaki [6] and have increased the electronic halfwidth Γ_e to 500 cm^{-1} since our measurements were made at room temperature. The frequencies in the ground and excited states were assumed to be equal and the summation over ν[3] was taken to ν = 8. Figure 1 shows the calculated and observed excitation profile of the ν_1 line of β-carotene. The experimental profile was standardized to the 992 cm^{-1} line of benzene. The agreement between the experimental points and the curves generated from the spontaneous resonance Raman effect is quite reasonable.

I. RESONANCE CARS SPECTRA OF HIGHLY FLUORESCENT COMPOUNDS

In an effort to illustrate the ease with which resonance enhancement can be applied to highly fluorescent compounds the CARS spectra of six highly fluorescent compounds have been obtained.

The compounds include four laser dyes (rhodamine 6G, rhodamine B, sodium fluorescein and coumarin 495) and two fused ring compounds (perylene and acridine orange). All of the above mentioned compounds have fluorescence quantum yields in the range of .6 to 1.0, excluding any possibility of study by ordinary spontaneous Raman techniques when pumped near their absorption maxima. The spectrum of each compound was obtained in its rigorous resonance region which allowed a relatively low concentration ($\sim 10^{-5}$M) to be used.

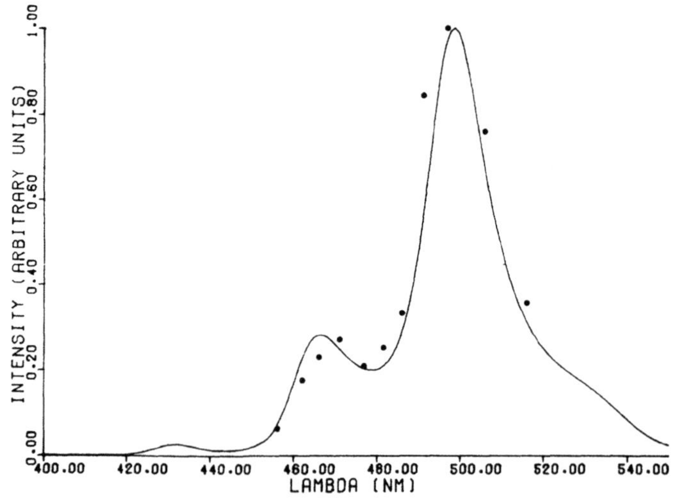

Figure 1. Observed (dotted line) and calculated (solid line) CARS excitation profile for the ν_1 vibrational mode of β-carotene.

Figures 2 and 3 show the CARS spectrum of rhodamine 6G and rhodamine B respectively near their absorption maxima (RG-530 nm; RB-555nm). The spectra obtained by inverse resonance Raman spectroscopy (IRRS) [13] agree qualitatively with those obtained using resonance enhanced CARS. However, a thousand fold greater concentration was necessary to obtain the IRRS spectra.

II. A RESONANCE CARS DETECTOR INCORPORATED INTO AN AUTOMATED HPLC SYSTEM HAVING MULTIPLE DETECTORS

The use of multiple detectors in chromatography for quantitative and qualitative analyses of unknown or complex samples is well known. The combined information gained from two or more detectors, which measure different physical or chemical properties of the solutes, can be much more useful than any of the individual detectors alone.

In high performance liquid chromatography (HPLC) one is often confronted with samples of unknown composition and/or of high complexity. It would, therefore, be very useful to have a HPLC system which would yield several types of information about the samples in the shortest amount of time. Such a system should yield

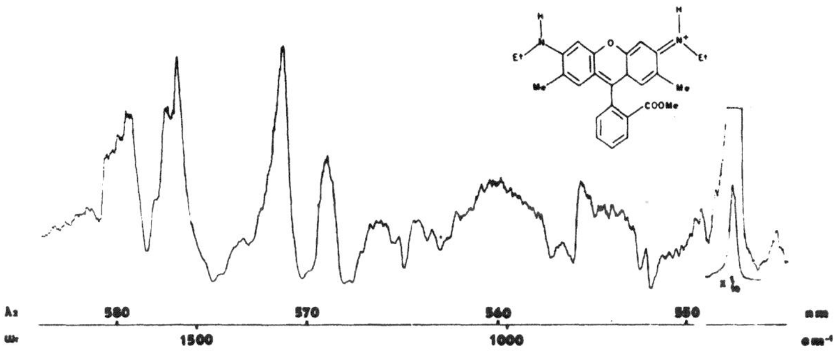

Figure 2. Coherent anti-Stokes Raman spectrum of rhodamine 6G pumped at 530 nm.

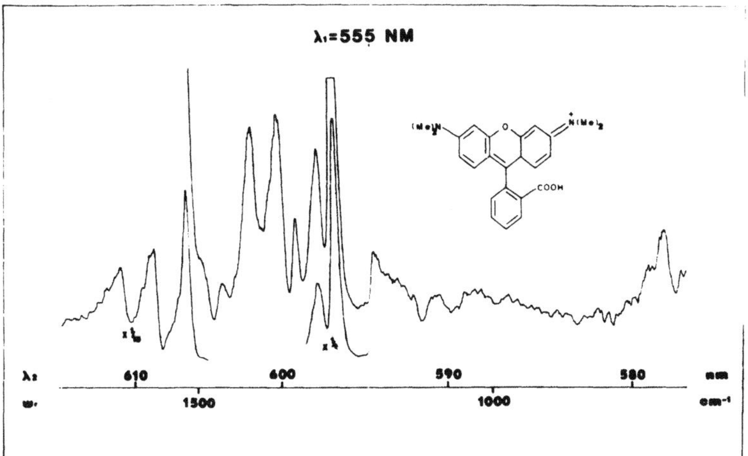

Figure 3. Coherent anti-Stokes Raman spectrum of rhodamine B.

qualitative as well as quantitative information and should be compatible with the multitude of solvents commonly used in HPLC. Klee et al. [7] have developed an automated HPLC system which has the capability for measuring UV/VIS absorption, fluorescence excitation-emission, and CARS spectra of the chromatographic effluent.

Non-resonant enhanced laser Raman has recently been described [8] as a detector for HPLC The authors stated that the choice of eluents was restricted due to interferences from the background signal of the eluent. Coherent anti-Stokes Raman spectroscopy (CARS) gives the same detailed structural information about sample components as does conventional Raman, and both types of Raman provide information that is complementary to that from infrared (IR) absorption. However, CARS is especially applicable to HPLC because it can be used easily with virtually any solvent systems including polar ones such as water and alcohols. This advantage over IR and conventional Raman is realized because resonance enhancement (pumping near to the 0 - 0 electronic transition of a solute) selectively increases the Raman emission of the solute without increasing the background signal due to the solvent. In addition, the CARS emission can also be discriminated from most of the background due to fluorescence [9].

A DEC LSI-11 microcomputer with 28 K words of memory and appropriate hardware for communication with stepper motors, analog devices, and an optical multichannel analyzer was used for experimental control, data acquisition, and data processing. Two Altex model 110A pumps (Altex Scientific, Berkeley, CA) and air-actuated Valco six- and eight-port rotary valves (Valco, Houston, TX) were used to deliver solvents, inject samples or stop eluent flow for stopped-flow acquisition of fluorescence excitation-emission and CARS spectra. The flow rate of each pump (and, therefore, the mobile phase composition) was controlled by 12-bit D/A converters. An Altex model 100 dynamic mixing chamber was used to mix solvents before the injector valve.

The absorption detection system, as illustrated in Fig. 4, consists of the following: a GCA-McPherson tungsten/deuterium light source module, an Altex 151-12 8μl flow-through cell, an ISA HR320 polychromator (Metuchen, NJ), and a PAR 1216 Vidicon multichannel analyzer (Princeton, NJ). The polychromator-Vidicon combination was used in conjunction with a Photochemical Research Associates model ALH-215 xenon arc lamp (London, Ontario, Canada), an ISA H-10 monochromator, and a 20 μl flow-through cell for detection of fluorescence excitation-emission spectra. The use of a 1.5 mm glass capillary for CARS detection has previously been described [10]. An ISA DH20A monochromator and refrigerated photomultiplier tube

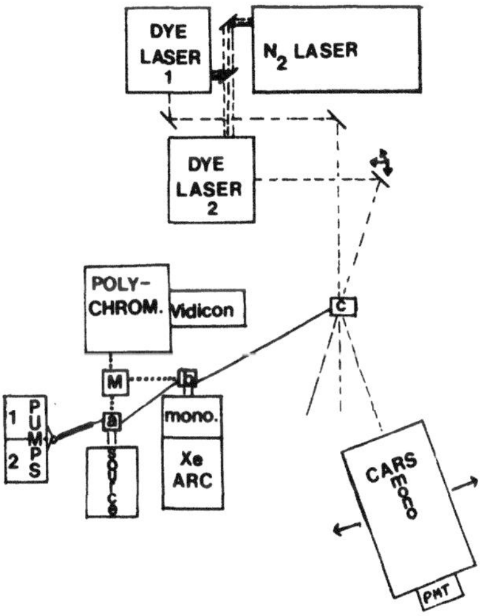

Figure 4. Block diagram of HPLC multiple detector system. Effluent from two HPLC pumps proceeds through: 8 µl absorption cell (a), 20 µl fluorescence cell (b), and CARS detection capillary (c). Mirror (m) allows passage of absorption continuum from UV/VIS source (source) when down and picks up fluorescence emission when up.

housing with a Hamamatsu R928 PMT were used for detecting CARS emissions. A NRG 0.5-5-150/B nitrogen laser (Madison, WI) was used to pump two dye lasers made at the University of Georgia instrument shop. The dye lasers have an estimated resolution of 1 cm^{-1}.

Chemicals

Six acid sulfonate rug dyes, four of whose structures are shown in Fig. 5, were used as model compounds. The structural information concerning the other dyes, A0128 and AR337, is proprietary. They correspond to DuPont Orange R, Conc. 200 %, and Red G "Merpacyl" dyes, respectively. The six dyes were chosen because they are non-volatile and are, therefore, better suited to separation by HPLC than by GC. They are also of environmental concern.

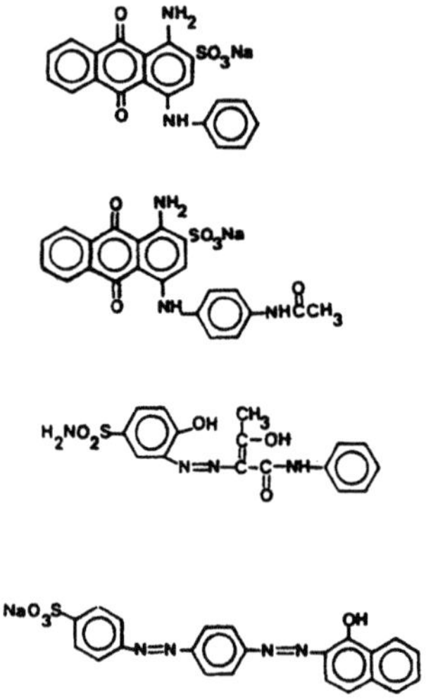

AB25

AB40

AY151

AR157

Figure 5. Molecular structure of four of the six acid sulfonate rug dyes used to assess instrument capabilities.

The software for control of the HPLC-multiple detector system includes several options. In addition to selecting the separation parameters, such as flow rate and gradient type, the user may designate the conditions under which complete fluorescence and/or UV/VIS spectra are to be stored for the eluting components. Spectra can be recorded either when the absorbance exceeds a given threshold value, or only once at each peak maximum. Spectra obtained using the former option, which includes both the leading and trailing edges of chromatographic peaks, can be used either to assess the purity of an individual peak (as in the case of coelution of two sample components) or to increase the signal-to-noise ratio (S/N) of the spectrum of a single component, by addition of successive spectra for that peak. (The same options are possible if fluorescence, rather than UV/VIS absorption, is used as the primary detector).

Fluorescence excitation wavelength and CARS pump wavelength are calculated in real-time based on the absorption data for the eluting components [10]. From a practical standpoint, the optimum pump wavelength cannot always be used, however, due to restrictions on available laser dyes. Since individual dyes can be scanned over only a 10-50 nm range, the suggested pump frequency does not always correspond to a good region for an available dye. In addition, it may be necessary to switch a laser dye for the probe in the middle of a scan in order to acquire CARS spectra for a frequency range greater than approximately 700 cm^{-1}.

Several options must, therefore, be considered in the selection of a practical pump wavelength. These options are important because the CARS signal depends on the square of the power of the pump beam and directly on the power of the probe beam [11]. By changing the pump wavelength from its theoretical optimum for resonance enhancement, one may be able to take advantage of a region of higher power for the probe laser.

Any changes in the software parameters for the CARS spectra to be acquired for eluted components, such as the selection of an alternate pump wavelength, are easily entered prior to the acquisition of individual spectra. After the parameters for obtaining a CARS spectrum have been determined, the program calculates the frequency of the probe laser and the signal based on the vibrational frequency to be measured. The monochromator, used for filtering the CARS emission from any background fluorescence and from scattered light due to the lasers, is scanned in accordance with the change in λ_3 during the acquisition of CARS spectra. The necessary change in the crossing angle of pump and probe beams as the CARS frequency changes is also corrected automatically [12].

The six rug dyes were baseline separated by the methanol/water TBAP gradient. Using the software option to store only one spectrum per peak, visible absorption spectra, from 400 to 650 nm, were taken on the fly every four to five seconds (limited by processor speed and run time of the software). A separation incorporating fluorescence detection on a similar chromatographic system has been reported elsewhere [9].

CARS spectra of three of the rug dyes are shown in Figs. 6-8. The structural information afforded by CARS allows unequivocal identification of the individual dyes providing that they are present in high enough concentration.

The utility of combined detectors used under favorable conditions is illustrated in Figs. 9 and 10. If the two red dyes were to coelute, the visible spectra would not allow identification of individual components. Fig. 10 shows, however, that CARS detection of the leading and trailing edges of this hypothetical peak should be able to differentiate between the individual components.

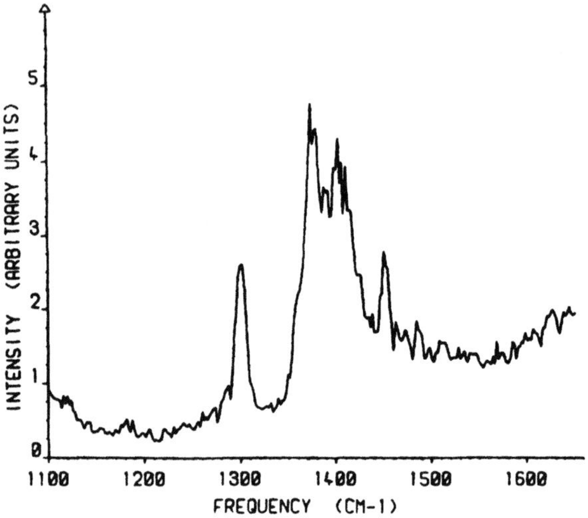

Figure 6. CARS spectrum of dye AR337 from 1100 to 1650 cm^{-1}. Dye was approximately $1.5x10^{-5}M$ in methanol and pump wavelength was 540 nm.

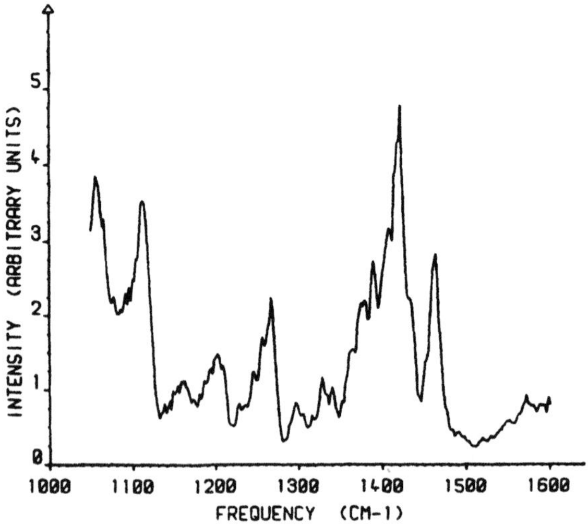

Figure 7. CARS spectrum of dye A0128 from 1100 to 1650 cm^{-1}. Dye was approximately $2x10^{-5}M$ in methanol and pump wavelength was 540 nm.

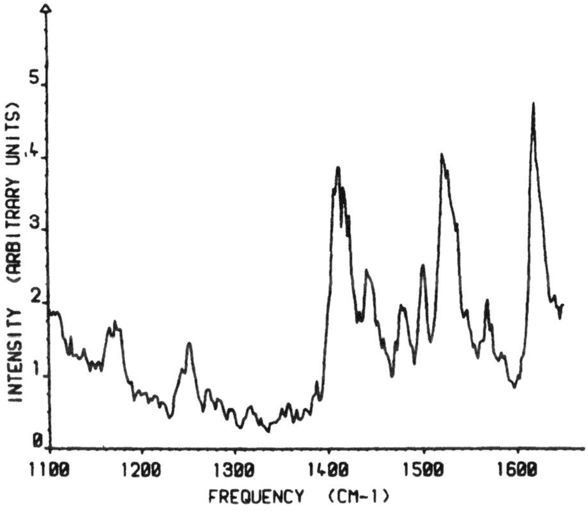

Figure 8. CARS spectrum of dye AR157 from 1100 to 1650 cm^{-1}. Dye was $3.0x10^{-5}M$ in methanol and pump wavelength was 545 nm.

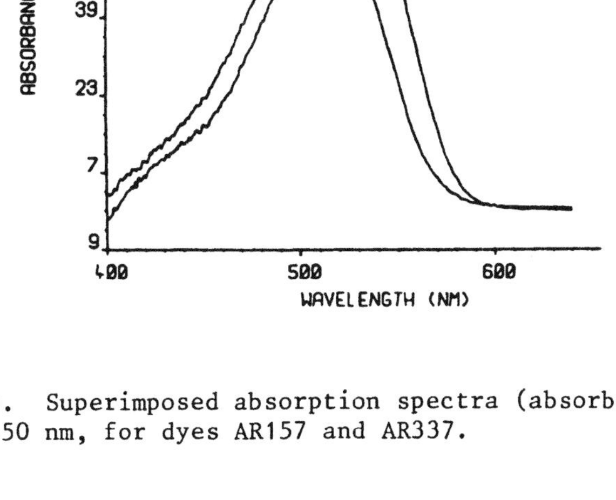

Figure 9. Superimposed absorption spectra (absorbance x 10^2) from 400 to 650 nm, for dyes AR157 and AR337.

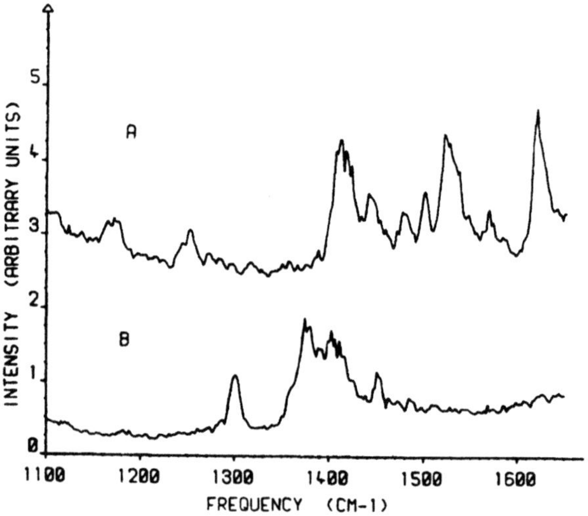

Figure 10. Superimposed CARS spectra from 1100 to 1650 cm^{-1} for dyes AR157 and AR337.

III. LOW FREQUENCY AND ZERO FREQUENCY RESONANCE ENHANCED CARS

In an earlier chapter a new phase-matching technique, Folded BOXCARS (FBCARS) was described. We have incorporated a FBCARS scanning algorithm along with optical modifications to our computer controlled CARS spectrometer and have obtained low and zero frequency resonance enhanced spectra of I_2 and β-carotene respectively.

Figure 11 is a resonance enhanced spectrum in the region of the 216 cm^{-1} peak of $5x10^{-3}$M I_2 in cyclohexane at a pump frequency of 540 nm. The peak is well defined and the stray light is less than 0.5 %.

Figure 12 is a zero frequency spectrum of benzene at a pump frequency of 510 nm. Figure 13 is a zero frequency resonance enhanced spectrum of $1x10^{-5}$M β-carotene in benzene. The latter peak exhibit a very large increase over the pure benzene peak.

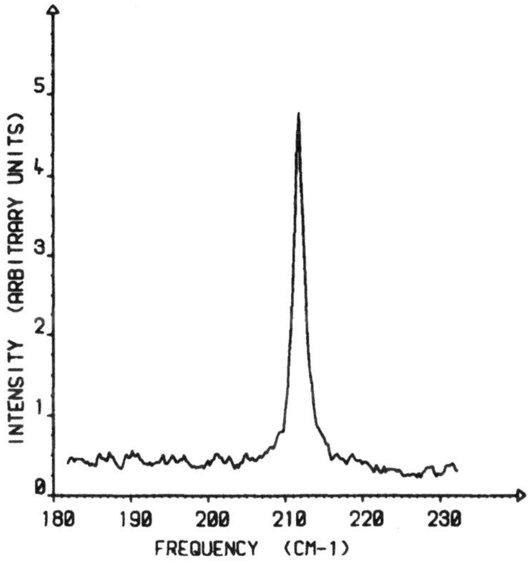

Figure 11. Resonance Folded BOXCARS spectrum of a 1×10^{-2}M solution of I_2 in cyclohexane. ω_1 is 540 nm.

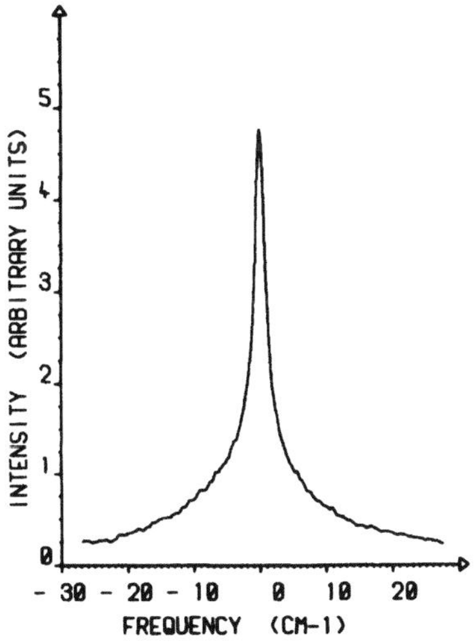

Figure 12. Low frequency spectrum of benzene with a pump frequency of 510 nm.

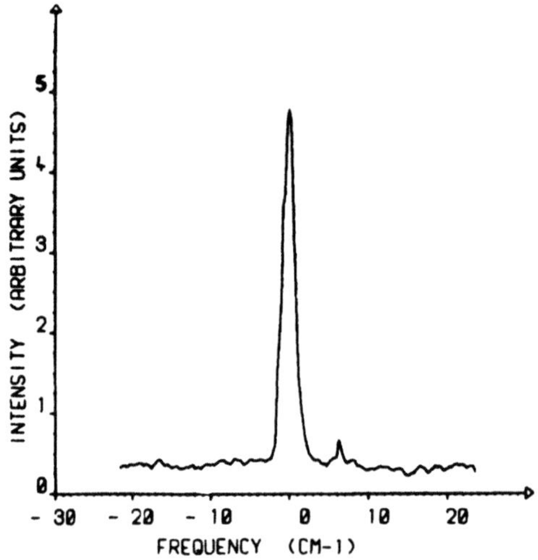

Figure 13. Resonance enhanced low frequency spectrum of $1x10^{-5}M$ β-carotene in benzene. ω_1 is 540 nm.

REFERENCES

1. Chemical Applications of Nonlinear Raman Spectroscopy (Harvey, A.B., ed.), 323, Academic Press, New York, 1981.

2. Chabay, I., Klauminzer, G.K. and Hudson, B.S., Appl. Phys. Lett, 28, 27 (1976).

3. Advances in Laser Chemistry (Zewail, A.H., ed.) Vol. 3, p. 277, Springer Series in Chemical Physics, 1978.

4. Horovitz and Carreira, to be published.

5. Carreira, L.A., Maguire, T.C. and Malloy, T.B., J. Chem. Phys. 69, 855 (1978).

6. Inagaki, F., Tasumi, M. and Miyazawa, T., J. Mol. Spectrosc. 50, 256 (1974).

7. Evaluation of an Automated HPLC System having Multiple Detectors, Kloe, M.S., Rogers, L.B., Carreira, L. A. and Azarraga, L, in Proc. Second Intl. Cong. of Anal. Techniques in Enviromental Chemistry.

8. D'Orazio, M. and Schimpf, U., Anal. Chem., 53(6), 809 (1981).

9. Boutilier, G.D., Irwin, R.M, Antcliff, R.R., Rogers, L.B., Carreira, L.A. and Azarraga, L., Appl. Spec., 35(6), 575 (1981).

10. Rogers, L.B., Stuart, J.D., Goss, L.P., Malloy, T.B. and Carreira, L.A., Anal. Chem., 49(7), 959 (1977).

11. Tolles, N.M., Nibler, J.W., McDonald, J.R. and Harvey, A.B. Appl. Spec., 31(4), 253 (1977).

12. Carreira, L.A., Rogers, L.B., Goss, L.P., Martin, G.W., Irwin, R.M., Von Wandruszka, R., Berkowitz, D.A., Chem. Biomed. Environ. Instru., 10(3), 249 (1980).

13. Carreira, L.A., Antcliff, R. and Horovitz, M.L., to be published.

14. Hudson, B., Heatherton, III, W., Kramer, S., Chabay, I. and Klauminzer, G.K., Proc. Natl. Acad. Sci. U.S.A., 73 3798 (1976).

BIOLOGICAL APPLICATIONS OF RESONANCE CARS

F. W. Schneider

Institut für Physikalische Chemie
Universität Würzburg
Würzburg, West Germany

Among the advantages of CARS regarding biological applications are its high energy conversion, its insensitivity towards fluorescence and its excellent time resolution (1). These strong points should make resonant CARS spectroscopy a preferred method to obtain Raman spectra of fluoresceing biological molecules. Surprisingly, a modest number of biological applications has been reported in the literature as compared with the much larger number of Raman and resonance Raman studies (2). A reason for this paucity of data may be the fact that CARS is not yet a routine method. It requires a judicious selection of biological systems whose spectra should be susceptible to a reasonable interpretation. This is often impossible to achieve with large molecules,a problem equally encountered in spontaneous Raman spectroscopy. Resonance enhancement is indispensible when CARS is applied to dilute aqueous solutions of biological molecules, whose concentrations should be at least 10^{-3} - 10^{-4} m/l in general.

A weak point of CARS becomes apparent at low signal strengths where dispersive line shapes may be observed at low concentrations of the biological molecules. This is due to the non-resonant background of the solvent (most often water in biological systems) which forms cross terms with the (resonant) real and (resonant) imaginary components of the susceptibility of the chromophore. It is the frequency dependence of the cross term involving the resonant real part that leads to dispersion-like curves. The second cross term (non-resonant background with imaginary part) becomes important if the pump and/or the anti-Stokes beam are tuned into an absorption band of the chromophore. This may lead to a negative band appearing in the nonresonant background which may look like a "anti-resonance". Thus it is possible that resonance CARS spec-

W. Kiefer and D. A. Long (eds.), Non-Linear Raman Spectroscopy and Its Chemical Applications, 445–459.

tra may show profiles which range from normal through dispersive to "negative" resonances, when the pump and/or anti-Stokes beams are scanned through an electronic absorption curve (3). Therefore it is advisable to employ sufficiently high solute concentrations whenever possible in order to obtain unique CARS spectra that approach Lorentz curves. Under such conditions the CARS amplitude is proportional to the square of the Raman amplitude.

The measurement of the degree of depolarisation offers another possibility to obtain information about the anisotropy of the molecular polarizability and on the symmetry of a vibrational mode. CARS offers here a greater flexibility than spontaneous Raman, since both pump and Stokes beams may be polarized independently. One keeps the plane of polarization of the Stokes beam constant and turns the plane of the pump beam to a parallel or perpendicular position. The CARS beam is polarized in the same plane as the Stokes beam. It has been shown that the degree of depolarization for CARS is equal to the square of the spontaneous Raman depolarization.

Surprisingly, radiation damage with biological molecules seems to be small, since the energy of the beams is normally in the microjoule and the average power is in the microwatt range. For comparison conventional Raman spectroscopy normally operates at much higher energies and higher average powers (~100 mW). To be safe, the use of a closed circuit flow system is advisable. In the following discussion of some biological systems we assume a basic familiarity with CARS.

CARS spectra of pure nucleic acids, proteins or lipids do not yet exist in the literature although their resonance enhancement in the UV range has now become routinely possible by the use of excimer lasers or frequency doubled dye lasers pumped by a Nd-YAG laser. A number of resonance CARS spectra of chromophors absorbing in the visible or near UV have been determined. Among the latter are porphyrines and hemes, flavines, acridine dyes, and carotenoids, which we would like to briefly discuss in the following.

Porphyrines and Hemes

The first biological chromophors studied by resonance CARS were cytochrome c and vitamin B_{12} by Spiro and coworkers (4). Ferrocytochrome c has two absorption bands at 550 nm (α band) and at 520 nm (β band). In order to make use of resonance enhancement of the pump beam as well as the CARS beam one may tune the pump beam into the α band of a 1 mM aqueous solution in a 1 mm cell and scan the Stokes beam through the range 590-605 nm. The anti-Stokes signal (505-525 nm) is then produced in the β band (Fig. 1).

The coherence length in liquids is approximately 1 mm, i.e. the incident beams must be crossed at a small angle to interact over a distance of about 1 mm to obtain an optimal CARS signal. Beam quality must be high and focussing tight in general.

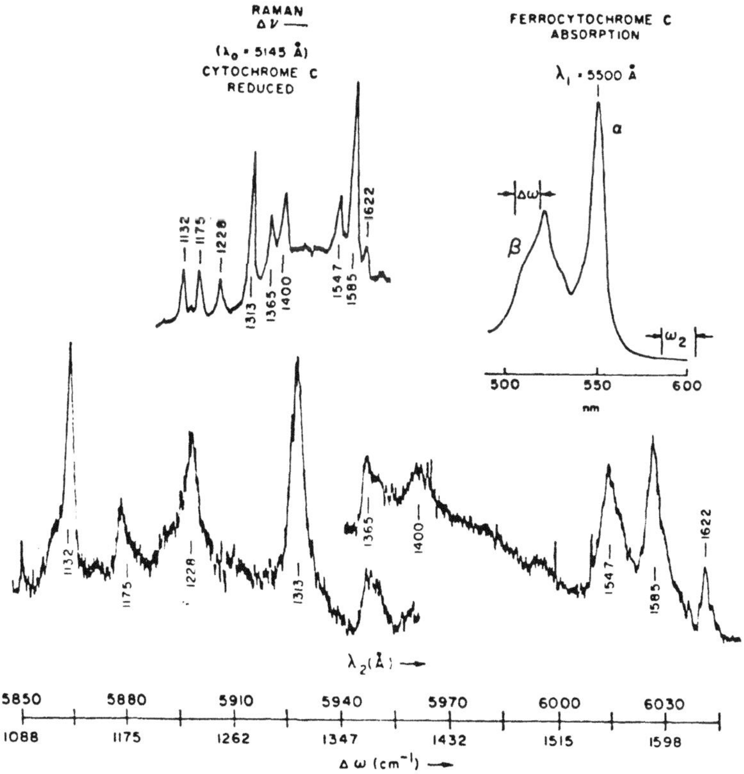

Fig. 1 Resonance CARS spectrum of ferrocytochrome c, 1 mM in H_2O. λ_1 = 5500 Å, λ_2 scan speed = 2 Å/min, laser repetition rate = 10 pulses per sec, 30 pulses averaged. The discontinuity in the spectrum ist due to the limiting tuning range of the apparatus. Insets: resonance Raman and visible absorption spectra of the same solution. (Adapted from ref. (4)).

It is noteworthy that the crossing angle of the two laser beams was kept constant over two relatively wide wavelength ranges in the above work. Optimal phase matching conditions could therefore not be achieved at all wavelength of the spectrum und relative line intensities cannot be used for any theoretical comparisons.

This state of affairs regarding intensities is characteristic for practically all solution CARS spectra unless an independently known substance is used as an internal intensity standard as has been done by Carreira et al (5) in the case of β-carotene with benzene (992 cm^{-1} band). On the other hand, the frequencies of the resulting CARS spectrum show good agreement with the independently known resonance Raman lines. It is advisable, however, to compare only those lines that have positive (Lorentz) profiles. At sufficiently high solute concentrations dispersive CARS line shapes give way to the normal Lorentzian bands and the above frequency problem disappears.

An interesting phenomenon arose when the authors tuned the pump beam through the low energy side of the absorption band towards longer wavelengths while scanning the Stokes beam in such a way as to maintain a constant frequency difference corresponding to a Raman line: a dispersive line shape was produced which eventually became completely negative at the longest off-resonance wavelength. A similar effect was observed with resonance CARS spectra of vitamin B_{12} whose maximum absorption also occurs at 550 nm: a Lorentzian line shape was obtained when the pump beam was practically resonant with the absorption band (555 nm) while a dispersive line shape was recorded at 580-595 nm and a completely negative CARS band originated when the pump wavelength was off-resonance at 615 nm (Fig.2). This change in band profile was successfully interpreted later on the basis of the expression given by Hudson et al (3).

In a kinetic study that shows the time resolution potential of CARS, Dallinger, Nestor and Spiro (6) claim evidence for the movement of iron out of the heme plane within 6 nsec of CO dissociation in CO hemoglobin (COHb). Here the pump beam (6μJ pulse energy) is tuned into the Q_o transition of COHb. Together with the Stokes beam, dissociation of the COHb complex occurs by photolysis during the pulse duration (~6 nsec). Calculations show that most of the COHb complexes are photolyzed. The authors concluded on the basis of the polarization properties of several CARS bands that the nanosecond spectrum produced corresponds to that of deoxyhemeglobin (Fig.3). Therefore the iron atom is assumed to have moved out of the heme plane during the laser pulse after CO dissociation has taken place. In later resonance Raman work (7) it was shown that a photointermediate is formed within about 30 psec during the excitation pulse. This short lived photointermediate is assumed to be a dissociative product of COHb whose iron atom is transiently in the porphyrin plane but has acquired a spin state of $S = 2$ as compared with $S = 0$ of the COHb ground state. More work is necessary to settle the problem of the time required for the iron atom and its imidazole ligand to move out of the heme plane.

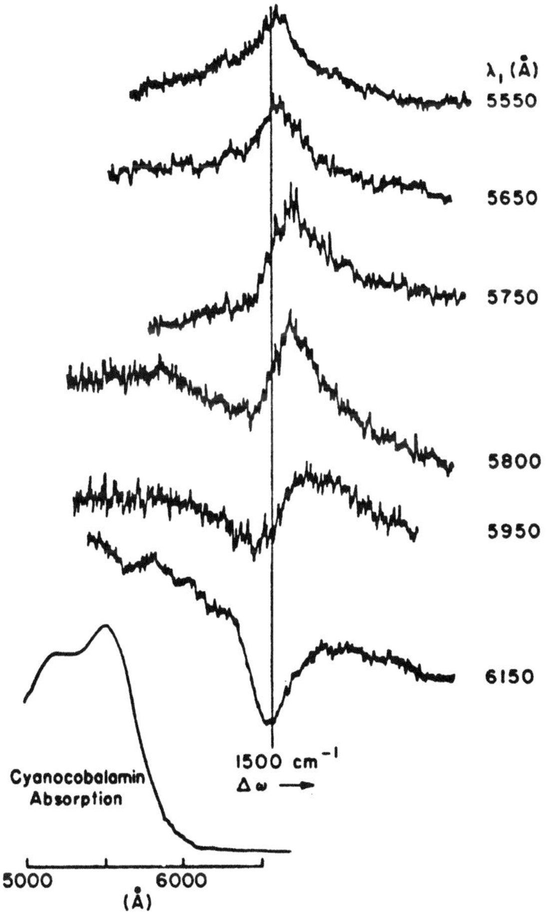

Fig. 2 Resonance CARS scans of the 1500 cm^{-1} band of cyanocobalamin, 1 mM in H_2O, as a function of λ_1 wavelength, λ_2 scan speed = 7.5 Å/min, laser repetition rate = 10 pulses per sec, 10 pulses averaged. (Adapted from ref. (4)).

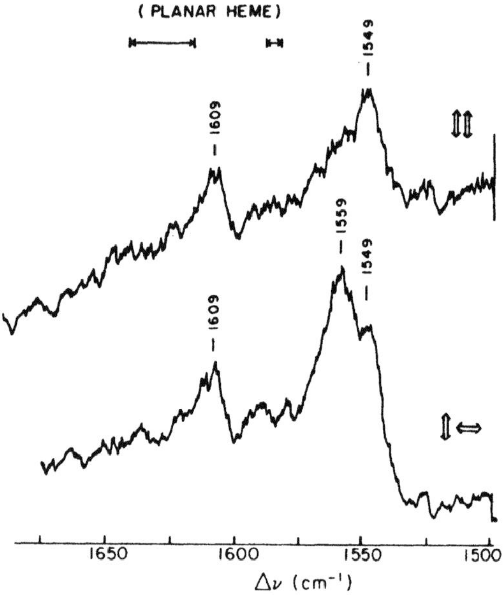

Fig.3 Polarized CARS spectrum of COHb: ⇕⇕ , dye lasers polarized in the same direction; ⇔, dye lasers polarized orthogonal to each other; λ_1 = 569 nm (pulse energy ~6 μJ); λ_2 = 622-629 nm (pulse energy ~20 μJ); [COHb] = 0.69 mM; laser repetition rate = 10 pulses; 100-pulse average; λ_2 scan rate = 0.75 nm/min. Sample in Teflon-stoppered 1 mm path length visible cell. (Adapted from ref. [6]).

Flavins

Flavins occur in the respiratory chain. They act as electron transfering molecules. The chromophore common to flavins is the isoalloxazine ring, which absorbs maximally at 450 nm (Fig.4) and fluoresces at 530 nm with a high quantum yield. Spiro and coworkers have reported the first vibrational frequencies of flavins by measuring the resonance CARS spectra of a number of flavin derivatives such as flavin adenine dinucleotide (FAD), FAD in D_2O, flavin bound to glucose oxidase, riboflavin binding protein in H_2O and D_2O, and flavodoxin [8-10].

Vibrational modes of the tricyclic isoalloxazine were observed at 1635, 1584, 1507, 1416, 1359 and 1255 cm^{-1} (Fig.4) by tuning the pump beam into the long wavelength shoulder at 480 nm and achieving resonance enhancement of the CARS beam at the flavin absorption band at ~450 nm. The isoalloxazine vibrational assign-

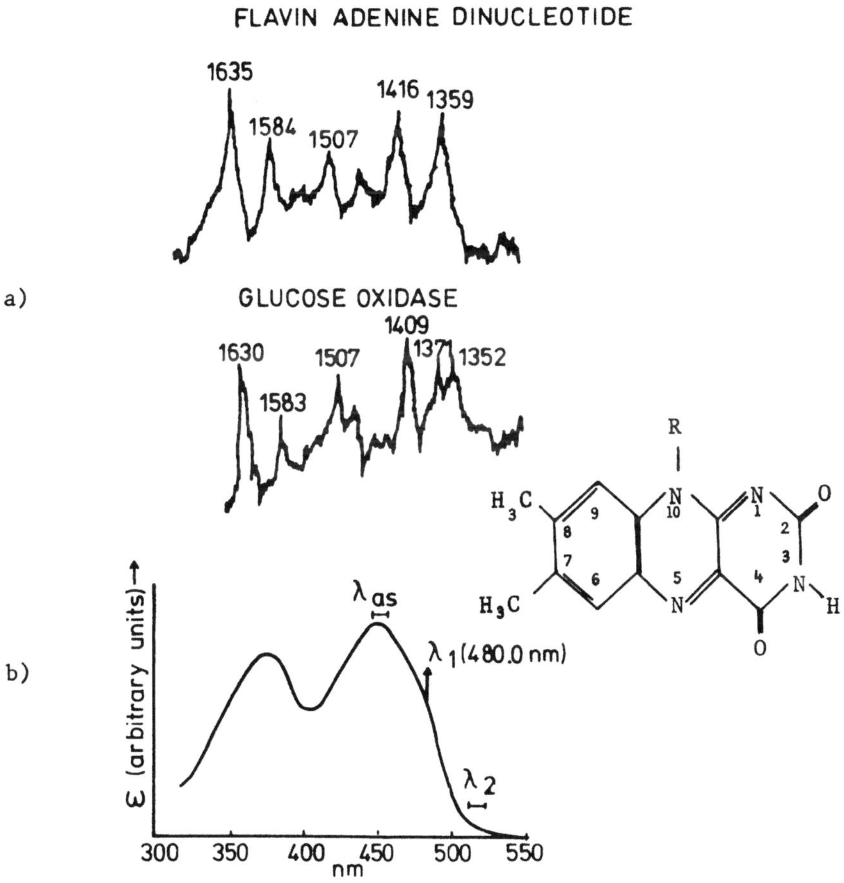

Fig. 4(a) CARS spectra of FAD and glucose oxidase at $\lambda_1 = 480.0$ nm, λ_2 scan speed was 0.6 nm/min, 10 pulses/sec, 30 pulses averaged. Laser pulse energies were ~10-25 μJ. (b) Structure of isoalloxazine ring in flavins, R is the ribose-phosphate-adenosine unit of FAD. Absorption spectrum of FAD with laser and CARS wavelengths (Adapted from ref. [8]).

ments were arrived at in analogy with those of uracil and also by the use of selective substitution of flavins modified by deprotonation at N(3), by CH substitution for N(1), N(3) and N(5) and by CH substitution at the 8 position [11]. When flavin is bound to glucose-oxidase, a frequency shift of 12 cm^{-1} has been observed in one component of the 1359 cm^{-1} band [8]. This shift was taken as evidence for hydrogen bonding of the N(3) proton to an unknown acceptor group of the protein. Furthermore the band at ~1255 cm^{-1} seems to be equally sensitive to H bonding. It shows a downward shift of 10 cm^{-1} in the riboflavin binding protein and disappears in glucose oxidase. More work is necessary to fully interpret these shifts. Evidence for the semiquinone form of flavodoxin was obtained for the first time by Dutta and Spiro (10).

Dutta and Spiro also carried out a line shape analysis of the 1635 cm^{-1} and 601 cm^{-1} vibrational bands of FAD in aqueous solution (12). The origin of the lower energy electronic transition (0-0) of FAD has been assigned from the low temperature spectrum of riboflavin to be at ~475 nm which represents a shoulder on the long wavelength side of the 450 nm band. The maximum absorption band at 450 nm is the envelope of the 0-1 vibronic transition. Satisfactory agreement with Hudson et al's (3) expression for the CARS intensities in the presence of strong interaction between non-resonant background and Raman scattering was obtained. The input parameters are the vibrational and electronic transition energies and widths and the ratio of the Raman to the background susceptibilities. In these calculations only the first term in the summation of the Franck-Condon products was used since higher terms were not necessary, i.e. in the vicinity of the electronic resonance only the ground state vibrational level of the electronically excited state contributes to resonance enhancement. The origin shift between equilibrium positions of ground and electronically excited states turns out to be relatively large for the 1635 cm^{-1} and 601 cm^{-1} band of FAD.

Acridines

It is well known that certain planar chromophores such as the strongly fluoresceing proflavine and acridine orange cations are able to intercalate between the base pairs of a DNA double helix thereby causing certain genetic effects. These dye cations tend to dimerize and self-aggregate in aqueous solution whereas their aggregating tendency is much less pronounced in alcoholic solvents.

Strong resonance enhancement of the CARS signal is achieved by tuning the pump beam into the monomer absorption band of the dye (13). Thus it is possible to detect dye concentrations as low as 10^{-4} m/l. The CARS experiments were carried out with two dye lasers that were pumped by a pulsed N_2 laser (1 MW, pulse width of 2.8 nsec) at 3371 Å. The authors also used a multiplex technique consisting of a narrow bandwidth pump laser and an untuned wide band Stokes laser. At a constant crossing angle an entire CARS spectrum of 300-500 cm^{-1} width is generated which is dispersed by a grating polychromator and detected by a SIT Vidicon targed of an optical multichannel analyzer system. The spectral resolution was ~0.3 A per channel. It is possible to obtain a CARS spectrum in one laser pulse with nanosecond time resolution. For signal to noise enhancement about 500 pulses were accumulated in the OMA system.

The resonance CARS spectra of acridine orange and proflavine monomer cations in methanol show great similarities in the range of 1200-1600 cm^{-1}. There are two strong polarized bands which may be attributed to the symmetric acridine ring modes. The latter are not signifincantly altered by methyl substitution in the amino groups. CARS bands of stretching vibrations in the 3000 cm^{-1} range are not observed, since they are not coupled to the electronic transition moment of the chromophore and therefore are not subject to resonance enhancement.

The frequencies of the intercalated DNA/dye complex in aqueous solution are quite similar to the resonance CARS frequencies of the free cation in methanol (Fig.5). This similarity also

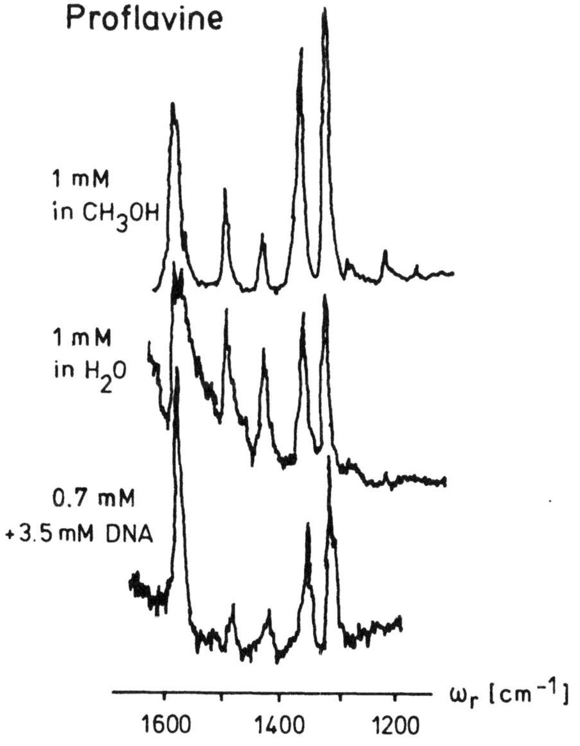

Fig. 5 Resonance CARS spectrum of proflavine in methanol, water and intercalated into DNA; λ_1 = 4900 A, λ_2 scanned; pump laser 7 KW, Stokes laser 5 KW, 30 pulses averaged (Adapted from ref.(11)).

extends to the relative intensities of the bands between aqueous complex and the acridine dissolved in methanol and not between complex and the aqueous solution of the acridine. This indicates that the intercalated dye is in a less polar environment than in aqueous solution. Thus intercalation does not affect the vibrational frequencies of the acridine ring appreciably although effects regarding the out-of plane vibrations might have been expected. On the other hand, intensity changes are observed upon intercalation that are similar to the changes in relative intensities between the dye monomers in aqueous and methanol solutions. Dipole-dipole interactions between the acridine and the base pairs of DNA or with the solvent water, respectively, may also play a role in determining intensity ratios in these systems.

Carotenoids

One of the most thoroughly studied biological molecules by resonant CARS is trans-β-carotene which is a non-fluorescent linear polyene and a constituent of visual pigments. Its resonance CARS spectrum consists of two lines at 1525 cm^{-1} (CC double bond stretching mode) and 1158 cm^{-1} (H_3C-C single bond stretch) together with some weak peaks at 1215, 1193 and 1005 cm^{-1} (5,19). Resonance enhancement may be achieved by tuning the pump laser into the electronic absorption which shows two partially overlapping bands at 453 nm and 480 nm, the latter being the origin of a π-π^*transition. The resulting CARS beam is also resonantly enhanced whereas the Stokes beam does not contribute to resonance enhancement. Energy conversion is particularly high for the carotenoids due to their strong π-electron delocalization. Thus very low concentration CARS spectra have been reported for β-carotene, ~5 x 10^{-7} m/l (5). In fact, saturation effects may occur even at moderate laser powers (~5-10 KW) (5). Therefore it has been necessary to lower the pump power to below 1 KW to obtain resonance CARS spectra in a standard melting point capillary. Carreira et al were able to experimentally determine the CARS excitation profiles of β-carotene for the two stretching modes. They compared the experimental excitation profiles with those calculated according to Albrecht theory (14) which was adapted for CARS by Hudson et al (3). Use was made of the origin shift parameters of the excited state of β-carotene which had been previously determined by Inagaki et al (15) from the excitation profiles for the spontaneous Raman effect. The origin shift parameter is a measure of the displacement of the excited electronic state along the normal coordinate for totally symmetric modes. In subsequent work Dutta et al (16) carried out a line shape analysis using an harmonic oscillator approximation for β-carotene which lead to similarly large shift parameters for the vibrational modes at 1525, 1155 and 1005 cm^{-1} (Fig.6). The relatively large origin shift parameters found for the excited state are consistent with the

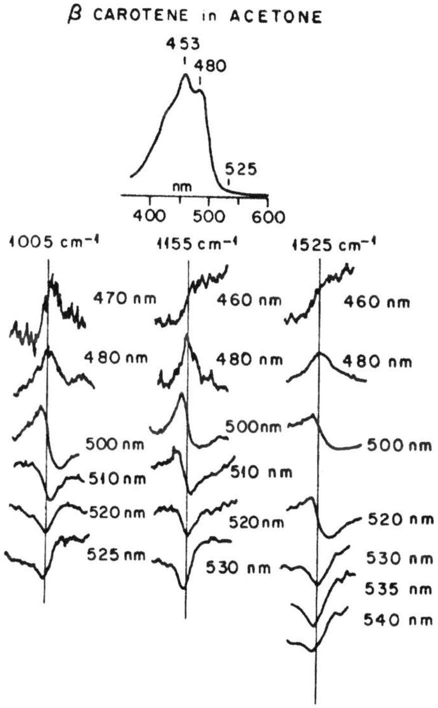

Fig. 6 β-carotene (1.5 x 10^{-5} M) in acetone: absorption spectrum with laser and CARS wavelengths; CARS profiles for the 1005, 1155 and 1525 cm^{-1} vibrations at various given pump wavelengths representing various resonance conditions. (Adapted from ref. [16])

pronounced vibrational structure displayed by the absorption spectrum. A Lorentzian CARS peak is observed near the absorption band at 480 nm, whereas a negative peak occurs ~1.3 vibrational quanta below 480 nm which is also consistent with an appreciable absorption band width (~725 cm^{-1}). However, when neighboring modes are included as "helping" modes in the calculations, the above single mode shift parameters are appreciably reduced as demanded by Warshel and Dauber [17] who calculated significantly lower shift parameters using semiemperical potential functions.

In a similar fashion CARS line shapes were calculated for ferrocytochrome c (aqueous solution) [16], Ni(II)-octaethylporphyrin [Ni(OEP)] (benzene solution) [18] and, earlier, for the 1635 cm^{-1} vibrational band of flavin adenine (FAD) [12]. In the case of Ni(OEP) the CARS and CSRS (Coherent Stokes Raman Scattering) line shapes of non-totally symmetric vibrations were measured and calculated. Here the effect of simultaneous resonance enhancement at the pump- and anti-Stokes frequencies in separate vibronic states has been clearly demonstrated. Therefore, on the basis of the above harmonic oscillator assumption, information on excited state parameters of a chromophore may also be obtained by a CARS line shape analysis. Earlier work on CARS line shapes was done by Lynch et al [20] and Lau et al [21].

Saturation effects may lead to deviations of the CARS signal from the theoretically required square dependence on the power of the pump beam as shown for β-carotene by Carreira et al [5]. However, this effect is only observed under electronic resonance conditions.

Retinal, a carotenoid, is the chromophor of bacteriorhodopsin which acts as a light driven proton pump in halobacterium halobium [22]. Bacteriorhodopsin converts photon energy into chemical free energy in the form of an electrochemical gradient that provides the driving force for the synthesis of ATP. Bacteriorhodopsin exists in two stable forms at room temperature: the light adapted (bR^{LA}) and dark adapted (bR^{DA}) bacteriorhodopsin. Both forms undergo a cyclic transformation whose intermediates have been well characterized for the light adapted form. The laser pulses act here in a two-fold manner, namely as initiators of the photochemical cycle and as the analyzing beams. Since the time scale of the CARS process is about 10^{-15} sec, a species with a longer life time is detectable in principle during a pulse of a longer half width at half maximum than 10^{-15} sec. In the case of the light adapted bacteriorhodopsin the first intermediate is formed within a picosecond and it decays in about 2 µsec at room temperature. Thus the time resolved CARS spectrum is a mixture of two species: bK^{LA}_{570} and bK^{LA}_{610} (Fig.7), whose composition depends on the wavelength and intensity of the laser beams [23]. The resonance CARS spectrum of this mixture shows good agreement with the known resonance Raman spectrum obtained by Marcus and Lewis [24] for bR_{570} at 77 K. Resonance CARS spectra of dark adapted bacteriorhodopsin bR^{DA} are significantly different to those of the light adapted form (Fig.7). In the former the 1530 cm^{-1} peak is broadened and separated into two well resolved maxima at 1528 cm^{-1} and 1547 cm^{-1}. This observation supports the finding that bR^{DA} is a mixture of the 13-cis and the all trans forms. The frequency separation of 20 cm^{-1} may be due to the decrease in electron delocalization in the 13-cis retinylidene chromophore. Further differences between the light and dark adapted nanosecond species exist in the fingerprint region (1100-1400 cm^{-1}).

Thus in the case of the carotenoids CARS has had the function of supporting the spontaneous resonance Raman results. Furthermore it was again demonstrated that CARS is a useful method to obtain time resolved Raman spectra of short lived intermediates.

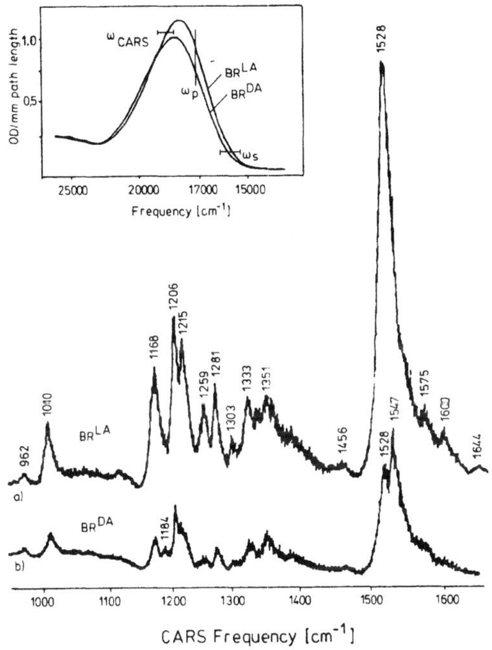

Fig. 7 Resonance CARS nanosecond spectra (scanned) of (a) light-adapted; (b) dark-adapted bacteriorhodopsin. The resonance conditions for the OD 11.2/cm suspension are shown in the inset. Pump frequency (ω_p) was 16936 cm^{-1} (= 5904 A). Stokes frequency (ω_s) was varied from 15260 to 16000 cm^{-1} (= 6550-6250 A). Pulse energies were 2.3 μJ at ω_p and 1.0 μJ at ω_s corresponding to 2.3 and 1.0 KW peak power at ω_p and ω_s, respectively. Spectral bandwidth was 1.0 cm^{-1} for ω_p and 0.5 cm^{-1} for ω_s. Intensity scales are different for both spectra (Adapted from ref. [23]).

The above applications of CARS show that certain phenomena like H-bond formation may be detected in principle even in complex biological systems. The main potential of CARS is its rejection of fluorescence. On the basis of its high energy conversion CARS is well suited for the detection of short-lived intermediates. This capability has been made use of recently in rapid photoisomerizations of vibrationally excited cycloheptatriene in the gas phase [25] or in vibrationally excited molecular fragments in the photodissociation of atmospheric gases [26].

REFERENCES

1. Hudson, B.S., Ann.Rev.Biophys.Bioeng., 6, 135 (1977)

2. Carey, P.R. and Salares, V.R., Raman and Resonance Raman Studies of Biological Systems, in Clark,R.J.H. and Hester, R.E. (eds.): Advances in Infrared and Raman Spectroscopy, Vol. 7, Chapter 1, Heyden, London, 1980.

3. Hudson, B.S., Hetherington III, W., Cramer, S., Chabay, I. and Klauminzer, G.K., Proc.Natl.Acad.Sci. USA,783,3798 (1976)

4. Nestor, J., Spiro, T.G. and Klauminzer, G.K., Proc.Natl.Acad. Sci. USA, 73, 3329 (1976)

5. Carreira, L.A., Maguire, T.C. and Malloy, T.B., Jr., J.Chem. Phys. 66, 2621 (1977)

6. Dallinger, R.F., Nestor, J.R. and Spiro, T.G., J.Am.Chem.Soc. 100, 6251 (1978)

7. Terner,J., Spiro, T.G., Nagumo, M., Nicol, M.F. and El-Sayed, M.A., J.Am.Chem.Soc. 102, 3238 (1980)

8. Dutta, P.K., Nestor, J. and Spiro, T.G., Proc.Natl.Acad.Sci. USA, 74, 4146 (1977)

9. Dutta, P.K., Nestor, J.R. and Spiro, T.G., Biochem.Biophys. Res.Comm., 83, 209 (1978)

10. Dutta, P.K. and Spiro, T.G., Biochemistry, 19, 1590 (1980)

11. Dutta, P.K., Spencer, R., Walsh, C. and Spiro, T.G., Biochim. Biophys.Acta, 623, 77 (1980)

12. Dutta, P.K. and Spiro, T.G., J.Chem.Phys. 69, 3119 (1978)

13. Tretzel, J. and Schneider, F.W., in .Balaban (ed.): Molecular Mechanism of Biological Recognition, Elsevier/North Holland, Biomedical Press, Amsterdam 1979; Chem.Phys.Lett., 59, 514 (1978)

14. Albrecht, A.C., J.Chem.Phys., 34, 1476 (1961)

15. Inagaki, F., Tasumi, M. and Miyazawa, T., J.Mol.Spectrosc., 50, 286 (1974)

16. Dutta, P.K., Dallinger, R. and Spiro, T.G., J.Chem.Phys. 73, 3580 (1980)

17. Warshel, A. and Dauber, P., J.Chem.Phys., 66, 5477 (1977)

18. Igarashi, R., Adachi, Y. and Maeda, S., J.Chem.Phys., 72, 4308 (1980)

19. Nitsch, W. and Kiefer, W., Sov.J.Quantum Electron., 7, 1460 (1977)

20. Lynch, Jr., R.T., Lotem, H. and Bloembergen, N., J.Chem.Phys., 66, 4250 (1977)

21. Lau, A., Pfeiffer, M. and Werncke, W., Optics Commun. 23, 59, (1977)

22. Oesterhelt, D. and Stoeckenius, W., Nature New.Biol., 233, 149 (1971); Oesterhelt, D., Meentzen, M. and Schumann, L., European J.Biochem., 40, 453 (1973); Oesterhelt, D. and Stoeckenius, W., Proc.Natl.Acad.Sci. USA, 70, 2853 (1973); Oesterhelt, D., Angew.Chem.Intern.Ed., B 15, 17 (1976)

23. Tretzel, J. and Schneider, F.W., Chem.Phys.Lett., 66, 475 (1979),and references therein.

24. Marcus, M.A. and Lewis, A., Biochemistry, 17, 4722 (1978)

25. Luther, K. and Wieter, W., J.Chem.Phys. 73, 4131 (1980)

26. Valentini, J.J., Moore, D.S. and Bomse, D.S., Chem.Phys.Lett., 83, 217 (1981)

SURFACE ENHANCED CARS ON METAL SURFACES AND METAL COLLOIDS

F. W. Schneider

Institut für Physikalische Chemie
Universität Würzburg
Würzburg, West Germany

Surface enhanced Raman scattering (SERS) from molecules on metal surfaces has been observed for a variety of systems almost exclusively by the method of spontaneous Raman scattering (1). In remarkable work a coherent Raman method has been used by Shen and coworkers (2,3) who investigated surface CARS of benzene adsorbed in multilayers on a silver surface. These authors interpret their findings as a four wave mixing process of plasmon waves which are confined to a surface layer of silver atoms of a few Ångstrom thickness in contact with bulk benzene. This process occurs in the metal/dielectric interface and it can be understood on the basis of ordinary bulk CARS with surface boundary conditions. The main optical trick in the experiment is to couple the CARS wave out of the surface without great loss of intensity. This has been achieved by the so-called Kretschmann geometry which uses a glass prism on to whose largest side a thin layer of silver has been deposited. The latter is in physical contact with the liquid sample to be investigated (Fig.1).

It may be worthwhile at this moment to briefly summarize the salient features of spontaneous SERS before discussing surface CARS in some detail. SERS has been observed for the first time by Fleischmann et al (4) who discovered a large enhancement of the Raman cross section of pyridine adsorbed at a silver electrode. A large number of SERS experiments by many authors followed this initial observation which unequivocally established SERS to be a reality for several metal surfaces. For extensive reviews see ref. (1). There are two main groups of surfaces that have been used: roughened or flat metal surfaces and colloidally dispersed metal particles providing a high surface to volume ratio. Similarly high enhancement factors of 10^4 - 10^6 over ordinary Raman scat-

W. Kiefer and D. A. Long (eds.), Non-Linear Raman Spectroscopy and Its Chemical Applications, 461–470.

tering were calculated for some of the latter sols on the assumption that a single monolayer of the adsorbed molecules is the cause for the enhancement.

For the interpretation of SERS two groups of theoretical models have been advanced. One approach includes models that deal with classical electromagnetic field enhancement due to resonances of surface plasmon waves caused by the shape of the surface (surface roughness) (5). Other but related models include image field enhancement (6) and Raman reflection (7). The second approach is quantum mechanical in nature. It involves exchange dipole mixing of the electronic states of the adsorbate with the electronic eigenstates of the surface (8). These charge transfer interactions between metal and adsorbate are considered to lead to resonant Raman enhancement. For molecules chemisorbed on a surface, a combination of the physical and chemical approach has been suggested to be responsible in principle (9).

Surface CARS

In the following Shen et al's work (2,3) on surface CARS and their experimental set-up is considered in some detail. For the surface CARS experiment the pump beam was provided by a Q-switched ruby laser, part of which also excited a dye laser system to obtain the tunable Stokes beam. The two pulsed beams were directed at a glass prism which was coated with a silver film on its large side. This metal film was in physical contact with the sample (e.g. benzene). In order to generate the CARS process in the metal/sample interface, the two beams have to be crossed in a fashion shown in figs.1 and 2. Their interaction length on the surface is calculated to be rather short, namely ~10 μm. The incident pump and Stokes beams may generate corresponding plasmon waves on the metal surface. Since the plasmon waves are confined in the xy-plane, their intensities may be significantly higher than those of the pump and Stokes beam themselves. It is the pump plasmon and Stokes plasmon waves that interact on the surface to generate the anti-Stokes plasmon wave. The phase matching condition must be fulfilled for the plasmon waves:

$$(k_a)_{//} = 2(k_1)_{//} - (k_2)_{//}$$

where the $(k_i)_{//}$ are the real components of the plasmon vectors in the xy-plane (fig.2) decribing the surface.

The CARS plasmon wave is coupled out of the surface by the prism and detected as a laser-like beam by a photomultiplier. The phase match is shown graphically by the parallelogram of the wave vector components in the xy-plane in fig. 2. The angles (incidence and xy-plane reflexion) between the wave vectors and the z-axis may be visualized from the figure for the pump (θ_1) and the Stokes

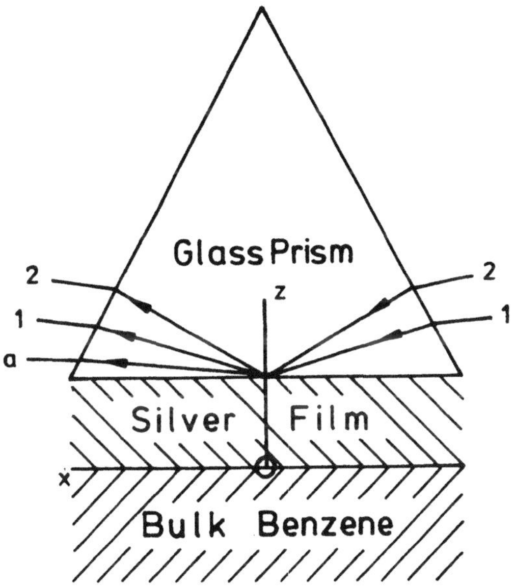

Fig. 1 Prism assembly for surface plasmons undergoing a CARS process in the xy-interface between silver and benzene. Pump beam 1 propagates only in the xz-plane (Adapted from ref. (2) and (3)).

(θ_2) beam propagating in the glass prism. θ_a is the corresponding angle for the anti-Stokes wave vector. For an optimal surface CARS signal, the following conditions must be met: the surface plasmon resonances must be optimally excited by the pump and Stokes fields, respectively, phase match of plasmon waves must be fulfilled and the nonlinear third order susceptibility of the sample must pass through a Raman resonance. Fig. 3 shows the surface CARS signal obtained for the 992 cm^{-1} band of benzene which was used by the authors as the dielectric medium in the experimental set-up described above. The beam energy densities at the crossing point were purposely kept low at 2.5 mJ/cm^{-2} and 25 mJ/cm^2 for the pump and Stokes beams, respectively, in order to prevent damage of the silver film by local heating.

The anti-Stokes power output from the glass prism has been calculated by Shen et al (2) as

$$P(\omega_a) = \frac{\varepsilon_g^{1/2} c}{2\pi} \int |E_a|^2 \, dA$$

where ε_g is the dielectric constant of the glass, $|E_a|^2$ is the intensity of the CARS beam (in glass) at frequency ω_a and A is its cross-sectional area. Their theoretical expression for the intensity $|E_a|^2$ is in very good agreement with experiment if the non-resonant contribution to the third order susceptibility is included in the calculations. The only adjustable parameter used was a scaling factor to normalize the experimental and theorectical amplitudes.

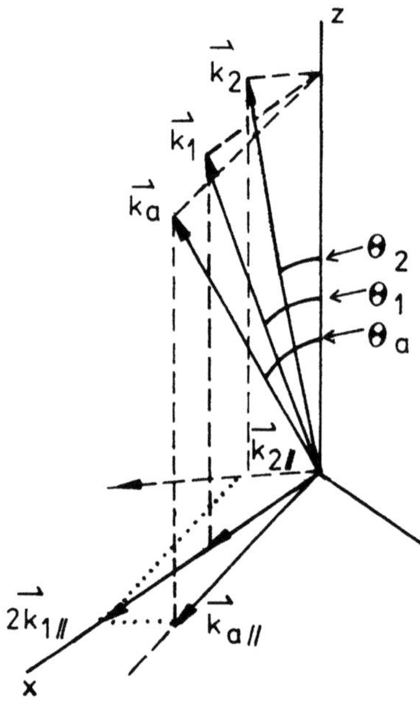

Fig. 2 Wave vectors in the glass prism; only components $\vec{k_i}$ // in the xy-plane must be phase matched (Adapted from ref. (2)).

In order to minimize the effects of unavoidable pulse variations in the laser beams, a bulk CARS experiment was done simultaneously with part of the pump and Stokes beams of the surface CARS set-up. This procedure produced the normalized surface CARS signal for benzene. The experimental anti-Stokes output at the resonance maximum agreed well with theoretical predictions of 2.5×10^5 photons/pulse. Higher CARS signals may be obtained in principle by using picosecond pulses and tighter focusing, since the CARS signal varies as $I_1^2 (\omega_1)$. t.A., where I_1 and I_2 are the pump and Stokes laser intensities at the pump (ω_1) and Stokes (ω_2) frequencies, respectively, t is the pulse duration and A is the cross sectional area of the beam at the interface. Shen et al were able to show by experiment that the surface CARS signal is very

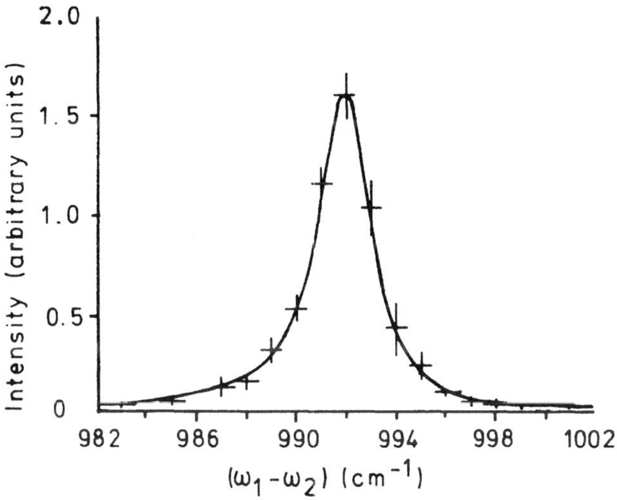

Fig. 3 The 992 cm^{-1} band of benzene by surface CARS. CARS intensity versus Raman frequency ($\omega_1 - \omega_2$), where ω_1 and ω_2 is the frequency of the pump and Stokes laser, respectively (Adapted from ref. [2,3]).

sensitive to the angle between the incident pump beam and the normal to the prism face, whereas the surface CARS signal is surprisingly insensitive towards a phase mismatch. A phase mismatch may be achieved by a controlled variation of the angle of incidence of the Stokes beam if the pump beam angle is kept constant. The latter insensitivity is due to the above mentioned short surface interaction length of ~10 μm which is equal to $1/K''_{//}$, where $K''_{//}$ is the imaginary component of the surface plasmon. It is interesting to note that the total surface CARS signal may originate from about 100 layers of benzene molecules, since the thickness of the benzene layer which leads to surface CARS is approximately 1000 Å (= $\lambda/2\pi$) as calculated on the basis of an exponentially decaying plasmon wave away from the interface. Thus Shen et al have demonstrated that the concept of surface plasmon interaction is consistent with the results of the surface CARS experiment. Any plasmon enhancement factors have not yet been reported for CARS.

Molecules Adsorbed on Metal Colloids

Metal colloids consist of microscopic metal spheres of diameters in the nanometer range. They are usually prepared in situ by reduction of Ag^+ with excess $NaBH_4$ in an aqueous solution in the case of silver colloids (sols). Adsorbate molecules may be added subsequently which causes the metal spheres to further grow or aggregate at a rate which depends on the type and concentration of the added molecules. Metal sols are characterized by a special type of scattering called Mie scattering which is due to resonant excitation of the dipolar and multipolar collective normal modes of the conduction electrons in the metal particles (11). The solutions are colored with extinction bands shifted to longer wavelengths as the particle diameter increases as shown and interpreted already in 1908 by Mie for gold sols (10). Creighton et al (11) were the first to observe surface enhancement by spontaneous Raman spectroscopy of the 1010 cm^{-1} line of pyridine adsorbed on silver and gold particles in aqueous solution. These authors found Raman enhancement of a factor ~100 over the same free pyridine concentration. Enhancement was greatest when the excitation wavelength was close to that of the Mie band. In other words, the Raman excitation spectrum was similar to the Mie spectrum at wavelengths between 475 - 625 nm for silver sols. On the basis of this result they concluded that Raman enhancement of adsorbed pyridine is a consequence of surface plasmon oscillations in a similar fashion as of pyridine adsorbed on roughened silver surfaces.

Kerker and coworkers (12) developed a theory which showed that Raman scattering by molecules adsorbed at the surface of a metal sphere may be enhanced by the order of up to 10^6 by the large local fields near the surface if the radius of the sphere is much less than the wavelength ($< 0.02\ \lambda$) and the relative refractive index of the sphere at either the exciting or the Raman wavelength approaches the value $\sqrt{2}\ i$. As the particle size becomes comparable with the wavelength the enhancement becomes smaller and covers a wider spectral range at longer wavelengths. Thus, for a particle radius of 5 nm the enhancement maximum is calculated to be 10^6 at 382 nm whereas for 50 nm the enhancement maximum is nearly 10^4 and it is shifted to longer wavelengths (~500 nm). The enhancement drops drastically with distance of the molecule form the surface. In order to obtain agreement between Kerker's calculations and Creighton et al's experiments one has to assume that the silver sols were quite polydisperse (diameter 5-500 nm) and partially aggregated to form particle clusters. On the other hand, Creighton et al report that their particles were roughly spherical with diameters in the range 1-50 nm, and for some preparations 1-10 nm as determined by transmission electron microscopy.

If plasmon excitation is responsible for the enhancement process, then one would expect practically no plasmon enhancement in a CARS experiment, since the plasmon CARS process cannot occur on a sphere whose diameter is much less than the required interaction length of ~10 μm on the surface. Any enhancement would then be expected to be of a different type such as due to charge transfer interaction leading to a "normal" resonance CARS effect.

To test this ascertion surface CARS of pyridine molecules adsorbed on silver particles was investigated (13) and compared with the results of Creighton et al's spontaneous Raman work. Similar conditions for the preparation of the silver sols were chosen. For the CARS measurements fresh silver sols were aged for 1-20 hours, depending on the concentration of the subsequently added pyridine to give sols whose extinction maxima occured at ~560 nm. Higher added pyridine concentrations cause the Mie extinction maxima to shift more rapidly to longer wavelengths as a function of time. Therefore it became necessary for higher pyridine concentrations (>40 mM) to use pure silver sols that were aged for a longer time. The reason for employing this particular wavelength is to take advantage of any resonance enhancement of the CARS beam which is generated around 560 nm when the pump wavelength is at about 590 nm and the Stokes beam is at a correspondingly higher wavelength. We were limited with our present laser set-up to these particular wavelengths as the maximal wavelengths of sufficient intensity to generate the CARS process. According to Kerker et al's theory (12) Mie scattering at ~500 nm should provide a plasmon enhancement of 10^3 - 10^4 for particles of ~50 nm diameter. This enhancement should be sufficiently strong to be readily detected under normal CARS conditions.

To obtain CARS spectra a multiplex method was employed which uses an untuned Stokes laser of wide band width (~300 cm^{-1}) in conjunction with a 0.3 m Jarrel-Ash spectrograph and SIT Vidicon OMA detection.

A CARS spectrum of a 0.1 M pyridine solution in water shows strong bands at 1003 cm^{-1} and 1035 cm^{-1}. Pyridine in silver sols showed similarly strong CARS bands at the same frequencies within an experimental error of $\pm$ 3 cm^{-1}. This surprising similarity of frequencies seems to indicate that - unlike spontaneous Raman of pyridine in silver sols (11) and of pyridine adsorbed on silver electrodes - CARS only detects the free pyridine present in the silver sols and not the much smaller amount of adsorbed pyridine. The concentration detection limit for free aqueous pyridine is about 40 mM in our CARS set-up even if several thousand pulses are averaged for S/N enhancement. This CARS signal becomes much smaller than the (constant) nonresonant background of the pyridine-silver sols as the pyridine concentration decreases. Therefore dispersive CARS bands are observed at the lowest pyridine concentration used. This is known to reduce the normal square dependence of the CARS intensity on concentration to an approximately linear one at low concentrations.

A comparison between the CARS spectra of free pyridine solutions and pyridine-silver sol solutions at 40 mM total pyridine concentration shows no apparent enhancement of the pyridine bands in the pyridine-silver sols. The ratio of the CARS signal to the nonresonant background did not depend on the wavelength of the pump beam in the range investigated (420-490 nm). At ~590 nm spontaneous Raman already produced a substantial enhancement (11). Importantly, solutions of 5 mM pyridine in silver sol (2.5×10^{-4} M) showed no CARS spectrum at all except for the nonresonant background. In Creighton et al's work on spontaneous Raman scattering the latter solution had a Raman spectrum whose 1010 cm^{-1} band was up to ~5 times more intense than the corresponding 1003 cm^{-1} band of a 100 mM aqueous pyridine solution depending on excitation wavelength thus establishing an apparent surface enhancement factor of ~10^2 for silver colloids. If one takes into account that only the adsorbed pyridine is responsible for enhancement then the enhancement factor of 10^2 must be multiplied by a factor greater than 5/0.25, since the amount of adsorbed pyridine is less than 0.25 mM. This leads to a real enhancement of $> 2 \times 10^3$ for pyridine adsorbed on silver particles. This enhancement factor as measured with spontaneous Raman is clearly absent in the surface CARS measurements.

A small CARS enhancement (of the order of <20) cannot be ruled out on the following basis: A total pyridine concentration of 40 mM represents a large excess over the silver concentration (2.5×10^{-4}) present. If it is assumed that the surface of all silver particles is completely covered with a monolayer of pyridine molecules, then more than 99.0 % of all pyridine molecules are present as free pyridine in solution. This means that less than 1 part in 100 is due to adsorbed pyridine. If the enhancement factor is assumed to be ~100 for the adsorbed pyridine then an intensity enhancement of less than a factor of 2 would be expected at 40 mM pyridine. This cannot be detected quantitatively at these low concentrations. However, if a lower total pyridine concentration of 5 mM is used similar to Creighton et al's experiments, the fraction of adsorbed pyridine becomes larger with an upper limit of about 5 %. Even lower enhancement factors of only about 20 should then be detectable in the CARS experiment. The fact that 5 mM pyridine in silver sol produced no CARS spectrum indicates that any CARS enhancement factor must be small i.e. less than ~20.

Thus the virtual absence of SERS in CARS of pyridine-silver sols may be taken as indirect evidence for a surface plasmon mechanism which can be readily detected by spontaneous Raman but not by CARS. Other mechanisms such as charge transfer interactions should lead to "normal" resonance enhancement in CARS, which, however, is too small to be detected.

In conclusion the measurement of surface CARS on solid metal surfaces is possible as shown by Shen's work. Surface enhancement by CARS on macroscopic surfaces has not yet been directly demonstrated. On the other hand surface CARS does not seem possible on silver colloids since the metal particles are too small to provide the minimal surface interaction length of ~10 μm necessary for CARS. Thus CARS in conjunction with spontaneous Raman may be diagnostic to distinguish between theoretical SERS models for the adsorbed molecules. The surface CARS results do not exclude a physical model of enhancement such as plasmon excitation as previously postulated by Creighton et al. In fact it is likely that plasmon enhancement is mainly operative in pyridine-silver sol solutions as detected by spontaneous Raman.

The surface CARS method holds promise for the study of thin films and adsorbed molecules whose absorption and fluorescence is unusually large.

REFERENCES

1. For reviews see:
Van Duyne, R.P., in: Chemical and Biochemical Applications of Lasers, Vol. 4, Ed. C.B.Moore, Academic Press, New York, 1978; Burstein, E., Chen, C.Y., Lundquist, S., in: Light Scattering in Solids, Eds. Birman, J.L., Cummins, H.Z., Rebane, K.K., Plenum,New York, 1979; Furtak, T.E., Reyes, J., Surf.Sci.,93, 351 (1980); Otto, A., Applications of Surf.Sci., 6, 309 (1980)

2. Chen, C.K., de Castro, A.R.B., Shen, Y.R., DeMartini, F., Phys.Rev.Letters, 43, 946 (1979)

3. Shen, Y.R., Chen, C.K., de Castro, A.R.B., Proc.VIIth Int. Conf. on Raman Spectroscopy, Ed. Murphy, W.F., North-Holland Publishing Company, p. 659 (1980)

4. Fleischmann, M., Hendra, P.J., McQuillan, A.J., Chem.Phys.Lett., 26, 163 (1974);
McQuillan, A.J., Hendra, P.J., Fleischmann, M., J.Electroanal. Chem., 65, 933 (1975);
Paul, R.L., McQuillan, A.J., Hendra, P.J., Fleischmann, M., J.Electroanal.Chem., 66, 248 (1975)
Fleischmann, M., Hendra, P.J., McQuillan, A.J., Paul, R.L., Reid, E.S., J.Raman Spectrosc., 4, 269 (1976)

5. Mokovits, M., J.Chem.Phys., 69, 4159 (1978);
McCall, S.L., Platzman, P.M., Wolff, P.A., Phys.Letters,77A, 381 (1981);
Hexter, R.M., Albrecht, M.G., Spectrochim.Acta,35A, 233 (1979)

6. King, F.W., Van Duyne, R.P., Schatz, G.C., J.Chem.Phys., 69, 4472 (1978);
Efrima, S., Metiu, H., Chem.Phys.Lett., 60, 59 (1978);
Schatz, G.C., Van Duyne, R.P., Surf.Sci., 101, 425 (1980) and references

7. Otto, A., Surface Sci., 75, L 392 (1978);
McCall, S.L., Platzmann, P.M., Bull.Am.Phys.Soc., 24, 340 (1979);
Maniv, T., Metiu, H., Chem.Phys.Lett., 79, 79 (1981)

8. Robinson, G.W., Chem.Phys.Lett., 76, 191 (1980);
Aussenegg, F.R., Lippitsch, M.E., Chem.Phys.Lett., 59, 214 (1978)

9. Persson, B.N.J., Chem.Phys.Lett., 82, 561 (1981)

10. Mie, G., Ann.Phys., 25, 377 (1908)

11. Creighton, J.A., Blatchford, C.G., Albrecht, M.G., J.Chem.Soc. Faraday II, 75, 790 (1979);
subsequent work: Wetzel, H., Gerischer, H., Chem.Phys.Lett. 76, 460 (1980);
Lippitsch, M.E., Chem.Phys.Lett., 74, 125 (1980);
Kerker, M., Siiman, O., Bumm, L.A., Wang, D.-S., Appl.Opt., 19, 3253 (1980);
Von Raben, K.V., Chang, R.K., Chem.Phys.Lett., 79, 465 (1981)

12. Kerker, M., Wang, D.-S., Chew, H., Appl.Opt., 19, 4159 (1980);
Wang, D.-S., Kerker, M., Phys.Rev. B, 24, 1777 (1981)

13. Schneider, F.W., Gehring, E., Wittmann, E., to be published

INVERSE RAMAN AND RAMAN GAIN SPECTROSCOPY

THEORY OF INVERSE RAMAN AND RAMAN GAIN SPECTROSCOPY

W. J. Jones

University College of Wales
Aberystwyth

Conventionally Raman spectroscopy has been regarded only as a scattering process in which a portion of the monochromatic radiation incident on a molecular sample is scattered inelastically. The differential scattering cross-section, $d\sigma/d\Omega$, relates to the power scattered into a given element of solid angle $d\Omega$, σ measuring the rate at which energy is removed from the incident beam by scattering relative to the rate at which energy crosses a unit area perpendicular to the direction of propagation of the incident beam [1]. σ and $d\sigma/d\Omega$ are usually related to a single molecule and a unit wavenumber interval and the differential Raman scattering cross-sections relate to the relevant transition polarisability components [1]. The observation of the stimulated Raman effect by Woodbury and Ng [2] and the interpretation of the effect by Hellwarth [3] shortly after the construction of the first laser, the ruby laser [4], provided an indication of the benefits that were to accrue as a result of the power density available from these radical new light sources. Shortly thereafter Jones and Stoicheff [5] using a "giant pulse" ruby laser demonstrated the process to which they gave the name Inverse Raman Effect. This process occurs when two coincident beams of monochromatic radiation are incident on a Raman medium, the frequency difference of the two beams coinciding with a Raman active transition in the molecules under study. In this process, when there are more molecules in the lower of the two vibrational states coupled by the process, the lower frequency radiation is amplified with energy being abstracted from the higher frequency radiation. Essentially the system behaves as an amplifier of Raman radiation [6] (Figure 1) and the process bears the same relationship to the stimulated

W. Kiefer and D. A. Long (eds.), Non-Linear Raman Spectroscopy and Its Chemical Applications, 473–485.

Raman effect as the laser amplifier bears to the laser oscillator.

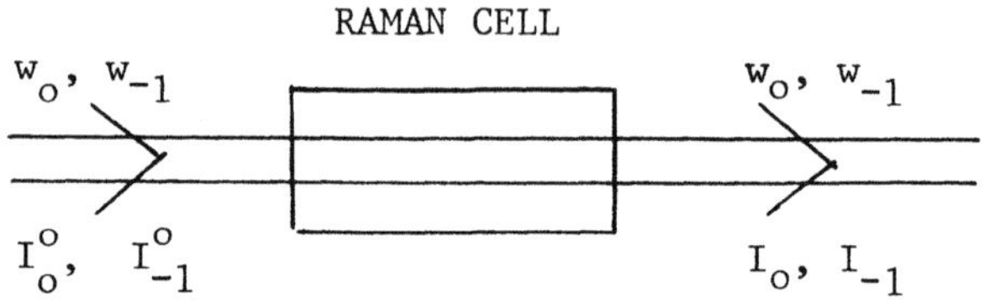

Fig. 1. The Raman Amplifier. w_o and w_{-1} are the frequencies of the two incident laser beams with $w_o > w_{-1}$. As a result of the interaction process occurring at resonance $I_{-1} > I_{-1}^o$ and $I_o < I_o^o$.

The term Raman Amplification Spectroscopy provides a general descriptive name for the overall process taking place, the terms Inverse Raman (Raman Loss), Raman Gain, Photoacoustic Raman Spectroscopy, relate specifically to the techniques employed for observing or enhancing the observable effect of the amplification process.

Initial studies of the phenomenon sought to record the absorption of photons from a continuum synchronised in time and space with intense pulses of monochromatic laser radiation, as in the initial study of Jones and Stoicheff [5], photographic detection being generally employed. These studies were successful but, because of the difficulty of recording small changes of intensity of the background continuum in the Inverse Raman process, the procedures developed were rapidly superseded by the related CARS process which, being largely background free, exhibited significantly greater potential for future development. With the advent of tunable laser sources in the visible region in recent years the potential for further development of Raman amplification spectroscopy has become apparent. In order to appreciate the value of this type of spectroscopy it is necessary to consider the theory in some detail.

The propagation of light waves in a medium is described [6] by the wave equation for the electric field of the radiation E

$$\nabla^2 E - \frac{n^2}{c^2}\frac{\partial^2 E}{\partial t^2} = \frac{4\pi}{c^2}\frac{\partial^2 P^{NL}}{\partial t^2} \quad (1)$$

where P^{NL} is the non-linear polarisation, n the refractive index and c the velocity of light. It is assumed that the electric field is described by a plane wave with varying amplitude and phase propagating in the z direction. For consideration of Raman amplification spectroscopy

$$E(z,t) = \tfrac{1}{2}\{E_o \exp i(k_o z - w_o t) + E_{-1} \exp i(k_{-1} z - w_{-1} t) + c.c.\} \quad (2)$$

k and w are the wave vector and frequency of the light field, respectively, the o and -1 right subscripts refer to the higher and lower laser frequencies, and c.c. represents the corresponding complex conjugates.

The non-linear polarisation is written as

$$P^{NL}(z,t) = \tfrac{1}{2}\{P_o^{NL}\exp(-\, i\, w_o t) + P_{-1}^{NL}\exp(-\, i\, w_{-1} t) + c.c.\} \quad (3)$$

From equations (1) - (3) one obtains

$$\frac{n_{-1}}{c}\frac{\partial E_{-1}}{\partial t} + \frac{\partial E_{-1}}{\partial z} = \frac{2\pi i w_{-1}}{c\, n_{-1}} P_{-1}^{NL}\exp(-\, i\, k_{-1}\, z), \quad (4)$$

with an equivalent expression for the higher frequency laser field E_o.

In the absence of this non-linear polarisation, homogeneous plane waves in the medium propagate independently of each other [7]. The non-linearity of the dielectric medium provides a coupling between different homogeneous plane waves, the lowest order non-vanishing term being a cubic function of the electric field amplitudes. This polarisation can be expressed as

$$P^{NL}(z,t) = \chi^{(3)}\ E(z,t).E(z,t)\ E(z,t) \tag{5}$$

χ, the third order non-linear susceptibility, is a complex fourth rank tensor with elements $\chi^{(3)}_{ijk\ell}$ which depend on the frequencies present. The i'th Cartesian component of the non-linear polarisation at the Stokes frequency is given by

$$P^{NL}_{-1i} = \chi^{(3)}_{ijk\ell}(-w_{-1},w_o,-w_o,w_{-1})\ E_{oj}.E^{*}_{ok}\ E_{-1\ell}\exp(ik_{-1}z) \tag{6}$$

The term $\chi^{(3)}_{ijk\ell}(-w_{-1},w_o,-w_o,w_{-1})$ describes a polarisation at the frequency w_{-1} which is proportional to the field at the same frequency E_{-1} and also to the intensity $|E_o|^2$ of a wave at frequency w_o. For this circumstance there is an exact momentum matching condition $k_{-1} - k_{-1} + k_o - k_o = o$. The subscripts i,j,k,ℓ denote the polarisation directions of the generated and incident electric fields E_o and E_{-1}. The expression for the non-linear polarisation at the Stokes frequency given by equation (6) leads to an exponential amplification of the Stokes wave according to

$$\frac{\partial E_{-1}}{\partial z} = -\frac{2\pi w_{-1}}{n_{-1}c}\chi^{(3)}_{ijk\ell}|E_o|^2\ E_{-1} \tag{7}$$

Since the intensity and the amplitude of the wave are related by the expression $I = \frac{nc}{8\pi}E^2$, we have that

$$\frac{\partial I_{-1}}{I_{-1}} = 2\frac{\partial E_{-1}}{E_{-1}}, \tag{8}$$

and the amplitude, as well as the intensity, of the Stokes wave are amplified according to the expressions

$$E_{-1} = E^o_{-1}\exp\left(\tfrac{1}{2}\ g\ I_o\ z\right) \tag{9a}$$

$$I_{-1} = I^o_{-1}\exp\left(g\ I_o\ z\right) \tag{9b}$$

where z is the distance traversed by the beams through the Raman sample and E^o_{-1} and I^o_{-1} represent the amplitude and intensity of the incident wave. The gain factor g is given by

$$g = - \frac{32\pi^2 w_{-1}}{n_{-1} n_o c^2} \chi^{(3)}_{ijk\ell}(-w_{-1}, w_o, -w_o, w_{-1}) \qquad (10)$$

In Raman amplification there are two contributions to the total non-linear susceptibility $\chi^{(3)}$ of equation (10): from non-resonant electronic processes, χ^{NR}, as well as from the Raman amplification process, χ^R, with which we are concerned. The Raman contribution to the non-linear susceptibility is derived from the differential equations describing the Raman active vibrations [6]. This Raman susceptibility is given by

$$\chi^R_{ijk\ell}(-w_{-1}, w_o, -w_o, w_{-1}) = \frac{N\Delta}{24\hbar} \left[\frac{\alpha^R_{ij} (\alpha^R_{k\ell})^*}{w_R - (w_o - w_{-1}) + i\Gamma/2} \right], \qquad (11)$$

where N is the number of molecules per unit volume, Δ the fractional population difference $(N_a - N_b)/N$ between the two states a and b involved in the Raman transition of frequency w_R, and Γ the full width at half maximum for the spontaneous Raman line. The Raman matrix elements, α^R_{ij}, relate to the differential spontaneous Raman scattering cross sections by the equations

$$|\langle a|\alpha^R_{ij}|b\rangle|^2 = \left(\frac{c}{w_{-1}}\right)^4 \left(\frac{d\sigma}{d\Omega}\right)_{ij} \qquad (12)$$

For liquids and solids the susceptibility is averaged over all orientations of the molecules.

In the usual experimental conditions employed in Raman Gain spectroscopy the pump (w_o) and probe (w_{-1}) lasers are polarised in the same direction, the relevant Raman susceptibility term being given by

$$\chi^R_{iiii}(-w_{-1}, w_o, -w_o, w_{-1}) = \frac{N\Delta}{24\hbar} \left(\frac{c}{w_{-1}}\right)^4 \left[\frac{(d\sigma/d\Omega)_{ii}}{w_R - (w_o - w_{-1}) + i\Gamma/2} \right] \qquad (13)$$

and $(d\sigma/d\Omega)_{ii}$ is related to the total Raman scattering cross section by the dipolarisation ratio ρ [6].

The real and negative imaginary parts of this Raman susceptibility are obtained, normalised to unity, by plotting χ^R_{iiii} in equation (13) as a function of the mismatch of the laser frequencies from the resonance condition ($w_R = (w_o - w_{-1})$). Such a plot is shown in Figure 2

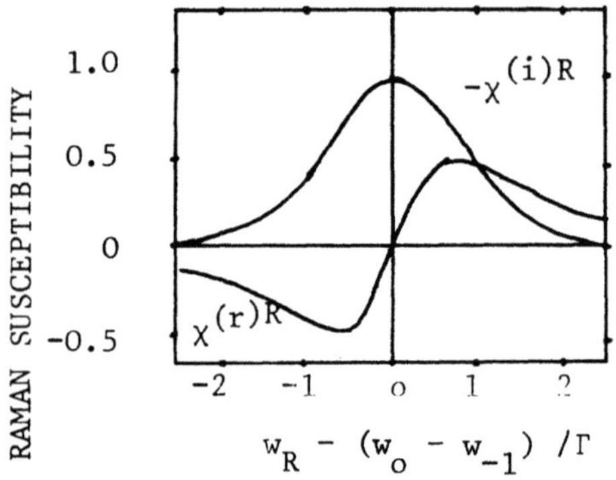

Fig. 2. Plots of the real part $\chi^{(r)R}$ and negative imaginary part $-\chi^{(i)R}$ of the third order Raman Susceptibility versus mismatch from the resonance condition $w_R = w_o - w_{-1}$. Γ is the line width of the Raman transition.

Bearing in mind the contribution from non-resonant terms,

$$\chi^{(3)}_{iiii}(-w_{-1}, w_o, -w_o, w_{-1}) = \frac{N\Delta}{24\hbar}\left(\frac{c}{w_{-1}}\right)^4 \left\{\frac{(d\sigma/d\Omega)_{ii}}{w_R - (w_o - w_{-1}) + i\Gamma/2}\right\} + \chi^{NR} \qquad (14)$$

The total third order Stokes susceptibility for parallel electric fields at w_o and w_{-1} is given by

$$\chi^{(3)}_{ijk\ell} = 6\,\chi^{(3)}_{iiii}(-w_{-1},w_o,-w_o,w_{-1})$$

$$= 6\left[\chi^{(r)} + i\,\chi^{(i)}\right]^{R} + 6\,\chi^{NR} \tag{15}$$

The term $[\chi^{(r)} + i\,\chi^{(i)}]^{R}$ is the complex resonant part of the third order Raman susceptibility.

The real part of $\chi^{(3)}_{ijk\ell}$ is given by [8]

$$\left[\chi^{(3)}_{ijk\ell}\right]^{(r)} = 6(\chi^{(r)} + \chi^{NR})$$

$$= \frac{N\Delta}{4\hbar}\left(\frac{c}{w_{-1}}\right)^{4}\left[\frac{w_R-(w_o-w_{-1})}{(w_R-(w_o-w_{-1}))^2 + \Gamma^2/4}\right]\left(\frac{d\sigma}{d\Omega}\right)_{ii}$$

$$+ 6\,\chi^{NR}, \tag{16}$$

and the imaginary part by

$$\left[\chi^{(3)}_{ijk\ell}\right]^{(i)} = \frac{-N\Delta}{8\hbar}\left(\frac{c}{w_{-1}}\right)^{4}\left[\frac{\Gamma}{(w_R-(w_o-w_{-1}))^2 + \Gamma^2/4}\right]\left(\frac{d\sigma}{d\Omega}\right)_{ii} \tag{17}$$

The imaginary part is due entirely to the Raman contribution and leads to the Lorentzian profile displayed in Figure 2. Since this term is negative g in equation (10) is positive and leads to an amplification of the Stokes wave at w_{-1} according to equation (9). It is this negative imaginary part of $\chi^{(3)}$ that is accessed in conventional Raman Gain spectroscopy. In terms of the imaginary component of the Raman susceptibility for parallel polarisations the Raman gain coefficient of equation (10) is given by

$$g = -\frac{192\pi^2 w_{-1}}{n_{-1}n_o c^2}\,\chi^{(i)R}_{iiii}(-w_{-1},w_o,-w_o,w_{-1}), \tag{18}$$

and in terms of the Raman scattering cross section for parallel polarisations of incident and scattered radiation from equations (10) and (17)

$$g = + \frac{4\pi^3 c^2 N\Delta}{hn_{-1}n_o w_{-1}^3} \left[\frac{\Gamma}{(w_R-(w_o-w_{-1}))^2 + \Gamma^2/4}\right] \left(\frac{d\sigma}{d\Omega}\right)_{ii} \tag{19}$$

At the peak of the Raman transition ($w_R = (w_o-w_{-1})$) the gain coefficient is given by

$$g = + \frac{16\pi^3 c^2 N\Delta}{\Gamma hn_{-1}n_o w_{-1}^3} \left(\frac{d\sigma}{d\Omega}\right)_{ii}, \tag{20a}$$

or in wavenumber (cm^{-1}) units, with $w = 2\pi c\bar{\nu}$ (and Γ also expressed in cm^{-1} units)

$$g = + \frac{N\Delta}{\pi hc^2 n_{-1}n_o \bar{\nu}_{-1}^3\Gamma} \left(\frac{d\sigma}{d\Omega}\right)_{ii} \tag{20b}$$

The gain coefficient shows that the enhancement in the intensity of the Stokes beam is linearly dependent on the spontaneous Raman scattering cross-section and, unlike the Raman scattering process, dependent on the population difference between the states involved in the Raman Amplification process. Since the spontaneous Raman line width, Γ, appears in the denominator of this expression it is clear that the process is ideal for the investigation of species displaying very narrow spectral lines in gases or low temperature solids. The spectra that are generated by this process using linearly polarised pump and probe lasers are conventional Raman spectra observed in forward scattering, depolarisation ratios being obtained by comparing spectra obtained with parallel and perpendicular polarisations of the two incident laser beams

$$\text{i.e.} \quad g_{11}/g_{\perp} = \left(\frac{d\sigma}{d\Omega}\right)_{ii} \Big/ \left(\frac{d\sigma}{d\Omega}\right)_{ij} \tag{21}$$

At the same time as there is observed an exponential increase at the Stokes frequency (equation (7)) there is an exponential loss at the higher frequency laser [7], governed by

$$\frac{\partial E_o}{\partial z} = - \frac{2\pi w_o}{n_o c} \left(- \chi^{(3)}_{ijk\ell}\right) |E_{-1}|^2 E_o \tag{22}$$

This is a consequence of the relationship

$$\chi(-w_{-1},w_o,-w_o,w_{-1}) = \chi^*(-w_o,w_{-1},-w_{-1},w_o),$$

which expresses the fact that the imaginary part of the Raman susceptibility changes sign on the interchange of the frequencies w_o and w_{-1}. The gain at w_{-1} is accompanied by a corresponding loss at w_o. The exponential loss at the anti-Stokes wave is governed by (cf equation 9(b))

$$I_o = I_o^o \exp(\ell I_{-1} z), \tag{23}$$

where the loss coefficient ℓ is related to the gain coefficient g by

$$\ell = w_o g/w_{-1} \tag{24}$$

For small signal gains (or losses), which is certainly the case in studies of Raman Amplification spectroscopy using continuous wave laser sources, the fractional increase in intensity is derived from (9b) for Raman Gain Spectroscopy as

$$\frac{\delta I_{-1}}{I_{-1}} = g I_o z \tag{25}$$

The intensity increase, δI_{-1}, at the Stokes frequency, here regarded as the probe laser, is modulated at any frequency impressed on the pump laser, w_o. [Methods of recording the gain at the Stokes frequency generally involve an amplitude or polarisation modulation of the pump beam w_o at a frequency f, the appearance of gain at frequency w_{-1} being detected by use of lock-in detection methods]. In this case heterodyne detection is intrinsically a part of the mixing process. Levenson and Eesley [9] have considered in detail methods for optimising the intensity of the local oscillator probe laser on the detector in order to maximise the signal to noise ratio attainable in the detection process. The intensity of radiation incident on a "square-law" detector in a Raman Gain experiment is

$$I_{-1} = \frac{nc}{8\pi}\left(E_{-1}^{LO} + E_{-1}^{s}\right)^2$$

$$I_{-1} = I_{-1}^{LO} + I_{-1}^{s} + \frac{nc}{4\pi}\left(E_{-1}^{s} \cdot E_{-1}^{LO*}\right)^{(r)} \qquad (26)$$

$$= I_{-1}^{LO} + I_{-1}^{s} + I_{-1}^{H}$$

where E_{-1}^{LO} and I_{-1}^{LO} are the amplitude and intensity of the background, local oscillator, radiation which adds coherently to the signal at the detector surface. The signal response generated is proportional, with proportionality constant K, to I_{-1} integrated over the detector surface. The term I_{-1}^{s} is generally too low to be detected under normal circumstances and the Raman signal of interest is produced as a heterodyne signal which is separated from the noise generated primarily by the local oscillator through the modulation impressed on the pump laser, w_o. The signal response is given by

$$S = KI_{-1}^{H} = \frac{ncK}{4\pi}\left(E_{-1}^{s} . E_{-1}^{LO\,*}\right)^{(r)}. \qquad (27)$$

The local oscillator and signal amplitudes in this expression are provided from the amplified signal emanating from the gain cell. E_{-1}^{LO} is very nearly equal to E_{-1}^{o} of equation (9a) (if it is assumed that all of this radiation falls on the detector) and from this same equation E_{-1}^{s} $(= E_{-1} - E_{-1}^{o})$ is found to be $\frac{1}{2}gI_oE_{-1}^{o}\, z$, giving the signal response of the detector as

$$S = K\, g\, I_o I_{-1}\, z \qquad (28)$$

The signal response of the detector is thus linearly dependent on the pump intensity, and on the probe intensity at the detector surface. Levenson and Eesley [9] considered the various terms contributing to the noise and identified the principal components as arising as a result of fluctuations in the probe laser intensity with mean square values

$$< \delta I_{-1}^2 > = k_{-1} I_{-1}^{o2}, \quad < \delta I_B^{\;2} > = k_B I_B^{\;2}, \qquad (29)$$

where $\sqrt{k_{-1}}$ is the root mean square fractional deviation of the probe laser intensity from its average value and $< \delta I_B^{\;2} >$ represents the mean square fluctuation of the intensity of the background which adds incoherently to the local oscillator wave: in the experimental methods generally employed for the study of Raman Amplification spectra this latter component can be neglected, although it acquires relevance in techniques such as Optical Heterodyne Raman Induced Kerr Effect Spectra [10] (OHD-RIKES). Additionally, as a result of the quantised nature of radiation, there is shot noise at the detector with a mean square value

$$< \delta I^2_{shot} > = i\,(I^o_{-1} + I_B), \tag{30}$$

where i is the intensity associated with a single photon. Considering the signal response generated by these intensity fluctuations the signal to noise ratio for this type of experiment as a whole can be written

$$\frac{S}{N} = \frac{g\, I_o\, I_{-1}\, z}{(k_{-1}\, I^{o2}_{-1} + \frac{1}{Q}\, i\, I^o_{-1} + \eta)^{\frac{1}{2}}} \tag{31}$$

The last term η in the denominator arises from external noise factors including thermal noise in the detector load that would arise in the absence of any optical signals, and Q is the detector quantum efficiency. Levenson [9,10] has considered the methods available for optimizing the local oscillator intensity to maximize the signal to noise ratios for various non-linear optical techniques.

The signal and noise contributions outlined here have been derived in terms of the incident intensities of the probe and pump beams. In this formulation the signal is linearly dependent on the interaction length z. In practice, however, it is found that for physically realisable path lengths that can be produced with collimated laser beams the intensity gain (or loss) is so low that it is preferable to generate gain in the vicinity of the focus of the laser beams. The interaction length is then very short but the intensity (power per unit cross sectional area) is extremely high and since the signal strength depends on the product intensities of the two laser beams this increased intensity more than compensates for the decrease in the length of the interaction zone. In practice it is found that the greater part of the gain (or loss) is produced in the vicinity of the focus of the two beams (assumed to be focussed together through the sample). The fractional gain at this focal region is given, from equations (18) and (25) (taking the negative imaginary part of $\chi^{(3)R}$) as

$$\frac{\delta I_{-1}}{I_{-1}} = \frac{384\pi^3\bar{\nu}_{-1}}{n_{-1}n_o c}\, \chi^{(i)R}_{iiii}\, I_o\, z, \tag{32}$$

with $I_o = P_o/\pi\omega_o^2$ and z as the confocal beam parameter $2\pi\omega_o^2/\lambda_o$. ω_o is the spot radius of the focussed pump beam of wavelength λ_o and the power of the pump beam is expressed in erg s^{-1} in this formulation. Owyoung and Percy [11] evaluated the negative imaginary part, $\chi_{iiii}^{(i)R}$ of the third order Raman susceptibility of the ν_2 mode of benzene as 15.9 x 10^{-14} cm^3 erg^{-1}. It is worth while evaluating the fractional change of intensity of a helium-neon probe laser beam under the influence of a pump beam of average power 50 mW in a sample of benzene liquid at the resonance condition. One assumes 2 mm diameter laser beams focussed by a lens of 20 mm focal length, the corresponding spot radius being 4.0 μm and the confocal beam parameter (z in equation 32) o.16 mm. With $\bar{\nu}_{-1}$ = 15,800 cm^{-1} and assumed values of n_{-1} and n_o of 1.46 the average signal gain is found to be 7.4 x 10^{-6}. The gain measured experimentally by Owyoung and Jones [12] at the peak of this Raman transition was 1.75 x 10^{-5}, implying that the most significant part of the signal (~ 40%) is produced in the focal region of the laser beams.

A more precise evaluation of the overall gain has been carried out by Owyoung and Jones [12] and by Barrett and Heller [8] by integrating the intensity dependent gain through the focal region of the beams (assuming gaussian beams) and a relationship is obtained which is independent of focussing (to a good approximation) and linearly dependent on pump power,

$$\frac{\delta P_{-1}}{P_{-1}} = \frac{384\pi^4}{\lambda^2 cn} \chi_{iiii}^{(i)R} P_o \tag{33}$$

providing the focal region of the two beams is entirely enveloped by the sample.

For a fixed sample and fixed pump and probe laser wavelength the gain or loss realizable is dependent only on the pump power, as is the signal to noise ratio when probe laser fluctuations dominate the noise contributions (i.e. k_{-1}, I_{-1}^{o2} greater than iI_{-1}^{o}/Q and η in equation (31)). The ultimate sensitivity limit, however, is attained when shot noise dominates all other noise contributions and the signal to noise ratio scales as $P_o P_{-1}^{\frac{1}{2}}$ (in Raman Gain). Under such circumstances improved signal to noise ratios may be obtained not only by increasing the pump power but also by increasing the probe power incident on the detector. However, this latter method of improving the signal to noise ratio tends to be of limited value because of the increasing difficulty of attaining shot noise limited performance with increasing probe power and because of the inability of detectors to handle powers substantially in excess of 10 mW. In practice the maximum probe powers useable with continuous wave lasers are typically in the region of 10 mW. At this power level, with a detector of 10%

quantum efficiency at 633 nm, the minimum signals detectable (with a 1:1 signal to noise ratio) correspond to a gain (or loss) of 2.10^{-8}. This sensitivity limit would lead to a signal to noise ratio of ~ 900:1 for the ν_2 vibration of benzene under the conditions stated. Methods that have been employed for approaching this theoretical noise limit and for generating the Raman signal will be discussed in detail in Part II.

REFERENCES

1. Long, D.A., *Raman Spectroscopy*, McGraw-Hill, 1977.

2. Woodbury, E.J. and Ng, W.K., Proc. IRE, 50, 2347 (1962).

3. Hellwarth, R.W., Appl. Opt., 2, 847 (1963).

4. Maiman, T.H., Nature, 187, 493 (1960); Phys. Rev., 123, 1145 (1961).

5. Jones, W.J. and Stoicheff, B.P., Phys. Rev. Lett., 13, 657 (1964).

6. Maier, M., Appl. Phys., 11, 209 (1976).

7. Bloembergen, N., Am. J. Phys., 35, 989 (1967).

8. Barrett, J.J. and Heller, D.F., J. Opt. Soc. Amer., 71, 1299 (1981).

9. Levenson, M.D. and Eesley, G.L., Appl. Phys., 19, 1 (1979).

10. Levenson, M.D., J. Raman Spectrosc., 10, 9 (1981).

11. Owyoung, A. and Percy, P.S., J. Appl. Phys., 48, 674 (1976).

12. Owyoung, A. and Jones, E.D., Opt. Lett., 1, 152 (1977).

APPLICATIONS OF INVERSE RAMAN AND RAMAN GAIN SPECTROSCOPY

W. J. Jones

University College of Wales
Aberystwyth

In the preceding paper the background theory relevant for the understanding of the Raman Amplification process was considered in some detail. It is necessary now to consider some of the approaches that have been adopted to maximize the signal to noise ratios attainable in this branch of spectroscopy. Before doing so, however, it is as well to be aware of the signal strengths available in some typical experiments using continuous wave lasers. We consider two instances: one the strong vibrational band of liquid benzene at 992 cm^{-1} considered in the previous paper; the other the J = 6 component of the vibrational Q branch in gaseous nitrogen. For the benzene sample we consider an incident average pump power of 50 mW and a probe power on the detector of 1 mW giving a fractional signal $\delta I_{-1}/I_{-1}$ of 1.75 x 10^{-5} and shot noise limited performance, expressed as a fraction of the probe laser power of 6.10^{-8}, leading to a signal to noise ratio of ~300 for a single pass of the laser beams through the sample cell. The resolution of the Raman Amplification process is determined by the linewidths and relative frequency stabilities of the lasers employed so that in principle the method is capable of extremely high resolution. In practice also the resolution of the individual rotational components of a vibrational Q branch of a small molecule presents no major difficulty - the major problems arise from the signal to noise ratios attainable. The scattering cross section for the vibrational Q branch of nitrogen gas at 2330 cm^{-1} is 4.4 x 10^{-31} $cm^2 Sr^{-1}$, the pressure broadened line width of a single rotational component at a pressure of 1 atmosphere being ~ 0.05 cm^{-1}. Since only ~ 10% of the molecules of nitrogen reside in the most populated rotational level, J = 6, the molecular number

W. Kiefer and D. A. Long (eds.), Non-Linear Raman Spectroscopy and Its Chemical Applications, 487–498.

density is $N\Delta = 2.6 \times 10^{18}$ molecule cm^{-3}. For the same powers of the HeNe and dye lasers as used in the liquid experiment the fractional increase of the probe laser power is estimated to be $\delta I_{-1}/I_{-1} = 8.10^{-9}$, so that for the same shot noise limit of 6.10^{-8} the signal to noise ratio estimated is only 0.13. It is apparent, therefore, that although these straightforward methods would be quite adequate for recording good quality liquid spectra they are far from being satisfactory for recording the very much weaker spectra produced by gaseous samples. In going from a liquid to a gaseous sample the number density, even at a pressure of 1 atmosphere, is decreased by ~ 250 and where resolved rotational structure is to be observed this decrease in the number density in a selected J state is even greater. For the Q branch of nitrogen the scattering cross section is more than an order of magnitude smaller than the cross section of ν_2 for benzene so that even the marked reduction in the line width in going from liquid to gas is not enough to produce a signal that is distinguishable from the noise. Clearly a marked improvement in the approach adopted is essential if the technique is to be of value in the investigation of gas phase spectra and even for liquid and solid samples, particularly if one wishes to have the facility for observing the spectra of surface adsorbed species, marked improvements in sensitivity are desirable. Owyoung [1] considered some of the factors influencing the signal to noise ratios attainable in this branch of optical spectroscopy using cw lasers. Utilising a 50 mW average dye laser power he obtained a single pass gain of 1.7×10^{-5} for the ν_2 mode of benzene and with a multipass liquid cell giving an increase of signal of 3 x he obtained a signal to noise ratio of 300 using a 30 second integration constant in his detection circuitry with 2 mW probe power on the detector. This signal to noise ratio was a factor of 45 from the shot noise limit, probe laser fluctuations being the primary source of noise limiting the sensitivity. He estimated for that particular system a capability of observing values of $\chi^{(i)R} = 5.5 \times 10^{-16}$ cm^3 erg^{-1} for a 1:1 signal to noise ratio and a 30 second time constant. As a lower limit on detection in bulk liquid samples, he estimated a 100 mW pump beam with a multireflection cell with 30 x gain and a 40 x improvement in probe laser noise would result in a detection threshold of $\chi^{(i)R} \sim 6.6 \times 10^{-19}$ cm^3 erg^{-1}, equivalent to a benzene concentration of 5 parts per million assuming detection on the 992 cm^{-1} mode.

Clearly one of the major problems encountered in recording good quality spectra lies in the difficulty of approaching the shot noise limit in the detection circuitry. Using lock in detection methods with modulated signals in the region of 10 kHz, a typical helium neon probe laser with a power of ~ 1 mW on the detector exhibits a fractional noise level in the region of 10^{-5} with a 1 second integration constant. Much of this noise arises from mode instabilities in the laser cavity. Use of a mode stabilized single frequency laser improves this figure by about an order of magnitude. A continuous wave dye laser, on the other hand, is very much worse than a helium neon laser because of these mode instabilities and the existence of bubbles and transient inhomogeneities in the dye jet. For such a laser it is not at all uncommon to observe a fractional noise level in the region of 10^{-3} - 10^{-5} for this same 1 Hz bandwidth at a frequency of 10 kHz. Careful attention to the flow of the dye solution can reduce this noise contribution significantly but rarely is it possible to approach the noise figure obtained with a single frequency HeNe laser. Many methods have been adopted for improving these measured noise figures, among the more successful [2, 3] being the introduction of electro-optic noise reduction systems (e.g. the Coherent Associates Model 307 Noise Reduction System - which serves to reduce probe laser fluctuations by a factor of ~ 50 in the frequency range up to ~ 100 kHz) and the use of differential amplifiers which subtract the outputs from two near-identical detectors on which are incident laser beams from the same laser source, only one of these laser beams carrying the signal of interest. More recently Levine and Bethea [4] measured the probe laser noise of a synchronously pumped mode-locked dye laser on a spectrum analyzer and found a broad minimum from about 3-20 MHz, with the noise at 10 MHz being approximately 3 orders of magnitude lower than at 2kHz. Use of modulation at 10 MHz allowed these authors to attain near shot limited performance in their studies with synchronously pumped mode-locked dye lasers [4,5].

Because of the difficulties of approaching shot noise limited performance, early studies of Raman Gain spectroscopy used a variation of the technique which has been termed Raman Induced Kerr Effect Spectroscopy (RIKES) [6,7,8]. RIKES takes advantage of the change in polarisation of the probe laser induced by the pump laser near a Raman resonance to reduce the local oscillator power on the detector without significantly decreasing the generated Stokes radiation, thereby decreasing markedly the significance of the probe laser noise. In general terms, however, though the technique is particularly valuable for use with noisy probe lasers the magnitude of the signal is reduced and the signal to noise ratio decreased.

As well as seeking methods of reducing probe laser noise, major efforts have been devoted to ways of increasing the magnitude of the signal generated. Three of these methods will be discussed in this paper, cw Raman Amplification Spectroscopy, the Quasi-cw Stimulated Raman Spectrometer of Owyoung and Raman Amplification Spectroscopy using picosecond lasers.

CONTINUOUS-WAVE RAMAN AMPLIFICATION SPECTROSCOPY

A typical cw Raman Amplification Spectrometer, used in some of the studies of gas phase spectra at Aberystwyth, is shown in Figure 1. The probe laser used in this Raman Gain equipment is a

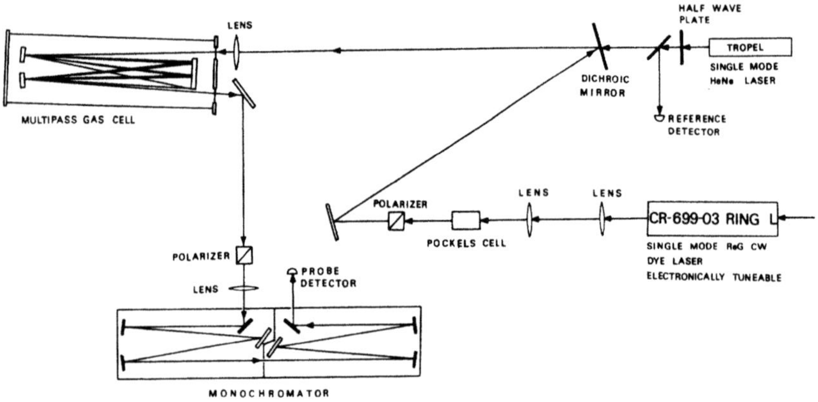

Fig. 1. Schematic diagram of the cw Raman Amplification (Gain) experiment (courtesy of Molecular Physics).

single frequency Tropel Model 100 laser exhibiting a long-term frequency stability of ~ 10 MHz, the pump laser being a Coherent Associates single frequency ring dye laser with an output power of ~ 500 mW when pumped with 6 W from an argon ion laser. The continuous scan range of the ring dye laser is 6 GHz between mode hops and the effective resolution of the instrument is ~ 30 MHz. This equipment was used to investigate the Q branches of the Fermi resonance diad $\nu_1/2\nu_2$ of carbon dioxide and for this purpose the dye laser had to be scanned near 585.2 and 581.7 nm for the 1285.4 cm^{-1} and 1388.15 cm^{-1} bands, respectively. The pump beam is amplitude modulated at a frequency of 20 kHz by means of a Pocckels cell and polariser, the horizontal component only being transmitted by the polariser. The pump and probe beams are combined at a dichroic beam splitter, the beam sizes and divergences being matched by means of a two lens telescope. In order to enhance the very weak signals impressed on the probe laser by the 200 mW average power of the pump laser, the combined beams are passed successively through a multi-reflection cell to increase the signal by a factor of ~ 20. In the study on CO_2 [9] with a probe power on the detector of 0.2 mW the noise in a 1 Hz bandwidth relative to the incident power was 4.10^{-7}, within a factor of 5 - 6 of the shot noise limit. To approach the shot limit more closely the polarisation of the probe laser was rotated by a half-wave plate to ~ $\pi/4$ in relation to the pump beam. A polariser in front of the double monochromator was used to reduce the local oscillator power to the 25 μW needed for shot limited performance. Calibration of the spectrum was obtained by transferring a small portion of the dye laser to a hollow cathrode lamp and three etalons with free spectral ranges of 1.6, 0.18 and 0.0083 cm^{-1}.

The spectrum in the vicinity of 1388 cm^{-1} measured with a 1 second time constant is shown in Figure 2, the fringes of the

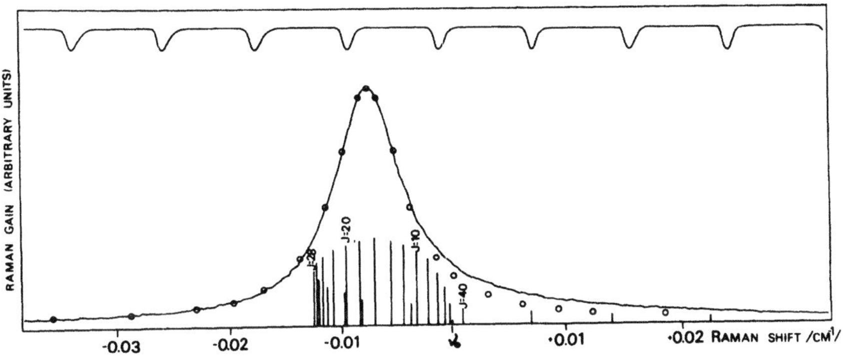

Fig. 2. The 1388.15 cm^{-1} Q branch of CO_2 at a resolution of 30 MHz (courtesy of Molecular Physics).

etalon with 250 MHz (0.0083 cm^{-1}) free spectral range being shown at the top of the diagram. The positions and relative intensities of the individual rotational transitions contributing to the total intensity of this $2\nu_2$ Raman band are indicated in stick form. The measured gain for the peak of this band with 200 mW average pump power was 1.3×10^{-4} and the signal to noise ratio better than 200.

The striking feature of the observed Q branch is its extreme narrowness, some 30 times less than the width of a single rotational S-branch transition in CO_2 at the pressure of 1 atmosphere used for this study. The reduction in the spectral line width of the Q branch from what might have been expected even for a single rotational transition is a consequence of motional narrowing, which occurs when the inelastic collision frequency exceeds the frequency separation of the Q branch components. Because the Q branch of $2\nu_2$ is so extremely sharp the gain observed is significantly higher than might have been expected for a normal Q branch with a total scattering cross section of only 0.2×10^{-30} cm^2 Sr^{-1} at 633 nm. By contrast the ν_1 band at 1286^{-1} is a more normal Q branch spectrum with a peak Raman gain of only 2.10^{-6} even though the total scattering cross section for this band is comparable with that of $2\nu_2$. This situation is quite understandable since the rotational constants for this ν_1 band lead to a more open rotational structure of the Q branch and the effects of motional narrowing are not so pronounced.

It is apparent then that under favourable circumstances good quality spectra of gases can be obtained with relatively low powers for pump and probe lasers. Owyoung, Patterson and McDowell [3] using a 514.5 nm single frequency argon ion pump laser, 1.5 mW of probe power from a scannable rhodamine 6G single mode dye laser, and a multireflection cell giving a gain of 50 x, were able to resolve the detailed rotational structure in the ν_1 mode of methane (with a total scattering cross section ~ 4 times greater than that of $2\nu_2$ of CO_2) at a total gas pressure of ~ 35 torr and a resolution of 25 MHz. By this means Owyoung et al were able to obtain detailed rotational constants from a band that it had previously not been possible to resolve using conventional techniques. The authors point out that with improved laser stabilisation the resolution could be extended to less than 1 MHz and the sensitivity to an order of magnitude better than reported in that paper. However, the authors were aware of the limitations of the cw techniques for the investigation of weak Raman spectra at high resolution and outlined briefly the advantages of hybrid cw-pulsed laser systems which were expected to lead to several orders of magnitude improvement in sensitivity with only a slight loss in resolving power. These ideas led to the development of The Quasi cw Stimulated Raman Spectrometer of Owyoung.

Many of the most important developments in the study of high resolution Raman spectra of gases have arisen as a result of the researches of Owyoung at the Sandia Research Laboratories, the quasi-cw spectrometer being the most sensitive instrument developed to date. The signal strength of a Raman amplification instrument depends on the product powers of the pump and probe beams employed

$$S^{cw} = C^{cw} P_o^{cw} P_{-1}^{cw} \quad (1)$$

so that high powers on the sample are desirable. Regrettably, however, pulsed lasers rarely have the narrow spectral line width necessary for the study of molecular spectra at high resolution. Owyoung [10] has overcome this problem by using the output from a single frequency ring dye laser with a cw power of > 50 mW and spectral line width of ~ 1 MHz but amplified to a train of pulses at a repetition rate of 10 pps using a series of dye laser amplifiers pumped by the frequency-doubled output from a Nd/YAG laser. By careful alignment and isolation of the amplifiers from each other and from the dye laser oscillator Owyoung is able to produce 10ns pulses with a minimum spectral line width of 75 MHz (transform limited) and peak power in the range 2 - 10 MW. This tunable pump laser is used with a fixed frequency probe laser in Inverse Raman experiments. Since the repetition rate is relatively low Owyoung is able to increase the probe power on the detector to ~ 100 mW by synchronising with the pump laser shaped pulses of ~ 100 μs duration (Fig. 3) from a single frequency argon ion laser (or dye) laser and yet avoid power saturation of the probe detector.

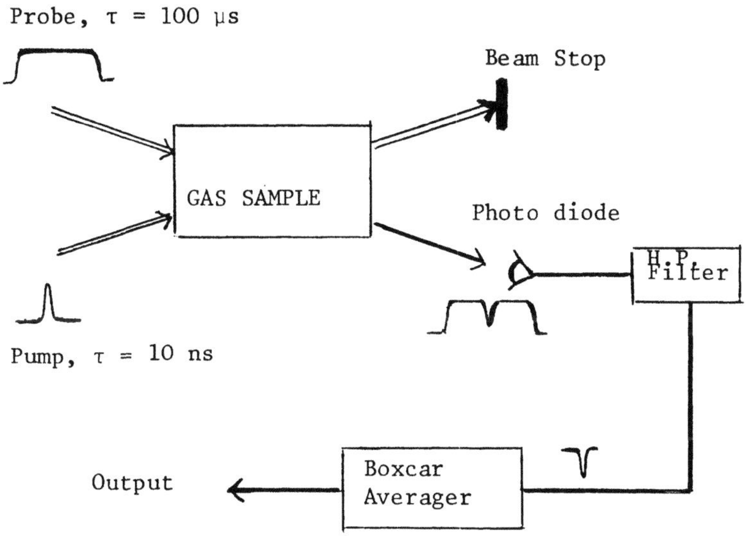

Fig. 3. A schematic diagram of the Inverse Raman Spectrometer of Owyoung.

The signal pulses arising from the Inverse Raman process are isolated by a high frequency pass filter and, following amplification by a radio frequency amplifier, are averaged using a Boxcar Averager. Owyoung points out that in addition to the gains which are realized through the increase in the pump power, the shot noise limit, which cannot be reached because of probe laser noise in cw gain studies, can be realized in the quasi-cw system. This attainment of shot noise limited performance arises because the probe laser noise spectrum is concentrated at lower frequencies whereas shot noise contributions are uniformly distributed. Thus, although the detection bandwidth is increased by ~ 10^7 over the cw case (> 10^7 Hz as compared with 1 Hz), laser noise contributions are not appreciably increased and the single pass sensitivity realized by the quasi-cw system is more than 10^4 times that observed in the cw system. It is worth realizing, however, that techniques such as the use of multireflection cells to enhance the sensitivity are less easy to use with the quasi-cw system so that the realistic improvement in performance is nearer 10^3. Nevertheless, the improvement in performance is quite dramatic and using this equipment Owyoung and his co-workers have been able to obtain beautifully resolved spectra and carry out detailed analysis of the Q branch structures of the totally symmetric bands of a variety of spherical top molecules at pressures of ~ 1 torr: $^{12}CH_4$, $^{13}CH_4$ [10,11] ; $^{28}SiH_4$ [12]; CF_4 [13].

RAMAN AMPLIFICATION SPECTROSCOPY USING PICOSECOND LASERS

An alternative approach to that adopted by Owyoung, but designed for the study of condensed phase species at lower resolution (~ 1 cm^{-1}), also takes advantage of the increase in signal strength from the use of pulsed, in place of continuous wave, lasers. Current generations of commercially available lasers include mode-locked ion lasers which provide a continuous train of pulses exhibiting high peak powers but comparatively low average powers. Typical pulse characteristics of the ion laser are a repetition rate of ~ 80 MHz and a pulse duration of ~ 200 ps. The peak power in the pulses is thus raised by a factor of ~ 60 over the acerage output power (which is typically in the region of > 1 W. When this laser is used to pump tunable dye lasers, a synchronously-pumped mode-locked dye laser can be produced in which the output dye laser pulses are synchronised with the output pulses from the ion laser but individual dye laser pulses are reduced in duration to ~ 20 ps (transform limited for a laser line width of ~ 1 cm^{-1}). The peak power in the 80 MHz pulses is thus raised ~ 600 x over the average dye laser output. With an average output power of ~ 100 mW the peak power is thus increased some three orders of magnitude over the powers typically employed in a cw Raman gain or Inverse Raman experiment. To appreciate the benefit of the use of such lasers one reverts again to equation (1) which shows how the signal generated depends on the product powers of the lasers employed. Since the average power of a train of laser pulses relates to the peak power by the expression

$$P^{av} = P^{pulse}, R. \tau, \tag{2}$$

where R and τ represent the pulse repetition rate and pulse duration, the pulsed signal generated from equation (1) (assuming both lasers to be pulsed) relates to the average power of the pump and probe lasers by

$$S^{pulse} = C^{pulse} \frac{P_o^{av}}{R_o \tau_o} \cdot \frac{P_{-1}^{av}}{R_{-1}\tau_{-1}} \tag{3}$$

C^{pulse} is distinguished from C^{cw}, the steady state gain or loss factor, since it does not necessarily follow that these two proportionality constants are identical. With such rapid repetition rate lasers it is possible to use phase sensitive detection methods and use detection bandwidths which are extremely small (< 1 Hz) in comparison with those used in the quasi-cw experiments of Owyoung. Under such circumstances the average signal power is relevant rather than the peak power S^{pulse} of equation (3). This average signal power depends on the pulse rate R_s and τ_s at which the signal pulses are generated.

$$S^{av} = C^{pulse} \frac{P_o^{av}}{R_o \tau_o} \cdot \frac{P_{-1}^{av}}{R_{-1}\tau_{-1}} \cdot R_s \tau_s \tag{4}$$

Using two synchronously pumped dye lasers, pumped by the same mode-locked ion lasers so that R_o and R_{-1} are identical, a signal pulse is generated each time a pump and probe laser pulse coincide. Equation (4) then reduces to

$$S^{av} = C^{pulse} P_o^{av} P_{-1}^{av} / R\tau, \tag{5}$$

where the pump and probe pulses are assumed to have the same time duration as well as, of course, the same repetition rate. The enhancement of the signal produced over the signal produced by the use of cw lasers is thus

$$\frac{S^{av}}{S^{cw}} = \frac{C^{pulse}}{C^{cw}} \cdot \frac{P_o^{av}}{P_o^{cw}} \cdot \frac{P_{-1}^{av}}{P_{-1}^{cw}} \cdot \frac{1}{R\tau} \tag{6}$$

For the same average powers employed using pulsed laser excitation as compared with cw excitation the enhancement in the signal is ~ 600 providing the pulsed gain (or loss) does not differ markedly from that attained in the cw experiment. At the shot noise limit the ultimate noise figure is determined by the number of photons incident on the detector in unit time i.e. on the average power incident on the detector rather than on any functions of the pulse power. As a result the enhancement in signal leads to a corresponding increase in the signal to noise ratio providing shot noise limited detection is attained.

In our experiments at Aberystwyth we are currently employing a synchronously-pumped mode-locked Rhodamine 6G dye laser as probe laser (R_o = 80 MHz, $\tau_o \simeq$ 25 ps) and a cavity-dumped synchronously-pumped mode-locked DCM dye laser as pump laser (R_{-1} = 4 MHz, $\tau_{-1} \simeq$ 25 ps). Since only one in 20 of the synchronously-pumped laser pulses coincide with each of the cavity-dumped pulses, signal is generated only at a rate of 4 MHz so that the enhancement over the cw system is again ~ $1/R_o\tau_o$.

Levine, Shank and Heritage [14] have considered the relationship between the pulsed gain in comparison with the cw gain as a function of the Raman relaxation time and the duration of the pump and probe laser pulses. For a sample such as benzene $C^{pulse}/C^{cw} \simeq 0.4$ so that the enhancement expected is ~ 250. In consequence of this dramatic signal enhancement Levine and Bethea [4,5] were able to obtain a signal to noise ratio of approximately 200,000 for the 992 cm^{-1} Raman band in bulk benzene. By introducing an additional low frequency modulation of the pump lasers these authors show how it is possible to reduce the large thermal background produced by absorbing substrates by a factor of $10^3 - 10^4$, thereby permitting the full use of the ultra high sensitivity which, seemingly, is sufficient to observe vibrational spectra of adsorbed monolayers.

APPLICATIONS OF RAMAN GAIN AND INVERSE RAMAN SPECTROSCOPY

The developments of the techniques of Raman Gain and Inverse Raman Spectroscopy discussed above are so striking in terms of the sensitivity attainable in what was once regarded as a relatively insensitive spectroscopic technique that it is difficult to appreciate the full potential for future applications. Nor would it be safe to assume that further dramatic improvements in sensitivity are unlikely. The very high resolution work of Owyoung has already shown how it is possible to obtain detailed structures of vibrational Q branches and no doubt this work is capable of extension to investigate the detailed structures of complete vibration-rotation bands although there will clearly be difficulties in such work because of the problems associated with scanning over very wide spectral regions at such high resolution. The sensitivity available has enabled the workers at the Sandia National Laboratories to record Q branch spectra of CH_4 and CD_4 cooled to > 13 K in a pulsed molecular jet [15] and to record sub-Doppler linewidth spectra of the Q_{01} (2) transition in deuterium using Inverse Raman techniques [16] analogous to sub-Doppler IR saturation spectroscopy. In this latter study it is shown that, by delaying the pulsed measurement from the pulsed saturation, the technique can be used to study velocity relaxation processes in homonuclear diatomics or other Raman-active molecules.

Rahn [17] has shown how it is possible to record sensitive, highly-reproducible Raman spectra resolved in time and space using single pump pulses from a Nd/YAG pumped dye laser and a single frequency argon ion probe laser at a resolution of 0.003 cm^{-1}. The author points out that the high sensitivity and excellent rejection of background radiation provide significant new capabilities for studies of turbulent combustion, reaction kinetics and other transient phenomena and for time averaged spectroscopic applications the technique permits the precise measurement of linewidths and line shapes, thus aiding studies of collisional processes.

The potential applications for the use of picosecond lasers at a lower resolution limit are equally exciting, particularly for the study of surface adsorbed species, gas-solid, gas-liquid and liquid-solid, in which fields the need for a sensitive probe of the nature of adsorbed species is extremely great.

REFERENCES

1. Owyoung, A., Opt. Comm., 22, 323 (1977).

2. Owyoung, A., IEEE J. Quant. Elec., QE14, 192 (1978).

3. Owyoung, A., Patterson, C.W. and McDowell, R.S., Chem. Phys. Lett., 59, 156, (1978).

4. Levine, B.F. and Bethea, C.G., IEEE J. Quant. Elec., QE16, 85 (1980).

5. Levine, B.F. and Bethea, C.G., Appl. Phys. Lett., 36, 245 (1980).

6. Heiman, D. et al, Phys. Rev. Lett., 36, 189 (1976).

7. Owyoung, A., Opt. Lett., 2, 91 (1978).

8. Levine, M.D., J. Raman Spectrosc., 10, 9 (1981).

9. Baran, J., Grofcsik, A. and Jones, W.J., Mol. Phys., to be published.

10. Owyoung, A. in *Laser Spectroscopy IV*, pp.175-187, (H. Walther and K.W. Roth, eds.), Springer-Verlag, Berlin, 1979.

11. McDowell, R.S., Patterson, C.W. and Owyoung, A., J. Chem. Phys., 72, 1071 (1980).

12. Owyoung, A., Esherick, P. and Robiett, A.G., J. Mol. Spectrosc., 86, 209 (1981).

13. Esherick, P., Owyoung, A. and Patterson, C.W., J. Mol. Spectrosc., 86, 250 (1981).

HIGH RESOLUTION INVERSE RAMAN AND RAMAN GAIN SPECTROSCOPY

P. Esherick and A. Owyoung

Sandia National Laboratories
Albuquerque, New Mexico 87185
United States

1. HISTORICAL BACKGROUND

Although spontaneous Raman spectroscopy has a history dating back to the discovery of the Raman effect in 1928, nonlinear Raman phenomena have only been reported since the introduction of laser sources in the past 20 years [1].

In particular, Jones and Stoicheff [2], in their report of the inverse Raman effect, recognized the potential utility of the technique in obtaining spectral data under conditions of high background luminescence or where time-resolved Raman data would be useful. Nevertheless spectroscopic applications of the technique were rather slow in coming, and it was not until several years later that stimulated Raman spectroscopy (SRS) demonstrated high spectral resolution capabilities in a gaseous medium [3].

In the decade that followed these early studies, very few investigations were conducted in which the spectroscopic capabilities of either the SRS or frequency mixing techniques were utilized. The studies by Lallemand and his co-workers at Harvard [3-5] provided an admirable example of the great potential of SRS as a high-resolution spectroscopic probe in H_2, yet this was a rather isolated case in which large Raman gain signals could be generated in a gaseous medium at high densities without appreciable pressure broadening. On the whole it is fair to say that most of the work in this period concentrated either on the potential application of coherent Raman effects as a means of nonlinear frequency conversion, or on the character of the nonlinear susceptibility

W. Kiefer and D. A. Long (eds.), Non-Linear Raman Spectroscopy and Its Chemical Applications, 499–517.

itself. In the latter case, the nonlinear dispersion which arose from the interference of Raman contributions to this susceptibility provided opportunities to determine Raman cross sections relative to two-photon cross sections and electronic hyperpolarizabilities [6-9]. The failure to exploit the spectroscopic capabilities of these phenomena during this period is at least partially attributable to the absence of tunable narrow-linewidth sources which facilitated later studies. It must also be acknowledged that even after the introduction of these sources, both frequency mixing studies [6-7] and the few stimulated Raman studies which were reported emphasized the determination of Raman cross sections and the character of the nonlinear dispersion rather than the general spectroscopic utility of the techniques [8-11].

2. STIMULATED RAMAN TECHNIQUES

Stimulated Raman techniques use a strong "pump" laser source to induce a resonant gain in a Raman active medium at a frequency which is downshifted from the pump source by the Raman transition frequency. Simultaneously, a corresponding absorption is induced at a frequency which is upshifted by that same amount. The process of using a "probe" laser source to measure the induced Raman gain profile has been termed "stimulated Raman gain spectroscopy" (Fig. 1a) whereas the probing of the absorption profile is referred to as "inverse Raman spectroscopy" [2] (Fig. 1b). Since the two processes are identical, except for an inversion in the signs of both the signal and the Raman frequency shift between the pump and probe beams, we shall use the term "stimulated Raman spectroscopy" (SRS) in describing both techniques.

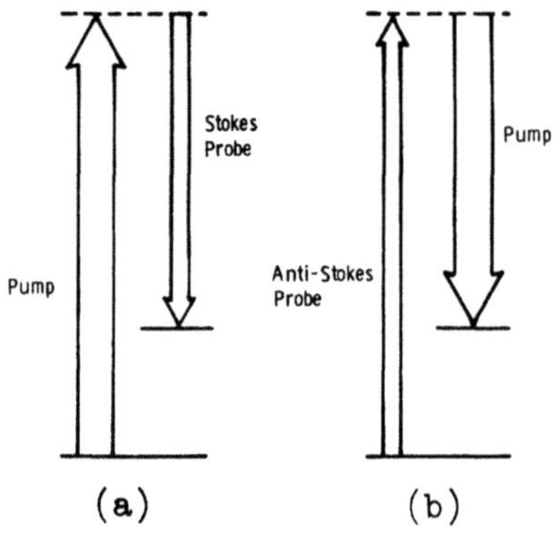

Fig. 1 Energy level diagrams for (a) stimulated Raman gain and (b) inverse Raman absorption processes.

The utility of the stimulated Raman processes arise primarily from two considerations: resolution and signal strength. Unlike spontaneous Raman spectroscopy, where resolving power is limited by spectrographs to 0.05-0.1 cm^{-1}, the resolution of the stimulated processes is limited only by the linewidths of the lasers used to induce and probe the absorption or gain profiles. Hence, Doppler-limited spectra are a realistic goal. Furthermore, in contrast to the 90-degree scattering geometry common to conventional light scattering studies [12], the colinear forward scattering geometry normally used in SRS studies minimizes the Doppler width.

The relations which govern the stimulated Raman processes are essentially quite simple and can be summarized by the gain equation, which in the plane wave approximation can be written in the form [13-15],

$$G \equiv \frac{\delta P(\omega)}{P(\omega)} \simeq \frac{N\lambda^3}{hc^2} \frac{d^2\sigma}{d\Omega d\frac{1}{\lambda}} I(\Omega)\ell \tag{1}$$

Here $\delta P(\omega)$ is the change induced in the probe power $P(\omega)$ at frequency ω by the pump beam of intensity $I(\Omega)$ [J sec^{-1} cm^{-2}] at frequency Ω, N is the difference in population between the lower and upper states, $d^2\sigma/d\Omega d(1/\lambda)$ [cm^2 $ster^{-1}$ $(cm^{-1})^{-1}$] is the spectrally resolved differential Raman cross section, λ is the vacuum wavelength of the probe beam, and ℓ is the path over which the beams interact in the medium.

Early in 1977 our involvement in studies of nonlinear refractivity and dispersion in liquids [9] caused us to re-examine the possibility of using SRS as a precision tool for Raman spectroscopy. In contrast to earlier studies where high-power lasers were used to induce changes in transmission sufficiently large ($\geqslant$ 10%) to be conveniently detected, it was our intention to use highly stable lasers and sensitive detection techniques to measure very small induced gains. Indeed, if shot noise limits of detection could be realized, then transmission changes of less than 10^{-7} would be measurable. This would imply that even very modest cw laser sources would be adequate to induce measurable transmission changes, sufficient for precision spectroscopic studies, even in some gaseous samples.

The first cw SRS studies were reported [14] late in 1977 and entailed the use of a 100-mW amplitude-modulated dye laser as the pump source and a single-mode HeNe laser as the probe. The 0.5 cm^{-1} linewidth of the pump laser was quite sufficient for demonstrating the feasibility of the technique in liquids and Raman spectra of a variety of liquid samples were obtained by observing the induced gain profile as a function of pump laser wavelength [13-15]. The true significance of these results, however, lies not in the spectral data obtained, but in the fact that these studies demonstrated that cw laser sources could be employed in an SRS study and that probe sources could be operated with near shot-noise-limited amplitude stability.

Subsequent gas-phase studies of diatomic molecules [13,16] established the ability of SRS to supply detailed lineshape and pressure broadening data. It was not until the publication [17] of the first fully resolved Q-branch spectrum of the ν_1 mode in methane (Fig. 2) that the significance of purely spectroscopic applications of SRS became apparent.

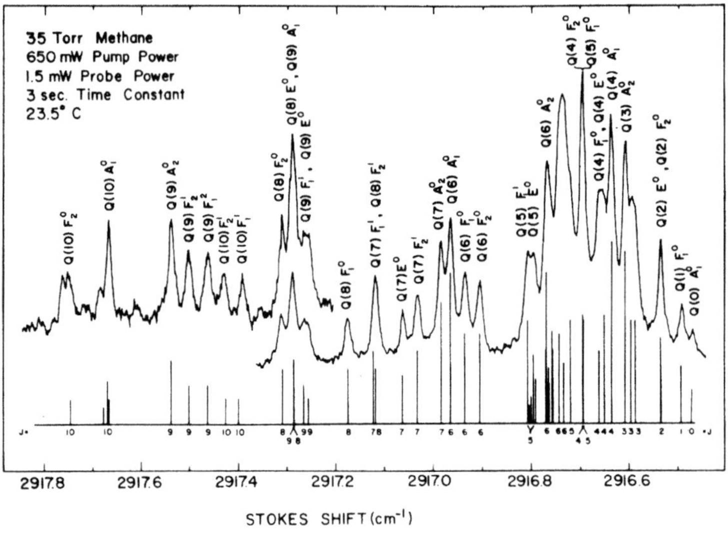

Fig. 2 Stimulated Raman gain spectrum of the ν_1 band of methane obtained using cw laser sources and lock-in amplifier detection techniques. The sample pressure was 35 Torr and sample was located in an optical cell that provided a multipass gain of 50X. The calculated positions and strengths of all lines having $J \leq 10$ are indicated beneath the spectrum.

Although the study of the ν_1 fundamental in methane is an indisputable demonstration of the potential utility of cw SRS for high-resolution Raman studies, it also serves to illustrate the limitations of this type of system. Although the methane cross section is quite large, it is spread out in a high-resolution spectrum and each individual resolved transition can be quite weak. Consequently, the spectrum is observable with cw lasers only with the aid of multipass techniques.

A. Pulsed Stimulated Raman Studies

In seeking to further improve the sensitivity of SRS it became apparent that the sensitivity scaled as $P(\Omega)\,\{P(\omega)/\Delta F\}^{1/2}$ where $P(\Omega)$ and $P(\omega)$ are the pump and probe power levels, respectively, and ΔF is the detection bandwidth. The largest gain will be realized by increasing the pump power $P(\Omega)$ to its practical limit. The probe power $P(\omega)$ may then be adjusted to avoid both saturation of the Raman transition and of the detector. Of course, the use of a pulsed pump source results in an increased detection bandwidth and a consequent increase in noise levels, but some of this will be compensated for in time averaging the signal over many pulses, thus effectively stretching the pulse duration.

Figure 3 shows a schematic representation of the pulsed laser version of the experimental apparatus detailed in Ref. 18. The pulsed pump laser source is a single-mode electronically scannable dye oscillator of ~ 1-MHz linewidth which has been pulse amplified up to 1 MW by three dye amplifiers pumped with a frequency-doubled Nd:YAG laser source. An electro-optic gate shutters the oscillator synchronously with the amplifier pump source, thus avoiding any possible thermal beam distortions that could result from absorption of the cw oscillator beam by the dye amplifiers.

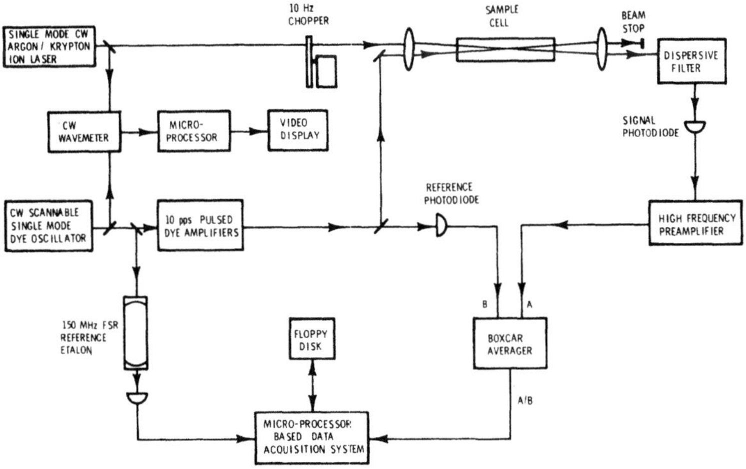

Fig. 3 Schematic representation of the apparatus developed for high-resolution pulsed stimulated Raman specroscopy experiments.

The system operates at a 10 pps repetition rate and emits pulses of ~ 60-MHz spectral linewidth (FWHM) and 12-nsec duration. The probe source is either an argon or krypton ion laser operating in a single longitudinal mode with nominally 0.6-watt output at 514.5 or 647.1 nm, respectively. To prevent premature saturation of the detector system, the probe source is gated to produce 100 μs pulses at 10 pps by a mechanical chopper which also sends signals to synchronously trigger the electro-optic gate and Nd:YAG laser. An interferometric cw wave meter [18-21] and scanning Fabry-Perot interferometers continuously monitor the spectral character and absolute frequencies of both laser sources.

The pump and probe beams are both sent through a 40-cm-focal-length lens that directs the two beams to a common focus in the gas sample cell. Here the stimulated Raman effect produces a small transient signal on the probe source as a result of the 12 ns pump. Two Pellin-Broca prisms and a grating-pinhole pair are used to isolate the probe signal from any residual pump radiation that is not eliminated by the beam stop.

The probe signal is detected by a fast silicon photo-diode. The transient electrical signal is isolated by high-pass filters, which eliminate the 100 μs gated waveform, leaving the 12 ns perturbation. This signal is then directed to a boxcar averager that operates with a 10 ns gate. Here a 12 ns signal derived from the pump laser output serves as a normalization source for improved stability in the signal averaging process.

In Fig. 4 we illustrate the improved sensitivity of pulsed SRS via spectra taken in the vicinity of the ν_1 fundamental in SF_6. In the 4-Torr spectrum (Fig. 4a), the 0.002 cm^{-1} instrumental bandwidth fully resolves the rotational structure in this Q-branch [22]. In contrast, the 100-Torr spectrum (Fig. 4b) of the same region is collisionally broadened to a point where the rotational structure is completely blended, leaving the broad rotational contours of the fundamental band and several underlying hotbands. Even in this latter spectrum, however, which required a scanning time of 20 minutes, the resolution is considerably better than the highest resolution (0.05 cm^{-1}) capabilities offered by spontaneous Raman techniques. Such a spectrum recorded on a high-resolution spectrographic instrument designed for such studies would require over 60 hours of exposure [23]. Due to the higher efficiency of photoelectric detection, however, a Raman spectrophotometer with comparable resolving power (0.05 cm^{-1}) could yield such a spectrum in about an hour [24,25]. The weak signal levels and limited resolving power obtained from conventional spectroscopic tools are characteristic of the spontaneous Raman methods when applied to very-high-resolution gas-phase studies.

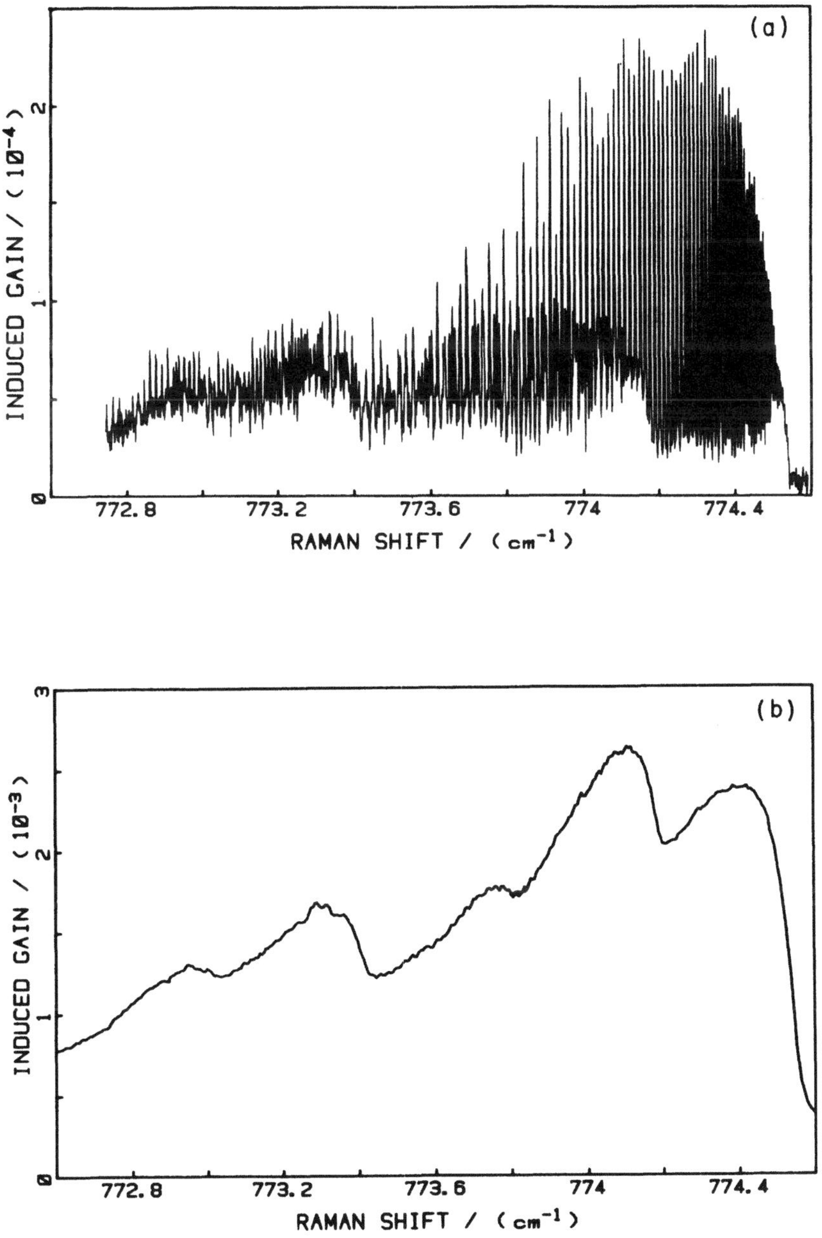

Fig. 4 Stimulated Raman spectra of the ν_1 fundamental of SF_6 taken with an instrumental resolution of 0.002 cm^{-1}. (a) 100 minute scan at 4 Torr, (b) 20 minute scan at 100 Torr.

3. APPLICATIONS

Our applications of stimulated Raman spectroscopy have progressed through three main areas [18]. Initial work exploited the ultra-high resolution capabilities of the technique in a number of fundamental studies of the spectroscopy of ν_1 modes in spherical-top molecules. Subsequent efforts, designed to take advantage of spatial resolution capabilities, led to the application of SRS to studies in supersonic molecular free-expansion jets. Our most recent experiments intended to illustrate a capability for temporal resolution, use SRS to probe the dynamics of infrared excitation in the prototype system: CO_2-laser-pumped SF_6.

A. Spectroscopy of Gas-Phase Molecules

Purely spectroscopic applications of SRS originated with the observation of the ν_1 spectrum of methane [17]. This work constituted the first occasion in which a Raman technique was successful in obtaining a fully resolved Q-branch spectrum of a complex molecule. Encouraged by the theoretical support and collaboration of Robin McDowell and Chris Patterson of Los Alamos National Laboratory, ν_1 spectra of several other light spherical-top molecules were obtained and analyzed [26-28]. These molecules constituted a special case for SRS for several reasons. First, the ν_1 hydrogen stretching vibration is a strong Raman mode, and it occurs at a large (2000 to 3000 cm^{-1}) Stokes shift that turns out to be experimentally convenient for available lasers. Second, in spherical-top molecules, only the Q-branch of the symmetric ν_1 mode appears. In spite of an increased interest in the spectroscopy of spherical-top molecules, these bands had never been previously resolved by Raman techniques, and thus they presented themselves as obvious subjects for detailed study using the high-resolution capabilities of SRS.

The most striking observation in the Q-branch spectra of the light spherical tops (CH_4, CD_4, and SiH_4) was the appearance of distinct splitting of the higher J rotational transitions. This added complexity, particularly for the methanes, makes it very difficult to visualize any progression in J at all.

For the totally symmetric ν_1 modes of spherical-top molecules this splitting is primarily due to tensor components of the centrifugal distortion terms that are either fourth- or sixth-rank in J. Using an expression given by Moret-Bailly [29] for the fourth-order Hamiltonian including these terms, all of the ν_1 spectra observed to date have been successfully analyzed and fit to obtain the relevant spectroscopic parameters.

Subsequent spectroscopic studies [22,30-31] investigated the ν_1 spectra of the heavy spherical-top molecules CF_4 and SF_6. In marked contrast to the lighter spherical tops, these molecules exhibit extremely regular spectra (Fig. 4a) with absolutely no evidence of tensor splitting of the individual rotational lines. In fact, their spectra can be fit readily by a simple rigid rotor expression $\nu(J) = \alpha + \Delta\beta J(J + 1)$ requiring only the two constants α and $\Delta\beta$.

Due to their smaller rotational B values, the spectra of these heavier molecules are spread out over many more rotational levels and are consequently over an order of magnitude weaker in peak signal strength. The individual rotational lines are more closely spaced than in the lighter molecules, and thus resolution becomes critically important. Although Doppler broadening is not as important as in the methanes, collision broadening can contribute significantly, as shown in Fig. 4b. The importance of maintaining high-sensitivity while developing SRS as a high-resolution spectroscopic technique is thus readily apparent from these studies.

B. Molecular Jet Studies

Over the past several years molecular beam techniques have been applied more and more frequently to laser spectroscopy problems. One of the major advantages is the simplification of gas phase spectra that results from the rotational and vibrational state cooling occurring in the molecular expansion. Recent laser excited fluorescence studies [32] are excellent examples of the dramatic improvement available in simplifying the electronic spectra of polyatomics; similar benefits should be realizable in Raman spectra.

In addition to spectral simplification, the adiabatic cooling in a free expansion also reduces the widths of velocity distributions and thus provides a means for improving the resolution of Doppler-limited spectra. However, unlike saturation spectroscopy techniques, which can produce sub-Doppler Raman linewidths in room temperature gases [33], molecular beam techniques are not really "sub-Doppler" in the same sense, but rather they are actually "reduced-Doppler" techniques.

Following the development of new high density pulsed molecular beam sources [34], Duncan and Byer [35] proposed that ultra-high-resolution, reduced-Doppler-width Raman spectra were now realizable with CARS. (It should be noted here that the development of such high-density molecular beams was crucial for the CARS approach with its quadratic dependence of signal on molecular density.) Motivated by Duncan and Byer's proposal, we began a collaborative effort with J. Valentini of Los Alamos

National Laboratory to obtain reduced-Doppler stimulated Raman spectra using a molecular beam system.

Our first molecular expansion SRS results [36] were obtained using a cw free-expansion jet of CH_4. Shown in Fig. 5, our results clearly demonstrated that Raman spectra could be observed, with linewidths less than the room temperature Doppler width, by using coherent Raman techniques in conjunction with molecular beam techniques.

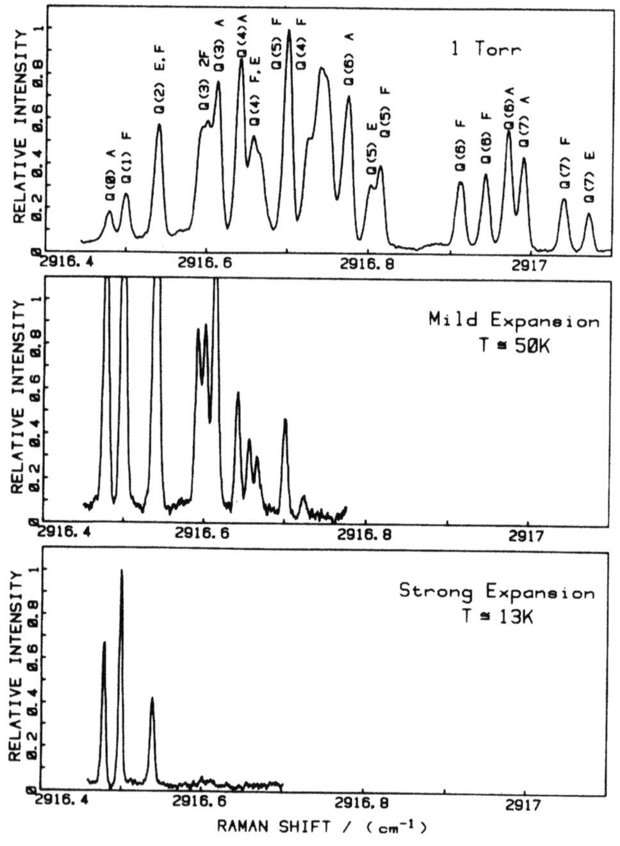

Fig. 5 Inverse Raman spectra of ν_1 fundamental of $^{12}CH_4$. (Top) Static gas, 1 Torr, 300K. (Center) cw molecular jet, 350 μm nozzle, 4.8 psia backing pressure, 1.8 mm downstream. (Bottom) cw molecular jet, 100 μm nozzle, 55 psia backing pressure, 4.5 mm downstream.

The room-temperature Doppler width for this transition in methane is 270 MHz, as illustrated in the top spectrum. Under mild expansion conditions (center spectrum) a temperature of 50°K is reached (1.8 mm downstream from the nozzle) that would give a minimum Doppler width of 110 MHz were it not for the relatively large angular sampling of the expansion at this point. The improvement in resolution, however, is still significant, particularly as seen in the well-resolved tensor split components of Q(3) and Q(4). Under stronger expansion conditions (bottom spectrum) the resolution improves even further, both because of the reduced temperature, 13°K in this case, and because at 4.5 mm downstream we are probing a smaller angular range of the expansion. In this particular example, a residual Doppler width of 105 MHz was measured which is almost entirely due to the angular spread of the free-expansion. Following these initial studies a pulsed valve was installed in the molecular beam apparatus which, by reducing the open-time duty cycle, allows for much higher molecular density operation while maintaining reasonable pumping speed requirements. The pulsed valve, designed by Valentini and Cross [37], is a modification of a commerical piezoelectric leak valve (Veeco, PV-10). This modified valve has been operating in our beam apparatus without failure since its installation.

The SF_6 spectrum shown in Fig. 6 illustrates the superb signal levels that can be achieved, even with weak transitions, using the pulsed free-expansion jet. Note that due to the reduced rotational and vibrational partition functions at reduced temperatures, the signal level in the pulsed molecular jet is actually stronger than that observed at 4 Torr under static cell conditions (Fig. 4a). Furthermore, because of the reduced kinetic temperature, we are able to achieve a reduction in collision linewidth for a given density.

The most promising future applications for molecular jets will take advantage of the reduced collisional relaxation rates that are present at low temperatures. This phenomenon makes it possible to study any number of transient species that would otherwise be very difficult to observe using spontaneous Raman techniques.

C. Time Resolved Studies

As a prototypical illustration of the capability of SRS to make time-resolved measurements on a transient specie, we have recently undertaken the study of the Raman spectroscopy of SF_6 during and following excitation of the ν_3 vibration by CO_2 laser radiation [38].

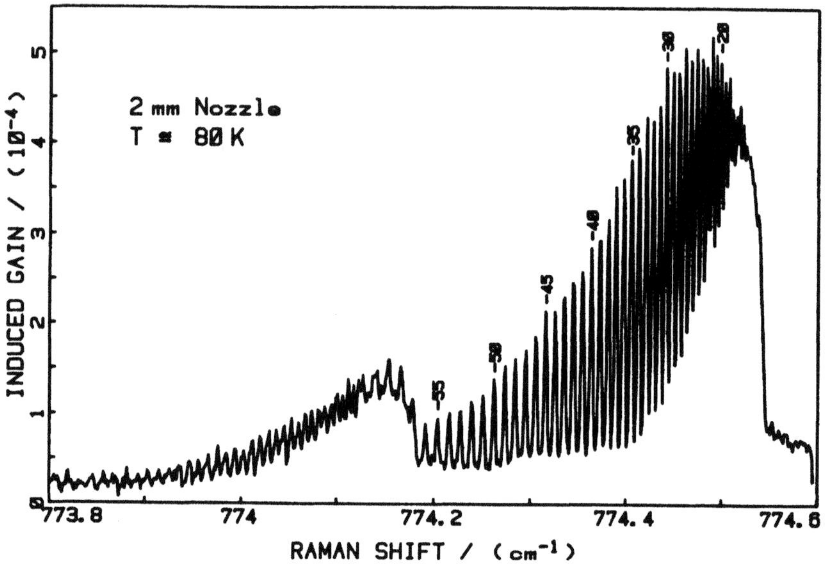

Fig. 6 Raman gain spectrum of the ν_1 band of SF_6 obtained in a pulsed free-expansion jet by probing 7 mm downstream from a 2 mm nozzle. Backing pressure = 86 psia. Instrumental resolution is ~70 MHz.

The energy diagram of Fig. 7a illustrates the use of the ν_1 band as a means of monitoring the rotationally specific removal of population from the ground state via CO_2 laser excitation in the ν_3 band. Experimentally this is achieved by operating the SRS system at 10 Hz and synchronizing a CO_2 laser to excite the sample at 5 Hz. Two alternately triggered boxcar averagers are then used to obtain both the unperturbed, "normal," spectrum and the "perturbed," CO_2 laser pumped spectrum.

The spectrum shown in Fig. 8 illustrates the selective depletion of the ν_1 fundamental when pumped by the P(18) line of the grating tuned CO_2 laser TEA laser. The spectrum was recorded at the peak of the 100 nsec CO_2 laser probe ($\pm$25 ns) at a fluence level of ~ 80 kW/cm^2. Comparison with the "normal" spectrum shows significant depletion in the J = 33 region where the P(18) line of CO_2 at 945.980 cm^{-1} overlaps the ν_3 P branch [39]. Additional scans taken at longer delay times (up to 1 μs) show a thermalization of the rotational states, which results in a depletion of the entire band and a filling of the "hole" near J = 33. At higher fluence levels (~ 600 kW/cm^2), more extensive depletion is seen over the entire band, even at short delay times.

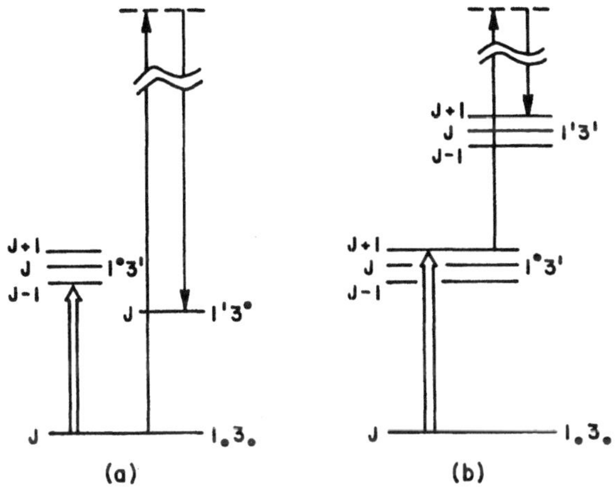

Fig. 7 Energy level schematic illustrating the use of SRS to probe SF_6 molecules excited by a CO_2 laser in the ν_3 band. (a) Monitoring of the ν_1 fundamental to measure ground state depletion. (b) Monitoring of the $\nu_1 + \nu_3 \leftarrow \nu_3$ hot band populated by the CO_2 laser.

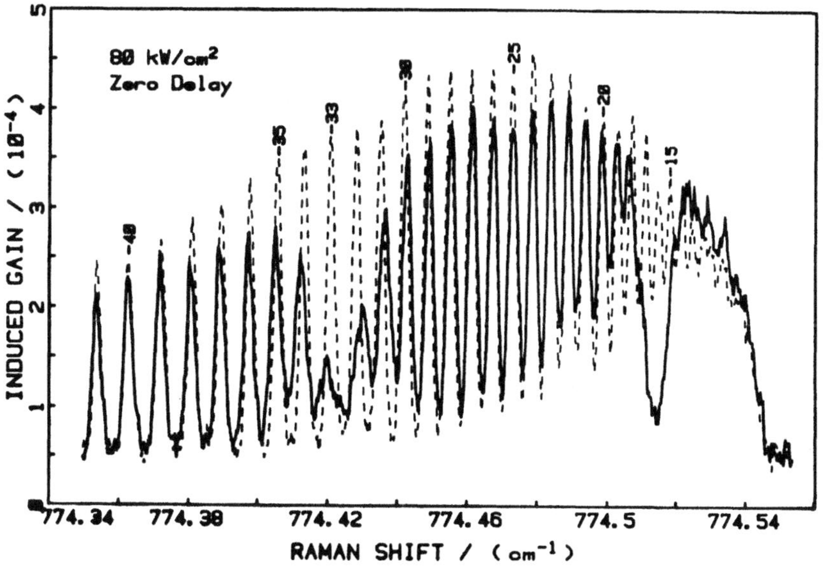

Fig. 8 Raman gain spectra of the ν_1 fundamental of SF_6 showing the effect of excitation of the ν_3 band near P(33) by the P(18) line of the CO_2 laser. The CO_2 laser "perturbed" spectrum (solid line) is compared to the "normal" spectrum (dashed line).

The lines around the region of strongest depletion are also found to be strongly pushed away from the frequency of the saturated line. This frequency shift arises from the near resonant interaction of the CO_2 laser intensity with the ν_3 P-branch transitions. The resultant up- and down-shifting of the ν_3 levels are consequently reflected in the ν_1 transitions through the Stark shift in their common ground states. Careful examination of Fig. 8 reveals that such a shift is also present in this spectrum, although not nearly as evident as it would be at higher fluences.

An unexpected feature of this spectrum is the appearance of a strong selective depletion of the band, in the vicinity of J = 16. Indeed, this observation could not be accounted for on the basis of direct excitation of the ground state via either one-, two-, or three-photon processes at the CO_2 laser frequency [40-41]. In addition, no effect on J = 16 levels was observed by Steinfeld, et al. [42], in an IR-IR double resonance study which employed a diode laser to monitor ground state populations via changes in the ν_3 absorption spectrum.

We have subsequently explained these observations on the basis of an AC Stark shift which affects only the upper level of the Raman transition. Specifically, the CO_2 P(18) line is very nearly resonant with the R(16) transition of $\nu_1 + \nu_3 \leftarrow \nu_1$. This results in spectral shifts of those lines adjacent to J = 16 in the ν_1 Raman spectrum. In a manner very similar to that observed at J = 33, the Stark shifts near J = 16 are in a direction away from the resonant frequency, such that the CO_2 laser essentially carves out a spectral depletion without actually affecting the ground state populations.

In addition to detecting depletion of the ground state populations, SRS techniques can also yield both spectroscopic and dynamic information on the excited species. This is diagramed in Fig. 7b where SRS is employed to monitor the CO_2 laser populated $\nu_1 + \nu_3 \leftarrow \nu_3$ hotband. The free-expansion jet becomes an essential part of this experiment because of the necessity to eliminate thermally populated "hotbands" which, at room temperature, would obscure the presence of the CO_2-laser-pumped species. Figure 9 illustrates the $\nu_1 + \nu_3 \leftarrow \nu_3$ spectrum, with band origin at 771.637 cm^{-1}, as recorded at short delays following excitation by the P(14) line of CO_2. The rotational structure in this band is not as well resolved as in the ν_1 fundamental due to the effects of Coriolis splitting of the ν_3 mode. The observed structure is reproducible, however, and can be analyzed using ΔB from the ν_1 fundamental spectrum and the value $\xi = 0.690$ obtained for the Coriolis coupling coefficient from the analysis of ν_3 [43]. Note that in addition to new spectral information, we are also able to observe specifically the nonthermal rotational state distribution that is created in the $v_3 = 1$ level during the CO_2-laser excitation process.

These preliminary results clearly demonstrate the applicability of SRS techniques to studying the dynamics of multi-photon excitation processes in molecular systems. They illustrate the potential capability for obtaining time resolved spectral information with SRS, and may lead the way to eventual Raman probing of molecular photo-fragments or other transient species.

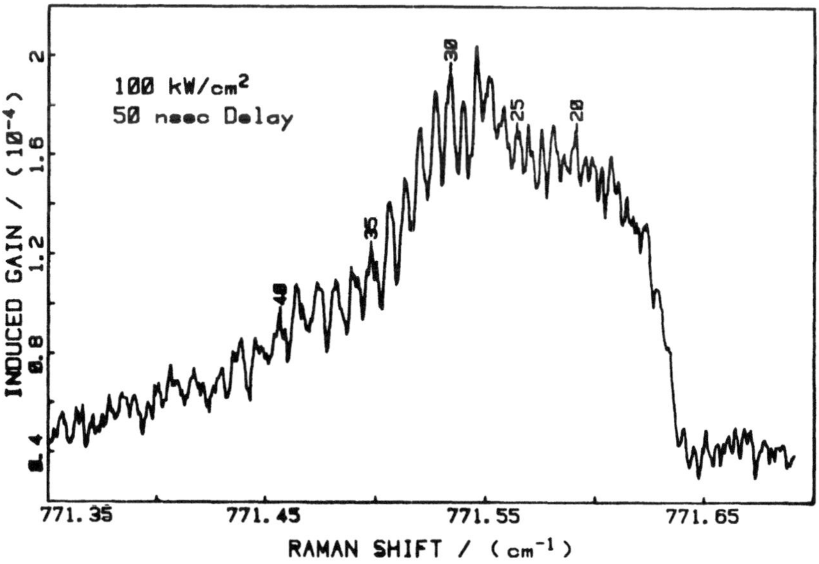

Fig. 9 Raman gain spectrum of the $\nu_1 + \nu_3 \leftarrow \nu_3$ transition in SF_6, observed during excitation of the ν_3 mode by a CO_2 laser. Tentative J assignments of the $\nu_1 + \nu_3$ spectrum are shown. The CO_2 P(14) line used here pumps the R(28) transition in ν_3 of SF_6. Note the nonthermal distribution of intensity in the region of J = 29.

4. CONCLUSIONS

The development of stimulated Raman spectroscopy over the past four years has clearly established the technique as a valuable tool in gas-phase Raman studies. In this brief article we have traced that development, detailed the capabilities of the stimulated Raman spectrometer, and surveyed a variety of applications of the system to problems in the gas phase. For a more thorough discussion of this topic, we refer the reader to Ref. 18.

SRS is clearly most attractive in ultra-high-resolution ($< 0.01\ cm^{-1}$) studies that are beyond the reach of spontaneous Raman spectroscopy. Collision broadening effects dictate that such studies be performed at low pressures in order to maximize sensitivity. Currently, sample pressures below approximately 5 Torr are required in order to achieve linewidths approaching our $0.002\ cm^{-1}$ instrumental resolution. Sensitivity considerations have thus limited our studies at these low pressures to relatively strong Raman scattering modes. It is our expectation that in the near future the introduction of higher-power sources and a multi-pas cell will significantly change this situation by increasing the overall sensitivity of SRS by at least two orders of magnitude. This should open up a much wider range of possible applications, which should, in turn, establish SRS as the standard for high-resolution Raman studies.

5. ACKNOWLEDGMENTS

We are pleased to acknowledge here the collaboration of C. W. Patterson and R. S. McDowell in the theoretical analysis of many of our spectra, and we thank them for many helpful discussions over the past few years. We are also grateful for the opportunity to collaborate with J. Valentini on the free-expansion jet studies and A. J. Grimley in the CO_2 laser pumped SF_6 experiments. Credit is particularly due to R. E. Asbill for his expert technical assistance throughout this research effort.

REFERENCES

1. Eckhardt, G., Hellwarth, R. W., McClung, F. J., Schwarz, S. E., Weiner, D., and Woodbury, E. J., Phys. Rev. Lett., 9, 455 (1962).

2. Jones, W. J., and Stoicheff, B. P., Phys. Rev. Lett., 13, 657 (1964).

3. Lallemand, P., Simova, P., and Bret, G., Phys. Rev. Lett., 17, 1239 (1966).

4. Bloembergen, N., Bret, G., Lallemand, P., Pine, A., and Simova, P., IEEE J. Quantum Electron., QE-3, 197 (1977).

5. Lallemand, P., and Simova, P., J. Mol. Spec., 26, 262 (1968).

6. Levenson, M. D., and Bloembergen, N., J. Chem. Phys., 60, 1323 (1974).

7. Lotem, H., Lynch, R. T., Jr., and Bloembergen, N., Phys. Rev. A, 14, 1748 (1976).

8. Owyoung, A., Opt. Comm., 16, 266 (1976).

9. Owyoung, A., and Peercy, P. S., Appl. Phys. Lett., 48, 674 (1977).

10. Reinhold, I., and Maier, M., Opt. Comm., 5, 31 (1972).

11. Kincaid, B. E., and Fontana, J. R., Appl. Phys. Lett., 28, 12 (1976).

12. Weber, A., in The Raman Effect Vol. 2 (A. Anderson, ed.) Marcel Dekker Inc., New York, NY, 1973, p. 543.

13. Owyoung, A., in Chemical Applications of Nonlinear Raman Spectroscopy (A. B. Harvey, ed.) Academic Press, New York, NY, 1981, p. 281.

14. Owyoung, A., and Jones, E. D., Opt. Lett., 1, 152 (1977).

15. Owyoung, A., Opt. Comm., 22, 323 (1977).

16. Owyoung, A., Opt. Lett., 2, 91 (1978).

17. Owyoung, A., McDowell, R. S., and Patterson, C. W., Chem. Phys. Lett., 59, 156 (1978).

18. Esherick, P., and Owyoung, A., in Advances in Infrared and Raman Spectroscopy, Vol. 9 (R. J. H. Clark and R. E. Hestor, eds.) Heyden & Son Ltd., London, 1982.

19. Petley, B. W., and Morris, K., Opt. and Quantum Electron., 10, 277 (1978).

20. Hall, J. L., and Lee, S. A., Appl. Phys. Lett., 29, 367 (1976).

21. Kowalski, F. V., Hawkins, R. T., and Schawlow, A. L., J. Opt. Soc. Am., 66, 965 (1976).

22. Kowalski, F. V., Teets, R. E., Demtroder, W., Schawlow, A. L., J. Opt. Soc. Am., 68, 1611 (1978).

23. Aboumajd, A., Berger, H., and Saint-Loup, R., J. Mol. Spec., 78, 486 (1979).

24. Fletcher, W. H., Rayside, J. S., and McLendon, W. B., J. Raman Spec., 7, 205 (1978).

25. Rubin, B., McCubbin, T. K., and Polo, S. R., J. Mol. Spec., 69, 254 (1978).

26. Owyoung, A., in Laser Spectroscopy IV (H. Walther and K. W. Rothe, eds.) Springer-Verlag, Berlin, 1979, p. 175.

27. McDowell, R. S., Patterson, C. W., and Owyoung, A., J. Chem. Phys., 72, 1071 (1980).

28. Owyoung, A., Esherick, P., Robiette, A. G., McDowell, R. S., J. Mol. Spectrosc., 86, 209 (1981).

29. Moret-Bailly, J., Cahiers Phys., 15, 237 (1961); J. Mol. Spectrosc., 15, 344 (1965).

30. Owyoung, A., and Esherick, P., in Lasers and Applications (W. O. N. Guimaraes, C. T. Lin, and A. Mooradian, eds.) Springer-Verlag, Berlin, 1981, p. 67.

31. Esherick, P., Owyoung, A., and Patterson, C. W., J. Mol. Spectrosc., 86, 250 (1981).

32. Smalley, R. E., Whartan, L., Levy, D. H., Accts. Chem. Res., 10, 139 (1977).

33. Owyoung, A., and Esherick, P., Opt. Lett., 5, 421 (1980).

34. Gentry, W. R., and Giese, C. F., Rev. Sci. Instrum., 49, 595 (1978).

35. Duncan, M. D., and Byer, R. C., IEEE J. Quantum Electron., QE-15, 63 (1979).

36. Valentini, J. J., Esherick, P., and Owyoung, A., Chem. Phys. Lett., 75, 590 (1980).

37. Cross, J. B., and Valentini, J. J., Rev. Sci. Instrum., 53, 38 (1982).

38. Esherick, P., Grimley, A. J., and Owyoung, A., in Laser Spectroscopy V (A. R. W. McKellar, T. Oka, and B.P. Stoicheff, eds.) Springer-Verlag, Berlin, 1981.

39. McDowell, R. S., Galbraith, H. W., Krohn, B. J., Cantrell, C. D., and Hinkley, E. D., Opt. Comm., 17, 178 (1976).

40. Patterson, C. W., private communication.

41. Patterson, C. W., Krohn, B. J., and Pine, A. S., Opt. Lett., 6, 39 (1981).

42. Dubbs, M., Harridine, D., and Steinfeld, J. I., private communication.

43. Brode, Ch. J., et al., in Laser Spectroscopy IV (H. Walther and K. W. Rothe, eds.) Springer-Verlag, Berlin, 1979, p. 142.

INVERSE RAMAN AND RAMAN GAIN SPECTRA OF LIQUIDS AND SOLUTIONS

M. D. Morris and C. E. Buffett

Department of Chemistry
University of Michigan
Ann Arbor, MI 48109, USA

1. INTRODUCTION

Although it is the oldest form of coherent Raman spectroscopy, inverse Raman spectroscopy is less widely practiced than CARS. The inverse Raman effect was first observed by Jones and Stoicheff [1]. That pioneering paper proposed the use of the inverse Raman effect for the study of short-lived species and suggested that rotational and electronic inverse Raman effects could be observed, as well as vibrational spectra. The possible utility of inverse Raman spectroscopy as a means of observing Raman spectra over an intense fluorescence background was very clearly stated.

The work of Jones and Stoicheff was cited frequently throughout the 1960's and gave rise to a small group of practitioners. However, the weakness of the effect and the lack of versatile instrumentation for observing it, prevented widespread investigation or application. Even less attention was paid to Raman gain spectroscopy, which remained virtually untouched during this period.

Inverse Raman spectroscopy was largely eclipsed by CARS in the early 1970's. The successes of CARS in generating high resolution gas phase spectra and solution spectra in luminescent media overshadowed important work in inverse Raman spectroscopy. In 1974 Yeung [2] and Werncke and co-workers [3] measured inverse Raman spectra from broadband dye lasers using Q-switched ruby lasers as pump sources. In both cases, the fluorescence rejection possibilities were pointed out. In fact, observation of Raman spectra of fluorescent media appears to have been a

W. Kiefer and D. A. Long (eds.), Non-Linear Raman Spectroscopy and Its Chemical Applications, 519–531.

major goal of Yeung, while the Werncke group stressed the ability to measure spectra on the nanosecond time scale as well as fluorescence rejection. Both groups used photographic detection, which offered limited sensitivity. However, the Werncke group used intra-cavity generation, affording a factor of about 10 improvement in detection limits.

Although its feasibility was well-established, inverse Raman spectroscopy remained decidedly less popular than CARS throughout the mid-1970's. The introduction of modulation techniques to inverse Raman and Raman gain spectroscopy by Owyoung [4,5] brought a further increase in sensitivity of these techniques. Owyoung's work also marked the beginning of widespread interest in inverse Raman and Raman gain spectroscopy as a serious alternative to CARS. At present, a variety of direct absorption and modulation schemes are used to obtain high quality inverse Raman spectra in condensed and gas phases. The same measurement techniques are applicable to Raman gain spectroscopy.

2. THEORY

The theory of inverse Raman and Raman gain spectroscopy is discussed in detail elsewhere in this volume. We present only a summary here. The inverse Raman absorption $\Delta P_1/P_1$, is defined by equation 1.

$$\frac{\Delta P_1}{P_1} = \frac{96\pi^2\omega_1\omega_2}{c^3 n} \mathrm{Im}\chi^{(3)}_{iiii}(-\omega_1,\omega_1,\omega_2,-\omega_2)P_2 2d \qquad (1)$$

Here, P_1, P_2 are the power densities in the pump and probe beams respectively, at frequencies ω_1 and ω_2, n is the refractive index of the sample and $\mathrm{Im}\chi^3$ is the imaginary part of the third-order susceptibility, which is proportional to the Raman cross-section and the concentration of the Raman-active molecule. The path length is taken to be 2d, where d is the distance from the center of the sample cell to one end. The center-to-end definition will be used later.

The same expression holds for Raman gain, if the pump beam is defined as the laser of higher optical frequency, and the probe as the lower frequency beam. The gain signal is an increase in pump intensity, rather than an attenuation. Otherwise, away from an electronic resonance, the techniques are equivalent, and the use of one rather than the other is determined by experimental convenience.

Equation 1 is the small signal approximation to an

exponentially increasing build-up of inverse Raman or Raman gain intensity. The small signal approximation is sufficient for gain or attenuation of less than about 5%. In practice, very large signals are not encountered except perhaps in neat solvents or gases at very high pressure. Thus, the signal should remain exactly analogous to a spontaneous Raman signal.

Both polarization components of the spontaneous Raman signal are observable in inverse Raman and Raman gain spectroscopy. If pump and probe laser are linearly polarized in the same plane, the parallel polarization component of the Raman signal is observed. If they are polarized 90° to each other, the perpendicular component is obtained.

Equation 1 assumes that power densities remain constant as the beams propagate through the sample. For this geometry, one would predict that indefinitely large signals could be obtained by the use of very long sample paths. In practice, almost all work is done using short path length cells, typically 1-10 cm, and focused beams. Under these conditions a limiting signal is reached when the path length reaches 50-75 confocal parameters [6].

The reason is that the power density is not constant in a focused beam, but increases to a limit at beam focus and then decreases again. For the assumption of a Gaussian beam, equation 2 holds.

$$P(d) = \frac{P}{\pi \omega_o^2} \frac{z_o^2}{(d^2 + z_o^2)} \qquad (2)$$

Here, P(d) is the power density of the beam at distance d from focus, P is the power of the beam and z_o is the confocal parameter of the lens/laser pair.

If both the pump and probe beams are assumed to be identical Gaussians, substitution of equation 2 for the power density - cell length product in equation 1 leads to the conclusion that inverse Raman or Raman gain signals should obey equation 3.

$$\frac{\Delta P_1}{P_1} = \frac{96\pi^2 \omega_1 \omega_2}{c^3 n} \mathrm{Im}\chi^{(3)}_{iiii}(-\omega_1,\omega_1,\omega_2,-\omega_2)P_2 \arctan \frac{d}{z_o} \qquad (3)$$

3. EXPERIMENTAL TECHNIQUES

Two general classes of experiment are in fairly widespread use. These are the intra-cavity absorption method and the modulation/demodulation method.

The Werncke group has developed the intra-cavity absorption method extensively [3, 7]. Inverse Raman absorption or Raman gain is measured from the output of a broad band dye laser. In most experiments the dye laser is pumped by the second harmonic of a ruby laser, while the residual ruby fundamental pumps the inverse Raman or Raman gain process. A photographic plate or an optical multichannel analyzer serves as the detector system.

The intracavity method has the advantage of experimental simplicity and an intrinsic multiplex advantage. Although intra-cavity absorption from a CW laser is enhanced by a factor of as much as 10^5, enhancements of only 5-10 are typical for short pulse systems. The enhancement factor is approximately equal to the number of cavity round trips a photon can make. The intra-cavity enhancement appears to compensate for the fact that the system operates with an unfocused beam, causing a decrease in sensitivity.

Although the Werncke work has been done with ruby lasers, the same configuration can be used with any short pulse lasers, such as Nd/YAG or excimer/narrow band dye/broad band dye. While the use of unfocused beams decreases the sensitivity of the experiment somewhat, it also simplifies the alignment greatly.

More commonly, modulation/demodulation schemes are used to achieve good sensitivity in inverse Raman or Raman gain experiments. These derive from experiments of Owyoung and co-workers, who used a modulated CW dye laser to observe Raman gain from a single mode He-Ne laser [4,5]. This experiment was reported as 'cw stimulated Raman spectroscopy.' Electro-optic modulation of the dye laser was used, and a lock-in amplifier was employed to extract the Raman signal. The sensitivity of the experiment was improved by multi-passing.

Modulation/demodulation achieves two goals simultaneously. First, it allows good sensitivity, since noise levels can approach the shot noise limit. Second, it allows measurements with laser-limited resolution, rather than monochromator-limited resolution. In solution, the first property is extremely important, while in the gas phase, both advantages are used. The beam quality and band width of a CW dye laser are superior to those obtainable with pulsed dye lasers. However, a pulsed dye laser is simpler to use than a CW device. With oscillator-amplifier systems, the beam profile can approach Gaussian. The band width of the typical pulsed dye laser is 0.1-1 cm^{-1} and, even with an etalon, band widths of less than 0.03 cm^{-1} are not commercially available. For condensed phases, Raman bandwidths are rarely less than 2 cm^{-1}, so that the available resolution is quite adequate. Thus, pulsed systems may be preferred for solution work.

The advantages of pulsed dye lasers and of modulation/demodulation measurement can be combined in a measurement scheme using a pulsed probe and a CW pump called ac-coupled inverse Raman by us [8, 9] and quasi-cw inverse Raman by Owyoung [10]. The experimental configuration used in our laboratories is outlined in Figure 1.

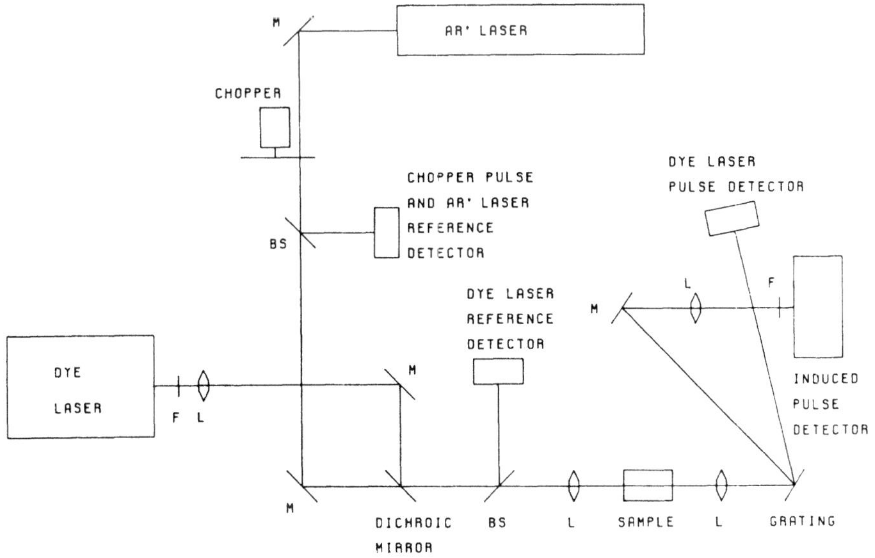

Fig. 1 Experimental configuration for AC-coupled inverse Raman spectroscopy. BS, 4% beamsplitter; F, dielectric filter; L, lens; M, mirror.

The beams from the argon ion laser and the pulsed dye laser are combined collinearly through a dichroic mirror and focused onto the sample cell with an achromat. After passage through the cell, the beams are re-collimated and then separated with a diffraction grating. The entire argon ion laser beam is focused onto a fast photodiode, which is capacitively coupled to a pulse amplifier.

Because the argon ion laser is a CW system, no signal goes through the coupling capacitor unless there is a change in the beam intensity. Thus, induced Raman signals appear as pulses, synchronous with the dye laser pulse, and free from the CW background signal. Gated integration serves to isolate these pulsed signals.

This modulation scheme provides better sensitivity than a direct subtraction measurement, since it does not depend on maintaining high common-mode rejection ratio in an RF difference amplifier [11, 12]. We have reported detection limits similar to those obtainable in spontaneous Raman spectroscopy, using a nitrogen laser pumped dye laser (pulse width 6 nsec, peak power 100 kw). Better sensitivity could be obtained with more powerful dye lasers, or dye lasers with better beam quality.

The use of a dichroic mirror as a beam combiner maximizes the signal from the system, by allowing colinear beam propagation. However, the dichroic mirror limits measurements to Raman shifts greater than about 600 cm^{-1}. A crossed beam geometry, or the use of counter-propagating beams, allows measurements of Raman shifts as small as the sum of the laser band widths [12].

For low CW laser powers, the mechanical chopper is not necessary. Its function is to limit the laser power incident on the photodiode in order to prevent damage to the device. In our lab we use a wheel rotating at about 30 Hz, with a notch which generates a pulse width of about 150 microsec. Our pulse amplifiers have essentially no frequency response below 100 kHz, so that the opening and closing of the chopper do not generate any signal. With the protective chopper, we have been able to use the same photodiode for over three years with no evidence of damage.

A single longitudinal mode laser is required for this experiment, but not for bandwidth. In the absence of an etalon, the mode beating signal generated by a typical argon ion laser is much larger than the desired inverse Raman signal. This signal, which occurs at a frequency of 150-200 MHz for typical cavity lengths, is sampled by the gated integrator, and generates artifacts which obscure the Raman bands.

Stone has used lengths of silica fiber to obtain long interaction lengths for inverse Raman measurements using a 20 milliwatt krypton ion laser and a Xenon arc lamp [13]. By using internal diameters of about 4 microns, power densities of about 0.1 megawatt/cm^2 are obtained. The small diameter, combined with path lengths of 1-20 meters, generates fairly intense spectra without recourse to modulation. Although the sensitivity of this technique is great, it has two disadvantages. First, it is limited to liquids whose refractive index is greater than that of quartz. Aqueous solutions can not be used. Second, the filling process is tedious, requiring as much as an hour for fibers of 10-20 meter length.

4. SPECTROSCOPIC PROPERTIES OF LIQUIDS AND SOLUTIONS

Figure 2 shows the inverse Raman spectrum of n-butyl acetate. The spectrum was taken in the rapid scan mode, as the average of 2 laser pulses, at 1 cm^{-1} intervals [14]. The lower wavenumber limit is set by the properties of the dichroic mirror, which switches from reflective to transmissive near 600 cm^{-1} to the red of the argon ion line. The upper wavenumber limit is the red end of the dye laser gain curve.

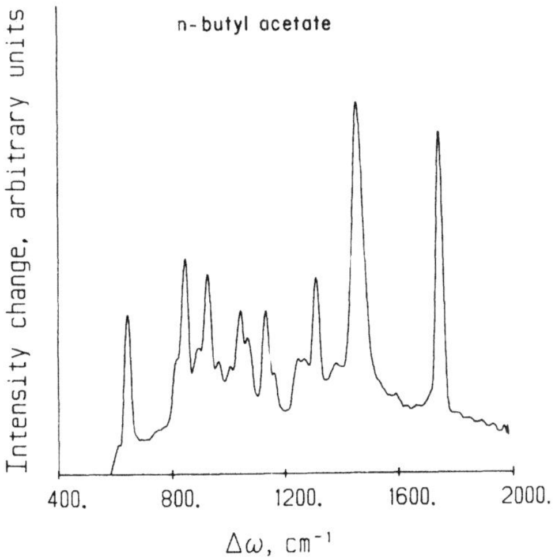

Fig. 2 Inverse Raman spectrum of neat n-butyl acetate, 2 laser pulses per data point.

Comparison of the spectrum to a spontaneous Raman spectrum of n-butyl acetate shows them to be identical. The example shown here illustrates two other features of inverse Raman spectroscopy. First, the signal/noise ratio is good. The scan rate is about 350 cm^{-1}/min, limited by the tuning rate of the grating. Second, despite the range of over 1000 cm^{-1} covered, no crossing angle adjustment is needed, unlike CARS or CSRS. In fact, colinear, or any other geometry can be maintained over the entire tuning range of the dye, allowing collection of spectra over a range of 2000 cm^{-1} or more. The linearity of inverse Raman signals with concentration is demonstrated by the data of Figure 3. Here, the height of the 1050 cm^{-1} band of the nitrate ion in water is plotted against concentration. Very good linearity is observed. Moreover, unlike CARS, IRS signals retain a Lorentzian band shape at all dilutions.

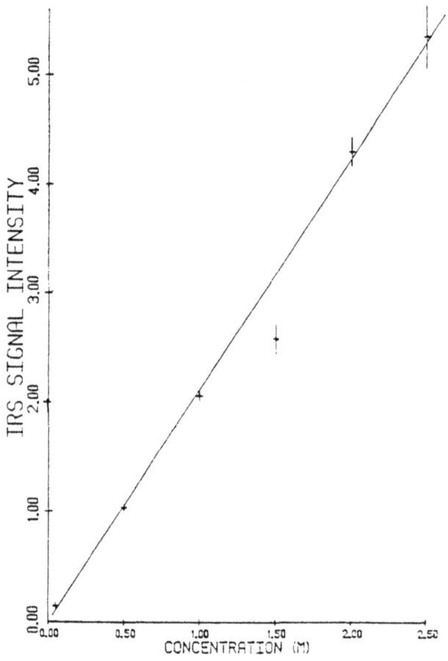

Fig. 3 Plot of NO_3^- 1052 cm^{-1} band height vs. concentration.

Figure 4 shows the IRS signal of 5 x 10^{-3} M nitrate in water. Although much smaller than the water 1650 cm^{-1} band, the nitrate signal remains Lorentzian.

Figure 4 demonstrates that the detection limits of inverse Raman spectroscopy are similar to those of normal Raman spectroscopy [14] under similar conditions. We have previously reported detection limits of 5 x 10^{-3} M NO_3^- using about 100 mW CW power and 250 microjoule pulses [9]. System improvements have lowered current detection limits to about 1 x 10^{-3} M, and some further improvement is possible without increasing laser power.

In principle, inverse Raman and Raman gain spectroscopy are

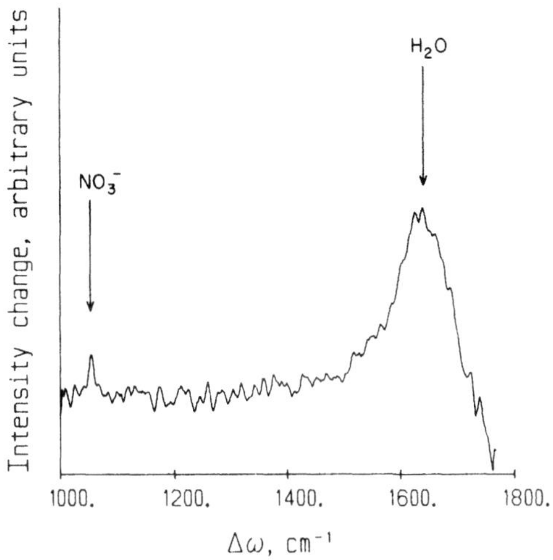

Fig. 4 Inverse Raman spectrum of 5×10^{-3} M NO_3^- in water. The water bending mode at 1645 cm^{-1} is clearly visible.

limited by the detector shot noise generated by the pump laser [15]. In solutions other factors set the observable detection limits. First, the solvent has a Raman signal which may be weak, but which is rarely identically zero. The desired signal must be subtracted from any solvent signal, and the noise associated with the solvent signal may ultimately be limiting. Second, all solvents have weak absorptions, typically 10^{-4} - 10^{-3} absorbance/cm. Thus, the solvent generates a small thermal lens signal, which can not be completely rejected. This signal may also limit low concentration measurements. Finally, the presence of small particles, which randomly scatter light, may generate small noise spikes, which can obscure small signals. In any case, in real samples it would appear difficult to approach the shot noise limit closer than perhaps a factor of two or three. However, the beam quality, peak power and repetition rate of currently available YAG or excimer laser pumped dye lasers should allow detection limits of 1×10^{-4} M NO_3^- or lower, without resonant enhancement. Detection limits for direct absorption techniques might be somewhat higher, limited by the dynamic range of array detectors.

Although there is no non-resonant background expected in inverse Raman or Raman gain spectroscopy, two-photon absorption can be limiting in some cases. Yeung has suggested that two-photon absorption will be a problem with high laser power densities [2]. Two photon absorption has been observed by Nestor as a weak, polarized band near the 790 cm^{-1} band of toluene [12]. In aqueous solutions, however, two photon absorption can be neglected.

Because inverse Raman signals are directly proportional to $Im\chi^3$, they are additive. There are no cross-term effects and no interferences between adjacent signals, as is often observed in CARS. As Figure 2 shows, closely spaced bands may not be completely resolved, but they appear as the simple sums of the individual band intensities.

Inverse Raman spectroscopy also preserves polarization properties of Raman spectra. Werncke and co-workers [3] have reported the inverse Raman lines of benzene to be polarized and depolarized as the spontaneous Raman lines. We have measured the depolarization ratio of the benzene 1178 cm^{-1} line to be 0.73, in good agreement with the 0.75 value expected for a depolarized line. We find the depolarization ratio of the 992 cm^{-1} line to be 3×10^{-2}, in adequate agreement with the literature value of 2×10^{-2} reported for spontaneous Raman spectroscopy [16].

5. SOME APPLICATIONS TO LIQUIDS AND SOLUTIONS

In the first observation of inverse Raman spectroscopy Stoicheff and Jones pointed out the potential use for fluorescence rejection. This use has remained a major area of application. Werncke and co-workers [7] have recorded the inverse Raman spectra of light oil and transformer oil. They demonstrate that very good spectra of the hydrocarbons are obtained, despite the presence of fluorescent components which almost completely mask the spontaneous Raman spectra.

More commonly, inverse Raman and Raman gain spectroscopy are used to obtain resonance-enhanced Raman spectra of luminescent molecules [17-21]. For this application, these techniques compare favorably with CARS in sensitivity. However, while Raman gain band shapes remain Lorentzian in the electronic resonance region, inverse Raman band shapes change [20, 21]. This phenomenon is discussed in detail elsewhere in this volume.

With crossed or counter-propagating beams, all forms of coherent Raman spectroscopy are useful for examining the region of small wavenumber shift. In principle, coherent Raman spectroscopy allows observation of wavenumber differences as small as the sum of the two laser bandwidths. For pulsed lasers this difference is 0.05-0.5 cm^{-1}. For cw lasers, it can be less than 0.001 cm^{-1}.

Little use has been made of this property, yet. However,

Nestor [12] has reported Raman gain spectra of several neat liquids down to about 0.5 cm^{-1}. A zero frequency band, attributed to bulk motion of the solvent, is observed.

5. DEPHASING MEASUREMENTS

Heritage has used Raman gain to measure the dephasing time for the CS_2 stretching mode at 656.5 cm^{-1} [22]. A pair of synchronized pulses from two independently tunable mode-locked cw dye lasers is used to generate a Raman gain signal. One of the lasers is mechanically chopped to generate a modulated signal. A second pair of pulses, delayed with respect to the first pair, probes the dephasing, which is measured by the intensity of the modulated gain on the second pair, as a function of delay time. This variation of the pump-probe experiment appears to be quite sensitive and is readily extended to any Raman-active vibration.

6. SOLIDS

Inverse Raman and Raman gain spectroscopy, like other coherent Raman spectroscopies, are inherently transmission techniques. As such, they have limited applicability to solids. There have been a few reports of measurements on optically transparent solids [23, 24]. Most notably, Heritage and Allara have demonstrated the Raman gain measurements are sensitive enough to measure monolayers of material on transparent substrates without the need for surface enhancement or electronic resonance enhancement.

Generation of inverse Raman or Raman gain spectra would appear to be an area ready for substantial progress. Photoacoustic or thermal lens detection would appear to be a useful method for observing such spectra, as would attenuated total reflection measurements, using sapphire reflectance elements. Although generation of spontaneous Raman spectra from evanescent waves is established as a method for depth profiling by Raman spectroscopy, there have been no reports of similar coherent Raman experiments. It should be noted that thermal artifacts would be formidable in both the photoacoustic and internal reflection schemes.

ACKNOWLEDGEMENTS

The authors wish to acknowledge the substantial experimental contributions of G. P. Ritz, J. P. Haushalter, K. Hillig and

G. R. Daigneault.

This work has been supported by the National Science Foundation, grant CHE 79-15185.

REFERENCES

1. Jones, W.J. and Stoicheff, B.P., Phys. Rev. Lett., 13, 657 (1964).

2. Yeung, E. S., J. Mol. Spectrosc., 53, 379 (1974).

3. Werncke, W., Klein, J., Lau, A., Lenz, K. and Hunsalz, G., Opt. Comm., 11, 159 (1974).

4. Owyoung, A. and Jones, E. D., Opt. Lett., 1, 152 (1977).

5. Owyoung, A., Opt. Comm., 22, 323 (1977).

6. Daigneault, G. R. and Morris, M. D., Appl. Spectrosc., 35, 591 (1981).

7. Werncke, W., Lau, A., Weigmann, H.-J., Buge, H.-G. and Reich, P., J. Raman Spectrosc., 6, 193 (1977).

8. Morris, M. D., Wallan, D. J., Ritz, G. P. and Haushalter, J. P., Anal. Chem., 50, 1976 (1978).

9. Haushalter, J. P., Ritz, G. P., Wallan, D. J. and Dien, K., Appl. Spectrosc., 34, 144 (1980).

10. McDowell, R. S., Patterson, C. W., and Owyoung, A., J. Chem. Phys., 72, 1980.

11. Nestor, J., J. Chem. Phys., 69, 1778 (1980).

12. Nestor, J., Appl. Spectrosc., 35, 81 (1981).

13. Stone, J., J. Chem. Phys., 69, 4349 (1978).

14. Buffett, C. E. and Morris, ., Appl. Spectrosc., 35, 203 (1981).

15. Owyoung, A., I.E.E.E. J. Quantum Electron., QE-14, 192 (1978).

16. Skinner, J. G. and Nilsen, W. G., J. Opt. Soc. Amer., 58, 113 (1968).

17. Werncke, W., Lau, A. Pfeiffer, M., Weigmann, H.-J., Hunsalz, G., and Lenz, K., Opt. Comm., 16, 128 (1976).

18. Werncke, W., Lau, A., Pfeiffer, M., Lenz, K., and Weigmann, H.-J., Sov. J. Quantum Electron., 7, 1464 (1977).

19. Haushalter, J. P., Buffett, C. E. and Morris, M. D., Anal. Chem., 52, 1284.

20. Haushalter, J. P. and Morris, M. D., Anal. Chem., 53, 21 (1981).

21. Schopfer, L. M., Haushalter, J. P., Smith, M., Milad, M. and Morris, M. D., Biochem., 20, 6734 (1981).

22. Heritage, J. P., Appl. Phys. Lett., 34, 470 (1979).

23. Grassl, H. and Maier, M., Opt. Comm., 30, 253 (1979).

24. Heritage, J.P. and Allara, D. L., Chem. Phys. Lett., 74, 507 (1980).

RESONANCE RAMAN GAIN AND RAMAN LOSS SPECTROSCOPY

L. A. Carreira and M. L. Horovitz

Department of Chemistry
University of Georgia
Athens, Georgia 30602

In accompanying chapters anti-Stokes Raman spectroscopy has been shown to be a powerful tool for the study of molecular species. However a few drawbacks have limited the applicability of the technique. First, the complex lineshapes exhibited by CARS require computer evaluation in order to obtain accurate frequencies. Second the nonresonant susceptibility may overshadow the Raman peak, limiting the detectability. Finally, the hardware required for scanning is complex and without the computer control previously mentioned, obtaining spectra is a very time-consuming procedure. Because of the drawbacks several processes have been proposed to obtain similar information with fewer problems: Rikes [1,2,3], Optical Heterodyne Detection (OHD) [4,5], Photo-Acoustic Raman Spectroscopy (PARS) [6] as well as Raman gain/loss spectroscopy [7].

The Raman gain/loss process is represented in Fig. 1 when the fields of two laser beams are crossed in a medium. There can be a stimulated gain in ω_2 $(\omega_2<\omega_1)$ or loss in ω_2 $(\omega_1<\omega_2)$. If we study the induced gain in ω_2, we will be doing stimulated Raman or Raman gain spectroscopy. Either way as the frequency difference between the two lasers approaches a Raman level $(\omega_1-\omega_2=\omega_r)$ the induced signal increases.

In this chapter we report the application of stimulated Raman and inverse Raman (Raman gain/loss) to obtaining near zero frequency Raman spectra of several compounds. This technique has the advantage of discriminating against Rayleigh scattering so that a scan may be taken through zero Raman shift. The lineshapes obtained in these studies are explained in light of the behaviour of the third order susceptibility, $\chi^{(3)}$.

W. Kiefer and D. A. Long (eds.), Non-Linear Raman Spectroscopy and Its Chemical Applications, 533–541.

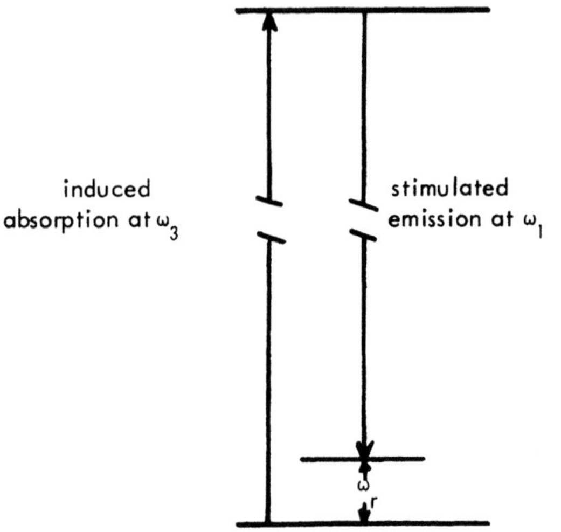

Figure 1. The Raman gain/loss process.

The experimental arrangement was similar to that introduced by Owyoung [8,9] which combined a single mode cw Argon laser and a pulsed dye laser to produce the Raman gain (loss) spectrum. In order to reduce stray light we incorporated the counter propagating plane-polarized beam configuration used previously [10,11]. The experimental set up is shown in Fig. 2. The single mode Argon ion laser (Spectra Physics 164-03) was chopped at 60 Hz and each pulse has a width of 100 μsec. The leading edge of each pulse drove a binary up-counter and every sixth pulse triggered a monostable which fired the N_2 laser (Molectron UV-1000). The N_2 laser pumped a dye laser (Molectron DL-200) to provide the pump beam. Since spectral filtering was useless at zero frequency difference, great care was taken to eliminate pump beam reflections from co-propagating with the chopped CW beam. All lenses and sample holders were tilted with respect to the laser beam axis and the beams were spatially filtered as shown in Fig. 2. Careful spatial filtering eliminated the necessity of a Faraday rotator [11] to prevent cw beam modulation by the pump beam. The large RF burst that accompanied the firing of the N_2 laser proved very troublesome and although RF shielding reduced the problem considerably it still interfered at low signal levels. To overcome this problem, the 5145 Å probe beam was focussed into a 100 meter optical fiber (Math Associates

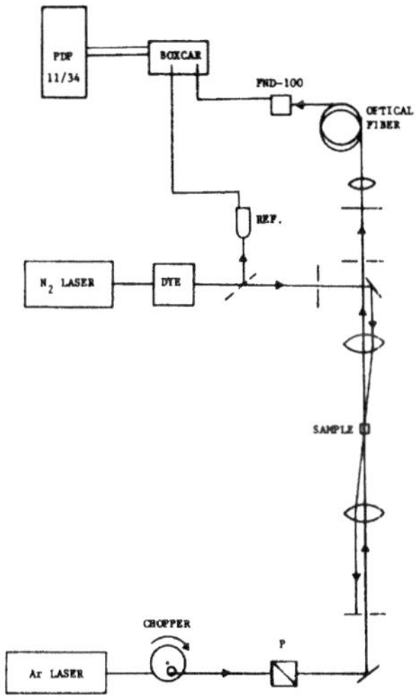

Figure 2. Raman gain/loss experimental setup.

QSF-300A). The resulting ~500 nsec optical delay allowed the signal to be measured well after the RF burst. Signal attenuation by the fiber at 5145 Å was only a factor of two over the 100 meters. The optical bandwidth for the fiber was calculated to be well over 200 MHz and no appreciable pulse broadening was observed. The optical fiber was positioned so that its output just filled the active area of a photodiode detector (EG and G, FND-100). The output of the photodiode was filtered to pass only the high frequency gain (loss) signal, amplified using two 10x amplifiers (ortec 9301) in series and averaged using a boxcar integrator (PAR 162-165). A PDP-11/34 computer controlled positioning of the beams, scanning of the pump laser and collection of the data.

The solvents used in this study were Baker reagent grade and were used without further purification. The β-carotene was obtained from Sigma Chemical Company.

Raman gain/loss phenomena are governed by the third-order susceptibility, $\chi^{(3)}(-\omega_2, \omega_2, \omega_1, -\omega_1)$. Carreira [12] have derived a lineshape equation assuming only one Raman level comes into play.

$$\mathrm{Im}\chi^{(3)} = \frac{A\delta}{\delta^2+\Gamma_r^2} - \frac{B}{\delta^2+\Gamma_r^2} \quad (1)$$

where

$$A = \frac{2\Gamma_e D}{(D^2+\Gamma_e^2)^2} \quad \text{and} \quad B = \frac{2\Gamma_e^2\Gamma_r}{(D^2+\Gamma_e^2)^2} \quad (2)$$

and $D = \omega_e-(\omega_1+\omega_2)/2$ where ω_e is the frequency of the electronic excited state with a dumping factor, $\Gamma_e \cdot \delta = \omega_1-\omega_2$.

The first term in equation (1) will produce a dispersive shaped line centered at zero Raman shift. The second term is a Lorentzian shape centered at a zero frequency. Away from an electronic absorption the A terms will dominate.

A preliminary study in this region has been done by Song et al. on carbon disulfide. We have reproduced this work to try our theory and fitting scheme. Figure 3 (bottom) shows the experimental data obtained along with our fit. As shown in Fig. 3 (top), the fit is composed of two lineshapes, the dispersive term at zero,

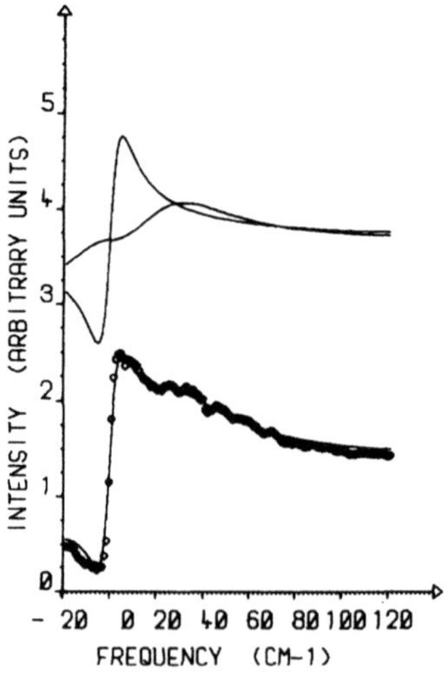

Figure 3. Experimental near zero frequency data (bottom) and lineshape components for theoretical fit of carbon disulfide (top).

and a Lorentzian Raman term at $\sim 34\ cm^{-1}$. The spectra of several solvents were then obtained and are shown in Figs. 4-9. A feature not seen in the CS_2 spectrum is seen in all of these spectra at $<1.0\ cm^{-1}$. That is, a very sharp ($\sim 1\ cm^{-1}$) Lorentzian term has been added in this region. This can best be displayed by showing the individual components of the benzene fit. In Fig. 4a we can

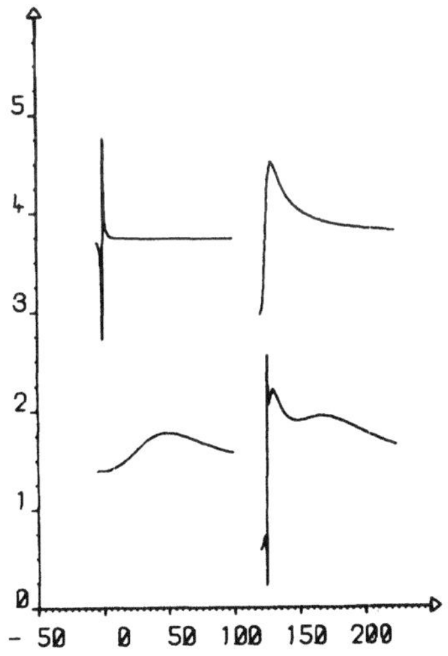

Figure 4. Individual components of the benzene fit.

see the first Raman component at .255 cm^{-1}, Fig. 4b shows the dispersive component about zero, Fig. 4c shows a second Raman band at 33.3 cm^{-1}, the addition of these components is shown in Fig. 4d. The original data overlayed with this three component fit is shown in Fig. 5. A possible explanation for this very sharp peak is Brillouin scattering. Jacobson and Shen [13] have observed Brillouin scattering with the RIKES technique in this frequency range. The spectra of carbon tetrachloride shown in Fig. 6 displays the sharp Raman peak at $\sim 0.5\ cm^{-1}$ and a very broad band centered at $\sim 40\ cm^{-1}$. Figs. 7,8 and 9 are more detailed spectra of this low frequency peak for carbon tetrachloride, acetone and ethanol, respectively.

Finally, we obtained the spectra of β-carotene near a one-photon resonance. Figure 10 shows a concentration study of β-carotene in benzene.

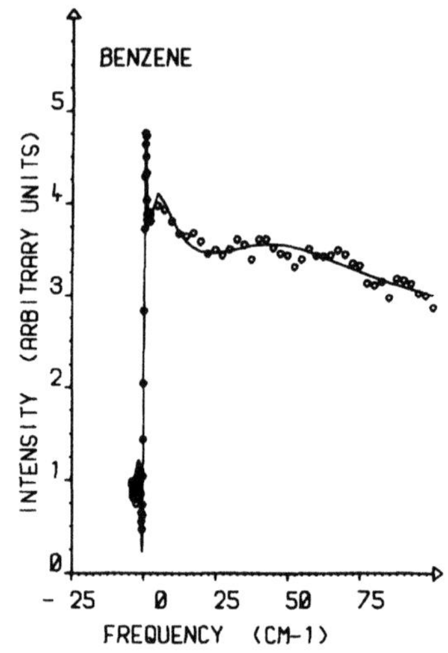

Figure 5. Experimental near zero frequency data for benzene overlayed with the three component fit.

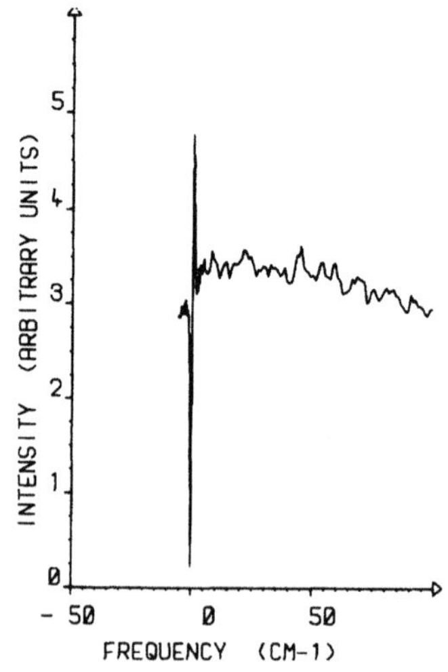

Figure 6. Experimental near zero-frequency data for carbon tetrachloride.

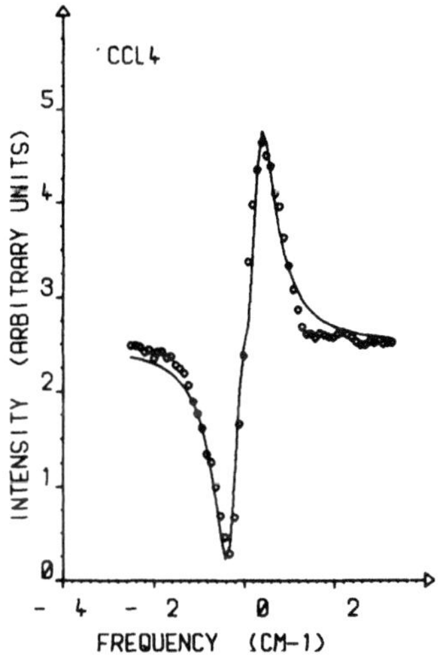

Figure 7. A more detailed spectrum of carbon tetrachloride overlayed with the three component fit.

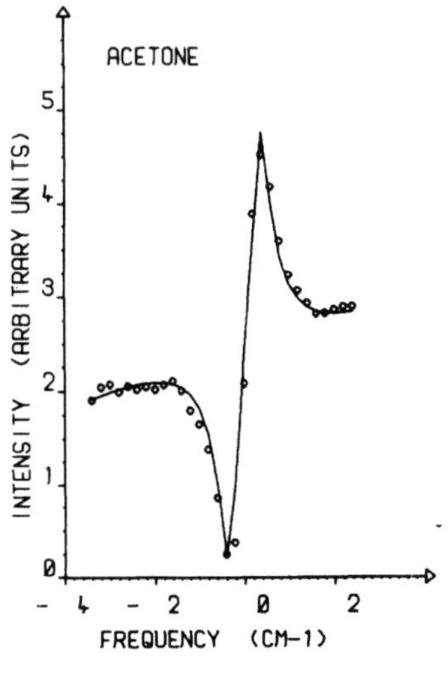

Figure 8. A detailed spectrum of acetone overlayed with the three component fit.

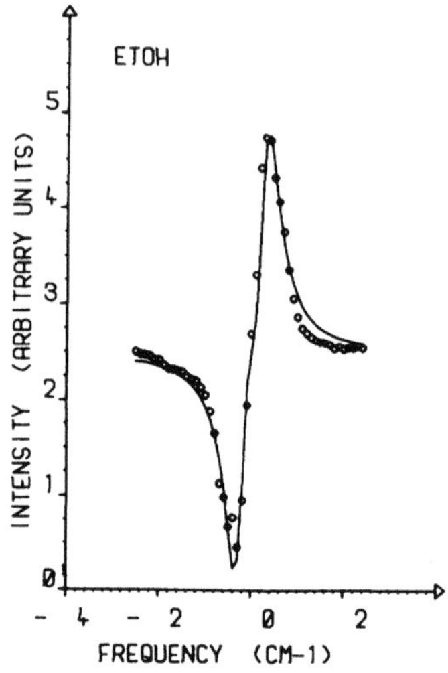

Figure 9. A detailed spectrum of ethanol overlayed with the three component fit.

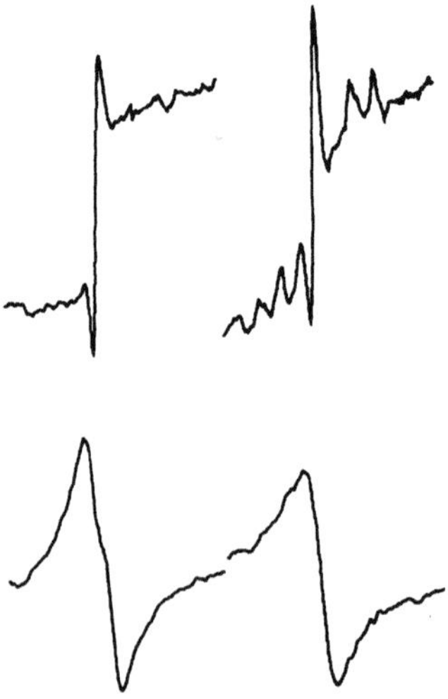

Figure 10. Near zero frequency spectra of β-carotene in benzene at the following concentrations: a) 1 x 10^{-7}M, b) 5 x 10^{-6}M, c) 1 x 10^{-5} M, d) 1 x 10^{-4}M.

At higher concentrations (Figs. 10c,d) the shape of the spectra change completely as gain and loss exchange sides at zero frequency. The change in line shape can be explained in two ways. The sign of D in equation (2) will change if the ω_e for β-carotene is such that $\omega_1 > \omega_e$. However, if one is seeing a vibration enhanced to the red of the 0 - 0 transition, then the line shape is due to a vibrational rather than an electronic effect. It is known for β- carotene that $\omega_1 < \omega_e$. Therefore, we are seeing a very low frequency heretofore unreported resonance enhanced signal due to a torsional vibration about β-carotene's double bonds.

REFERENCES

1. Heiman, D., Hellwarth, R.W., Levenson, M.D. and Martin, G., Phys. Rev. Lett., 36,189 (1976).

2. Dutta, N.K., J. Appl. Phys., 51, 40 (1980).

3. Levenson, M.D. and Song, J.J., J. Opt. Soc. Am., 66, 641 (1976).

4. Eesley, G.L., Levenson, M.D. and Tolles, W.M., IEEE J. Quant. Elec., QE-14, 45 (1978).

5. Levenson, M.D. and Eesley, G.L., Appl. Phys., 19, (1979).

6. Chemical Applications of Nonlinear Raman Spectroscopy (Harvey, A.B., ed.) 89, Academic Press, New York, 1981.

7. Chemical Applications of Nonlinear Raman Spectroscopy (Harvey, A.B., ed.) 281, Academic Press, New York, 1981.

8. Owyoung, A. and Percy, D.S., J. Appl. Phys. 48, 674 (1977).

9. Owyoung, A. and Toney, E.D., Opt. Lett. 1, 152 (1977).

10. Nestor, J.R., Appl. Spec. 35, 81 (1981).

11. Scurpano, M.A.F., Lee, T.H. and Song, J.J., Opt. Lett. 6, 193 (1981).

12. Carreira, L.A., Antcliff, R. and Horovitz, M.L., to be published.

13. Jacobson, A. and Shen, Y.R., Appl. Phys. Lett. 34, 464 (1979).

APPLICATIONS OF RESONANCE INVERSE RAMAN AND RAMAN GAIN SPECTROSCOPY TO BIOLOGICAL SYSTEMS

M. D. Morris and R. J. Bienstock

Department of Chemistry
University of Michigan
Ann Arbor, MI 48109, USA

1. INTRODUCTION

Resonance Raman spectroscopy ranks among the most powerful probes of biochemical molecules currently available. Few, if any, other techniques provide the combination of fairly high resolution, applicability to aqueous solutions and good sensitivity. However, spontaneous Raman spectroscopy is inapplicable to highly luminescent media and many systems of biochemical interest are fluorescent. Among the important fluorescent chromophores are the flavins, the bile pigments and aromatic amino acids. In addition, many important probe molecules are luminescent, including almost all of those which intercalate with nucleic acids.

Coherent Raman spectroscopy has established itself as an important supplement to spontaneous Raman techniques for generation of Raman spectra of luminescent systems. Certainly, CARS is the most familiar of the coherent Raman spectroscopies, with well known advantages and problems. Inverse Raman spectroscopy has recently emerged as a useful alternative to CARS for certain applications. Away from an electronic resonance inverse Raman spectroscopy provides background-free Raman spectra which are exactly analogous to spontaneous Raman spectra. Using modern modulation/demodulation techniques, inverse Raman spectroscopy provides sensitivities equal to those achievable by spontaneous Raman spectroscopy [1].

Several groups have reported resonance and near resonance inverse Raman and Raman gain spectra. Lin and co-workers observed near resonance spectra of nitroaniline [2] by direct

W. Kiefer and D. A. Long (eds.), Non-Linear Raman Spectroscopy and Its Chemical Applications, 543–559.

absorption. Using intra-cavity absorption or gain from a broadband dye laser, Werncke and co-workers have observed both resonance inverse Raman and resonance Raman gain spectra of various fluorescent dye molecules [2,3].

Our group has employed modulation techniques to obtain resonance inverse Raman spectra at high sensitivities. We have used ac-coupled inverse Raman spectroscopy [5] to obtain resonance Raman spectra of acridines, flavins, bile pigments and carotenes at concentrations of 10^{-6} - 10^{-4} M.

Much of the interest in resonance inverse Raman spectroscopy has arisen from the assumption that the spectra would, like other inverse Raman spectra, be Lorentzian and background free. As such, it was assumed that inverse Raman spectroscopy would preserve the CARS advantage of luminescence rejection, but would yield more easily interpretable spectra and be usable at lower concentrations. There are, however, certain complications. Inverse Raman spectra, under electronic resonance conditions, are observable at low concentrations in luminescent solutions. They are fairly easy to interpret. But, they are not always Lorentzian [6, 7].

2. THE RESONANCE INVERSE RAMAN EFFECT

Near an electronic resonance, inverse Raman signals differ dramatically from spontaneous Raman signals. Band shapes change from positive Lorentzians to dispersive shapes to inverted Lorentzians. These band shapes are due to the interaction of the electromagnetic fields of the pump and probe laser beams with the sample. This interaction is described by the expansion of the third order susceptibility tensor [8-10]. Figure 1 shows the interactions possible in a resonance inverse Raman effect. The three-level system, 1a, is appropriate on the red side of resonance. Here only ground state vibrations occur in the inverse Raman spectrum. Neglecting the effects of stimulated emission, we have shown that equation 1 describes this case [7].

$$\chi_I^{(3)} = 6NLK \frac{\Gamma_R(\omega_e-\omega_1)^2-\Gamma_R\Gamma_e^2 + 2\Gamma_e(\omega_e-\omega_1)[\omega_R-(\omega_1-\omega_2)]}{[(\omega_e-\omega_1)^2+\Gamma_e^2]^2[(\omega_R-(\omega_1-\omega_2))^2+\Gamma_R^2]} \quad (1)$$

Here, $\chi_I^{(3)}$, is the imaginary part of the third order susceptibility, with polarization subscripts omitted. Transition moments and constants are lumped into K. N is the number density of the sample, and frequencies and damping constants are defined by the transitions in Figure 1a.

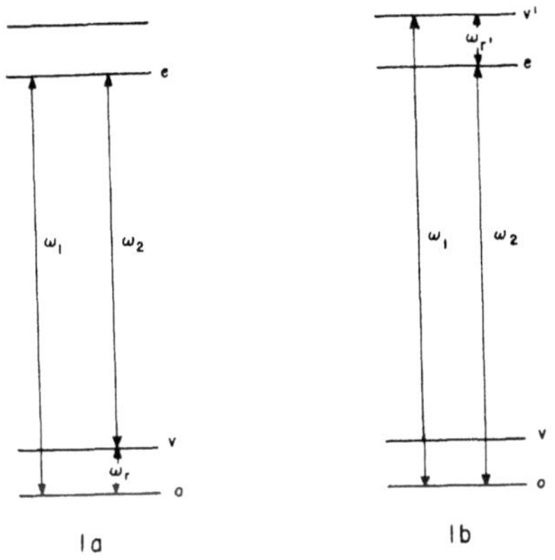

FIG. 1. Energy level diagram for resonance inverse Raman spectroscopy; a. three-level system ; b. four-level system.

Equation 1 is the leading term of a sum over excited states. For highly structured narrow band absorption spectra, it is generally adequate to retain only this leading term. Equation 1 also neglects stimulated electronic transitions [6], which contribute a broad, structureless background.

For the four level system depicted in Figure 1b, one must add another set of terms, analogous to equation 1, in $\omega_R{}'$, the excited state vibrational frequency. These contributions may dominate when both lasers are well inside the electronic absorption band [10,11].

The band shapes predicted by equation 1 are shown schematically in Figure 2. Well below electronic resonance, one predicts the usual Lorentzian bands. As resonance is approached, the band shape becomes dispersive and, ultimately, an inverted Lorentzian. Figure 3 demonstrates this behavior for acridine orange.

The non-Lorentzian band shapes obtained near an electronic resonance are an added complication in resonance inverse Raman spectroscopy. If absorption from a narrow band pump laser is used, all bands will have the same shape, since band shape depends only on detuning of the pump laser from the electronic origin. If, however, absorption from a broad band pump laser is employed, it is possible to have band shapes systematically change across the spectrum.

Resonance Raman gain spectroscopy, for the three-level case, has a frequency factor of somewhat different form from equation

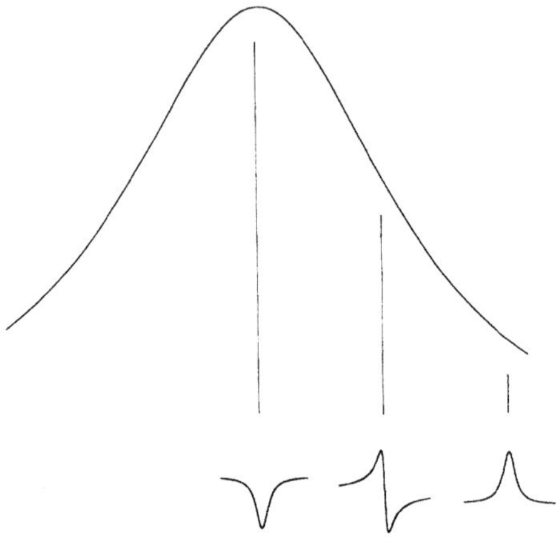

Fig. 2. Idealized resonance inverse Raman band shapes.

1. Equation 2 describes the Raman gain case.

$$\chi_I^{(3)} = 6NLK \frac{-[(\omega_e-\omega_1)^2 + \Gamma_e^2]\Gamma_r}{[(\omega_e-\omega_1)^2+\Gamma_e^2]^2[\omega_r+(\omega_2-\omega_1)]^2+[(\omega_e-\omega_1)^2+\Gamma_e^2]^2\Gamma_r^2} \qquad (2)$$

Equation 2 predicts Lorentzian behavior in the electronic resonance region.

We have verified the pump frequency dependence of inverse Raman band shapes in the resonance region [7]. Equation 1 predicts that band shapes depend only upon pump laser frequency. If a narrow band laser is employed as the pump, as in AC-coupled inverse Raman spectroscopy, all of the resonance enhanced bands in a spectrum will have the same shape. This behavior is fortunate, since it simplifies the evaluation of spectra.

3. EXPERIMENTAL PROBLEMS

The resonance enhanced experiment presents some problems not encountered in non-resonance work [5-7]. How bothersome these are depends upon the way in which the experiment is performed.

The most serious experimental problem in AC-coupled inverse Raman experiments is thermal blooming [7]. Thermal blooming of even pure water is readily observed. In a sample which may have absorbance 0.1 - 1, the thermal lens effect is very strong. Although inverse Raman spectroscopy is most commonly employed

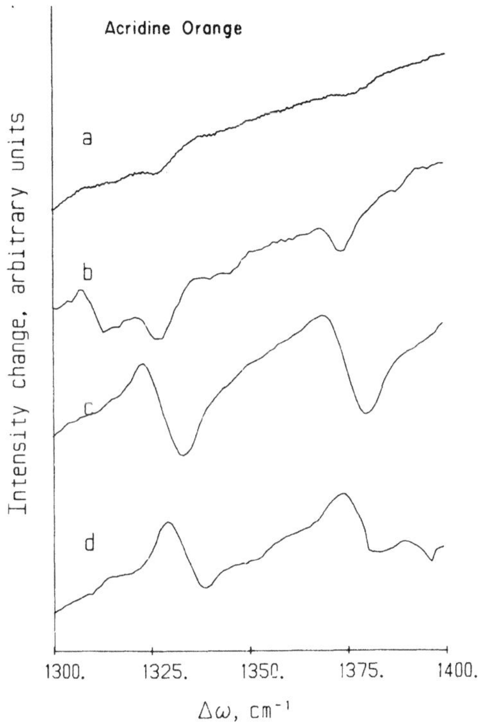

Fig. 3. Band shapes for resonance inverse Raman spectra of acridine orange. a. ω_1=20,492 cm^{-1}; b. ω_1=20,141 cm^{-1}; c. ω_1=19,932 cm^{-1}; d. ω_1=19,436 cm^{-1}.

with luminescent systems, these do not have unit quantum efficiency. Consequently, thermal lens effects remain important.

A thermal lens generated with a pulsed laser has a rise time determined by the focused beam size and the speed of sound in the medium. In aqueous solutions, thermal lens rise times will typically be in the range 10-50 nsec, depending upon the focal length of the lens used. Fall times are determined by thermal diffusion and are typically 10-100 msec.

It is possible to minimize the thermal lens signal by careful experimental design. In principle, thermal lens signals would be undetectable if there were no limiting apertures in the optical system. The one limiting aperture which must exist is the detector itself. Because the response of a photodiode is limited by its junction capacitance, which is proportional to area, there is a maximum practical size to this device. This is about 0.05 cm^2 with commercially available silicon photodiodes.

It is also necessary to maintain good beam colinearity, since the thermal lens effects are more easily minimized when they appear as beam expansion rather than as beam deflection. With these precautions, thermal lens effects are tolerable, even in systems with relatively low quantum efficiency for fluorescence.

The high power densities obtained with focused laser pulses can cause local photochemical damage to many potentially interesting samples, such as the bile pigments and many of the acridines. We have found that the simplest way to minimize this problem is to place the sample cell on a motorized stage to constantly expose fresh solution to the laser. Translation is a more satisfactory motion than stirring. Stirring agitates small particles present in even filtered biochemical samples and often produces very noisy spectra.

In systems with singlet lifetimes comparable to or longer than the laser pulse duration, it is possible to generate a significant excited state population, 1 to 10 per cent of the ground state. Under these conditions, the transmission of the CW laser radiation increases. Since that increase is synchronous with the pulse, it is captured by the gated detection system. The effect is to generate a sloping background signal, which resembles the absorption spectrum [6]. Because this background will generally be structureless, it is fairly easy to subtract it by the conventional techniques. It is observed largely because of the sensitivity of ac-coupled measurements.

In addition, ordinary stimulated emission occurs, and generates a signal which more or less follows the absorption spectrum. This signal, too, is of importance only because of the sensitivity of the measurements.

What is actually observed as a background in resonance enhanced measurements is the resultant of all of these processes. The measurement system detects only pulsed transmission changes, independent of their source. In our experiments, these background processes typically contribute a signal between two and five times larger than the largest signal peaks, a relatively minor problem.

Although resonance Raman signals would, in principle, be most sensitive at exact resonance, we find it useful to work somewhat on the red side of resonance. Typically, we position our argon ion laser about two half bandwidths from the absorption maximum. This approach has the effect of minimizing the artifacts associated with exact resonance because it places the pulsed dye laser in a region of low absorption. Moreover, the system behaves as a three-level, not a four-level system. Consequently, the observed spectra are true ground state resonance Raman spectra. They are also only insignificantly perturbed from Lorentzian band shapes, so that one can evaluate them in the usual way. These advantages more than compensate for the loss, if any, of sensitivity.

The artifacts associated with the AC-coupled measurement do not complicate direct absorption measurements from broad band pulsed lasers [3,4]. In this case, thermal lensing and ground state depletion simply change the absolute magnitude of the signal. Stimulated emission, or, indeed, ordinary absorption cause preferential transmission of some wavelengths, and generate a sloped background. Since the intensity of the pulse follows the dye laser gain curve anyway, this does not seem to add a new complication. However, if the pulsed dye laser output is positioned very close to resonance, it is quite possible for Raman band shapes to be a function of wave number shift. This problem could complicate the measurement of weak bands.

4. FLAVINS

Flavoproteins are a very versatile group of oxidation-reduction enzymes. Almost 200 of these enzymes have been isolated. In most cases, but not all, the natural substrate is known. Despite close to a century of research in this field, there is vigorous controversy about the mechanisms of their reactions [13], and even about the classification of the different subgroups [12].

What is perhaps most remarkable about the flavoproteins is that almost all of them contain the same co-enzyme, 7,8-dimethyl isoalloxazine. In the great majority of cases, this group is non-covalently bound to the protein active site. Two groups are common at position 10, a ribose-phosphoric acid chain and a ribose-pyrophosphate-adenosine chain. The first molecule is called flavin mononucleotide (FMN) and the second is flavin adenine dinucleotide (FAD). In some enzymes, however, the flavin is linked covalently to the protein, usually through the 8-methyl group as 8-α-histidylflavin or 8-α-cysteinylflavin.

There are only a few known naturally occurring flavins which do not contain the basic 7,8-dimethyl isoalloxazine structure. These include roseoflavin (8-dimethylaminoflavin), 6-hydroxyflavin , 8-hydroxyflavin and 8-OH, 5-deazaflavin. The functions of these compounds are unclear at present, but it is thought that roseoflavin is an antibiotic, produced by Streptomyces sp. to disrupt the flavo-enzyme functions of other bacteria.

There are several schemes for classifying the types of

reactions catalyzed by flavoproteins. Although there is disagreement on the classifications and the mechanisms [12,13], it is known that all flavins cycle between the fully oxidized, or quinone state, the semiquinone state generated by one electron reduction and/or the dihydroquinone state generated by two-electron reduction.

It is widely understood that the protein matrix must be a major factor in determining the behavior of any flavoprotein. To some extent, the protein controls the stereochemistry, limiting the substrates which will be a good fit at the active site. However, the protein must also control the redox potential of the enzyme, since this is also known to vary widely among the flavoproteins. Moreover, the protein must also influence the choice of reactive site on the isoalloxazine, since there are several possible reactive centers [12]. These two considerations are not completely independent.

The difficulty of preparing suitable crystals has limited the penetration of x-ray spectroscopy in this field. Moreover, flavoproteins may contain hundreds of amino-acid residues, so that solving crystal structures is exceedingly difficult in any case. Although there have been some elegant NMR studies done, two factors limit success in this field. First, the isoalloxazine ring system contains protons directly bonded only at three sites. Proton NMR spectroscopy is of limited value. Carbon-13 NMR is much more useful, but the flavin concentration in all but a handful of flavoproteins can not be made much higher than 10^{-4} M. Because flavin concentrations are so low, natural abundance ^{13}C spectroscopy is impractical. Spectroscopy of isotopically enriched flavins is possible, but even this work has been largely limited to free flavins, because of the concentration constraints [14].

For these reasons, resonance Raman spectroscopy has emerged as the only structural probe broadly applicable to flavins and flavoproteins. However, the isoalloxazine ring system is highly luminescent. Quantum yields are typically 0.2-0.3. This fluorescence is quenched efficiently by some protein, notably riboflavin binding protein. But other proteins leave quantum yields undiminished or only slightly reduced. Over the past four years several groups have begun systematic investigations of the Raman spectroscopy of flavins. The first critical review of results in this area has already appeared [15]. Spontaneous Raman spectroscopy has been used to study flavins inserted in the fluorescence quenching riboflavin binding protein. However, study of free flavins and of most enzymes, requires coherent Raman spectroscopy.

Spiro and co-workers conducted the first CARS studies of

flavins [16]. They presented resonance CARS spectra for FAD and glucose oxidase and provided tentative interpretations. The Spiro group has also presented the CARS spectra of 8-aminoriboflavin and 8-hydroxyriboflavin.

Our interest has been in the spectroscopy of flavins containing electron-donating groups in the 8-position. Typical donor groups include $-NH_2$, $-N(CH_3)_2$, $-O^-$ and $-S^-$. The dimethylamino and hydroxy (O^-) are naturally occurring, while the sulfide-substituted flavin appears to be extremely useful as a probe of electron distribution in flavins imbedded in flavoproteins [12].

The 8-thio substituted flavin is used to probe electron distribution changes as red shifts in the absorption spectra of the modified flavoprotein [12]. We have recently shown [17,18] that resonance Raman spectroscopy of the modified flavoproteins provides a more detailed picture of electronic distribution than does absorption spectroscopy.

Raman spectra of flavins 8-substituted with electron donating groups are substantially different from those of the standard chromophore, or flavins substituted with electron withdrawing groups [17,18,19]. The resonance inverse Raman spectrum of $8\text{-}O^-$ riboflavin and $8\text{-}NH_2$ riboflavin, shown as Figure 4, illustrate this point. Most notably, the intense band at 1560-1565 cm^{-1} is displaced about 20 cm^{-1} with respect to its position in normal flavins (1580-1585 cm^{-1}). Similarly, other bands are displaced, typically to lower frequency. The thio substituted compounds show this behavior even more strongly. As we have shown, band positions in the thio derivatives are protein-dependent [18].

We interpret these shifts as evidence for the charge redistribution pictured in Figure 5. The normal flavin structure most closely resembles the benzenoid form. However, if there is a strongly electron donating group in position 8, and/or if the protein carries a positive charge near the site of position 1, the structure can shift toward the quinonoid form. The amino and hydroxy substituted flavins are presumably intermediate between the two limiting forms, although not very far from the benzenoid form.

Similarly, the inverse Raman spectrum of roseoflavin (8-dimethylamino riboflavin) shows it to be slightly perturbed from the benzenoid form, as shown in Figure 6. Similar results have been obtained in CARS for 8-methylamino riboflavin [20] and by spontaneous Raman spectroscopy for fluorescence quenched 8-methoxyriboflavin [21].

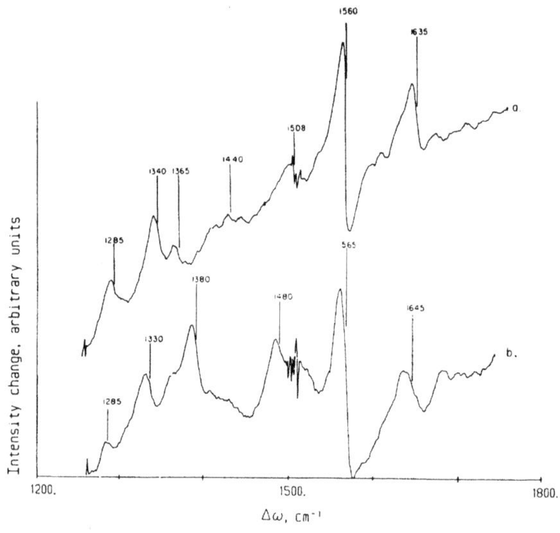

Fig. 4. Inverse Raman spectra of substituted Flavins; a. 8-OH Riboflavin, b. 8-NH_2 Riboflavin

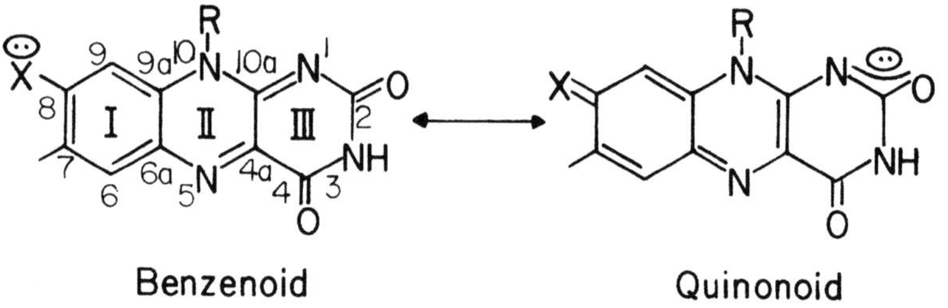

Fig 5. Limiting forms of the charge distributions on flavins electron donating 8-substituents.

The significance of these findings is that they suggest a means by which flavin reactivity can be controlled in enzymes. Studies with substituted flavins show clearly that the protein charge density near ring III varies among the oxidases [17,18]. The resulting shifts in electron density in the flavin cause changes in the redox potential and, perhaps, even in the active site. Although this effect is easily observed with substituted

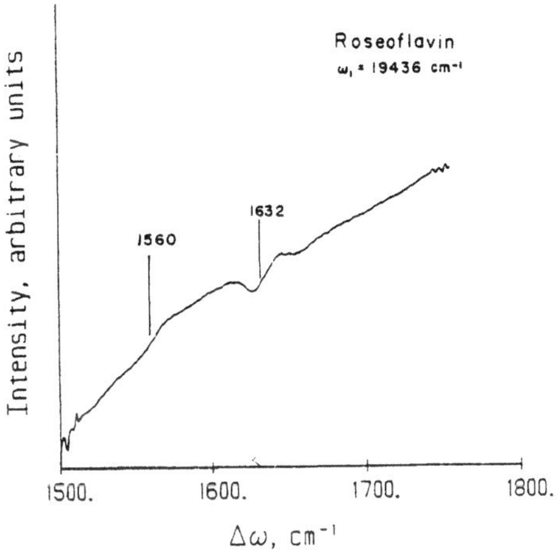

Fig 6. Inverse Raman spectrum of roseoflavin.

compounds, it is smaller with the native flavins themselves. However, Spiro and co-workers [22] have observed that the 1580 cm^{-1} band shifts to about 1575^{-1} in glucose oxidase. This observation is consistent with the charge redistribution hypothesis. At present, however, there is no large and systematic body of resonance Raman data on native and substituted enzymes to verify or disprove this theory.

5. BILE PIGMENTS

The bile pigments are linear tetrapyrroles formed by the breakdown of hemoglobin in humans and animals [23,24]. Members of this class of compounds also function as photosynthetic pigments in certain plants [25]. In human metabolism the most important bile pigments are bilirubin and its glucuronic ester. The bilirubin structure is shown as Figure 7.

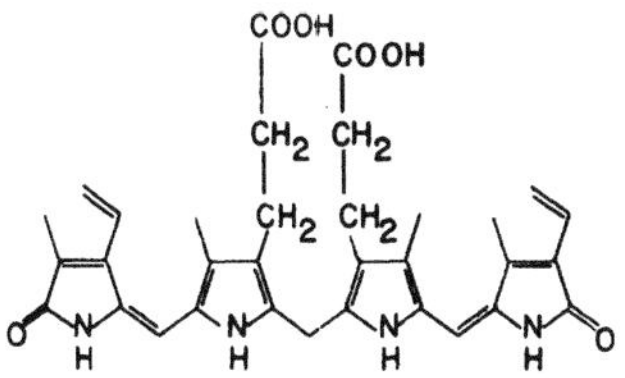

Fig. 7. Bilirubin structure.

The linear structure of Figure 7 is misleading. The two dipyrrole rings function as independently conjugated ring systems. In crystalline bilirubin and in solution they are internally hydrogen bonded to give a bent, hydrophobic structure. In the body, free bilirubin is always complexed to albumin. Bilirubin diglucuronide is not tightly internally hydrogen bonded and, unlike bilirubin, is quite water soluble.

Build-up of bilirubin and bilirubin diglucuronide in the body is the cause of jaundice and is indicative of liver or kidney malfunction. For this reason, the bilirubin measurements are common in hospital laboratories. However, the complications of internal hydrogen bonding and albumin complexation make automation of these measurements rather difficult. Thus, information about bile pigment structures and chemistry is of potentially great practical significance.

Bilirubin and its diglucuronide are sufficiently luminescent that spontaneous resonance Raman spectroscopy is extremely difficult. It is possible to find only one or two of the strongest bands. Inverse Raman spectroscopy, however, allows observation of bile pigment resonance Raman spectra at biochemically realistic concentrations. Figure 8 shows the resonance inverse Raman spectrum of bilirubin in chloroform at 1.3×10^{-4} M Placement of the argon ion laser at 514.5 nm preserves Lorentzian line shapes but preserves adequate sensitivity for biochemical studies. Attempts to observe spontaneous Raman spectra at this excitation frequency have been completely unsuccessful.

The major bands in the spectrum are the pyrrole ring breathing mode at 1190 cm^{-1} and the exo C=C stretch at 1613 cm^{-1}. The lactam carbonyl stretch is also visible at 1640 cm^{-1}.

In DMSO the lactam carbonyl stretch occurs at 1636 cm^{-1}. For bilirubin diglucuronide, the lactam carbonyl stretch occurs at 1637 cm^{-1} in DMSO and at 1639 cm^{-1} in water. These data demonstrate that hydrogen bonding involving the lactam carbonyls can not be very strong, since the lactam C=O stretch varies over only 4 cm^{-1} in the several media studied. This structure is plausible for bilirubin, based on the x-ray structure of the solid. Esterification of the propionic acid side chains eliminates this possibility in the diglucuronide.

6. CONJUGATED POLYENES

Because conjugated polyenes are only weakly fluorescent, measurement of their spontaneous resonance Raman spectra is not unduly difficult. In fact, the spectroscopy of polyenes has been

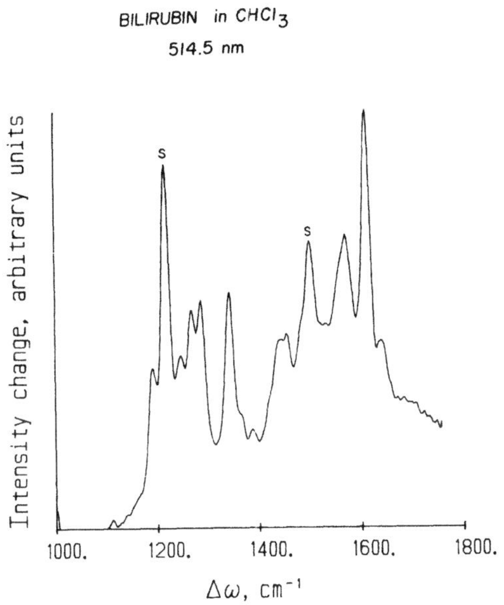

Fig. 8. Resonance inverse Raman spectrum of Bilirubin, 1.33×10^{-4} M in $CHCl_3$. $CHCl_3$ bands are marked, s.

one of the major areas of biochemical resonance Raman spectroscopy [26]. The inverse Raman work with these molecules has consisted largely of demonstrations aimed at verifying properties of inverse Raman spectroscopy [7,27]. However, Kubota and co-workers have obtained the inverse Raman spectrum of neurosporene obtained by extraction from the bacterium rhodopseudomonas sphaeroides. The spectrum appears to be identical to the spontaneous Raman spectrum. They were able to obtain measurements in single dye laser pulses and suggest that inverse Raman spectroscopy may prove advantageous for time-resolved work.

7. ACRIDINES

Acridines are, in general, highly luminescent. Members of this family have been used as fluorescent cell stains by biologists. Several substituted acridines are commonly employed by biochemists and biophysicists because they form strong complexes with nucleic acids. This complexation, called intercalation, occurs by insertion of the acridine between the coils of the nucleic acid helix. It is of academic interest as a means of perturbing DNA conformation and is, therefore, an important probe of DNA structure. The intercalation reaction is also the means by which an important group of anti-tumor molecules acts. This relation to cancer chemotherapy gives intercalation wide practical interest as well.

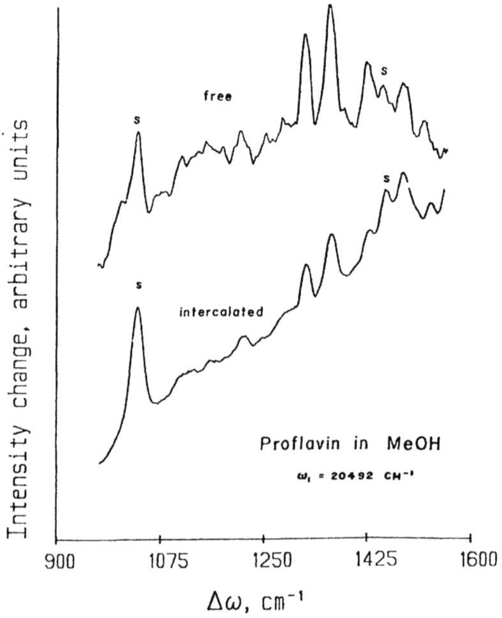

Fig. 9. Proflavin-DNA inverse Raman spectrum. a. Free proflavin. b. Proflavin intercalated with DNA. Solvent bands are marked, s.

In some cases, it has been possible to obtain spontaneous Raman spectra of intercalating molecules and their DNA complexes under resonance or near resonance conditions. Spontaneous Raman spectroscopy has been particularly successful for adriamycin, a hydroxyanthraquinone, whose absorption and emission spectra are sufficiently displaced [28-30]. For the intercalating acridines, coherent Raman spectroscopy has been more appropriate. Tretzel and Schneider have reported CARS studies of free and intercalated proflavin [28,31]. They observe intensity reductions in some of the ring modes of intercalated proflavin. This Raman hypochromism is characteristic of ring stacking interactions.

We have observed similar hypochromism in the inverse Raman spectroscopy of the proflavin-DNA system, as shown in Figure 9. In particular, we find that the 1370 cm^{-1} and 1430 cm^{-1} are more strongly attenuated than the other bands in the aromatic ring mode region. This suggests that the interaction of proflavin, with DNA may be partially localized on the ring system, presumably on the outer rings.

8. CONCLUSIONS

Resonance inverse Raman spectroscopy and Raman gain spectroscopy are both fairly new techniques. There is need for improved experimental configurations as well as for further theoretical progress. At present, these forms of coherent Raman spectroscopy are promising tools for the biochemist. In the future, they could rank among the more important probes of biochemical structures.

ACKNOWLEDGEMENTS

The authors wish to thank L.M. Schopfer, J.P. Haushalter, T.I. Chen, K. Hillig and C. Buffett for their very important contributions to this work.

This work has been supported by the National Science Foundation, grant CHE 79-15185.

REFERENCES

1. Haushalter, J.P., Ritz, G.P., Wallan, D.J., Dien ,K., and Morris, M.D., Appl. Spectrosc., 34, 144 (1980).

2. Lin, S.H., Reid, E.S., and Tredwell C.J., Chem. Phys. Lett., 29, 389 (1974).

3. Werncke, W. , Lau, A., Pfeiffer, M., Weigmann, H.-J., Hunsalz, G., and Lenz, K., Opt. Comm., 16, 128, (1976).

4. Wernke, W., Lau, A., Pfeiffer, M., Lenz, K., and Weigmann, H.-J., Sov. J. Quantum Electron., 7, 1464 (1977).

5. Morris, M.D., Wallan, D.J., Ritz, G.P., and Haushalter, J.P., Anal. Chem., 50, 1796, (1978).

6. Haushalter, J.P., Buffet, C.E., and Morris, M.D., Anal. Chem., 52, 1284, (1980).

7. Haushalter, J.P., and Morris, M.D., Anal. Chem., 53,2, (1981).

8. Bloembergen,N., Lotem, H., and Lynch, R.T., Indian J. of Pure and Appl. Phys., 16, 151, (1978).

9. Lotem, H., Lynch, R.T., and Bloembergen, N., Phys. Rev. A,14, 1748, (1976).

10. Druet, S.A.J., Taran, J.P.E., and Borde, C.J., J. Physique,40,, 819, (1979).

11. Druet, S.A.J., Taran, J.P.E., and Borde, C.J., J. Physique,41, 183, (1980).

12. Massey, V., and Hemmerich, P., Biochem. Soc. Trans., 8, 246, (1980).

13. Bruice, T.C., Acc. Chem. Res., 13, 256, (1980).

14. Muller, F., Van Schagen, G.G., and Van Berkal, W.J.H., in Yagi, K., and Yamano, T., eds., Flavins and Flavoproteins, 6th International Symposium, Elsevier, New York, 1981.

15. Muller, F., Photochem. Photobiol.,34, 735, (1981).

16. Dutta, P.K., Nestor, J.R., and Spiro, T.G., Proc. Nat. Acad., U.S.A.,74, 4146 (1977).

17. Schopfer, L.M., Haushalter, J.P., Smith, M., Milad, M., and Morris, M.D., Biochem.,20, 6734, (1981).

18. Schopfer, L.M., and Morris, M.D., in Flavins and Flavoproteins, VII, Massey, V., and Williams, C., eds., Elsevier, New York, in press.

19. Schopfer, L.M., and Morris, M.D., Biochem.,19, 4932, (1980).

20. Dutta, P.K., Spencer, R., Walsh, C., and Spiro, T.G., Biochem. Biophys. Acta,623, 77-83 (1980).

21. Nishina, Y., Shiga, K., Horiike, K., Tojo, H., Kasai, S., Yanase, K., Matsui, K., Natari, H., and Yamano, T., J. Biochem. (Tokyo),88, 403 (1980).

22. Dutta, P.K., Nestor, J.R., and Spiro, T.G., Biochem. Biophys. Res. Comm.,83, 209 (1978).

23. Schmid, R., and McDonagh, A.F., in The Porphyrins, Volume 6, (Dolphin, D., ed.), Academic, New York, 1979.

24. McDonagh, A.F., in The Porphyrins, Volume 6, (Dolphin, D., ed.), Academic, New York, 1979.

25. Lightner,D.A., in The Porphyrins, Volume 6, (Dolphin, D., ed.), Academic, New York, 1979.

26. Warshel,A., Ann. Rev. Biophys. Bioeng.,6, 273 (1977).

27. Kubota, K., Kuroda, S., and Koyama, Y., Biopolymers,20, 2701 (1981).

28. Hillig, K. W., and Morris, M.D., Biochem. Biophys. Res. Comm., 71, 1228 (1976).

29. Mainfatt, M., Bernard, L., and Theophonides, T., J. Raman Spectrosc.,4, 68 (1981).

30. Angeloni, L.L., Smulevich, G., and Marzocchi, M.P., Spectr. Chim. Acta., 38A, 213 (1982).

31. Tretzel, T., and Schneider, F.W., in Molecular Mechanisms of Biological Recognition, Balaban, M., ed., Elsevier, Amsterdam, 1979, pp. 235-240.

PHOTOACOUSTIC RAMAN SPECTROSCOPY AND RAMAN INDUCED KERR EFFECT

THEORY OF PHOTOACOUSTIC RAMAN SPECTROSCOPY

J. J. Barrett and D. F. Heller

Allied Corporation
Warren, New Jersey 07060

1. DESCRIPTION OF PROCESS

Photoacoustic Raman spectroscopy is a nonlinear spectroscopic technique based upon the selective population of a given energy state of a system by the process of coherent Raman amplification (stimulated Raman scattering). The conditions required for this process are: (1) the transition involving the initial and final energy levels must be Raman-active, i.e., the transitions must involve a change in the molecular polarizability, and (2) the frequency difference of the two incident laser beams must be adjusted to equal the frequency of this Raman-active transition. Raman-active transitions can occur for molecules that have no infrared spectrum, since the occurrence of a Raman spectrum depends on the change in polarizability of the molecule and not on the presence of a transition dipole moment. This makes PARS a particularly attractive analytical technique for studying molecules which have no infrared spectrum. The photoacoustic Raman scattering (PARS) process is shown schematically in Figure 1. Two incident laser beams with frequencies ω_p and ω_s interact with two energy states (labelled $|a\rangle$ and $|b\rangle$) of a molecule (Fig. 1a). During or following stimulated Raman scattering (SRS), collisional relaxation of the excited molecules produces a pressure change in the sample which is detected by a microphone. The laser beam at the frequency ω_p is referred to as the pump beam and the second laser beam at the frequency ω_s is referred to as the Stokes beam. If E_a and E_b are the energies associated with the levels $|a\rangle$ and $|b\rangle$, then the frequency ω_o (in wavenumbers), for the Raman transition between those energy states is equal to $(E_b-E_a)/hc$, where h is Planck's

W. Kiefer and D. A. Long (eds.), Non-Linear Raman Spectroscopy and Its Chemical Applications, 563–583.

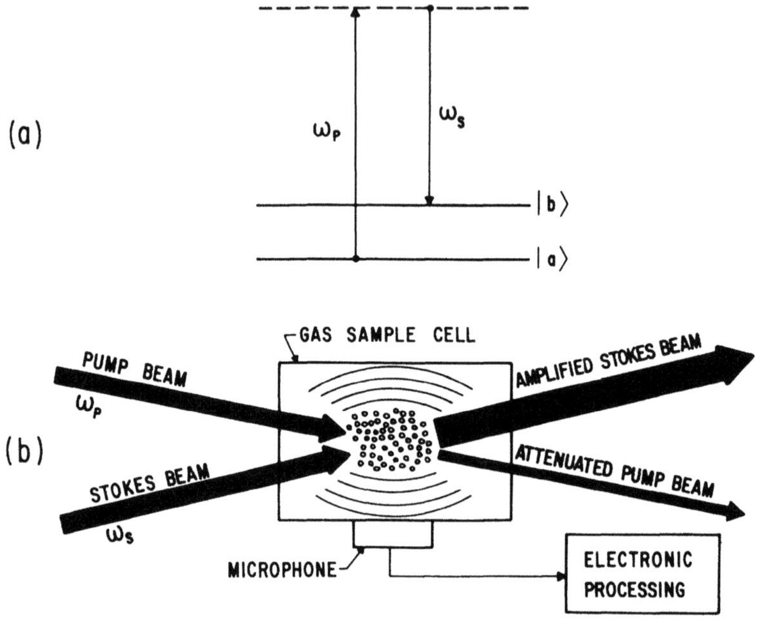

Figure 1. Schematic representation of the photoacoustic Raman scattering (PARS) process.

(a) An energy level diagram for a Raman type process involving the interaction of pump and Stokes photons at frequencies ω_p and ω_s, respectively, with the energy states $|a\rangle$ and $|b\rangle$ of a molecular system. The horiztonal dashed line represents an intermediate virtual state of the interaction.

(b) Schematic representation of the arrangement for generating a PARS signal in a gas. The pump and Stokes beams are overlapped spatially and temporally in the gas sample. When the frequency difference $\omega_p - \omega_s$ equals a Raman frequency of the gas, amplification of the Stokes beam and attenuation of the pump beam occurs and the molecular population of the upper energy level $|b\rangle$ is increased. Relaxation of these excited molecules by V-T processes results in the generation of an acoustic wave which is detected by a microphone.

constant and c is the speed of light. When the frequencies of the incident lasers are adjusted such that $\omega_p - \omega_s = \omega_o$, a nonlinear interaction between the ω_p and ω_s beams occurs through the third-order nonlinear susceptibility $\chi^{(3)}(\omega_s)$ of the sample. This nonlinear process results in the amplification of the Stokes beam and the attenuation of the pump beam. One molecule is transferred from level $|a\rangle$ to level $|b\rangle$ for each Stokes photon which is generated by this interaction. An excess population in level $|b\rangle$ is produced and the translational energy of the molecules in the gas sample is increased by collisional deactivation (V,R→T energy transfer) of this excess population. A pressure change which is the manifestation of the increase in translational energy is detected by a microphone as depicted in Fig. 1b. The theoretical aspects of photoacoustic Raman spectroscopy will be discussed in the next two sections.

2. THEORETICAL APPROACH

The magnitude of the pressure change in a gaseous sample produced by the photoacoustic Raman technique may be determined by calculating the number of molecules which are excited to an upper energy state. Since the number of Stokes photons which are produced by the stimulated Raman process is equal to the number of molecules which are deposited in the upper energy state, we begin by calculating the number of photons which are generated in the amplified Stokes wave. In the following sections we derive the theoretical expressions associated with PARS. A kinetic model, which considers the rates of various processes for populating and depopulating the upper energy level, is also discussed.

A. Stokes Amplification Process

The irradiance of the Stokes wave after traversing a distance z through a sample may be written as

$$I_s(z) = I_s(0)e^{g_s z} \tag{1}$$

where $I_s(0)$ is the initial irradiance of the Stokes laser beam at the cell entrance (z=0). The quantity g_s is the Stokes gain coefficient which is equal to [1]

$$g_s = -\frac{16\pi^2 k_s^o}{c n_s^2 n_p} \chi'' I_p. \tag{2}$$

where k_s^o ($=\omega_s n_s /c$) is the initial magnitude of the Stokes wave vector before amplification (at z=0); n_p and n_s are the refractive indices of the sample at the pump and Stokes frequencies, respectively; I_p is the irradiance of the pump laser beam and χ'' is the imaginary part of the third order nonlinear susceptibility at the Stokes frequency. In general, the third order susceptibility is equal to the sum of a complex resonant term and a nonresonant term, vis., $\chi^{(3)}(\omega_s) = (\chi' + i\chi'')_R + \chi_{NR}$ where the prime and double prime denote real and imaginary parts, respectively.

An explicit expression for the stimulated Raman scattering susceptibility $\chi^{(3)}(-\omega_s, \omega_s, \omega_p, -\omega_p)$ can be calculated from the response of a polarizable oscillator to the driving (pump and Stokes) fields. The equation of motion for a driven-damped classical oscillator in terms of the (displacement coordinate $\underset{\sim}{q}$ and resonance frequency ω_o) is [2,3]

$$\frac{\partial^2 \underset{\sim}{q}}{\partial t^2} + \Gamma \frac{\partial \underset{\sim}{q}}{\partial t} + \omega_o^2 \underset{\sim}{q} = \frac{1}{m} \frac{\partial}{\partial \underset{\sim}{q}} \{\underset{\sim}{P} \cdot \underset{\sim}{E}\} \tag{3}$$

where Γ is the radiative damping rate, E is the driving field and $\underset{\sim}{P}$ is the electronic polarizability of the oscillator, i.e.,

$$\underset{\sim}{P} = \underset{\approx}{\alpha} \cdot \underset{\sim}{E} + 1/2\, \underset{\approx}{\underset{\sim}{\beta}} \cdot \underset{\sim}{E}\underset{\sim}{E} + \ldots \tag{4}$$

For the case of interest, stimulated Raman scattering, only the first term in Eq. (4) is needed and Eq. (3) becomes

$$\frac{\partial^2 \underset{\sim}{q}}{\partial t^2} + \Gamma \frac{\partial \underset{\sim}{q}}{\partial t} + \omega_o^2 \underset{\sim}{q} = \frac{1}{m} \left(\frac{\partial \alpha}{\partial \underset{\sim}{q}} \right) E^2 . \tag{5}$$

For near resonant driving ($\omega_p - \omega_s \approx \omega_o$) only those terms in E^2 which have Fourier components near ω_o need be considered, i.e.,

$$E^2 \approx \frac{1}{2} E_p E_s^* \exp i[(k_p - k_s)z - (\omega_p - \omega_s)t] . \tag{6}$$

The above equations of motion [Eqs. (5) and (6)] are now easily solved for $\underset{\sim}{q}(t)$ by Fourier transform methods. Defining

$$\underset{\sim}{q}(\omega) = \frac{1}{2\pi} \int_{-\infty}^{\infty} e^{i\omega t} \underset{\sim}{q}(t) dt \tag{7}$$

we obtain

$$-\omega^2 \tilde{q}(\omega) - i\omega\Gamma\tilde{q}(\omega) + \omega^2\tilde{q}(\omega)$$

$$= \frac{1}{2m}\left(\frac{\partial\alpha}{\partial\tilde{q}}\right)\left\{E_p E_s^* e^{i(k_p-k_s)z}\,\delta(\omega-\omega_p+\omega_s) + E_p^* E_s e^{i(k_s-k_p)z}\,\delta(\omega-\omega_s+\omega_p)\right\} \tag{8}$$

for the transformed equation of motion. Thus,

$$\tilde{q}(\omega) = \frac{\left(\frac{\partial\alpha}{\partial \tilde{q}}\right)E_p E_s^* e^{i(k_p-k_s)z}\,\delta(\omega-\omega_p+\omega_s)}{2m[\omega_o^2-\omega^2-i\omega\Gamma]} \tag{9}$$

since only the near resonance term contributes.

Back transforming

$$\tilde{q}(t) = \int_{-\infty}^{\infty} e^{-i\omega t}\tilde{q}(\omega)d\omega \tag{10}$$

we arrive at our result

$$\tilde{q}(t) = \frac{\left(\frac{\partial\alpha}{\partial \tilde{q}}\right)E_p E_s^* e^{i(k_p-k_s)z}\, e^{-i(\omega_p-\omega_s)t}}{2m\left\{\omega_o^2-(\omega_p-\omega_s)^2-i(\omega_p-\omega_s)\,\Gamma\right\}}\,. \tag{11}$$

The nonlinear (Raman) polarization induced in a quantized ensemble of N such oscillators is

$$\tilde{P}^{NL} = N\Delta\tilde{q}^*\left(\frac{\partial\alpha}{\partial \tilde{q}}\right)\tilde{E} \tag{12}$$

where $\Delta = (N_a-N_b)/N$ is the fractional population difference between the upper and lower states. We note that the nonlinear contributions are explicitly carried by $\tilde{q}$. The ith component of the nonlinear polarization at the Stokes frequency P_{si}^{NL} is readily determined from the complex conjugate of Eq. (11) and the pump component of E. Its cartesian components are

$$P_{si}^{NL} = \frac{N\Delta}{4m\{\omega_o^2-(\omega_p-\omega_s)^2 + i(\omega_p-\omega_s)\Gamma\}} \qquad (13)$$

$$\cdot \sum_{jkl} \left(\frac{\partial\alpha}{\partial \tilde{q}}\right)_{ij} \left(\frac{\partial\alpha}{\partial \tilde{q}}\right)^*_{kl} E_{pj}E^*_{pk}E_{sl}e^{ik_s z} .$$

The quantity P_{si}^{NL} is related to the third order susceptibility $\chi^{(3)}$ through

$$P_{si}^{NL} = 3\sum_{jkl}\left\{\chi^{(3)}_{ijkl}(-\omega_s,\ \omega_p,\ \omega_p,\ -\omega_a)E_{pj}E_{pk}E^*_{pl}\ \ e^{i(2k_p-k_a)z}\right. \qquad (14)$$

$$\left. + 2\chi^{(3)}_{ijkl}\ (-\omega_s,\ \omega_p,\ -\omega_p,\ \omega_s)E_{pj}E^*_{pk}E_{sl}e^{ik_s z}\right\}.$$

Only the second term in Eq. (14) leads to amplification at the Stokes frequency, i.e., to stimulated Raman scattering. Using Eq. (13) we can identify $\chi^{(3)}$ in Eq. (14)

$$\chi^{(3)}_{ijkl}(-\omega_s,\ \omega_p,\ -\omega_p,\ \omega_s) = \frac{N\Delta\left(\frac{\partial\alpha}{\partial \tilde{q}}\right)_{ij}\left(\frac{\partial\alpha}{\partial \tilde{q}}\right)^*_{kl}}{24m\{\omega_o^2-(\omega_p-\omega_s)^2 + i(\omega_p-\omega_s)\ \Gamma\}} \qquad (15)$$

This expression can be reduced to a more standard form by noting that $\alpha(Q) = \alpha(Q_o) + (\partial\alpha/\partial Q)_o Q + \ldots$ Taking harmonic oscillator matrix elements we obtain the relation

$$\left(\frac{\partial\alpha}{\partial \tilde{q}}\right)_{ij} = \left(\frac{2m\omega_o}{\hbar}\right)^{1/2}\alpha_{ij} \qquad (16)$$

where the matrix elements of α are given by

$$\langle a|\ \alpha_{ij}\ |b\rangle = \frac{1}{\hbar}\sum_g\left\{\frac{\langle a|\ Q_i\ |g\rangle\langle g|\ Q_j\ |b\rangle}{\omega_{ga} + \omega_s} + \frac{\langle a|\ Q_j\ |g\rangle\langle g|\ Q_i\ |b\rangle}{\omega_{ga} - \omega_p}\right\} . \qquad (17)$$

Here Q_i is the ith component of the electric dipole moment operator, $|a\rangle$ and $|b\rangle$ are the initial and final states, and the summation associated with the index g is over all intermediate states. By noting that near resonance $\omega_p-\omega_s \approx \omega_o$ ($>> \Gamma$) and $\omega_o^2 - (\omega_p-\omega_s)^2 \approx [\omega_o-(\omega_p-\omega_s)](2\omega_o)$, we obtain

$$\chi^{(3)}_{ijkl}(-\omega_s, \omega_p, -\omega_p, \omega_s) = \frac{N\Delta}{24\hbar}\left[\frac{\alpha_{ij}\alpha^*_{kl}}{\omega_o-(\omega_p-\omega_s) + i\Gamma/2}\right]. \quad (18)$$

Of course, no real system is a simple two level system and thus the total third order susceptibility also contains other contributions from nonresonant states. These contributions are collectively labelled χ_{NR} and

$$\chi^{(3)}_{ijkl}(-\omega_s, \omega_p, -\omega_p, \omega_s) = \chi_{NR} + \chi \text{ [from Eq. (18)]} \quad (19)$$

The differential spontaneous Raman scattering cross-section, $d\sigma/d\Omega$, is related to the Raman polarization matrix elements by the equations

$$\overline{|\langle a|\ \alpha_{ij}\ |b\rangle|^2} = (c/\omega_s)^4\left(\frac{d\sigma}{d\Omega}\right)_{ij} \quad (20)$$

where the horizontal bar indicates an average over all orientations of the molecules. Combining Eq. (18) through Eq. (20) yields

$$\chi^{(3)}_{1111}(-\omega_s, \omega_s, \omega_p, -\omega_p) = \chi_{NR} + \frac{N\Delta c^4}{24\hbar\omega_s^4}\left[\frac{1}{\omega_o-\omega_p+\omega_s+i\Gamma/2}\right]\left(\frac{d\sigma}{d\Omega}\right) \quad (21)$$

where ω_o is the Raman transition frequency and Γ is the full width at half maximum (FWHM) for the spontaneous Raman line.

The total third order Stokes susceptibility for parallel electric fields at ω_p and ω_s, is

$$\begin{aligned}\chi^{(3)}(\omega_s) &= 6\chi^{(3)}_{1111}(-\omega_s, \omega_s, \omega_p, -\omega_p) \\ &= 6[\chi' + i\chi'']_R + 6\chi_{NR}\end{aligned} \quad (22)$$

where the term in brackets is the complex resonant part of the third order susceptibility. The real part of $\chi^{(3)}(\omega_s)$ is

$$[\chi^{(3)}(\omega_s)]' = 6(\chi' + \chi_{NR}) \tag{23}$$

$$= \frac{N\Delta c^4}{4\hbar\omega_s^4}\left[\frac{\omega_o-\omega_p+\omega_s}{(\omega_o-\omega_p+\omega_s)^2 + \Gamma^2/4}\right]\left(\frac{d\sigma}{d\Omega}\right) + 6\chi_{NR}$$

and the imaginary part of $\chi^{(3)}(\omega_s)$ is

$$[\chi^{(3)}(\omega_s)]'' = \chi'' = -\frac{N\Delta c^4}{8\hbar\omega_s^4}\left[\frac{\Gamma}{(\omega_o-\omega_p+\omega_s)^2 + \Gamma^2/4}\right]\left(\frac{d\sigma}{d\Omega}\right). \tag{24}$$

At resonance, $\omega_p-\omega_s = \omega_o$ and the peak value of $\chi^{(3)}(\omega_s)$ is

$$\chi^{(3)}(\omega_s)_{\text{Peak}} = -i\,\frac{N\Delta c^4}{2\hbar\omega_s^4\Gamma}\left(\frac{d\sigma}{d\Omega}\right) + 6\chi_{NR}. \tag{25}$$

Since the imaginary part of the Stokes susceptibility (χ'') is a negative quantity [Eq. (24)], g_s is positive [cf. Eq. (2)] and the Stokes wave will be amplified according to Eq. (1). Using Eq. (24) to eliminate χ'' from Eq. (28) yields

$$g_s = \frac{4\pi^3 N\Delta c^2}{h\, n_p n_s \omega_s^3}\left[\frac{\Gamma}{(\omega_o-\omega_p+\omega_s)^2 + \Gamma^2/4}\right]\left(\frac{d\sigma}{d\Omega}\right)I_p. \tag{26}$$

At resonance ($\omega_p-\omega_s=\omega_o$), the peak value of g_s is

$$\left\{g_s\right\}_{\text{Peak}} = \frac{16\pi^3 N\Delta c^2}{h\, n_s n_p \omega_s^3\Gamma}\left(\frac{d\sigma}{d\Omega}\right)I_p. \tag{27}$$

Often, it is convenient to express the frequency factors ω_s and Γ(FWHM) in wavenumber units [cm^{-1}]. Since ω(angular units) = $2\pi c\omega(cm^{-1})$, Eq. (27) may be re-written as

$$\left\{g_s\right\}_{\text{Peak}} = \frac{N\Delta}{\pi h c^2 n_s n_p \omega_s^3\Gamma}\left(\frac{d\sigma}{d\Omega}\right)I_p \tag{28}$$

where ω_s and Γ are in wavenumber units. In section C, we will use these results for the Raman amplification process to derive an expression for the PARS pressure change.

B. The Quantum Mechanical Two-Level System

The exact dynamics of a two-level system in the presence of the pump and Stokes fields can be determined by standard techniques. We consider the system shown in Fig. 1a where the upper level $|b\rangle$ undergoes some dephasing (or relaxation) process with rate Γ_b. The hamiltonian for this system is

$$H = H^o - \underset{\approx}{\alpha}:\underset{\sim}{E}\underset{\sim}{E} \tag{29}$$

where H^o is the hamiltonian for the zero-order states $|a\rangle$ and $|b\rangle$ and $\underset{\approx}{\alpha}$ is the polarizability. The equations of motion for the time-dependent probability amplitudes $C_a(t)$ and $C_b(t)$ to be in states $|a\rangle$ and $|b\rangle$ respectively are

$$\dot{C}_a(t) = -i/\hbar\, V_{ab}C_b e^{-i\omega t} \tag{30a}$$

$$\dot{C}_b(t) = -i/\hbar\, V_{ba}C_a e^{+i\omega t} - 1/2\,\Gamma_b C_b(t) \tag{30b}$$

where

$$\omega \equiv \omega_b - \omega_a \tag{31}$$

$$V_{ab} = V^*_{ba} = -\underset{\approx}{\alpha}_{ba}:\underset{\sim}{E}\underset{\sim}{E} \tag{32}$$

with α_{ba} being the (Raman scattering) matrix element of the polarizability connecting states $|a\rangle$ and $|b\rangle$. For near resonant driving, the only important contributions to E^2 are at frequency $\nu = \omega_p - \omega_s$. Therefore, we may write

$$V_{ab} = -\alpha_{ba} E_p E_s \cos[(k_p - k_s)z - \nu t]. \tag{33}$$

The resultant equations of motion [Eq. (30)] can be solved analytically in the rotating wave approximation to give

$$P_b(t) = \left|C_b(t)\right|^2$$

$$= \left|\frac{\alpha_{ba}E_pE_s}{\hbar\,\Omega_R}\sin(\Omega_R t/2)\right|^2 e^{-\Gamma_b t} \tag{34}$$

where Ω_R the (complex) Rabi frequency is given by

$$\Omega_R = \left[(\omega_{ab}-\nu)^2 + \Gamma_b^2/4 - i(\omega_{ab}-\nu)\Gamma_b + \left(\frac{\alpha_{ba}E_pE_s}{\hbar}\right)^2\right]^{1/2}. \quad (35)$$

The dephasing rate

$$\Gamma_b = \Gamma_b^D + \Gamma_b^R \quad (36)$$

is written as the sum of pure dephasing (Γ_b^D) and pure relaxation (Γ_b^R) contributions. The above description ignores repopulation of the state $|a\rangle$ through Γ_b^R or subsequent relaxation processes. The rate that energy is dissipated by the system is

$$\frac{\partial E}{\partial t} = -\hbar\omega_{ab}\,\Gamma_b^R\,P_b(t) \quad (37)$$

and $(-\partial E/\partial t)$ is the rate at which energy is made available to other (e.g., acoustic) processes. In most molecular systems, Γ_b is dominated by collisional relaxation processes at pressures above a few torr. Assuming $\Gamma_b^D > \mathrm{Re}\,\Omega_R > \Gamma_b^R$ coherent effects become unimportant in the overall energy dissipation dynamics and the dynamics of the energy flow can be modelled by simple kinetic equations. For simplicity we use this latter approach in the following section to show how PARS signal magnitudes can be obtained for any system and how these signals are expected to vary with laser intensity, etc. Under conditions where coherence effects become important, e.g., at high driving field intensities and small collision rates, recourse to Eqs. (34-37) can be taken.

3. Magnitude of PARS Signals

The magnitude of the photoacoustic Raman signals may be deduced from the change in internal translational energy of the sample due to the Stokes amplification process. One pump photon is annihilated for every Stokes photon which is generated by the nonlinear interaction and a molecule is transferred from the lower to the upper energy state associated with the Raman transition. The internal translational energy change is found by computing the number of Stokes photons which are produced in the amplification process and equating this number to the number of vibrational quanta which are formed.

The acoustic pressure wave is produced by molecules in the upper energy state which lose their vibrational energy by means of collisions. These collisional processes increase the translational energy of the gas molecules which results in heating of the gas. The change in thermal energy produces a pressure change for a gas at constant volume in an enclosed cell. If the input lasers are modulated at a rate which is slow compared to the vibrational-to-translational (V-T) relaxation rate, then the temperature and hence the gas pressure will vary at the modulation frequency. This modulated pressure wave is the sound wave which is detected in all photoacoustic processes.

We begin with an analysis which illustrates the basic features of the acoustic generation process. Following this, a more detailed description of the process will be presented.

A. A Simple Quasiequilibrium Model

The amplified Stokes irradiance is

$$\begin{aligned} \Delta I_s &\equiv I_s(z) - I_s(0) \\ &= I_s(0)[e^{g_s z} - 1]. \\ &\simeq I_s(0) g_s z \end{aligned} \tag{38}$$

for small values of the exponential gain factor, $g_s z$.

The energy in the amplified Stokes wave is

$$E_s = \Delta I_s \, A_s T = \Delta \eta_s (\hbar \omega_s) \tag{39}$$

where A_s is the cross-sectional area of the Stokes beam (assuming exact spatial overlap of the pump and Stokes beams), T is the interaction time of the pump and Stokes laser beams, and $\Delta\eta_s$ is the number of Stokes photons produced by the amplification process during the time interval T. For chopped cw laser beams, T is equal to the "on" time during one chopper cycle whereas for pulsed laser beams, T is equal to the duration of the laser pulses. In Eq. (39), the Stokes frequency is expressed in radians/sec.

For the sake of this discussion let us assume that the Raman transition involved in the amplification process occurs in a two level gaseous system between the ground and first vibrational state. Let N_0 and N_1 be the molecular population densities in the ground and first vibrational state, respectively, and let N be the total population density equal to the sum of N_0 and N_1. Solving Eq. (39) for $\Delta\eta_s$ and equating this to the total population increase ΔN_1 in the first vibrational state yields

$$\Delta N_1 = \Delta \eta_s = \Delta I_s A_s T/\hbar\omega_s. \tag{40}$$

Substituting for ΔI_s [Eq. (38)], we obtain

$$\Delta N_1 = I_s(0) g_s z A_s T/\hbar\omega_s. \tag{41}$$

Using Eq. (41), it is possible to calculate directly the increase in the number of molecules in the first vibrational level brought about by the nonlinear interaction of the ω_p and ω_s laser beams. The quantity ΔN_1 is directly proportional to the Raman gain coefficient g_s which in turn is proportional to only the imaginary part of the third order nonlinear susceptibility χ'' [Eq. (2)]. Because of this, the real nonresonant susceptibility does not contribute to the PARS signal and therefore background signals due to χ_{NR} will be absent. In the CARS technique, the nonresonant susceptibility is a limiting factor in the trace analysis of mixtures. Since the PARS technique does not depend on the nonresonant susceptibilities, it is more suitable for trace analysis.

The excess molecules in the first vibrational level, ΔN_1, collide with other molecules in the gas and with the walls of the sample cell. As a result of these collisions, the vibrational energy ($\hbar\omega_o$) of these excited molecules is converted into translational kinetic energy and the excited molecules return to the ground state. Therefore, the internal translational energy, U, of the gas is increased by an amount ΔU equal to

$$\Delta U = \Delta N_1(\hbar\omega_o). \tag{42}$$

Using the gas law $pV = (\gamma-1)U$, where γ is the ratio of specific heats (C_p/C_v) of the gas, we can relate the pressure change, Δp, in the sample cell to the change in translational energy by the equation

$$\Delta p = (\gamma-1)\Delta U/V \tag{43a}$$

or

$$\Delta p = [(\gamma-1)\Delta N_1 \hbar\omega_o]/V \tag{43b}$$

where the pressure change, Δp, is in units of dynes/cm^2 (microbars) when ΔU and V are expressed in units of ergs and cm^3, respectively. Substituting Eq. (41) into Eq. (43b) yields

$$\Delta p = [(\gamma-1) I_s(0) g_s z A_s T(\omega_o/\omega_s)]/V \tag{44}$$

which is the pressure increase in a cell of volume V. The acoustic power, P_a, due to the pressure change Δp is equal to $V\Delta p/T$, i.e.,

$$P_a = (\gamma-1)(\omega_o/\omega_s) g_s z P_s(0) \tag{45}$$

where $P_s(0)$ is the total incident power of the Stokes (ω_s) laser beam. In the next section a more exact expression for the acoustic pressure change will be derived using a rate equation approach.

B. Rate Equation Approach

The dynamics of the photoacoustic Raman process may be described by considering the time dependence of the number of molecules per unit volume (N_1) in the upper energy level. It is possible to define a Raman transition probability, k_r, in terms of the exponential form of the Raman gain [Eq. (38)]. However, the use of the linear approximation, which considers small values of the Raman gain factor $g_s z$, provides particularly simple expressions for the rate constant k_r. The error introduced by this linear approximation is not very large even for values of $g_s z$ approaching unity. The various processes for populating and depopulating the upper energy level are illustrated in Figure 2. In the incoherent limit, which arises when molecular dephasing processes (e.g., due to "weak" collisions) are faster than the Raman excitation processes, the level populations are described by simple kinetic equations. For simplicity we consider a two-level model. Hence, the sum of N_0 and N_1 equals the total molecular population density, N. For real molecular systems, the situation is more complex since many other rotational and vibrational energy states are usually present with significant populations at room temperature. The analysis of such systems is straightforward although somewhat more complicated.

The Raman amplification process populates level 1 at a rate $k_r N_0$ (per unit volume) and depopulates it at a rate $k_r N_1$ where N_0 and N_1 are the number of molecules per unit volume in levels 0 and 1, respectively. At the peak of a Raman transition, the transition probability, k_r, is equal to

$$k_r = \left\{ \frac{C_o}{\hbar \omega_s} \right\} I_s I_p \tag{46}$$

where

$$C_o = \frac{16\pi^3 c^2}{h\, n_s n_p \omega_s^3 \Gamma} \left(\frac{d\sigma}{d\Omega} \right) \tag{47}$$

and k_r is in units of [sec^{-1}]. The quantity, k_r, is not a constant since it depends on the irradiances of the pump (I_p) and Stokes (I_s) beams. These irradiances will vary spatially within the interaction (or source) volume of the laser beams depending on such factors as the degree of focusing, spatial laser modes, etc. The total upper state population change per

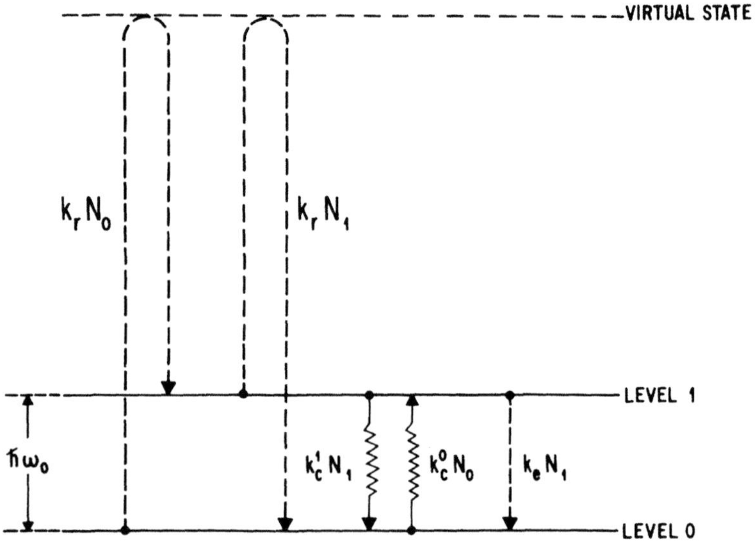

Figure 2. An energy level diagram for a two level molecular system depicting the transition rates for various processes. The quantities k_r, k_c, and k_e represent the transition probabilities (units of sec^{-1}) for Raman, collisional, and emission processes, respectively.

unit time due to the nonlinear Raman interaction may be found by integrating the quantity $k_r(N_0-N_1)$ over the source volume. The rate of radiation induced transitions from the upper to the lower state is equal to $(k_r+k_e)N_1$ where k_e^{-1} is the radiative lifetime of the upper state for spontaneous emission of radiation. The quantities $k_c{}^1N_1$ and $k_c{}^0N_0$ are the collision-induced transition rates from the upper to lower state and from the lower to upper state, respectively. For a gas in thermal equilibrium, the collision rate from the upper to the lower state, $k_c{}^1N_1{}^e$, is equal to the reverse collision rate $k_c{}^0N_0{}^e$, and since $N_1^e = N_0^e \exp(-\hbar\omega_o/kT)$, we have

$$k_c^0 = k_c^1 \exp(-\hbar\omega_o/kT) \tag{48}$$

where $\hbar\omega_0$ is the energy difference between levels 0 and 1 and the quantities N_0^e and N_1^e represent the equilibrium population densities for levels 0 and 1, respectively, before laser excitation. The rate equation for the population in level 1 may be written as

$$\frac{dN_1}{dt} = k_r N_0 + k_c^0 N_0 - k_r N_1 - k_c^1 N_1 - k_e N_1. \tag{49}$$

In this treatment, we ignore the coherent effects introduced into the level populations by the applied laser fields. These coherent effects become increasingly important at high fields and may be treated by replacing the kinetic equations we use here by their quantum mechanical counterparts (as in Section 2.B). The rate of thermal heating of the gas will not be greatly altered by the kinetic (incoherent driving) approximation used in this discussion. Collisions of gas molecules with the walls of the photoacoustic cell will be neglected in this analysis. Such collisions serve to carry away the thermal energy which is deposited in the gas by the Raman amplification process. As a result, the gas sample will eventually reach a higher average temperature whose magnitude depends on the laser irradiances (i.e., the rate heat is deposited in the gas) and on the rate energy is being lost to the walls of the photoacoustic cell. Generally, while these losses are not significant for ballistic acoustic wave propagation characteristic of pulsed laser excitation, they can be important for the case of cw laser excitation. We may express Eq. (49) in terms of the population density deviation from equilibrium, i.e.,

$$\frac{dF_1}{dt} = C_1 - C_2 F_1 \tag{50}$$

where $F_1 \equiv N_1 - N_1^e$, $F_0 \equiv N_0 - N_0^e$ and $F_0 + F_1 = 0$. The constants C_1 and C_2 in Eq. (50) are equal to

$$C_1 = N_0^e \left\{ k_r - (k_r + k_e) e^{-\frac{\hbar\omega_0}{kT}} \right\} \tag{51a}$$

and

$$C_2 = 2k_r + k_c\left\{1 + e^{-\frac{h\omega_o}{kT}}\right\} + k_e \tag{51b}$$

where we let $k_c \equiv k_c^1$ for simplicity. The solution of Eq. (50) is

$$F_1(t) = \frac{C_1}{C_2}\left\{1 - e^{-C_2 t}\right\} \tag{52}$$

where we have assumed the boundary condition $F_1(\text{at } t=0) = 0$. The assumption $k_c >> k_e$ is generally valid and for vibrational energy levels above several hundred wavenumbers, the approximations $\exp(-\hbar\omega_o/kT) << 1$ and $N_o^e \approx N$ can be made without a significant effect on the final result. By use of these approximations, Eq. (52) takes the simple form

$$F_1(t) = \left[\frac{Nk_r}{2k_r + k_c}\right]\left\{1 - e^{-(2k_r + k_c)t}\right\}. \tag{53}$$

Let us assume that the gas sample is in thermodynamic equilibrium at each point within the photoacoustic cell. Therefore, the quasi-equilibrium pressure change in a cell of volume V_c due to a change in internal energy is

$$\frac{\partial p}{\partial t} = \left(\frac{\gamma-1}{V_c}\right)\int_{V_f} H dV \tag{54}$$

where H is the rate heat is produced per unit volume in the interaction region of the laser beams and the integration is over the volume of the interaction region (source). The quantity H is equal to

$$H = \hbar\omega_o k_c F_1(t). \tag{55}$$

Using Eq. (53) and the fact that the value of k_r varies spatially within the source volume, we obtain

$$\frac{\partial p}{\partial t} = \left(\frac{\gamma-1}{V_c}\right)\hbar\omega_o N k_c \int_{V_f}\left(\frac{k_r}{2k_r+k_c}\right)\left\{1 - e^{-(2k_r+k_c)t}\right\} dV. \tag{56}$$

The solution of Eq. (56) for the pressure change from equilibrium (at t=0) is

$$p(t) = \left(\frac{\gamma-1}{V_c}\right)\hbar\omega_o N k_c \int_{V_f} \left(\frac{k_r}{2k_r+k_c}\right)\left\{t + \frac{1}{2k_r+k_c}\left[e^{-(2k_r+k_c)t} - 1\right]\right\} dV. \tag{57}$$

Let us consider the form of Eq. (57) in two limiting cases for which $(2k_r + k_c)t \gg 1$.

Case I: $k_c \gg k_r$. For this case, there is no population build-up in level 1, since molecules which are "pumped" into level 1 by the Raman amplification process are very quickly relaxed by the fast V-T rate. Therefore, Eq. (57) may be written as

$$p(t) \approx \left(\frac{\gamma-1}{V_c}\right)\hbar\omega_o N t \int_{V_f} k_r dV = \left(\frac{\gamma-1}{V_c}\right)\hbar\omega_o N t K_r \tag{58}$$

where K_r is the total Raman transition rate integrated over the volume of the interaction region. Equation (58) is essentially the same as Eq. (43b) which was derived using the quasiequilibrium model. In Eq. (58), the pressure increases linearly with excitation time and is directly proportional to NK_r, which is the rate molecules are entering level 1. After the lasers are switched off at some time $t=\tau$, the maximum pressure value, $p(\tau)$, quickly decays. Therefore, the observed pressure changes follow closely the temporal variations of the excitation laser sources when the laser excitation time τ is much longer than the collisional relaxation time $(k_c)^{-1}$.

Case II: $k_r \gg k_c$. For this case, molecules are entering level 1 faster than they can be removed by the V-T process and the population in level 1 builds up. The condition $k_r \gg k_c$ (Case II) would usually apply to the experimental situation involving excitation of the sample by high power (ca. 1 MW) lasers. For this case, if the collisional dephasing is slow [i.e., if the dephasing rate is less than or about equal to the Raman transition probability (k_r)] then the time dependence of our model system is more appropriately given by the two level Rabi-like solutions rather than by our kinetic equations. However, when coherent effects are not important in determining the energy which is finally given over to translations (e.g., $\tau_{pulse} \ll k_c^{-1}$), the kinetic approach is acceptable. The pressure change [Eq. (57)] now has the form

$$p(t) \approx \frac{1}{2}\left(\frac{\gamma-1}{V_c}\right)\hbar\omega_o N t k_c V_f \tag{59}$$

which is independent of the Raman transition probability k_r. When the lasers are switched off at some time $t=\tau$, the excess population (re equilibrium population) in level 1, and hence the pressure, will decay at a rate governed by the collisional transition probability k_c. Therefore, for $t>\tau$,

$$F_1(t>\tau) = F_1(\tau)e^{-k_c t} \tag{60}$$

and the pressure change is

$$p(t>\tau) \approx \frac{1}{2}\left(\frac{\gamma-1}{V_c}\right)\hbar\omega_o N\tau k_c V_f \, e^{-k_c(t-\tau)} \quad . \tag{61}$$

Therefore, the complete solution for Case II is

$$p(t) = \begin{cases} p_1(t) & \text{for } t\leq\tau \\ p_2(t) & \text{for } t>\tau \end{cases} \tag{62}$$

where $p_1(t)$ and $p_2(t)$ are defined by Eq. (59) and Eq. (61), respectively.

Numerical Estimate:

In order to get a rough idea of the magnitude of the photoacoustic Raman effect in a gas, numerical calculations were made using typical values of the relevant parameters for a gaseous sample. The following conditions were assumed:

Pump wavenumber:	$\omega_p/2\pi c = 20\ 000\ \text{cm}^{-1}$
Stokes wavenumber:	$\omega_s/2\pi c = 19\ 000\ \text{cm}^{-1}$
Raman linewidth:	$\Gamma/2\pi c = 0.1\ \text{cm}^{-1}$
Raman cross section:	$(d\sigma/d\Omega) = 10^{-30}\ \text{cm}^2$
Gas temperature:	$T = 300$ K
Specific heat ratio:	$\gamma = C_p/C_v = 1.3$
Collisional relaxation rate:	$k_c = 10^6\ \text{sec}^{-1}$
Molecular number density:	$N = 2.5 \times 10^{19}$ molecules cm^{-3}

Let us assume that the focal regions of the pump and Stokes beams are exactly overlapped and have uniform irradiances throughout the focal region. The major portion of the PARS signal will be generated at this common focus where the laser photon densities are the greatest. We will assume that the focal region is approximated by a small circular cylinder of cross sectional area 2.9×10^{-6} cm^2, a length (confocal parameter) of 0.11 cm and a volume, V_f, of 3.2×10^{-7} cm^3. These dimensions of the focal region are approximately what one would obtain for a 5 mm diameter Gaussian laser beam focused with a 15 cm focal length lens.

Using Eq. (47), we obtain

$$C_o = 7.8 \times 10^{-38} \text{ cm}^4 \text{ sec/erg.}$$

Case I: $k_c \gg k_r$

This situation will be realized when the pump and Stokes lasers are at low power levels (e.g., cw laser sources) and when the collisional relaxation rate k_c out of level 1 is fast. If we assume that the power of the pump and Stokes lasers are both equal to 3 watts, then the laser irradiances in the focal region are

$$I_p = I_s = 10^{13} \text{ erg sec}^{-1} \text{ cm}^{-2}.$$

Substituting numerical values into Eq. (46) yields

$$k_r = 2.2 \text{ sec}^{-1}$$

for the stimulated Raman transition rate. The peak value of the Stokes gain [Eq. (28)] is

$$g_s = 2.0 \times 10^{-5} \text{ per cm}$$

and the product of the interaction length and the Stokes gain is 2.1×10^{-6}. For lasers which are amplitude modulated (chopped) at 500 Hz, the excitation time, τ, during one chopper cycle is 10^{-3} sec. From Eq. (58), the total number of molecules in the focal volume which are excited to level 1 during one chopper cycle is $N\tau V_f k_r = 1.8 \times 10^{10}$ molecules. This corresponds to the excitation of one molecule for every 450 molecules in the focal region. This result obtained using Eq. (58) is in exact agreement with the upper state population change [Eq. (41)] which was derived using the simple model. The total acoustic energy (heat) generated during one chopper cycle is 1.1×10^{-3} ergs which was calculated using Eq. (58) as the product of the pressure change and the cell volume.

Case II: $k_r >> k_c$

This condition will apply for the case of excitation by high peak power lasers or when the collisional relaxation rate is slow. For this estimate, we will assume that the peak laser powers are $P_p = P_s = 1$ MW and that the laser pulse duration, τ, is 10 nsec.

The irradiances of the pump and Stokes lasers in the focal region are:

$$I_p = I_s = 3.4 \times 10^{18} \text{ erg sec}^{-1} \text{ cm}^{-2}.$$

The stimulated Raman rate constant [Eq. (46)] is

$$k_r = 2.5 \times 10^{11} \text{ sec}^{-1}$$

and the peak value of the Stokes gain [Eq. (28)] is

$$g_s = 6.7 \text{ per cm}.$$

The product of the Stokes gain and the interaction length (confocal parameter) is equal to 0.74. Thus, even for this case of high power laser excitation, the error due to the linear approximation of the Raman gain relative to the exponential form of the Raman gain is equal to only 32%. From Eq. (59), the number of molecules in the focal volume which contribute to the generation of the acoustic pressure pulse is equal to 0.5 $N\tau k_c V_f = 4 \times 10^{10}$ molecules. At the end of the laser pulse, the number of molecules in the upper energy level as given by Eq. (53) is equal to 0.5 N, i.e., half of the total number of molecules are in the upper state. Therefore, for high laser powers, the populations in the lower and upper states tend to equalize and the transition becomes saturated. For this situation, the stimulated Raman process is driving upper state molecules to the lower state before they can give up their energy by collisions. Therefore, the PARS signal will build up to some maximum level and, after saturation occurs, no further pressure change and consequently no further increase in the PARS signal will be generated.

4. CONCLUSION

In this paper, we have developed a useful theory of photoacoustic Raman spectroscopy. Equations governing the Raman gain process were derived and an expression for the third order Stokes susceptibility was obtained from the equation of motion for a driven-damped classical oscillator. An outline of a quantum mechanical treatment describing the dynamics of a two level system in the presence of the pump and Stokes fields was presented. A simple quasiequilibrium model was described and the pressure change in a gaseous sample cell due to the nonlinear Raman interaction was derived. Equations for the pressure change were also derived using a kinetic model which explicitly accounted for the various rates molecules were entering and leaving the upper energy state of the Raman transition. Two limiting cases of this kinetic model were identified and numerical estimates using typical values for the various parameters were presented.

REFERENCES

1. J. J. Barrett, "Photoacoustic Raman Spectroscopy of Gases," in Chemical Applications of Nonlinear Raman Spectroscopy, (A.B. Harvey, Editor) Academic Press, New York (1981), pp. 89-169.

2. M. Maier, W. Kaiser, and J. A. Giordmaine, "Backward Stimulated Raman Scattering," Phys. Rev., 177, pp. 580-599, (1969).

3. M. Maier, "Applications of Stimulated Raman Scattering," Appl. Phys., 11, pp. 209-231, (1976).

APPLICATIONS OF PHOTOACOUSTIC RAMAN SPECTROSCOPY

J. J. Barrett, D. R. Siebert and G. A. West

Allied Corporation
Warren, New Jersey 07060

1. INTRODUCTION

Photoacoustic Raman spectroscopy (PARS) is a nonlinear spectroscopic technique which has been developed recently [1]. The first experimental demonstration of the PARS technique was accomplished using low power continuous wave laser excitation of a gaseous sample [2]. Significant improvements in the experimental capability of PARS were realized by the use of pulsed laser excitation [3]. The application of the PARS technique to the problem of trace analysis of gaseous mixtures has resulted in a trace detection capability of about one part per million [4]. Recent improvements [5] in the design of the photoacoustic sample cell should make it possible to achieve even lower detection limits. The PARS technique has been used successfully to study pure rotational Raman transitions in gases [6-7]. One of the characteristic features of pure rotational PARS spectra is the absence of the Rayleigh component at zero frequency shift. In pure rotational Raman spectra recorded by conventional optical techniques, Rayleigh scattering, which is much more intense than Raman scattering, can sometimes interfere with the observation of low lying rotational states. Care must be taken to reduce stray light in the experimental arrangement (sample cell, spectrometer, etc.) as much as possible in order to be able to successfully record a spectrum. Since the PARS technique detects only the energy that is deposited in the gaseous sample, it is not affected by problems with Rayleigh scattering.

The use of acoustic methods to detect the absorption of light in gases was first reported by Bell [8] and Tyndall [9] about

W. Kiefer and D. A. Long (eds.), Non-Linear Raman Spectroscopy and Its Chemical Applications, 585–602.

one hundred years ago. Photoacoustic detection for spectroscopic applications had not received much attention until high brightness laser sources became available in the 1960's. The potential of acoustic techniques for detecting laser absorption was first explored by Kerr and Atwood [10]. Later work [11-16] demonstrated the extremely high sensitivity of acoustic detection. This high sensitivity has resulted in the ability to observe very weak vibrational overtone transitions [17] in gases using visible tunable dye lasers.

These photoacoustic techniques are based on the direct absorption of light in the sample. However, the PARS technique is fundamentally different because the light is not absorbed by the sample. Population of the upper energy level of a transition is accomplished by a stimulated Raman process rather than by direct absorption. This stimulated Raman excitation requires the simultaneous illumination of the sample by two laser beams whose frequencies differ by a Raman transition frequency. This experimental arrangement is the same as that required for other nonlinear Raman techniques such as coherent anti-Stokes Raman scattering (CARS) [18], Raman-induced Kerr effect spectroscopy (RIKES) [19], and stimulated Raman gain spectroscopy (SRGS) [20]. These nonlinear techniques all measure an optical signal by means of a photodetector whereas the PARS technique employs a microphone to sense the energy that is deposited in the sample after the passage of the laser beams.

2. EXPERIMENTAL ARRANGEMENTS

A. CW Laser Excitation

The experimental arrangement for cw PARS experiments [2] is shown schematically in Figure 1. Two argon ion lasers were employed in this experimental arrangement. The 514.5 nm line from one of these lasers provided the pump frequency ω_p with a maximum power of 6.6 W. This laser was modulated at 573 Hz with a mechanical light chopper and then directed onto the focusing mirror M_4 by means of the prisms, P_1 and P_3, and other auxiliary mirrors. The mirror M_4, which had a 30 cm radius of curvature, focused the pump beam to a point about 15 cm from the mirror surface. Mirror M_5 retro-reflected the ω_p beam so that two passes through the focal region were achieved. A halfwave plate was used to rotate the vertical polarization of the argon ion lasers by 90° so that Brewster angle optical components with a fixed beam height above the optical table could be used. The second argon ion laser was used to pump a Rhodamine 6G jet stream dye laser. The argon laser beam entered the dye laser cavity through the special dichroic mirror M_1. Mirrors M_1 and M_5 formed the optical cavity of the dye laser. A three element

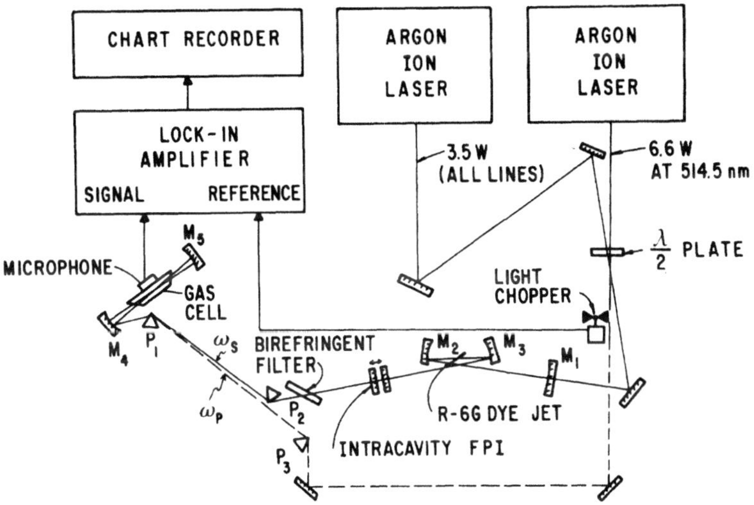

Figure 1. Schematic representation of the experimental arrangement used for obtaining photoacoustic Raman spectra with continuous wave (cw) laser sources.

birefringent filter was used for coarse tuning of the dye laser frequency, while fine tuning and line narrowing (<0.1 cm^{-1} FWHM) were obtained by use of an intracavity Fabry-Perot interferometer with piezoelectric scanning. Careful alignment of the ω_p and ω_s beams in the photoacoustic cell was necessary to insure that the focal regions for these beams were spatially overlapped. A photoacoustic cell with Brewster angle windows was placed in the dye laser cavity at the focus midway between mirrors M_4 and M_5. This photoacoustic cell used four miniature electret microphones mounted on the cell walls to detect the photoacoustic signals from the gas sample. The electrical signals from these microphones were detected by a lock-in amplifier tuned to the 573 Hz modulation frequency. The electrical output from the lock-in amplifier was displayed on a chart recorder. A discussion of the experimental results obtained with this apparatus is presented in Section 3.

B. Pulsed Laser Excitation

Since the PARS technique is a nonlinear process whose signals scale with the product of the excitation time and the powers in the pump and Stokes beams, a significant improvement in magnitude of the PARS signal can be realized by use of pulsed rather than cw laser sources. A schematic diagram of the

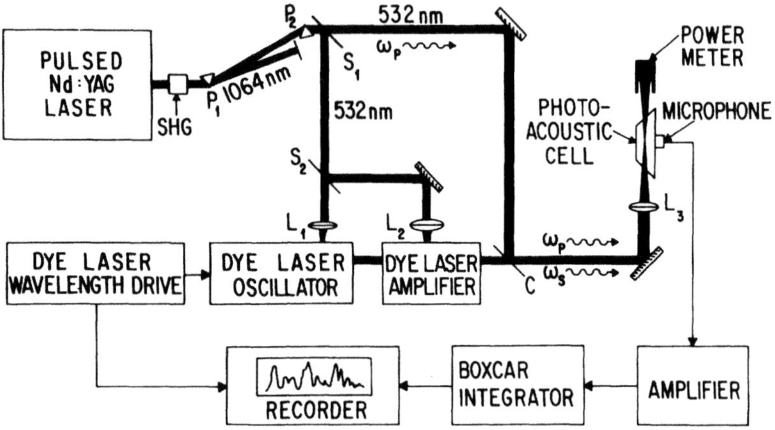

Figure 2. Schematic representation of the experimental arrangement used for photoacoustic Raman studies with pulsed laser sources.

experimental arrangement for obtaining PARS spectra with pulsed lasers is shown in Figure 2. A high power Q-switched Nd:YAG laser system provided the primary source of coherent radiation for this experiment. This laser, which is capable of generating 700 mJ in a 10 nsec pulse (~70 MW peak power) at 1.06μm, is operated at a pulse repetition rate of 10 Hz. The 1.06μm radiation is frequency doubled in an angle-tuned KD*P crystal to produce 200 mJ (~20 MW peak power) at 532 nm. A prism P_1 spatially separates the 1.06μm beam from the 532 nm beam and a beam splitter S_1 divides the 532 nm beam into two beams, one for pumping a dye laser and the other for producing the PARS signal in the photoacoustic cell PC. A dichroic beam combiner C was used to superimpose the ω_p and ω_s beams and a lens L_3 focused these beams at a common point in the photoacoustic cell. Within the limits of experimental error, the magnitudes of the observed PARS signals were independent of the degree of focusing for lenses whose focal lengths varied between 10 cm to 33 cm. An energy meter was used to measure the energies of the pump and Stokes beams after passing through the photoacoustic cell. The signal from the microphone in the photoacoustic cell was amplified (usually with a gain of about 100) and then analyzed using a boxcar integrator (PAR Model 162). The output from the boxcar integrator was displayed on a chart recorder. Spectra were obtained by scanning the wavelength of the dye laser while recording the boxcar output on the chart recorder.

Typical waveforms of the photoacoustic signals obtained using the pulsed laser apparatus is shown in Figure 3. This

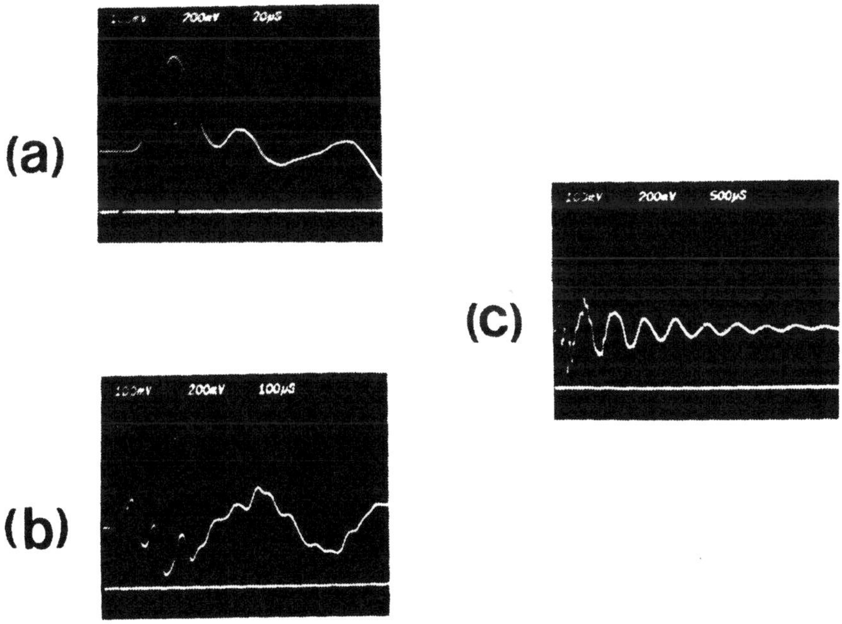

Figure 3. Oscilloscope photographs of the photoacoustic Raman signals in 6.7 kPa of carbon dioxide for pulsed laser excitation. The three photographs (a), (b), and (c) were obtained using the same input signal from the microphone but with different oscilloscope sweep rates of 20μsec/cm, 100μsec/cm, and 500μsec/cm, respectively. The lower trace in each photograph displays the gate apertures of the boxcar integrator.

data was obtained for pure CO_2 gas at a pressure of 6.7 kPa (50 torr) for three different horizontal sweep rates of the oscilloscope. The frequency difference between the pump and Stokes lasers was equal to 1388 cm^{-1} which is the Raman shift for the ν_1 band in CO_2. The pump beam at 532 nm and the Stokes beam at 574 nm had pulse energies of 15 mJ and 10 mJ, respectively, and peak powers of approximately 1.5 MW and 1.0 MW, respectively, assuming nominal pulse durations of 10 nsec. The photoacoustic cell was about 13.5 cm in length with Brewster angle entrance and exit windows. The distance from the focal region to the microphone was 0.75 cm. The electrical signal from the electret microphone was amplified 100 times by a low-noise amplifier. In Fig. 3a, the initial (ballistic) acoustic signal (upper trace) is shown with its peak amplitude of 650 mV occurring about 53 μsec after the lasers were fired. The initial acoustic wave reached the microphone about 27 μsec after the laser pulses passed through the cell but the maximum electrical signal from the microphone (due to this acoustic wave) occurred approximately 26 μsec later. This delay time of 26 μsec is due to the high frequency cut-off of the microphone. Figures 3b and 3c show larger portions of the acoustic waveforms. The damped oscillatory waveform of Fig. 3c is due to successive reflections of the initial acoustic pulse at the windows of the photoacoustic cell. In general, the pulsed acoustic waveforms for all gases are similar to those shown in Fig. 3. Waveform variations are observed for different photoacoustic cells due to geometrical effects and for different gas samples due to variations in the speed of sound.

3. EXPERIMENTAL RESULTS

A. Continuous Wave Laser Excitation

The first observation of a photoacoustic signal produced by a Raman transition in a gas was made using low power cw laser sources with the experimental arrangement shown in Figure 1. Although the photoacoustic signals produced by cw laser excitation are rather small, the amplitude stability, narrow linewidths (ca. 0.1 cm^{-1}) and well-defined Gaussian spatial modes of the cw laser sources make it possible to characterize accurately the experimental parameters, thereby permitting comparison of the experimentally observed PARS signals with theoretical predictions.

The photoacoustic Raman signals generated in a sample of methane gas at a pressure of 106.7 kPa (800 torr) are shown in Figure 4. The Raman active symmetric stretch vibrational mode of CH_4 (at 2916.7 cm^{-1}) was investigated by illuminating the CH_4 gas sample with two focused cw laser beams which were spatially overlapped. The ω_p pump beam was provided by the 514.5 nm [ω_p (vacuum) = 19429.8 cm^{-1}] line of an argon ion laser and the ω_s Stokes beam was generated by a tunable R-6G dye laser. When the dye laser was tuned to 605.4 nm, the frequency difference of the lasers was equal to the Raman transition frequency (2916.7 cm^{-1}) and a photoacoustic Raman (PARS) signal was observed. These PARS signals are shown in the lower part of Fig. 4 for repetitive scans of the (ω_p-ω_s) frequency difference across the CH_4 Raman band. The wavelength of the dye laser was monitored by use of a high-resolution spectrometer set to pass the CH_4 Stokes line at 605.4 nm as shown in the upper part of Fig. 4. The small lines between the main peaks are wavelength

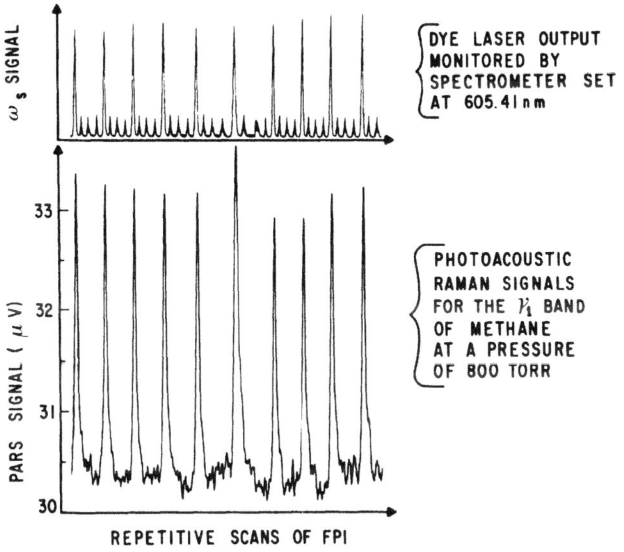

Figure 4. Experimental results of the cw PARS experiments in methane gas.

markers spaced ca. 0.5 cm^{-1} which were generated by passing some dye laser light through a solid etalon. Each time the dye laser was tuned to the CH_4 Stokes line a PARS signal was observed. The signal disappeared when either the ω_p or ω_s beam was blocked and the magnitude of the PARS signal varied linearly with gas pressure. The PARS background signal was due to a small amount of light from the ω_p beam striking a baffle in the photoacoustic cell.

B. Pulsed Laser Excitation

Large gains in the magnitude of the observed PARS signals have been reported [3] using high peak power pulsed laser sources. In comparison to cw laser excitation, the amount of energy deposited in a gas via the stimulated Raman process can be greater by factors of the order of 10^6. As a result of this large improvement in the magnitude of the PARS signal, the usefulness of PARS as a general spectroscopic technique capable of analyzing a wide variety of molecular species has been verified experimentally. In certain cases, the PARS technique may be the preferred method of analysis over other analytical techniques.

A PARS spectrum of the ν_1 vibrational region in CH_4 gas at a pressure of 6.7 kPa (50 torr) is shown in Fig. 5. The ν_1 Q branch

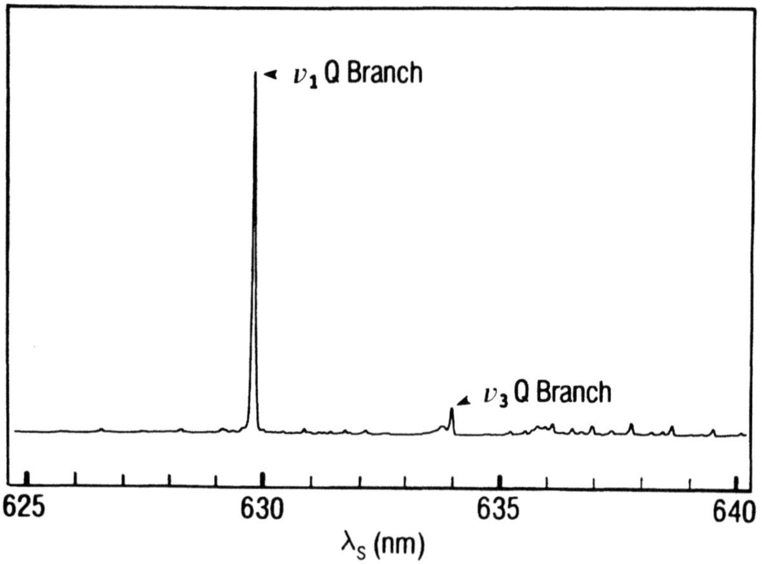

Figure 5. Vibrational PARS spectrum of CH_4 gas at a pressure of 6.7 kPa. The laser powers at the pump and Stokes wavelengths were 1.2 MW and 0.5 MW, respectively, with linewidths of 0.3 cm^{-1} and 0.5 cm^{-1}, respectively.

with a Raman shift of 2917 cm^{-1} ($\omega_s \sim$ 630 nm) and the ν_3 Q branch with a Raman shift of 3019 cm^{-1} ($\omega_s \sim$ 634 nm) are the two prominent features of the spectrum. The dye laser, which was operated using cresyl violet laser dye, had a linewidth of about 0.5 cm^{-1}. The peak powers in the pump and Stokes beams were 1.2 MW and 0.5 MW, respectively, with corresponding linewidths of 0.3 cm^{-1} for the ω_p beam and 0.5 cm^{-1} for the ω_s beam. The rich spectrum shown in Fig. 6 was generated by keeping all the experimental parameters the same as in Fig. 5 except for an increase in amplifier gain by a factor of twenty-five. The rotational O, P, R, and S branches along with the $2\nu_2$ Q branch are now clearly visible with a flat, essentially noise-free baseline. The time required to obtain the complete spectrum shown in Fig. 6 was 50 minutes.

The linearity of the PARS signal with gas pressure was investigated for the ν_1 (1388 cm^{-1}) band of CO_2. Spectral data for the ν_1 band for various gas pressures in the range 6.7 to 100 kPa (50 to 750 torr) are shown in Fig. 7. A plot of the peak amplitude of the CO_2 PARS signals for the ν_1 band as a function of gas pressure is shown in Fig. 8. The linearity of this plot indicates that the pressure changes in the sample, which are produced by the photoacoustic Raman process, are linear in molecular number

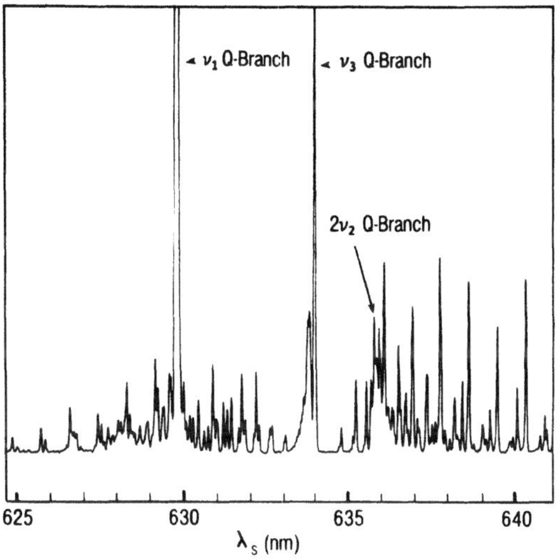

Figure 6. Vibrational PARS spectrum of CH_4 under the same experimental conditions as indicated in Figure 5 with the exception of an increased microphone amplifier gain by a factor of twenty-five.

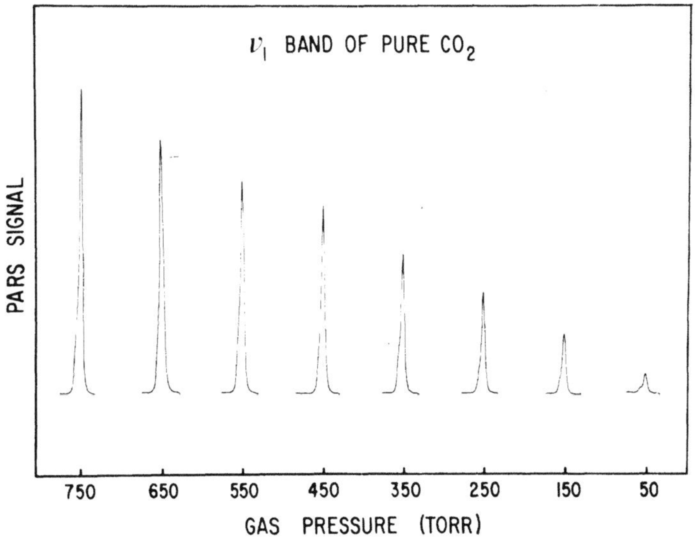

Figure 7. Spectral traces of the ν_1 band of CO_2 gas for different gas pressures. The data shown in this figure were obtained with laser powers of 1.1 MW and 0.34 MW at the pump and Stokes wavelengths, respectively.

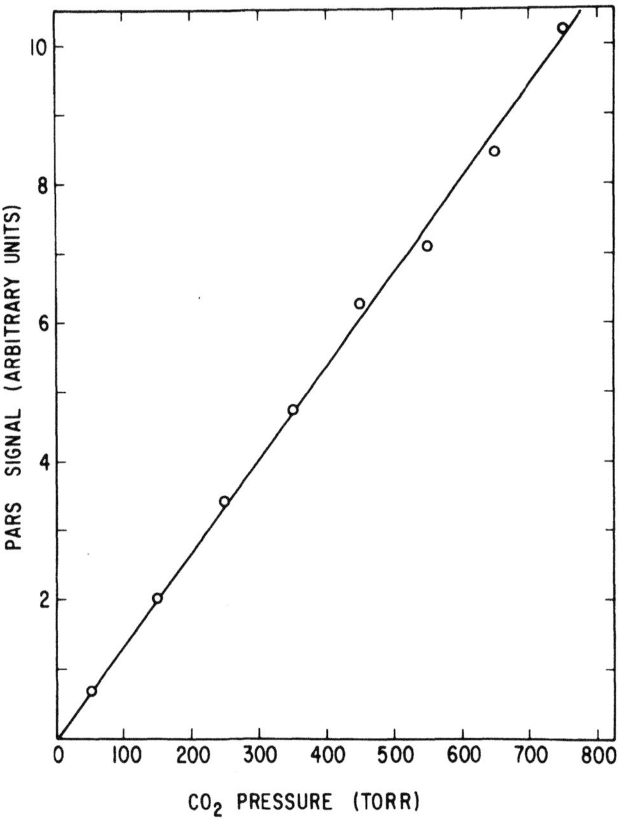

Figure 8. Plot of the peak spectral amplitude versus gas pressure for the data presented in Figure 7.

density as predicted by theory. A PARS spectrum of vinyl chloride (CH_2H_3Cl) is shown in Fig. 9. The gas pressure was 6.7 kPa and the peak laser powers in the pump and Stokes beams were 0.9 MW and 0.3 MW, respectively. The scan of Fig. 9 covers the region from about 1300 cm^{-1} to 1665 cm^{-1} which includes the vibrational ν_3 band at 1369 cm^{-1} and the vibrational ν_4 band at 1608 cm^{-1}. This spectrum has not been corrected for variations in dye laser power over the wavelength region of the scan.

B. PARS Spectra of Gaseous Mixtures

The application of the PARS technique to the analysis of gases in mixtures has been investigated recently by Siebert et al [4]. The usefulness of PARS for studying gas mixtures was realized shortly after the initial pulsed experiments were

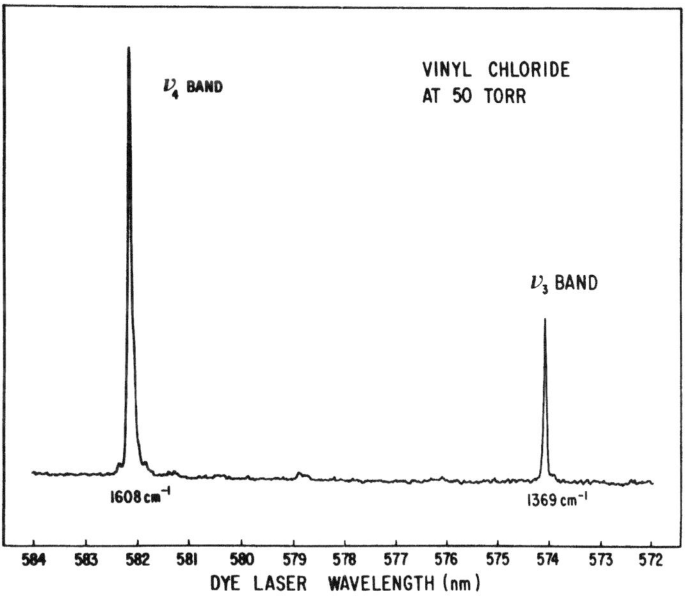

Figure 9. The photoacoustic Raman spectrum of the vibrational bands of vinyl chloride ($H_2C{=}CHCl$). Laser powers used to obtain this spectrum were 0.9 MW and 0.3 MW in the pump and Stokes beams, respectively.

carried out. A typical PARS spectrum of a gaseous mixtures is shown in Fig. 10 for a mixture of 26.7 kPa (200 torr) of SF_6 and 53.3 kPa (400 torr) of H_2. The $J=1\rightarrow3$ and $2\rightarrow4$ pure rotational transitions in H_2 at 587 cm^{-1} and 814 cm^{-1}, respectively, were observed along with the ν_1, ν_2, and ν_5 bands in SF_6 at 774 cm^{-1}, 642 cm^{-1}, and 525 cm^{-1}, respectively.

A more important aspect of the analysis of gaseous mixtures deals with the identification and quantitative measurement of trace quantities of a minor component of the mixture. Since the PARS signal is derived from the imaginary part of the third order nonlinear susceptibility tensor, contributions from the non-resonant susceptibility, which restrict the usefulness of coherent anti-Stokes Raman scattering (CARS) to trace detection levels on the order of 10 ppm, do not pose an equivalent problem for PARS. If the PARS technique is to be useful for trace analysis in mixtures, there must be a predictable relationship, (preferably linear) between gas concentration and the PARS signal magnitude. For the case of ethylene, a linear relationship does not exist between gas pressure and PARS signal

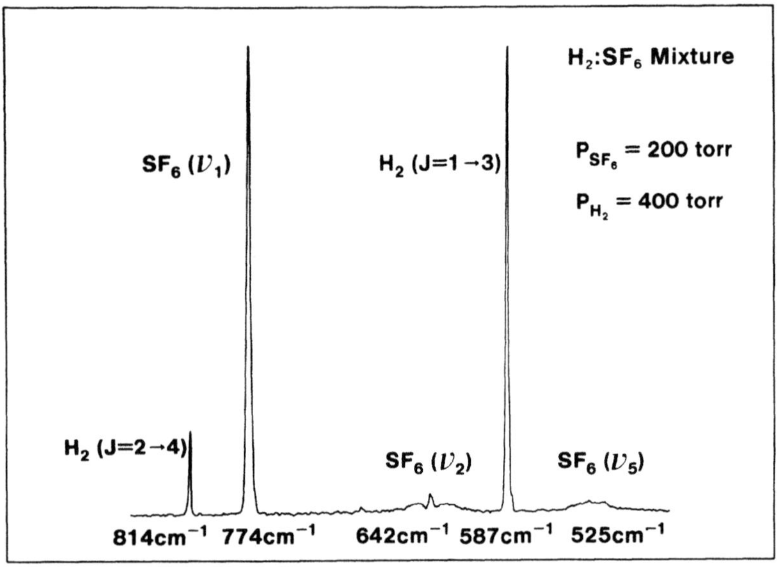

Figure 10. Photoacoustic Raman spectrum of a mixture of hydrogen and sulfur hexafluoride.

amplitude in the pure gas over the pressure range 0.67 to 100 kPa. This departure from linearity is believed due to pressure broadening of the Raman transition. In all gaseous mixtures studied up to now, a linear relationship does exist between the PARS signal amplitude and gas concentration for concentration values of one percent or less. An example of this linear behavior is shown in Fig. 11 for carbon dioxide in nitrogen. In Fig. 11, the partial pressure of CO_2 varied over the range 0.067 to 1.3 kPa. The total pressure of the gas mixture was equal to 100 kPa. A concentration of 6.3 ppm of CO_2 in N_2 was observed with a signal-to-noise ratio of 2.5. Therefore, the detection of CO_2 with a signal-to-noise ratio of one implies that a concentration of 2.5 ppm of CO_2 in 100 kPa of N_2 can be measured by our present apparatus. All measurements were obtained using a single pass of the pump and Stokes beams through the photoacoustic cell.

At present, the trace detection capability of PARS is about one part per million of CO_2 in air or nitrogen at atmospheric pressure. In order to lower the trace detection limit below one ppm, it will be necessary to increase the amount of energy deposited in the gas relative to the background noise due to the microphone/amplifier combination. One way of increasing the deposited

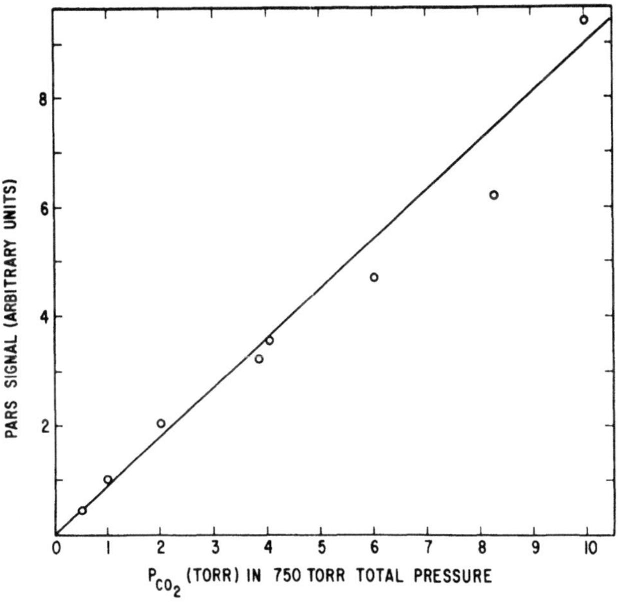

Figure 11. A plot of the PARS signal amplitude versus CO_2 pressure over the range of 0.067 to 1.3 kPa (0.5 to 10 torr).

energy is to pass the laser beams through the photoacoustic cell many times. If schemes for enhancing the magnitude of the PARS signal are successful, then trace concentration measurements in the parts per billion range would be possible which would make PARS competitive with other trace analytical techniques (e.g., GC/MS and FTIR).

C. Pure Rotational PARS

The pure rotational PARS spectrum of CO_2 is shown in Fig. 12. The laser powers in the pump and Stokes beams were 3.3 MW and 0.12 MW, respectively, and the CO_2 pressure was 80 kPa. The spectral linewidths of the pump and Stokes beams were about 0.3 cm^{-1} and 0.5-1.0 cm^{-1}, respectively. The striking feature of this spectrum is the absence of a strong Rayleigh component at the pump wavelength (532 nm). For other Raman methods, such as spontaneous Raman scattering, CARS, RIKES, SRGS, etc., the presence of intense Rayleigh scattering makes it difficult to observe pure rotational Raman transitions, especially if the Raman shifts are very small. This absence of the Rayleigh component makes the PARS technique particularly well suited for studying pure rotational Raman spectra.

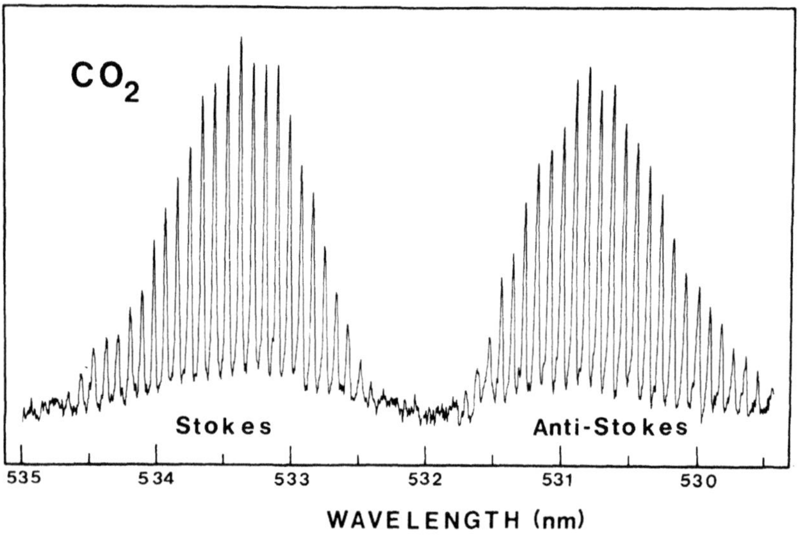

Figure 12. The pure rotational photoacoustic Raman spectrum of CO_2 gas at a pressure of 80 kPa. The spacing between adjacent rotational lines in this spectrum is about 3.1 cm^{-1}.

A computer model has been used to calculate the rotational PARS spectrum of CO_2. The results of this computer model are shown in Fig. 13 for CO_2. The upper part of this figure displays an experimentally generated PARS spectrum for CO_2 with a resolution estimated to be about 0.3 cm^{-1}. Although there is some indication of the presence of the odd J rotational lines of the 01^10 vibrational level, the resolution is not sufficient to completely resolve them. In the lower part of Fig. 13, the computer generated CO_2 PARS spectrum is shown. Good qualitative agreement exists between the experimental and computed spectra. The small discrepancies are due to laser power fluctuations during the scan and uncertainties concerning the laser linewidths and profiles.

The main advantages of the rotational PARS technique are its high sensitivity and resolution capability. In addition, it offers the unique feature of being able to observe very small Raman shifts with no interference due to Rayleigh scattering. Interferometric measurements of the input laser frequencies will result in the ability to study rotational Raman spectra with an accuracy far greater than that which can be achieved by other Raman techniques.

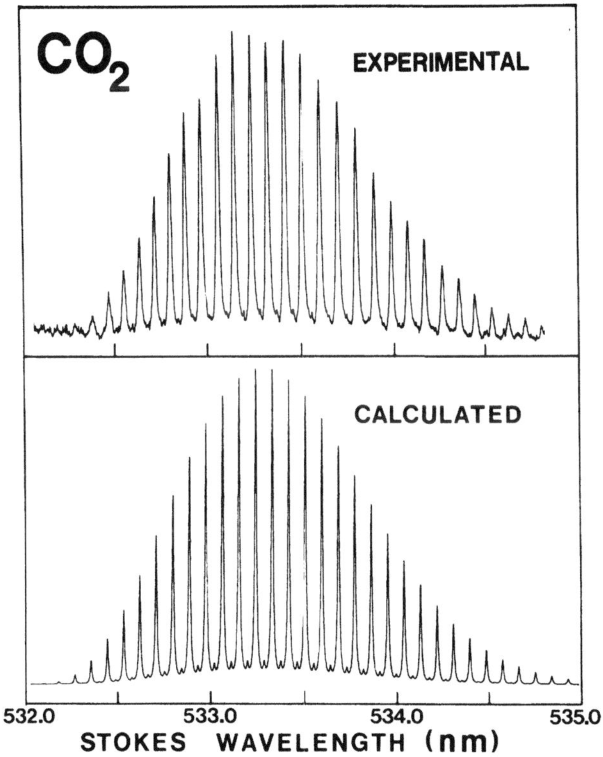

Figure 13. Comparison of the experimental and computer-generated pure rotational PARS spectra of CO_2. The experimental spectrum was obtained using laser powers of 5.2 MW and 0.12 MW in the pump and Stokes laser beams, respectively.

4. CONCLUSION

Photoacoustic Raman spectroscopy differs from all other nonlinear Raman techniques in the way the desired signal is detected. The PARS technique involves the acoustic detection of a pressure change in the sample whereas other nonlinear Raman techniques are concerned with the measurement of an optical signal. This fundamental difference in detection schemes provides a new dimension to the experimental methods which have been used previously to obtain Raman spectra. The use of acoustic measuring techniques provides a direct way for detecting the energy deposited in the sample. Other techniques such as stimulated Raman gain spectroscopy (SRGS) or conventional infrared absorption spectroscopy are based upon the measurement of small changes in the energy of the optical radiation passing through the samples. Since it is difficult to accurately measure small changes in a large background signal, these techniques are limited in their sensitivity. Photoacoustic absorption spectroscopy, which is also based on the direct observation of energy deposited in the sample, has demonstrated extremely high sensitivity in its ability to measure gas concentrations below one part per billion (ppb). A major factor responsible for this high sensitivity is the inherent sensitivity of acoustic detection methods. This same measurement sensitivity capability, which is intrinsic to photoacoustic Raman spectroscopy, provides a bright outlook regarding the usefulness of PARS as a sensitive analytical technique.

REFERENCES

1. J. J. Barrett, "Photoacoustic Raman Spectroscopy of Gases," in Chemical Applications of Nonlinear Raman Spectroscopy, (A.B. Harvey, Editor) Academic Press, New York (1981), pp. 89-169.

2. J. J. Barrett and M. J. Berry, "Photoacoustic Raman Scattering in Gases," in the Proceedings of the Sixth International Conference on Raman Spectroscopy, Bangalore, India, September 4-9, 1978, E. Schmid, R. Krishnan, W. Kiefer, and H. Schrotter, Eds., Heyden, London (1978), Vol. 1, pp. 466-467; "Photoacoustic Raman Spectroscopy using cw Laser Sources," Appl. Phys. Lett., **34**, pp. 144-146,(1979).

3. G. A. West, D. R. Siebert, and J. J. Barrett, "Gas Phase Photoacoustic Raman Spectroscopy using Pulsed Laser Excitation," J. Appl. Phys., **51**, pp. 2823-2828, (1980).

4. D. R. Siebert, G. A. West, and J. J. Barrett, "Gaseous Trace Analysis using Pulsed Photoacoustic Raman Spectroscopy," Appl. Optics, **19**, pp. 53-60, (1980).

5. Preliminary results have been reported by J. J. Barrett at the Second Chemical Congress of the North American Continent, Las Vegas, NV, August 24-29, 1980 and at the 1980 Federation of Analytical Chemistry and Spectroscopy Societies (FACSS) Meeting, Philadelphia, PA, September 28 - October 3, 1980.

6. G. A. West and J. J. Barrett, "Pure Rotational Stimulated Raman Photoacoustic Spectroscopy," Opt. Letters, **4**, pp. 395-397, (1979).

7. G. A. West and J. J. Barrett, "Photoacoustic Pure Rotational Raman Spectroscopy," in The Proceedings of the VII International Conference on Raman Spectroscopy, North-Holland Publishing Company, Amsterdam (1980), pp. 696-697.

8. A. G. Bell, Proc. Am. Assoc. Adv. Sci., "On the Production and Reproduction of Sound by Light," 29, pp. 115-136, (1880); Philos. Mag., "Upon the Production of Sound by Radiant Energy," 11, pp. 510-528, (1881).

9. J. Tyndall, "Action of an Intermittent Beam of Radiant Heat upon Gaseous Matter," Proc. Roy. Soc. London, 31, pp. 307-317, (1881).

10. E. L. Kerr and J. G. Atwood, "The Laser Illuminated Absorptivity Spectrophone: A Method for Measurement of Weak Absorptivity in Gases at Laser Wavelengths," Appl. Opt., 7, pp. 915-921, (1968).

11. L. B. Kreuzer, "Ultralow Gas Concentration Infrared Absorption Spectroscopy," J. Appl. Phys., 42, pp. 2934-2943, (1971).

12. L. B. Kreuzer and C. K. N. Patel, "Nitric Oxide Air Pollution: Detection by Optoacoustic Spectroscopy," Science, 173, pp. 45-47, (1971).

13. L. B. Kreuzer, N. D. Kenyon, and C. K. N. Patel, "Air Pollution: Sensitive Detection of Ten Pollutant Gases by Carbon Monoxide and Carbon Dioxide Lasers," Science, 177, pp. 347-349, (1972).

14. L. B. Kreuzer, "Laser Optoacoustic Spectroscopy--A New Technique of Gas Analysis," Anal. Chem., 46, pp. 239A-244A, (1974).

15. C. K. N. Patel, E. G. Burkhardt, and C. A. Lambert, "Spectroscopic Measurements of Stratospheric Nitric Oxide and Water Vapor," Science, 184, pp. 1173-1176, (1974).

16. E. Kritchman, S. Shtrikman, and M. Slatkine, "Resonant Optoacoustic Cells for Trace Gas Analysis," J. Opt. Soc. Am., 68, pp. 1257-1271, (1978).

17. K. V. Reddy, R. G. Bray, and M. J. Berry, "Dye Laser-Induced Photochemistry," in Advances in Laser Chemistry, (A. H. Zewail, Editor), Springer-Verlag, Berlin (1978), pp. 48-61.

18. M. W. Tolles, J. W. Nibler, J. R. McDonald, and A. B. Harvey, "A Review of the Theory and Application of Coherent Anti-Stokes Raman Spectroscopy (CARS)," Appl. Spectroscopy, 31, pp. 253-271, (1977).

19. D. Heiman, R. W. Hellwarth, M. D. Levenson, and G. Martin, "Raman-Induced Kerr Effect," Phys. Rev. Lett., 36, pp. 189-192, (1976).

20. A. Owyoung, "Coherent Raman Gain Spectroscopy Using CW Laser Sources," IEEE J. Quant. Electronics, QE-14, pp. 192-203, (1978).

RAMAN INDUCED KERR EFFECT

H.W. Schrötter

Sektion Physik der LMU München,
München, West Germany

1. INTRODUCTION

When a Raman active medium is irradiated simultaneously with an elliptically polarized pump laser beam of frequency ω_L and a linearly polarized Stokes laser beam as probe of frequency ω_S, a Stokes component is produced which is perpendicularly polarized to the incident probe beam. This perpendicular component shows resonance for $\omega_L-\omega_S = \omega_j$, where ω_j is a Raman active eigen frequency of the medium, and can therefore be used to observe the Raman spectrum.

The observed line shape near resonance and the background signal off resonance are determined by the direction and shape of the polarization ellipse of the pump beam and its orientation with respect to the polarization direction of the probe beam on one hand and the resonant and non-resonant components of the non-linear susceptibility on the other hand. For a circularly polarized pump beam a background free Raman spectrum is observed when the probe laser is tuned. This Raman induced Kerr effect (RIKE) has been first demonstrated by Heiman et al. (1) in organic liquids. In relation to CARS the advantage of RIKE is that no phase matching is required. The first investigations of RIKE in a crystal were made by Bergmann et al. (2, 3) in $LiIO_3$.

RIKE with an elliptically polarized pump beam can be used for measuring the ratio of the optical Kerr coefficient to the Raman scattering tensor elements and of the electronic contribution to the Kerr constant in liquids (4, 5) and in crystals (6). The resonant and nonresonant parts of the third order non-linear

W. Kiefer and D. A. Long (eds.), Non-Linear Raman Spectroscopy and Its Chemical Applications, 603–611.

susceptibility of benzene and bromobenzene were determined by Bergmann et al. (7).

2. THEORY

The theory of RIKE has recently been reviewed by Eesley (8) and we follow his notation in this summary.

A. Pure RIKE Spectroscopy

Two collinear laser beams propagating in z-direction are described as the electric fields

$$\vec{E}(\omega_1) = E_x(\omega_1)\cdot\frac{\vec{x}}{x} + iE_y(\omega_1)\frac{\vec{y}}{y} \tag{1}$$

with circular polarization ($|E_x|=|E_y|$) and

$$\vec{E}(\omega_2) = E_y(\omega_2)\frac{\vec{y}}{y} \tag{2}$$

with linear polarization, where $\vec{x}/x$ and $\vec{y}/y$ are the unit vectors in the respective directions. The nonlinear polarization density in x-direction, orthogonal to $\vec{E}(\omega_2)$, at frequency ω_2 is then

$$P_x(\omega_2) = \frac{3}{8}i\left[\chi_{1212}(-\omega_2,\omega_2,-\omega_1,\omega_1) - \right. \tag{3}$$

$$\left. -\chi_{1221}(-\omega_2,\omega_2,-\omega_1,\omega_1)\right]\cdot E_y(\omega_2)\cdot\left|E(\omega_1)\right|^2 ,$$

where $\chi_{1211}=\chi_{1222}=0$, valid for isotropic media, and

$|E(\omega_1)|=|E_x(\omega_1)|\cdot\sqrt{2} = |E_y(\omega_1)|\cdot\sqrt{2}$ has been used.

$\chi^{(3)}$ can be expressed as the sum of the nonresonant, the real resonant, and the imaginary resonant parts of the third order susceptibility

$$\chi_{ijkl} = \chi^{NR}_{ijkl} + \chi'_{ijkl} + i\chi''_{ijkl} \tag{4}$$

and therefore, when χ^{NR}_{1212} and χ^{NR}_{1221} are approximately equal, the nonresonant background is cancelled in a RIKE experiment with circularly polarized pump beam.

The signal field produced by the nonlinear polarisation at resonance $(\omega_1-\omega_2=\omega_j)$ is obtained as

$$E_x(\omega_2) = -\frac{3}{8}\cdot L\cdot\frac{\omega_2}{n_2}\sqrt{\frac{\mu_o}{\varepsilon_o}}\left[\chi_{1212}-\chi_{1221}\right] E_y(\omega_2)\left|E(\omega_1)\right|^2 \tag{5}$$

where L is the interaction length of the lasers, n_2 the refractive index at the signal frequency ω_2, and μ_o the permeability of free space. It is to be noted that eq. (5) contains no phase matching factor as for CARS.

The imaginary part of the nonlinear susceptibility at resonance is related to the peak differential scattering cross section of the linear Raman effect $d^2\sigma/d\Omega d\omega_2$ per unit volume,

$$\chi''_{1212} - \chi''_{1221} = \frac{16\pi^4}{3}\ \frac{n_1\varepsilon_o^2 c^4}{n_2 h\omega_1\omega_2^3}\ (1-3\rho_j)\cdot \tag{6}$$

$$\cdot\ \frac{d^2\sigma_{jxx}}{d\Omega d\omega_2}\cdot\left[\exp(-\frac{h\omega_j}{2\pi kT}) - 1\right] \ ,$$

where n_1 is the refractive index at ω_1, ρ_j the depolarization ratio of the Raman band ω_j, h Planck's and k Boltzmann's constant, and T the temperature.

The intensity of the signal has been calculated (8) to be

$$I_x(\omega_2) = \frac{9}{16}\ \frac{\omega_2^2\ L^2}{n_1^2 n_2^2}\ (\frac{\mu_o}{\varepsilon_o})^2\ \left|\chi_{1212}-\chi_{1221}\right|^2\ I_y(\omega_2) I^2(\omega_1). \tag{7}$$

Because of the cancellation of the nonresonant background a symmetric lineshape is obtained.

B. Optically heterodyned RIKE

As can be seen from eqs. (6) and (7), the RIKE intensity is proportional to the square of the Raman scattering cross section, as in CARS, and weak Raman lines are suppressed. This situation can be overcome by introduction of a local oscillator of frequency ω_2 in a technique termed optically heterodyned RIKE spectroscopy (9). The local oscillator beam has the same polarization as the expected RIKE signal and its electric field E_{LO} adds coherently to the RIKE signal field $E_R = E_x$ of eq. (5). Thus the net detector intensity of an ideal system will be

$$I_D = \frac{c}{2} n_2 \varepsilon_o |E_{LO}(\omega_2) + E_R(\omega_2)|^2 = \tag{8}$$

$$= I_{LO} + I_R + \frac{c}{2} n_2 \varepsilon_o (E_R E^*_{LO} + E^*_R E_{LO}).$$

If the local oscillator intensity I_{LO} is much greater than the RIKE intensity I_R, the heterodyne cross-term

$$I_H = n_2 c \varepsilon_o \, \mathrm{Re}(E_R E^*_{LO}) \tag{9}$$

will also be larger than the RIKE intensity I_R. From eqs. (5) and (6) it is seen that it is linear in the Raman scattering cross section and an improved signal-to-noise ratio is obtained for weak Raman lines.

3. EXPERIMENTAL ARRANGEMENTS

A schematic diagram of a RIKE spectrometer with additional provision for OHD-RIKE is shown in Fig. 1. A tunable dye laser with frequency ω_1 is used as pump laser; its beam is made circularly polarized by a quarter wave plate. The beam of the probe laser with fixed frequency ω_2 must be linearly polarized to one part in 10^6 for a pure RIKE experiment. It can be made collinear with the pump laser beam by a dichroic mirror M_2 and the y-component of the probe beam is rejected after interaction with the sample S by a Glan-Thompson prism G. The frequencies ω_1 and ω_2 are separated by a dispersive element like a prism P and a suitable aperture. The detector D may be a diode or a photomultiplier or, if the probe laser is operated broadband and a spectrometer is used for dispersion, a photographic plate or an optical multichannel detector.

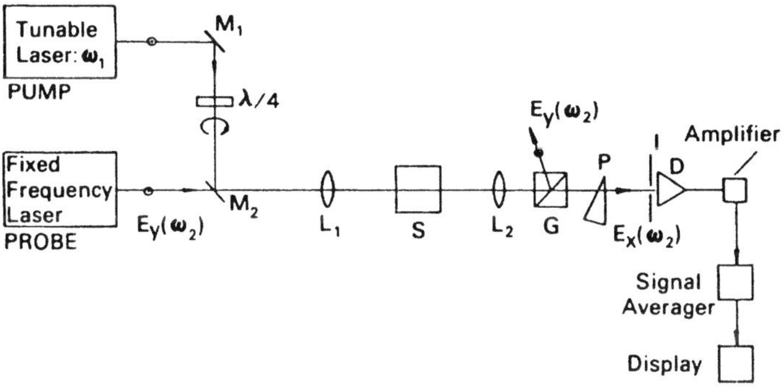

Fig. 1 Schematic diagram of RIKE spectrometer (8). M_1 100% mirror, M_2 dichroic mirror, $\lambda/4$ quarter wave plate, S sample, L_1, L_2 lenses, G Glan-Thompson prism, P dispersing element (prism), I aperture, D detector.

Only slight modifications are necessary to perform an OHD-RIKE experiment. When the Glan-Thompson prism G is rotated slightly from the position of minimum probe transmission, the transmitted component of the ω_2 beam serves as a local oscillator wave in phase with the probe wave and for a circularly polarized pump beam a spectrum with dispersion-like line shapes is obtained. For a linearly polarized pump beam the lineshapes of the linear Raman spectrum are reproduced.

Alternately the probe beam can be made slightly elliptically polarized by insertion of a strained glass plate in its path before the sample (9). In this case the local oscillator wave has the opposite phase as the probe wave and the produced signal shows the linear Raman line shape for a circularly polarized probe and the dispersion-like line shape for a linearly polarized probe.

4. EXPERIMENTAL RESULTS

The results of RIKE and OHD-RIKE spectroscopy have recently been reviewed by Eesley (8) and Levenson (10). Spectra were published for benzene (1, 9), pyridine (5), sodium benzoate in water (9), diphenyloctatetraene (11), calcite (11), diamond (12), lithium iodate (2, 3), and hydrogen (13). For benzene in carbon tetrachloride a spectrum of a 0.1 molar solution was recorded and the detection limit calculated to be at a 0.001 molar concentration (9). At this point it may be mentioned, however, that the linear Raman spectrum of a 0.0005 molar solution of benzene in carbon tetrachloride was recorded as early as 1967 by Murphy and Bernstein (see Ref. (14), p. 99).

Bergmann and Schubert (7) have investigated the change of the line shapes in RIKE spectra of benzene and bromobenzene upon the change from circular to elliptical polarization of the pump beam. Fig. 2 shows the shapes of the 992 cm^{-1} band of benzene

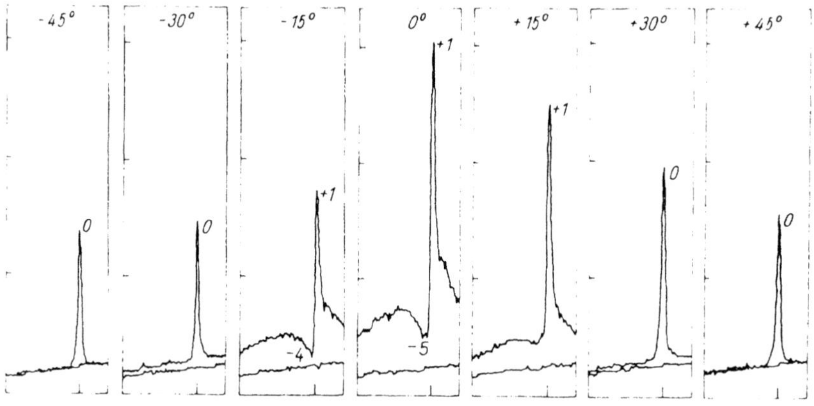

Fig. 2 RIKE spectra of the 992 cm^{-1} band of liquid benzene for different polarizations of pump beam (7).

and Fig. 3 those of the 1000 cm^{-1} and 1020 cm^{-1} bands of bromobenzene. An angle of 45° of the polarization direction of the incident pump laser beam indicates that it is converted to a circularly polarized beam after passing through a quartz plate.

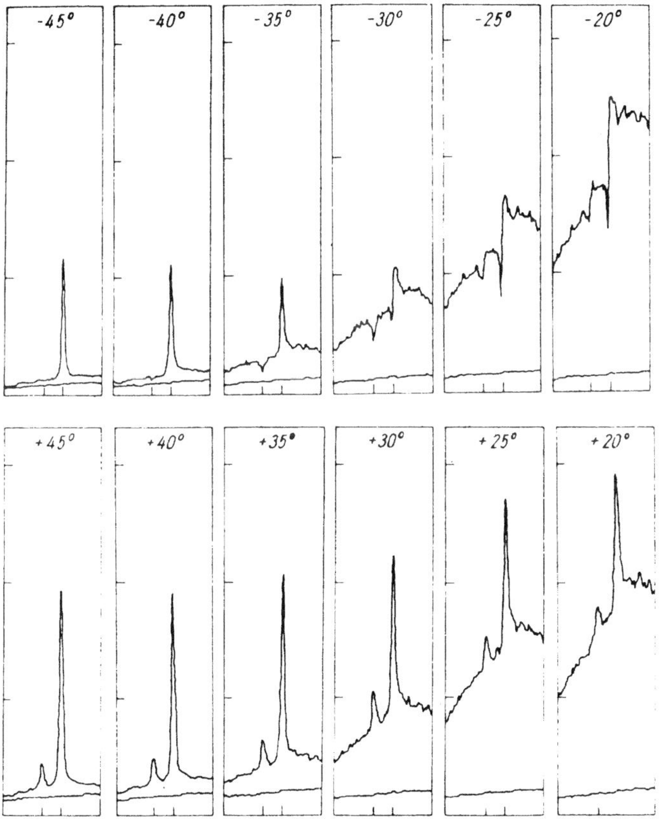

Fig. 3 RIKE spectra of 1020 cm^{-1} (weak line) and 1000 cm^{-1} (strong line) bands of bromobenzene for different polarizations of pump beam (7).

Scarparo et al. (15) have used counterpropagating pump and probe beams, the polarization direction of which was turned by 22.5° by a Faraday rotator, and have obtained a considerable improvement of the signal-to-noise ratio over a conventional OHD-RIKE experiment.

5. CONCLUSION

This chapter has been kept rather short in view of the excellent reviews published recently (8, 10) which should be consulted for further details.

REFERENCES

1. Heiman, D., Hellwarth, R.W., Levenson, M.D., and Martin, G., Phys. Rev. Lett. 36, 189 (1976).

2. Bergmann, J., Kneipp, K., and Ponath, H.E., phys. stat. sol. (b) 80, K 55 (1977).

3. Bergmann, J., Kneipp, K., and Ponath, H.E., Kvant. Elektron. 4, 3570 (1977).

4. Levenson, M.D., and Song, J.J., J. Opt. Soc. Am. 66, 641 (1976).

5. Song, J.J., and Levenson, M.D., J. Appl. Phys. 48, 3496 (1977).

6. Bergmann, J., Kneipp, K., and Ponath, H.E., phys. stat. sol. (b) 86, 215 (1978).

7. Bergmann, J., and Schubert, M., Exp. Tech. Phys. 27, 37 (1979).

8. Eesley, G.L., Coherent Raman Spectroscopy, Pergamon Press, Oxford 1981.

9. Eesley, G.L., Levenson, M.D., and Tolles, W.M., IEEE J.QE. 14, 45 (1978).

10. Levenson, M.D., in Chemical Applications of Nonlinear Raman Spectroscopy (Harvey, A.B., ed.), Academic Press, New York 1981, p. 205; J. Raman Spectrosc. 10, 9 (1981).

11. Eesley, G.L., Journ. Quantit. Spectrosc. Rad. Transf. 22, 507 (1979).

12. Eesley, G.L., and Levenson, M.D., Opt. Lett. 3, 178 (1978).

13. Owyoung, A., Opt. Lett. 2, 91 (1978).

14. Schrötter, H.W., in Raman Spectroscopy (Szymanski, H.A., ed.), Vol. 2, Plenum Press, New York 1970, p. 69.

15. Scarparo, M.A.F., Song, J.J., Lee, J.H., Cromer, C., and Levenson, M.D., Appl. Phys. Lett. 35, 490 (1979).

SUBJECT INDEX

AUTHOR INDEX

REFERENCE AUTHOR INDEX

First number indicates page on which reference is cited. Numbers in brackets are reference numbers. Underlined numbers give the page on which the complete reference is listed.

GPSR Compliance

The European Union's (EU) General Product Safety Regulation (GPSR) is a set of rules that requires consumer products to be safe and our obligations to ensure this.

If you have any concerns about our products, you can contact us on ProductSafety@springernature.com

In case Publisher is established outside the EU, the EU authorized representative is:

Springer Nature Customer Service Center GmbH
Europaplatz 3
69115 Heidelberg, Germany

Batch number: 10815968

Printed by Printforce, the Netherlands